T0321908

CAMBRIDGE TRACTS IN MATHEMATICS

---

191 Malliavin Calculus for Lévy Processes and Infinite-Dimensional Brownian Motion

CAMBRIDGE TRACTS IN MATHEMATICS

A complete list of books in the series can be found at www.cambridge.org/mathematics.
Recent titles include the following:

154. Finite Packing and Covering. By K. Böröczky, Jr
155. The Direct Method in Soliton Theory. By R. Hirota. Edited and translated by A. Nagai, J. Nimmo, and C. Gilson
156. Harmonic Mappings in the Plane. By P. Duren
157. Affine Hecke Algebras and Orthogonal Polynomials. By I. G. Macdonald
158. Quasi-Frobenius Rings. By W. K. Nicholson and M. F. Yousif
159. The Geometry of Total Curvature on Complete Open Surfaces. By K. Shiohama, T. Shioya, and M. Tanaka
160. Approximation by Algebraic Numbers. By Y. Bugeaud
161. Equivalence and Duality for Module Categories. By R. R. Colby and K. R. Fuller
162. Lévy Processes in Lie Groups. By M. Liao
163. Linear and Projective Representations of Symmetric Groups. By A. Kleshchev
164. The Covering Property Axiom, CPA. By K. Ciesielski and J. Pawlikowski
165. Projective Differential Geometry Old and New. By V. Ovsienko and S. Tabachnikov
166. The Lévy Laplacian. By M. N. Feller
167. Poincaré Duality Algebras, Macaulay's Dual Systems, and Steenrod Operations. By D. Meyer and L. Smith
168. The Cube-A Window to Convex and Discrete Geometry. By C. Zong
169. Quantum Stochastic Processes and Noncommutative Geometry. By K. B. Sinha and D. Goswami
170. Polynomials and Vanishing Cycles. By M. Tibăr
171. Orbifolds and Stringy Topology. By A. Adem, J. Leida, and Y. Ruan
172. Rigid Cohomology. By B. Le Stum
173. Enumeration of Finite Groups. By S. R. Blackburn, P. M. Neumann, and G. Venkataraman
174. Forcing Idealized. By J. Zapletal
175. The Large Sieve and its Applications. By E. Kowalski
176. The Monster Group and Majorana Involutions. By A. A. Ivanov
177. A Higher-Dimensional Sieve Method. By H. G. Diamond, H. Halberstam, and W. F. Galway
178. Analysis in Positive Characteristic. By A. N. Kochubei
179. Dynamics of Linear Operators. By F. Bayart and É. Matheron
180. Synthetic Geometry of Manifolds. By A. Kock
181. Totally Positive Matrices. By A. Pinkus
182. Nonlinear Markov Processes and Kinetic Equations. By V. N. Kolokoltsov
183. Period Domains over Finite and p-adic Fields. By J.-F. Dat, S. Orlik, and M. Rapoport
184. Algebraic Theories. By J. Adámek, J. Rosický, and E. M. Vitale
185. Rigidity in Higher Rank Abelian Group Actions I: Introduction and Cocycle Problem. By A. Katok and V. Niţică
186. Dimensions, Embeddings, and Attractors. By J. C. Robinson
187. Convexity: An Analytic Viewpoint. By B. Simon
188. Modern Approaches to the Invariant Subspace Problem. By I. Chalendar and J .R. Partington
189. Nonlinear Perron–Frobenius Theory. By B. Lemmens and R. Nussbaum
190. Jordan Structures in Geometry and Analysis. By C.-H. Chu
191. Malliavin Calculus for Lévy Processes and Infinite-Dimensional Brownian Motion. By H. Osswald
192. Normal Approximations with Malliavin Calculus. By I. Nourdin and G. Peccati

Malliavin Calculus for Lévy Processes and Infinite-Dimensional Brownian Motion

An Introduction

HORST OSSWALD
Universität München

CAMBRIDGE
UNIVERSITY PRESS

University Printing House, Cambridge CB2 8BS, United Kingdom

One Liberty Plaza, 20th Floor, New York, NY 10006, USA

477 Williamstown Road, Port Melbourne, VIC 3207, Australia

314-321, 3rd Floor, Plot 3, Splendor Forum, Jasola District Centre, New Delhi - 110025, India

103 Penang Road, #05-06/07, Visioncrest Commercial, Singapore 238467

Cambridge University Press is part of the University of Cambridge.

It furthers the University's mission by disseminating knowledge in the pursuit of education, learning and research at the highest international levels of excellence.

www.cambridge.org
Information on this title: www.cambridge.org/9781107016149

First published 2012

A catalogue record for this publication is available from the British Library

ISBN 978-1-107-01614-9 Hardback

To
Ruth
Christine, Silas, Till
Fabian
and in memoriam
Horst

Contents

Preface *page* xv

PART I THE FUNDAMENTAL PRINCIPLES

1 Preliminaries 3

2 Martingales 9

2.1 Martingales and examples 9
2.2 Stopping times 12
2.3 The maximum inequality 13
2.4 Doob's inequality 14
2.5 The σ-algebra over the past of a stopping time 16
2.6 L^p-spaces of martingales and the quadratic variation norm 17
2.7 The supremum norm 19
2.8 Martingales of bounded mean oscillation 19
2.9 $\left(L^1_{[}\right)'$ is BMO_2 21
2.10 $(L^1_{\sim})'$ is BMO_1 24
2.11 B–D–G inequalities for $p=1$ 28
2.12 The B–D–G inequalities for the conditional expectation for $p=1$ 30
2.13 The B–D–G inequalities 31
2.14 The B–D–G inequalities for special convex functions 32
Exercises 36

3 Fourier and Laplace transformations 37

3.1 Transformations of measures 37
3.2 Laplace characterization of $\mathcal{N}(0,\sigma)$-distribution 38
3.3 Fourier and Laplace characterization of independence 39

3.4 Discrete Lévy processes and their representation 42
3.5 Martingale characterization of Brownian motion 45
Exercises 46

4 Abstract Wiener–Fréchet spaces 50
4.1 Projective systems of measures and their limit 50
4.2 Gaussian measures in Hilbert spaces 52
4.3 Abstract Wiener spaces 54
4.4 Cylinder sets in Fréchet spaces generate the Borel sets 57
4.5 Cylinder sets in Fréchet space valued continuous functions 61
4.6 Tensor products 62
4.7 Bochner integrable functions 64
4.8 The Wiener measure on $C_{\mathbb{B}}$ is the centred Gaussian measure of variance 1 66
Exercises 70

5 Two concepts of no-anticipation in time 71
5.1 Predictability and adaptedness 71
5.2 Approximations of the Dirac δ-function 73
5.3 Convolutions of adapted functions are adapted 75
5.4 Adaptedness is equivalent to predictability 76
5.5 The weak approximation property 77
5.6 Elementary facts about L^p-spaces 78
Exercises 81

6 †Malliavin calculus on real sequences 82
6.1 Orthogonal polynomials 82
6.2 Integration 84
6.3 Iterated integrals 85
6.4 Chaos decomposition 86
6.5 Malliavin derivative and Skorokhod integral 88
6.6 The integral as a special case of the Skorokhod integral 89
6.7 The Clark–Ocone formula 90
6.8 Examples 91
Exercises 94

7 Introduction to poly-saturated models of mathematics 95
7.1 Models of mathematics 96
7.2 The main theorem 102
Exercises 105

8 Extension of the real numbers and properties **107**
8.1 ${}^*\mathbb{R}$ as an ordered field 107
8.2 The *extension of the positive integers 107
8.3 Hyperfinite sets and summation in ${}^*\mathbb{R}$ 109
8.4 The underspill and overspill principles 110
8.5 The infinitesimals 110
8.6 Limited and unlimited numbers in ${}^*\mathbb{R}$ 111
8.7 The standard part map on limited numbers 112
Exercises 113

9 Topology **115**
9.1 Monads 115
9.2 Hausdorff spaces 116
9.3 Continuity 117
9.4 Compactness 117
9.5 Convergence 118
9.6 The standard part of an internal set of nearstandard points is compact 119
9.7 From S-continuous to continuous functions 120
9.8 Hyperfinite representation of the tensor product 121
9.9 †The Skorokhod topology 124
Exercises 128

10 Measure and integration on Loeb spaces **130**
10.1 The construction of Loeb measures 130
10.2 Loeb measures over Gaussian measures 133
10.3 Loeb measurable functions 135
10.4 On Loeb product spaces 137
10.5 Lebesgue measure as a counting measure 138
10.6 Adapted Loeb spaces 142
10.7 $\boldsymbol{S}$-integrability and equivalent conditions 143
10.8 Bochner integrability and $\boldsymbol{S}$-integrability 145
10.9 Integrable functions defined on $\mathbb{N}^n \times \Lambda \times [0, \infty[^m$ 149
10.10 Standard part of the conditional expectation 153
10.11 Witnesses of $\boldsymbol{S}$-integrability 155
10.12 Keisler's Fubini theorem 157
10.13 $\boldsymbol{S}$-integrability of internal martingales 160
10.14 $\boldsymbol{S}$-continuity of internal martingales 160
10.15 On symmetric functions 165
10.16 The standard part of internal martingales 166
Exercises 170

PART II AN INTRODUCTION TO FINITE- AND INFINITE-DIMENSIONAL STOCHASTIC ANALYSIS

Introduction **175**

11 From finite- to infinite-dimensional Brownian motion **177**

11.1 On the underlying probability space 177
11.2 The internal Brownian motion 179
11.3 $\boldsymbol{S}$-integrability of the internal Brownian motion 181
11.4 The S-continuity of the internal Brownian motion 182
11.5 One-dimensional Brownian motion 182
11.6 Lévy's inequality 183
11.7 The final construction 186
11.8 The Wiener space 190
Exercises 194

12 The Itô integral for infinite-dimensional Brownian motion **196**

12.1 The $\boldsymbol{S}$-continuity of the internal Itô integral 196
12.2 On the $\boldsymbol{S}$-square-integrability of the internal Itô integral 203
12.3 The standard Itô integral 204
12.4 On the integrability of the Itô integral 207
12.5 $\mathcal{W}_{C_{\mathbb{H}}}$ is generated by the Wiener integrals 208
12.6 †The distribution of the Wiener integrals 209
Exercises 210

13 The iterated integral **211**

13.1 The iterated integral with and without parameters 211
13.2 The product of an internal iterated integral and an internal Wiener integral 216
13.3 The continuity of the standard iterated integral process 218
13.4 The $\mathcal{W}_{C_{\mathbb{H}}}$-measurability of the iterated Itô integral 219
13.5 $I_n^M(f)$ is a continuous version of the standard part of $I_n^M(F)$ 221
13.6 Continuous versions of internal iterated integral processes 222
13.7 Kolmogorov's continuity criterion 224
Exercises 228

14 †Infinite-dimensional Ornstein–Uhlenbeck processes 229
14.1 Ornstein–Uhlenbeck processes for shifts given by Hilbert–Schmidt operators 231
14.2 Ornstein–Uhlenbeck processes for shifts by scalars 239
Exercises 246

15 Lindstrøm's construction of standard Lévy processes from discrete ones 247
15.1 Exponential moments for processes with limited increments 248
15.2 Limited Lévy processes 251
15.3 Approximation of limited processes by processes with limited increments 256
15.4 Splitting infinitesimals 256
15.5 Standard Lévy processes 257
15.6 Lévy measure 260
15.7 The Lévy–Khintchine formula 264
15.8 Lévy triplets generate Lévy processes 265
15.9 Each Lévy process can be divided into its continuous and pure jump part 266
Exercises 270

16 Stochastic integration for Lévy processes 271
16.1 Orthogonalization of the increments 271
16.2 From internal random walks to the standard Lévy integral 275
16.3 Iterated integrals 278
16.4 Multiple integrals 282
16.5 The σ-algebra generated by the Wiener–Lévy integrals 283
Exercises 286

PART III MALLIAVIN CALCULUS

Introduction 291

17 Chaos decomposition 293
17.1 Admissible sequences 293
17.2 Chaos expansion 296
17.3 A lifting theorem for functionals in $L^2_{\mathcal{W}}(\widehat{\Gamma})$ 299
17.4 Chaos for functions without moments 300
17.5 Computation of the kernels 300

17.6 The kernels of the product of Wiener functionals 303
Exercises 304

18 The Malliavin derivative 306
18.1 The domain of the derivative 306
18.2 The Clark–Ocone formula 308
18.3 A lifting theorem for the derivative 309
18.4 The directional derivative 310
18.5 A commutation rule for derivative and limit 312
18.6 The domain of the Malliavin derivative is a Hilbert space with respect to the norm $\|\cdot\|_{1,2}$ 313
18.7 The range of the Malliavin derivative is closed 314
18.8 A commutation rule for the directional derivative 315
18.9 Product and chain rules for the Malliavin derivative 315
Exercises 318

19 The Skorokhod integral 319
19.1 Decomposition of processes 319
19.2 Malliavin derivative of processes 322
19.3 The domain of the Skorokhod integral 323
19.4 A lifting theorem for the integral 324
19.5 The Itô integral is a special case of the Skorokhod integral 325
Exercises 327

20 The interplay between derivative and integral 328
20.1 The integral is the adjoint operator of the derivative 328
20.2 A Malliavin differentiable function multiplied by square-integrable deterministic functions is Skorokhod integrable 330
20.3 The duality between the domains of D and δ 332
20.4 $L^2_{\mathcal{W}\otimes\mathcal{L}^1}(\widehat{\Gamma\otimes\nu},\mathbb{H})$ is the orthogonal sum of the range of D and the kernel of δ 333
20.5 Integration by parts 334
Exercises 334

21 Skorokhod integral processes 335
21.1 The Skorokhod integral process operator 335
21.2 On continuous versions of Skorokhod integral processes 336
Exercises 338

22 Girsanov transformations **339**
22.1 From standard to internal shifts 341
22.2 The Jacobian determinant of the internal shift 342
22.3 Time-anticipating Girsanov transformations 343
22.4 Adapted Girsanov transformation 347
22.5 †Extension of abstract Wiener spaces 348
Exercises 350

23 Malliavin calculus for Lévy processes **352**
23.1 Chaos 352
23.2 Malliavin derivative 356
23.3 The Clark–Ocone formula 357
23.4 Skorokhod integral processes 358
23.5 Smooth representations 360
23.6 A commutation rule for derivative and limit 362
23.7 The product rule 362
23.8 The chain rule 366
23.9 Girsanov transformations 368
Exercises 374

APPENDICES EXISTENCE OF POLY-SATURATED MODELS

Appendix A. Poly-saturated models **379**
A.1 Weak models and models of mathematics 379
A.2 From weak models to models 380
A.3 Languages for models 381
A.4 Interpretation of the language 382
A.5 Models closed under definition 383
A.6 Elementary embeddings 384
A.7 Poly-saturated models 386

Appendix B. The existence of poly-saturated models **388**
B.1 From pre-models to models 388
B.2 Ultrapowers 390
B.3 Elementary chains and their elementary limits 393
B.4 Existence of poly-saturated models with the same properties as standard models 395

References 398
Index 404

Preface

The aim of this book is to give a self-contained introduction to Malliavin calculus for Lévy processes $L : \Omega \times [0,\infty[\to \mathbb{B}$, where $\mathbb{B}$ is a finite-dimensional Euclidean space or a separable Fréchet space, given by a countable sequence of semi-norms. We only take for granted that the reader has some knowledge of basic probability theory and functional analysis within the scope of excellent books on these fields, for example, Ash [5] or Billingsley [13] and Reed and Simon [98] or Rudin [101].

The most important Lévy processes are Brownian motion and Poisson processes, where Brownian motion is continuous and Poisson processes have jumps.

In Chapter 6 we will study Malliavin calculus for discrete stochastic processes $f : \left(\mathbb{R}^d\right)^{\mathbb{N}} \times \mathbb{N} \to \mathbb{R}^d$. The probability measure on $\left(\mathbb{R}^d\right)^{\mathbb{N}}$ is the product of a Borel probability measure on $\mathbb{R}^d$. For simplicity let us set $d = 1$; later on we accept $d = \infty$. In an application we obtain calculus for abstract Wiener spaces over l^2, the space of square summable real sequences. By using suitable extensions of $\mathbb{R}$ and $\mathbb{N}$, we obtain calculus for abstract Wiener spaces over arbitrary separable Hilbert spaces in the same manner, where we only identify two spaces if there exists a canonical, i.e., basis independent, isomorphic isometry between them.

In the short but very crucial Chapter 7 we extend $\mathbb{R}$ and $\mathbb{N}$ to ${}^*\mathbb{R}$ and ${}^*\mathbb{N}$ in such a way that the elements of ${}^*\mathbb{R}$ and ${}^*\mathbb{N}$ can be handled as though they were the usual real numbers and positive integers, respectively. In ${}^*\mathbb{N}$ there exist **infinitely large** positive integers H, which means that $n < H$ for all $n \in \mathbb{N}$. We take an infinitely large $H \in {}^*\mathbb{N}$ and use $T := \left\{ \frac{i}{H} \mid i \in {}^*\mathbb{N}, i \leq H^2 \right\}$ instead of $\frac{{}^*\mathbb{N}}{H}$. Then T is like a finite set and can be seen as an infinitely fine partition of $[0,\infty]$. It follows that there is no great difference between T and $[0,\infty[$. Our fixed sample space now is $\Omega_d := \left({}^*\mathbb{R}^d\right)^T$. It only depends on the dimension d of the Lévy process we have in mind. If d is infinite, we can take Ω_ω, where ω is again an infinitely large number in ${}^*\mathbb{N}$. It turns out that each d, ∞-dimensional Lévy process L lives on $\Omega := \Omega_d,\ \Omega_\omega$ (see Theorem 15.8.1 for $d = 1$ and Theorem 11.7.7 for ∞-dimensional Brownian motion), i.e., L is a mapping from $\Omega \times [0, \infty[$ into $\mathbb{R}^d$, where $\mathbb{R}^d$ is a Fréchet space in case $d = \infty$.

Moreover, since T is infinitely close to the continuous time line $[0, \infty[$, processes $f : \Omega \times [0, \infty[\to \mathbb{R}^d$ are infinitely close to processes $F : \Omega \times T \to {}^*\mathbb{R}^d$, where ${}^*\mathbb{R}^d = {}^*\mathbb{R}^\omega$ in the infinite-dimensional case. This relation 'infinitely close' will be studied and applied in the whole book in great detail.

I hope that this short Chapter 7 may help to achieve my most cherished objective to convince my gentle readers that there is no reason to fear model-theoretical reasoning in mathematics.

The choice of the sample space Ω implies that we can study finite- and infinite-dimensional Lévy processes simultaneously. Although Ω can be handled as though it were a finite-dimensional Euclidean space (even in the infinite-dimensional case), Ω is very rich. In particular, each right continuous function from $[0, \infty[$ into $\mathbb{B}$ is a path of each Lévy process $L : \Omega \times [0, \infty[\to \mathbb{B}$. The proof of this fact is simple, but the result may be surprising and seems to be inconsistent.

This inconsistency disappears by observing that, in the case of Brownian motion, the set of non-continuous functions $f : [0, \infty[\to \mathbb{B}$ or the set of functions not starting in 0 is a nullset with respect to the image measure of the probability measure on Ω by L (it is the Wiener measure).

In the case of Poisson processes, the set of functions that are not increasing or fail to be counting functions or have only a finite range is a nullset.

In both cases the set of right continuous functions not having left-hand limits is also a nullset. It follows that we may assume that all Lévy processes L are almost surely surjective mappings from Ω onto the space D of **càdlàg** functions $f : [0, \infty[\to \mathbb{B}$, i.e., f is right continuous and has left-hand limits. Indeed, from each càdlàg function $f : [0, \infty[\to \mathbb{B}$ we can explicitly construct an $X \in \Omega$ with $L(X, \cdot) = f$.

One aim is to construct Brownian motion $b_{\mathbb{B}} : \Omega \times [0, \infty[\to \mathbb{B}$, where $\mathbb{B}$ is a separable Fréchet space with metric d, generated by a sequence $(|\cdot|_i)_{i \in \mathbb{N}}$ of separating semi-norms $|\cdot|_i$ on $\mathbb{B}$. What is a $\mathbb{B}$-valued Brownian motion? According to the finite-dimensional situation $\mathbb{B} = \mathbb{R}^m$, it is required that the components are one-dimensional Brownian motions, running independently on orthonormal axes; axes are elements of an orthonormal basis of $\mathbb{R}^m$. A new question arises immediately: what does 'orthogonality' mean in infinite-dimensional Fréchet spaces $\mathbb{B}$? Here is an answer: there exists a Hilbert space $\mathbb{H} \subseteq \mathbb{B}$ with norm, say $\|\cdot\|$, such that $(\mathbb{B}, d)$ is the completion of $(\mathbb{H}, d)$ and such that $\varphi \upharpoonright \mathbb{H}$ is continuous with respect to $\|\cdot\|$ for each φ in the topological dual $\mathbb{B}'$ of $\mathbb{B}$. Then

$\mathbb{B}'$ is a dense subspace of $\mathbb{H}' = \mathbb{H}$, with respect to $\|\cdot\|$ (see Section 4.3). So, orthogonality can be defined for all elements of $\mathbb{B}'$. Now $b_{\mathbb{B}} : \Omega \times [0, \infty[\to \mathbb{B}$ is a Brownian motion (BM), provided that $\varphi \circ b_{\mathbb{B}}$ is a one-dimensional BM for all $\varphi \in \mathbb{B}'$ with $\|\varphi\| = 1$, and $\varphi \circ b_{\mathbb{B}}$ and $\psi \circ b_{\mathbb{B}}$ are independent, provided that φ is orthogonal to ψ. So, axes of $\mathbb{B}$ are orthogonal elements of the dual space of $\mathbb{B}$. Since orthogonal sets in $\mathbb{R}^m$ can be identified with orthogonal sets in $(\mathbb{R}^m)'$, the notion of infinite-dimensional Brownian motion is literally a generalization of the finite-dimensional concept.

The Fréchet space $\mathbb{B}$ is called an **abstract Wiener space** over the so-called **Cameron–Martin space** $\mathbb{H}$. A famous result, due to Leonhard Gross [41], tells us that each Gaussian measure on the algebra of cylinder sets of $\mathbb{H}$ can be extended to a σ-additive measure $\gamma_{\mathbb{B}}$ on the Borel algebra of the described extension $\mathbb{B}$ of $\mathbb{H}$. Here is a nice example: The space of real sequences, endowed with the topology of pointwise convergence, is an abstract Wiener–Fréchet space over l^2.

There are many, quite different, abstract Wiener spaces over l^2. However, it will be seen that for any two abstract Wiener spaces $\mathbb{B}$ and $\mathbb{D}$ over the same Hilbert space $\mathbb{H}$ and for all $p \in [0, \infty]$ the L^p-spaces $L^p(\mathbb{B}, \gamma_{\mathbb{B}})$ and $L^p(\mathbb{D}, \gamma_{\mathbb{D}})$ can be identified, because there exists a canonical (i.e., basis independent) isomorphic isometry between them.

Our aim is to study:

- **in the finite-dimensional case**: Malliavin calculus on the space D of càdlàg functions, endowed with probability measures, generated by a large class of Lévy processes;
- **in the infinite-dimensional case**: Malliavin calculus on the space $C_{\mathbb{B}}$ of continuous functions from $[0, \infty]$ into $\mathbb{B}$, endowed with the Wiener measure. The reason why we take the space $C_{\mathbb{B}}$ instead of $\mathbb{B}$ (both are Fréchet spaces) is the following: in analogy to the classical Wiener space we want to use the timeline $[0, \infty[$ in order to be able to define the notions 'no time-anticipating' and 'Itô integral'. We replace in the classical Wiener space $C_{\mathbb{R}}$ the set $\mathbb{R}$ of real numbers by separable Fréchet spaces $\mathbb{B}$.

Since our sample space Ω is finite-dimensional in a certain sense, Ω is much simpler to handle than the space D of càdlàg functions or the space $C_{\mathbb{B}}$. Therefore, we may work on Ω without any loss of generality. In addition, Ω is even richer than $C_{\mathbb{B}}$. Because of the choice of Ω, Malliavin calculus reduces to finite-dimensional analysis. In particular, the Malliavin derivative is the usual derivative (see Example 18.3.2).

This book is an extension of a two-semester course on the Malliavin calculus, given in winter 2000/2001 and in summer 2001 at the University of Munich. It is organized as follows:

The results in Chapters 2, 3, 4 and 5 are well known and serve as the basis for the whole book.

In order to make preparations for the general Malliavin calculus we sketch a simple example in Chapter 6, following [88]. Although we work on $\mathbb{N}$ and on powers of $\mathbb{N}$, Malliavin calculus is obtained for Poisson processes and Brownian motion with values in abstract Wiener spaces over l^2. As we have already mentioned, techniques of Chapter 6 will be used later by enlarging $\mathbb{N}$ to a rich set which can be treated as though it were a finite set.

Models of mathematics (so called poly-saturated models), in which finite extensions of $\mathbb{N}$ exist, are introduced in Chapter 7.

Chapters 8 and 9 present some well-known applications of poly-saturated models to the real numbers and elementary topology.

Chapter 10 gives a detailed introduction to the well-established Loeb spaces with special regard to martingale theory.

Chapters 11, 12, 13 and 16 contain stochastic integration for infinite-dimensional Brownian motion and certain finite-dimensional Lévy processes, following [83], [84], [85], [89], [90], [91]. The results are partially an extension of results due to Cutland [23] and Cutland and Ng [24] for the special case of finite-dimensional Brownian motion.

In an application of Chapters 11, 12 and 13 we construct path-by-path continuous solutions to certain Langevin equations in infinite dimension, following [87] (see Chapter 14). Using this construction, one can easily see that each continuous function is a path of the solutions.

Following the work of Lindstrøm [67], we will show in Chapter 15 that each Lévy triplet can be satisfied by a Lévy process, which is infinitely close to a Lévy process, defined on a finite timeline, finite in the extended sense. Moreover, we will prove the well-known result that each Lévy process can be divided into a constant, a continuous Lévy martingale and a pure jump Lévy martingale.

Finally, Chapters 17 through 23 treat Malliavin calculus for infinite-dimensional Brownian motion and for a large class of finite-dimensional Lévy processes, following [83], [84], [85], [90], [89], [91], [92]. Again the results are partially extensions of results due to Cutland and Ng [24] for finite-dimensional Brownian motion.

In the appendices at the end of the book the reader can find a detailed proof of the existence of poly-saturated models of mathematics, following [86].

In order to avoid technical difficulties, without being seriously less general, and in order to make the book easier to read, we only take two dimensions d, namely $d = 1$ and $d = \infty$.

We start with $d = \infty$, thus with Malliavin calculus on abstract Wiener spaces. The case $d = 1$ seems to be much simpler, but additional difficulties appear in connection with Lévy processes more general than Gaussian processes.

Each chapter ends with exercises, which we try to keep as close as possible to the subject of that chapter. A † at headings of chapters or sections indicates topics that may be omitted on first reading.

Acknowledgement: I am very grateful to Ralph Matthes for many helpful comments. I also express my sincere thanks to my colleagues and friends Josef Berger, Erwin Brüning, Cornelius Greither and Martin Schottenloher, who have read large parts of the manuscript and have sent me corrections, criticisms and many other useful comments. I wish to thank Cambridge University Press for their kindness and help in the production of the book. In particular, my thanks go to the copy-editor, Mairi Sutherland, for her careful reading of the manuscript and for many queries which led to an improvement of the text.

PART I

The fundamental principles

1

Preliminaries

In order to fix the terminology, let us start with some well-established basic facts from functional analysis and measure theory. The reader is referred to the books of Reed and Simon [98] or Rudin [101] and Ash [5], Halmos [43] or Billingsley [13] for details.

We study Fréchet spaces, because the archetype of an abstract Wiener space is the space of real sequences, endowed with the topology of pointwise convergence. It is a Fréchet space and an abstract Wiener space over 'little' l^2.

Let $\mathbb{N}$ be the set of positive integers and let $\mathbb{N}_0 := \mathbb{N} \cup \{0\}$. We identify $n \in \mathbb{N}_0$ with the set $\{1, \ldots, n\}$, thus $0 = \emptyset$. For elements a, b of a totally ordered set, we set

$$a \wedge b := \min\{a, b\}, \quad a \vee b := \max\{a, b\}.$$

The following notation is important: if f is a binary relation, then $f[A]$ is the set of all second components of pairs in f, where the first components run through A, i.e.,

$$f[A] := \{y \mid \exists x \in A\, ((x, y) \in f)\}.$$

In particular, if f is a function, then $f[A] := \{f(x) \mid x \in A\}$ and $f^{-1}[B] := \{x \mid f(x) \in B\}$. Note that, in contrast to $f[A]$, $f(A)$ denotes the value of A if A is an element of the domain of f (see the notion 'image measure' below).

A semi-metric d fulfils the same conditions as a metric, except that from $d(a, b) = 0$ it need not follow that $a = b$. Let $(|\cdot|_i) := (|\cdot|_i)_{i \in \mathbb{N}}$ be a separating sequence of semi-norms on a real vector space $\mathbb{B}$; **separating** means: if $x \neq 0$, then there exists an $i \in \mathbb{N}$ with $|x|_i \neq 0$. A neighbourhood base of an element a in $\mathbb{B}$ in the locally convex topology $\mathcal{T}_{(|\cdot|_i)}$, given by $(|\cdot|_i)$, is the family of sets of the form

$$U_{\frac{1}{m}}(a) := \left\{ x \in \mathbb{B} \mid \max_{j \leq m} |x - a|_j < \frac{1}{m} \right\},$$

with $m \in \mathbb{N}$. It follows that a sequence $(a_k)_{k\in\mathbb{N}}$ converges to a in the topology $\mathcal{T}_{(|\cdot|_i)}$ if and only if $\lim_{k\to\infty} |a_k - a|_i = 0$ for all $i \in \mathbb{N}$. The topology $\mathcal{T}_{(|\cdot|_i)}$ is generated by a **translation invariant** metric d, i.e., $d(a,b) = d(a+c, b+c)$, where

$$d(a,b) = \sum_{i=1}^{\infty} \frac{1}{2^i} \cdot \frac{|a-b|_i}{1+|a-b|_i}.$$

This metric d is called the **metric generated by** $(|\cdot|_i)_{i\in\mathbb{N}}$. Note that we have $d(a+b,0) \leq d(a,0) + d(b,0)$. The space $(\mathbb{B}, d) = (\mathbb{B}, (|\cdot|_i))$ is called a **Pre-Fréchet space.** It is called a **Fréchet space**, provided $\mathbb{B}$ is complete. We always assume that Fréchet spaces are separable. The space $(\mathbb{B}, d)$ is complete as a metric space if and only if it is **complete as a locally convex space** $(\mathbb{B}, (|\cdot|_i))$, i.e., if (a_n) is a Cauchy sequence for each semi-norm $|\cdot|_i$, then there exists an $a \in \mathbb{B}$ such that $\lim_{n\to\infty} |a_n - a|_i = 0$ for all $i \in \mathbb{N}$.

The **topological dual** of a locally convex space $\mathbb{B}$ over $\mathbb{R}$ is denoted by $\mathbb{B}'$. It is the space of all linear and continuous functions $\varphi : \mathbb{B} \to \mathbb{R}$.

Fix a Hilbert space $\mathbb{H}$ over $\mathbb{R}$ with scalar product $\langle\cdot,\cdot\rangle : \mathbb{H} \times \mathbb{H} \to \mathbb{R}$ and norm $\|\cdot\|$ given by $\|a\| := \sqrt{\langle a,a\rangle}$. For subsets $A, B \subseteq \mathbb{H}$, we set $A + B := \{a + b \mid a \in A \text{ and } b \in B\}$. The **Cauchy–Schwarz inequality** says that $|\langle a,b\rangle| \leq \|a\| \cdot \|b\|$ for all $a, b \in \mathbb{H}$.

By the Cauchy–Schwarz inequality, we have $(\langle a,\cdot\rangle : \mathbb{H} \to \mathbb{R}) \in \mathbb{H}'$ for each $a \in \mathbb{H}$. Vice versa, by the **Riesz lemma**, for each $\varphi \in \mathbb{H}'$ there exists a (unique) $a_\varphi \in \mathbb{H}$ such that $\varphi = \langle a_\varphi, \cdot\rangle$ (see Theorem II.4 in [98]). It is a common practice to identify φ and a_φ, thus $\mathbb{H}' = \mathbb{H}$.

We say that $a \in \mathbb{H}$ is **orthogonal to** $b \in \mathbb{H}$ if the distance between a and b is equal to the distance between a and $-b$; in this case we shall write $a \perp b$. Note that $a \perp b$ if and only if $\langle a,b\rangle = 0$. Let $S \subseteq \mathbb{H}$. We will write $a \perp S$ if $a \perp b$ for all $b \in S$ and set $S^\perp := \{x \in \mathbb{H} \mid x \perp S\}$. Since $\langle\cdot,b\rangle$ is linear and continuous for all $b \in \mathbb{H}$, $S^\perp$ is a closed linear subspace of $\mathbb{H}$.

Let $\mathcal{E}(\mathbb{H})$ denote the set of all finite-dimensional subspaces of $\mathbb{H}$. If $A \subseteq \mathbb{H}$, then **span**A denotes the linear subspace of $\mathbb{H}$, generated by A.

A subset $S \subseteq \mathbb{H}$ is called an **orthonormal set** (ONS) in $\mathbb{H}$ if $\|a\| = 1$ and $a \perp b$ for all $a, b \in S$ with $a \neq b$. An ONS $\mathfrak{B}$ in $\mathbb{H}$ is called an **orthonormal basis** (ONB) of $\mathbb{H}$ if $\mathfrak{B}$ is **maximal**, i.e., there does not exist an ONS S which is a strict extension of $\mathfrak{B}$. Each Hilbert space $\mathbb{H}$ has an ONB (see Theorem II.5 in [98]).

We will often use the so-called **projection theorem** (see Theorem II.3 in [98]), which tells us that, if G is a closed subspace of $\mathbb{H}$, then each $x \in \mathbb{H}$ can be composed of a sum $x = a + b$ with $a \in G$ and $b \in G^\perp$; the pair (a, b) is uniquely determined by x; the mapping $f:\mathbb{H}\to G$ that assigns to x the element a is called the **orthogonal projection** from $\mathbb{H}$ onto G and is denoted by $\pi_{\mathbb{H}G}$.

In what follows let us assume that $\mathbb{H}$ is an infinite-dimensional Hilbert space and **separable**. This means there exists a countable **dense** subset $D\subseteq\mathbb{H}$, i.e., for each $a\in\mathbb{H}$ and each $\varepsilon>0$ there exists a $d\in D$ with $\|d-a\|<\varepsilon$. It follows that each separable $\mathbb{H}$ has a countable ONB $\mathfrak{E}:=(\mathfrak{e}_i)_{i\in\mathbb{N}}$ (see Theorem II.5 in [98]). Now each a is an infinite linear combination of elements in $\mathfrak{E}$, i.e., $a=\sum_{i=1}^{\infty}\langle a,\mathfrak{e}_i\rangle\,\mathfrak{e}_i$ (see Theorem II.6 in [98]). The $\langle a,\mathfrak{e}_i\rangle$ are called the **Fourier coefficients** of a with respect to $\mathfrak{E}$.

One form of the **Hahn–Banach theorem for locally convex spaces** $\mathbb{B}$, given by a separating sequence $(|\cdot|_i)$ of semi-norms, is (see Theorem 3.3 in Rudin [101]): Fix $i\in\mathbb{N}$ and a subspace M of $\mathbb{B}$. Then each linear map $f:M\to\mathbb{R}$ with $|f(\cdot)|\le|\cdot|_i$ on M can be extended to a linear mapping $g:\mathbb{B}\to\mathbb{R}$ such that $|g(\cdot)|\le|\cdot|_i$ on $\mathbb{B}$. In the special case of normed spaces we obtain the following result. Let $(\mathbb{B},|\cdot|)$ be a normed space and fix $a\in\mathbb{B}$ with $a\neq0$. Then there exists a $\varphi\in\mathbb{B}'$ such that $|\varphi|:=\sup_{|x|\le1}|\varphi(x)|\le1$ and $\varphi(a)=|a|$.

We also need some elementary facts from measure theory (see Ash [5] or Billingsley [13] for details). The **power set** of a set Λ, i.e., the set of all subsets of Λ, is denoted by $\mathcal{P}(\Lambda)$. The symmetric difference of sets A,B is denoted by

$$A\,\Delta\,B:=(A\setminus B)\cup(B\setminus A).$$

An **algebra** on a set Λ is a subset of $\mathcal{P}(\Lambda)$, closed under finite unions and under complements, and containing Λ. An algebra on Λ is called a σ**-algebra** on Λ if it is closed under countable unions. If $\mathcal{D}\subseteq\mathcal{P}(\Lambda)$, then the intersection $\sigma(\mathcal{D})$ of all σ-algebras $\mathcal{S}\supseteq\mathcal{D}$ is again a σ-algebra, which is called the σ**-algebra, generated by** $\mathcal{D}$. For two subsets $\mathcal{X},\mathcal{Y}\subseteq\mathcal{P}(\Lambda)$ let us set

$$\mathcal{X}\vee\mathcal{Y}=\sigma(\mathcal{X}\cup\mathcal{Y}).$$

The **Borel** σ**-algebra** on a topological space Λ is generated by the set of open sets in Λ and is denoted by $\mathcal{B}(\Lambda)$. The elements of $\mathcal{B}(\Lambda)$ are called the **Borel sets** of Λ.

Fix $a,b\in\mathbb{R}\cup\{-\infty\}$. **Right intervals** in $\mathbb{R}$ are sets of the form $]a,b]=\{x\in\mathbb{R}\mid a<x\le b\}$ or of the form $]a,\infty[=\{x\in\mathbb{R}\mid a<x<\infty\}$. Fix $n\in\mathbb{N}$. **right rectangles** in $\mathbb{R}^n$ are sets of the form $J_1\times\ldots\times J_n$, where $J_1,\ldots,J_n$ are right intervals in $\mathbb{R}$. Note that the set $\mathcal{R}(\mathbb{R}^n)$ of finite unions of pairwise disjoint right rectangles is an algebra on $\mathbb{R}^n$. This set also generates the Borel σ-algebra on $\mathbb{R}^n$.

Let $\mathcal{A}$ be an algebra on a set Λ. A function $\mu:\mathcal{A}\to[0,\infty]$ is called a **finitely additive measure on** $\mathcal{A}$ if $\mu(\emptyset)=0$ and $\mu(A\cup B)=\mu(A)+\mu(B)$ for all $A,B\in\mathcal{A}$

with $A \cap B = \emptyset$. A finitely additive measure μ on $\mathcal{A}$ is called a **measure** if it is σ**-additive**, i.e., $\mu\left(\bigcup_{n\in\mathbb{N}} A_n\right) = \sum_{n=1}^{\infty} \mu(A_n)$ for all pairwise disjoint $A_n \in \mathcal{A}$ such that $\bigcup_{n\in\mathbb{N}} A_n \in \mathcal{A}$. In this case $(\Lambda, \mathcal{A}, \mu)$ is called a **measure space**, provided that $\mathcal{A}$ is a σ-algebra. A measure μ on $\mathcal{A}$ is called **finite** if $\mu(\Lambda) < \infty$, and μ is called a **probability measure** if $\mu(\Lambda) = 1$, in which case $(\Lambda, \mathcal{A}, \mu)$ is called a **probability space**. The measure μ is called a **Borel measure** if $\mathcal{A}$ is a Borel σ-algebra.

Since we are working only with finite measures with the exception of the σ-finite Lebesgue measure on $]0, \infty]$, the Lévy measure on $\mathbb{R}$ and the counting measure on $\mathbb{N}$, it is always assumed that measures μ are σ**-finite**, i.e., there exists a sequence $(A_n)_{n\in\mathbb{N}}$ in $\mathcal{A}$ with $\Lambda = \bigcup_{n\in\mathbb{N}} A_n$ and $\mu(A_n) < \infty$. Here we have to be careful, because σ-finite measures, restricted to a σ-subalgebra of $\mathcal{A}$, and image measures (see below) of a σ-finite measure need not to be σ-finite.

A set $N \in \mathcal{A}$ is called a μ**-nullset** if $\mu(N) = 0$. The set of all μ-nullsets is denoted by $\mathcal{N}_\mu$. A measure space $(\Lambda, \mathcal{A}, \mu)$ is called **complete** if each subset of a μ-nullset belongs to $\mathcal{A}$, whence it is a μ-nullset.

Fix a measure space $(\Lambda, \mathcal{A}, \mu)$. The **Borel–Cantelli lemma** will be often used (see Ash [5] 2.2.4): Let $(U_i)_{i\in\mathbb{N}}$ be a sequence of sets in $\mathcal{A}$ such that $\sum_{i\in\mathbb{N}} \mu(U_i) < \infty$. Then

$$\mu\left(\bigcap_{n\in\mathbb{N}} \bigcup_{n\le i\in\mathbb{N}} U_i\right) = 0.$$

Two functions F, G, defined on Λ, are identified if $G = F$ μ-a.e. (μ-almost everywhere), i.e., $\mu\{X \mid F(X) \neq G(X)\} = 0$. Sometimes let us write μ-a.s. (μ-almost surely) instead of μ-a.e., in case μ is a probability measure. We are interested in L^p-spaces with $p \in [1, \infty]$. For $p \in [1, \infty[$ let $L^p(\mu)$ be the space of all real random variables F such that $|F|^p$ is μ-integrable and set $\|F\|_p := (\mathbb{E}|F|^p)^{\frac{1}{p}}$. Let $L^\infty(\mu)$ be the set of all μ-a.e. bounded random variables and for $F \in L^\infty(\mu)$ set $\|F\|_\infty := \inf\{c > 0 \mid |F| \le c\ \mu\text{-a.e.}\}$. All these L^p-spaces are Banach spaces with norm $\|\cdot\|_p$. We will write L^p instead of $L^p(\mu)$ if it is clear which measure we mean.

Let $(\Lambda, \mathcal{A}, \mu)$ be a probability space. The elements in $\mathcal{A}$ are called **events.** The **expected value** of a real-valued random variable f, i.e., the integral $\int_\Lambda f d\mu$, if it exists, is denoted by $\mathbb{E}_\mu(f)$ or simply $\mathbb{E}(f)$ or $\mathbb{E}f$ if it is clear which measure μ we mean. The **conditional expectation** of a μ-integrable random variable f with respect to a sub-σ-algebra $\mathcal{C}$ of $\mathcal{A}$ is denoted by $\mathbb{E}^{\mathcal{C}}f$ or $\mathbb{E}^{\mathcal{C}}_{\mu}f$ if it is not clear which measure we mean. It is the μ-a.s. uniquely determined

$\mathcal{C}$-measurable function g such that for all $C \in \mathcal{C}$

$$\int_C g\, d\mu = \int_C f\, d\mu.$$

In Ash's book [5] Section 6.5 the reader can find an arrangement of all properties of the conditional expectation we need; in particular, we use Jensen's inequality

$$\left(\mathbb{E}^{\mathcal{C}}|f|\right)^p \le \mathbb{E}^{\mathcal{C}}\left(|f|^p\right) \ \mu\text{-a.s.}$$

over and over again, where $p \in [1, \infty[$ and $|f|^p$ is integrable.

Let $(\Lambda, \mathcal{A}, \mu)$ be a measure space, let $\mathcal{A}'$ be a σ-algebra on a set Λ' and let $f : \Lambda \to \Lambda'$ be $(\mathcal{A}, \mathcal{A}')$**-measurable**, i.e., $f^{-1}[B] \in \mathcal{A}$ for all $B \in \mathcal{A}'$. The measure μ_f, defined on $\mathcal{A}'$ by $\mu_f(B) := \mu\left(f^{-1}[B]\right)$, is called the **image measure of** μ **by** f. If $\mathcal{A}'$ is a Borel σ-algebra, then f is simply called $\mathcal{A}$**-measurable**.

We have the following **transformation rule** (see Bauer [6], 19.1 and 19.2): Let $f : \Lambda \to \mathbb{R}^n$ be $\mathcal{A}$-measurable. Then we have, for all Borel-measurable $g : \mathbb{R}^n \to \mathbb{R}^d$,

$$\int_{\mathbb{R}^n} g\, d\mu_f = \int_\Lambda g \circ f\, d\mu,$$

provided that at least one integral exists.

Assume that $f : \Lambda \to \mathbb{R}_0^+$ is $\mathcal{A}$-measurable. The set function $f\mu : \mathcal{A} \to [0, \infty]$ is defined by setting

$$f\mu(B) := \int_B f\, d\mu = \int_\Lambda \mathbf{1}_B \cdot f\, d\mu.$$

Here $\mathbf{1}_B$ is the **indicator function** of B, i.e., $\mathbf{1}_B(x) = 1$ for $x \in B$ and $\mathbf{1}_B(x) = 0$ for $x \notin B$. By the monotone convergence theorem, $f\mu$ is a measure on $\mathcal{A}$. We say that the measure $f\mu$ has **density** f **with respect to** μ. The previous equality can be extended from $\mathbf{1}_B$ to all $\mathcal{A}$-measurable functions $g : \Lambda \to \mathbb{R}$:

$$\int_\Lambda g\, df\mu = \int_\Lambda g \cdot f\, d\mu,$$

provided that at least one integral exists.

We call a measure μ **absolutely continuous to** a measure ν, where μ and ν are defined on the same domain, if each ν-nullset is also a μ-nullset. Both measures are called **equivalent** if μ is absolutely continuous to ν and ν is absolutely continuous to μ.

The product measure of measures $\mu_1, \ldots, \mu_k$ is denoted by $\mu_1 \otimes \ldots \otimes \mu_k$.

In the following chapter we will use extensions of the previously mentioned **Riesz lemma** for σ-finite measures. The dual space of $L^p(\mu)$ with $1 \leq p < \infty$ is $L^q(\mu)$, where $\frac{1}{p} + \frac{1}{q} = 1$ and $q = \infty$ for $p = 1$, in the following sense: for each $\varphi \in (L^p(\mu))'$ there exists an $\psi \in L^q(\mu)$ such that $\varphi(f) = \int_\Lambda \psi \cdot f d\mu$ for all $f \in L^p(\mu)$.

2

Martingales

In this chapter a detailed introduction to martingale theory is presented. In particular, we study important Banach spaces of martingales with regard to the supremum norm and the quadratic variation norm. The main results show that the martingales in the associated dual spaces are of bounded mean oscillation. The Burkholder–Davis–Gandy (B–D–G) inequalities for L^p-bounded martingales are very useful applications. All results in this chapter are well known; I learned the proofs from Imkeller's lecture notes [47]. We also need the B–D–G inequalities for special Orlicz spaces of martingales.

In this chapter we study martingales, defined on standard finite timelines. Later on the notion 'finite' is extended and the results in this chapter are transferred to a finite timeline, finite in the extended sense. We obtain all established results also for the new finite timeline. Then we shall outline some techniques to convert processes defined on this new finite timeline to processes defined on the continuous timeline $[0,\infty[$ and vice versa. The reader is referred to the fundamental articles of Keisler [53], Hoover and Perkins [46] and Lindstrøm [64].

From what we have now said it follows that we only need to study martingales defined on a discrete, even finite, timeline.

2.1 Martingales and examples

Fix a countable set I, totally ordered by $<$ with smallest element $\square$. If not otherwise determined, we assume that I is finite and $H = \max I$. The set I can be viewed as a **timeline**. Choose an arbitrary object $H^+ \notin I$, define $t < H^+$ for all $t \in I$ and set $\inf \emptyset := H^+$. For each $t \in I$ we set $I_t := \{s \in I \mid s \leq t\}$.

Fix a complete probability space $(\Lambda, \mathcal{C}, \mu)$. Then $\left(\Lambda, \mathcal{C}, \mu, (\mathcal{C}_t)_{t\in I}\right)$ is called an **adapted probability space** if $(\mathcal{C}_t)_{t\in I}$ is a **filtration on** $\mathcal{C}$, i.e., $\mathcal{C}_t$ is a

σ-subalgebra of $\mathcal{C}$ and $\mathcal{C}_s \subseteq \mathcal{C}_t$ for $s \leq t$. The events in $\mathcal{C}_t$ represent the **state of information at time** $t \in I$. We tacitly assume that each $\mathcal{C}_t$ contains all μ-nullsets.

A property $P(X)$ about elements $X \in \Lambda$ holds **almost surely** if the set $\{P \text{ fails}\} := \{X \in \Lambda \mid P(X) \text{ fails}\}$ is a μ-nullset. As it is a common practice, we will write $\{P\}$ instead of $\{X \in \Lambda \mid P(X) \text{ holds}\}$ if P is a property on elements of Λ.

For $F : I \to \mathbb{R}$ and $t \in I$ the difference $\Delta F_t := F(t) - F(t^-)$ is called the **increment of** F **to** t. Here t^- is the immediate predecessor of t if $t > \Box$. Set $F(\Box^-) := 0$ and assume that $\Box^- < t$ for all $t \in I$. So $\Delta F_\Box = F_\Box$ is the first 'jump' of F. As usual, we write F_t instead of $F(t)$. Set $\mathcal{C}_{\Box^-} := \{\emptyset, \Lambda\} \vee \mathcal{N}_\mu$. Then $\mathbb{E}^{\mathcal{C}_{\Box^-}} F = \mathbb{E} F$ μ-a.s. for all $F \in L^1$.

A process $F : \Lambda \times I \to \mathbb{R}$ is called μ-p-**integrable** if $F_t \in L^p(\mu)$ for all $t \in I$. If F is μ-1-integrable, then we simply say F is μ-**integrable** or simply **integrable**. A process $M := (M_t)_{t\in I}$ is called a $(\mathcal{C}_t)_{t\in I}$-μ-**martingale** if the following conditions are fulfilled.

(a) $(M_t)_{t\in I}$ is $(\mathcal{C}_t)_{t\in I}$-**adapted**, i.e., M_t is $\mathcal{C}_t$-measurable for all $t \in I$.
(b) M is integrable.
(c) $\mathbb{E}^{\mathcal{C}_s} M_{s^+} = M_s$ μ-a.s. if $s^+ \in I$ is the immediate successor of $s \in I$.

If under (c) we have "$\geq$" instead of "$=$", then M is called an $(\mathcal{C}_t)_{t\in I}$-μ-**submartingale**. By Jensen's inequality, $|M|^p$ with $1 \leq p < \infty$ is a $(\mathcal{C}_t)_{t\in I}$-μ-submartingale if M is a $(\mathcal{C}_t)_{t\in I}$-μ-martingale and M is p-integrable.

If we understand $M_t(X)$ as the result of the chance X at time $t \in I$, then condition (a) means that the result at time t does not depend on what will happen after time t. Condition (c) means that, under the present state $\mathcal{C}_t$ of information, the expected result at the future time t^+ is identical to the achieved result at the present time t.

Let us drop $(\mathcal{C}_t)_{t\in I}$ or μ in the phrases martingale or submartingale if it is clear which filtration or measure we mean. We call $F : \Lambda \times I \to \mathbb{R}$ a **canonical martingale** if F is a $\left(\mathcal{C}_t^F\right)_{t\in I}$-martingale, where $\left(\mathcal{C}_t^F\right)_{t\in I}$ is the **filtration generated by** F, i.e.,

$$\mathcal{C}_t^F = \left\{(\Delta F_s)_{s\in I_t}^{-1}[B] \mid B \text{ is a Borel set in } \mathbb{R}^{I_t}\right\}.$$

Examples 2.1.1 Let $N : \Lambda \to \mathbb{R}$ be μ-integrable.

(i) $\left(\mathbb{E}^{\mathcal{C}_t} N\right)_{t\in I}$ is a martingale.
(ii) $\mathbf{1}_A \cdot M$ is a martingale if $A \in \mathcal{C}_\Box$ and M is a martingale.
(iii) $M : (\cdot, t) \mapsto \mathbf{1}_A(\cdot) \cdot \mathbf{1}_{[s,H]}(t) \cdot \left(\mathbb{E}^{\mathcal{C}_t} N - \mathbb{E}^{\mathcal{C}_s} N\right)$ is a martingale if $s \in I$ and $A \in \mathcal{C}_s$.

(iv) Fix $s \in I$ and suppose that N is $\mathcal{C}_s$-measurable. Set for all $t \in I$

$$M_t := \begin{cases} N & \text{if } s = \square \\ \mathbf{1}_{[s,H]}(t) \cdot \left(N - \mathbb{E}^{\mathcal{C}_{s^-}} N\right) & \text{otherwise.} \end{cases}$$

Then $M := (M_t)_{t \in I}$ is a martingale in both cases.

(v) Fix an integrable process K. Set

$$M_t := \sum_{\square < s \leq t} \left(\mathbb{E}^{\mathcal{C}_s} K_s - \mathbb{E}^{\mathcal{C}_{s^-}} K_s\right).$$

Then $(M_t)_{t \in I}$ is a martingale with $M_\square = 0$.

We have the following simple characterization of martingales, where the proof of '(a)$\Rightarrow$(b)' uses induction on t.

Proposition 2.1.2 *Fix a $(\mathcal{C}_t)_{t \in I}$-adapted sequence $(M_t)_{t \in I}$ of random variables such that $M_H \in L^1$. Then the following conditions (a), (b), (c) and (d) are equivalent.*

(a) *$(M_t)_{t \in I}$ is a martingale.*
(b) *$\mathbb{E}^{\mathcal{C}_s} M_t = M_s$ μ-a.s. for all $s, t \in I$ with $s < t$.*
(c) *$\mathbb{E} \mathbf{1}_A M_s = \mathbb{E} \mathbf{1}_A M_t$ for all $s, t \in I$ with $s < t$ and all $A \in \mathcal{C}_s$.*
(d) *$\mathbb{E} \mathbf{1}_A (M_{s^+} - M_s) = 0$ for all $s, s^+ \in I$ and all $A \in \mathcal{C}_s$.*

The **quadratic variation** $[M] : \Lambda \times I \to \mathbb{R}$ of a martingale is defined by

$$[M]_t := \sum_{s \leq t} (\Delta M_s)^2 = \sum_{s \leq t} (M_s - M_{s^-})^2.$$

We often use the following properties of martingales:

Lemma 2.1.3 *Fix martingales M, N and $s, t, r \in I$ with $s \leq t < r$ and suppose that $M_H, N_H \in L^2$. Then*

(a) *$\mathbb{E}^{\mathcal{C}_s} (M_r - M_t)^2 = \mathbb{E}^{\mathcal{C}_s} \left(M_r^2 - M_t^2\right)$ μ-a.s.*
(b) *$\mathbb{E} (M_s \cdot N_s) = \mathbb{E} \left(\sum_{i \leq s} ((M_i - M_{i^-}) \cdot (N_i - N_{i^-}))\right)$.*
This equality remains true if $M_H \in L^1$ and $N_H \in L^\infty$.
(c) *$\mathbb{E} M_s^2 = \mathbb{E} [M]_s$.*
(d) *$\mathbb{E}^{\mathcal{C}_s} \sum_{i \in [s,t]} (M_i - M_{i^-})^2 = \mathbb{E}^{\mathcal{C}_s} (M_t - M_{s^-})^2$ μ-a.s.*

Proof (a) follows from the fact that μ-a.s.

$$\mathbb{E}^{\mathcal{C}_s} (M_r \cdot M_t) = \mathbb{E}^{\mathcal{C}_s} \mathbb{E}^{\mathcal{C}_t} (M_r \cdot M_t) = \mathbb{E}^{\mathcal{C}_s} \left(\mathbb{E}^{\mathcal{C}_t} M_r \cdot M_t\right) = \mathbb{E}^{\mathcal{C}_s} \left(M_t^2\right).$$

(b) Since $\mathbb{E}(M_s \cdot N_s) = \mathbb{E}\left(\sum_{i \le s} (M_i - M_{i^-}) \cdot \sum_{j \le s} (N_j - N_{j^-})\right)$, the result follows from the fact that for $i < j$

$$\mathbb{E}\left((M_i - M_{i^-}) \cdot (N_j - N_{j^-})\right) = \mathbb{E}\left((M_i - M_{i^-}) \cdot \mathbb{E}^{\mathcal{C}_{j^-}}(N_j - N_{j^-})\right) = 0.$$

(c) follows from (b), and (d) follows from (a). □

For a fixed filtration $(\mathcal{C}_t)_{t \in I}$ and a fixed probability measure μ the set of $(\mathcal{C}_t)_{t \in I}$-$\mu$-martingales constitutes a linear space over $\mathbb{R}$.

Later on we shall use martingales on the continuous timeline $[0, \infty[$. It will be seen that there is a very close relationship between martingales on $[0, \infty[$ and martingales on a finite timeline, finite in the sense of poly-saturated models.

The **product σ-algebra** of $\mathcal{C}$ and the Lebesgue σ-algebra $\mathrm{Leb}[0, \infty[$ on $[0, \infty[$, augmented by all $\mu \otimes \lambda$-nullsets, is denoted by $\mathcal{C} \otimes \mathrm{Leb}[0, \infty[$. A $\mathcal{C} \otimes \mathrm{Leb}[0, \infty[$-measurable function $g : \Lambda \times [0, \infty[\to \mathbb{R}$ is called a **stochastic process**, or a **process** for short. A **c-filtration** on $\mathcal{C}$ is a $[0, \infty[$-sequence $(\mathfrak{f}_t)_{t \in [0, \infty[}$ of σ-algebras $\mathfrak{f}_t \subseteq \mathcal{C}$ such that $\mathfrak{f}_t \subseteq \mathfrak{f}_s$ for $t \le s$. We agree that a c-filtration $(\mathfrak{f}_t)_{t \in [0, \infty[}$ always fulfils the so called **usual (Doob–Meyer) conditions**, i.e.,

(a) $\mathfrak{f}_0$ contains all μ-nullsets and
(b) $(\mathfrak{f}_t)_{t \in [0, \infty[}$ is **right continuous**, i.e., $\mathfrak{f}_t = \bigcap_{s > t} \mathfrak{f}_s$.

A measurable set $B \in \mathcal{C} \otimes \mathrm{Leb}[0, \infty[$ is called **strongly $(\mathfrak{f}_t)_{t \in [0, \infty[}$-adapted** if for each $s \in [0, \infty[$ the section

$$B(\cdot, s) := \{x \in \Lambda \mid (x, s) \in B\} \in \mathfrak{f}_s.$$

Let $\mathfrak{A}_0$ be the set of strongly $(\mathfrak{f}_t)$-adapted sets. Note that $\mathfrak{A}_0$ is a σ-algebra. Set $\mathfrak{A} := \mathfrak{A}_0 \vee \mathcal{N}_{\mu \otimes \lambda}$. A function $f : \Lambda \times [0, \infty[\to \mathbb{R}$ is called **$(\mathfrak{f}_t)_{t \in [0, \infty[}$-adapted** (**strongly $(\mathfrak{f}_t)_{t \in [0, \infty[}$-adapted**) if f is $\mathfrak{A}$-($\mathfrak{A}_0$-)measurable.

A strongly $(\mathfrak{f}_t)_{t \in [0, \infty[}$-adapted process m is called an **$(\mathfrak{f}_t)_{t \in [0, \infty[}$-martingale** if m_t is integrable for each $t \in [0, \infty[$ and if for all $s, t \in [0, \infty[$ with $s < t$

$$\mathbb{E}^{\mathfrak{f}_s}(m_t) = m_s \, \mu\text{-a.s.}$$

Let us now return to martingales defined on an ordered finite timeline I.

2.2 Stopping times

Stopping time techniques provide a powerful tool to truncate martingales without losing the martingale property.

Define $\overline{I} := I \cup \{H^+\}$ and $\mathcal{C}_{H^+} := \mathcal{C}_H$. A function $\tau : \Lambda \to I \cup \{H^+\}$ is called a $(\mathcal{C}_t)_{t \in I}$-**stopping time** if for each $t \in I$, $\{\tau \leq t\} \in \mathcal{C}_t$. Note that, if τ is a $(\mathcal{C}_t)_{t \in T}$-stopping time, then $\{\tau = t\} \in \mathcal{C}_t$ for each $t \in \overline{I}$. We drop $(\mathcal{C}_t)_{t \in I}$- if it is clear which filtration we mean. Here is a perfect example of a stopping time:

Example 2.2.1 Let $A : \Lambda \times I \to \mathbb{R}$ be $(\mathcal{C}_t)_{t \in I}$-adapted. Define for $c \in \mathbb{R}$ and $X \in \Lambda$

$$\tau(X) := \inf \{t \in I \mid |A_t(X)| \geq c\}.$$

(Recall that $\inf \emptyset = H^+$.) Then τ is a stopping time.

Proof We have for each $t \in I$

$$\{\tau \leq t\} = \{|A_{\square}| \geq c\} \cup \ldots \cup \{|A_t| \geq c\} \in \mathcal{C}_t.$$

□

Proposition 2.2.2 *Let M be a martingale and let τ be a stopping time. Then the truncated process*

$$M^\tau : \Lambda \times I \to \mathbb{R}, (X,t) \mapsto M(X, \tau(X) \wedge t)$$

is a martingale.

Proof For all $t \in I$, M_t^τ is $\mathcal{C}_t$-measurable, because for all $c \in \mathbb{R}$

$$\{M_t^\tau \leq c\} = \bigcup_{i=\square}^{t} \{\tau = i \wedge M_i \leq c\} \cup \{\tau > t \wedge M_t \leq c\} \in \mathcal{C}_t.$$

In order to prove the martingale property, fix $t \in I$ with $t < H$ and $A \in \mathcal{C}_t$. Then, by Proposition 2.1.2 (d),

$$\mathbb{E}\mathbf{1}_A \left(M_{t^+}^\tau - M_t^\tau\right) = \mathbb{E}\mathbf{1}_{A \cap \{\tau > t\}} \left(M_{t^+} - M_t\right) = 0.$$

□

2.3 The maximum inequality

We use the maximum inequality in order to prove Doob's inequality.

Proposition 2.3.1 (Doob [33]) *Fix $p \in [1, \infty[$ and a non-negative submartingale N such that $N_H \in L^p$. Then for each $c \geq 0$*

$$c \cdot \mu \left\{ \max_{s \in I} N_s^p \geq c \right\} \leq \mathbb{E}\mathbf{1}_{\left\{\max_{s \in I} N_s^p \geq c\right\}} N_H^p.$$

Proof We define a stopping time τ by

$$\tau := \inf\left\{t \in I \mid N_t^p \geq c\right\}.$$

Recall that $\inf \emptyset = H^+$. Set $F := \max_{s \in I} N_s^p$. Then we obtain

$$\begin{aligned} c \cdot \mu\{F \geq c\} &= \mathbb{E}\mathbf{1}_{\{F \geq c\}} c \leq \mathbb{E}\mathbf{1}_{\{\tau \in I\}} N^p(\cdot, \tau(\cdot) \wedge H) \\ &= \mathbb{E}N^p(\cdot, \tau(\cdot) \wedge H) - \mathbb{E}\mathbf{1}_{\{\tau = H^+\}} N_H^p \\ &= \mathbb{E}N^p(\cdot, \tau(\cdot) \wedge H) - \mathbb{E}N_H^p + \mathbb{E}\mathbf{1}_{\{\tau \leq H\}} N_H^p. \end{aligned}$$

We estimate the first summand. Recall that $\mathcal{C}_{H^+} := \mathcal{C}_H = \mathcal{C}$. Since $\left\{\tau = H^+\right\}$ belongs to $\mathcal{C}$, we obtain

$$\begin{aligned} \mathbb{E}N^p(\cdot, \tau(\cdot) \wedge H) &= \sum_{i=1}^{H^+} \mathbb{E}\mathbf{1}_{\{\tau = i\}} N_{i \wedge H}^p \\ &\leq \sum_{i=\square}^{H^+} \mathbb{E}\mathbf{1}_{\{\tau = i\}} \left(\mathbb{E}^{\mathcal{C}_i} N_H\right)^p \leq \sum_{i=\square}^{H^+} \mathbb{E}\mathbf{1}_{\{\tau = i\}} \mathbb{E}^{\mathcal{C}_i} N_H^p \\ &= \sum_{i=\square}^{H^+} \mathbb{E}\mathbb{E}^{\mathcal{C}_i} \mathbf{1}_{\{\tau = i\}} N_H^p = \sum_{i=\square}^{H^+} \mathbb{E}\mathbf{1}_{\{\tau = i\}} N_H^p = \mathbb{E}N_H^p. \end{aligned}$$

Therefore,

$$c \cdot \mu\{F \geq c\} \leq \mathbb{E}\mathbf{1}_{\{\tau \leq H\}} N_H^p = \mathbb{E}\mathbf{1}_{\left\{\max_{s \in I} N_s^p \geq c\right\}} N_H^p.$$

This proves the result. □

2.4 Doob's inequality

The next two very useful results are due to Doob. The results tell us that for $p > 1$ the L^p bounded martingales can be identified with the L^p-spaces, independent on the length of I (see also Remark 2.7.3 below).

Lemma 2.4.1 (Doob [33]) *Fix $p \in]1, \infty[$ and $\mathcal{C}$-measurable $F, G : \Lambda \to \mathbb{R}_0^+$ such that $G \in L^p$. Moreover, let us assume that*

$$c \cdot \mu(\{c \leq F\}) \leq \int_{\{c \leq F\}} G d\mu \quad \textit{for each positive } c \in \mathbb{R}. \tag{1}$$

Then

$$\|F\|_p \leq \frac{p}{p-1}\|G\|_p.$$

Proof We may assume that $\|F\|_p > 0$. First let F be bounded in $\mathbb{R}$. By the transformation rule, Fubini's theorem and Hölder's inequality, we obtain

$$\begin{aligned}
\int_\Lambda F^p d\mu &= \int_{[0,\infty[} y^p d\mu_F(y) = \int_{[0,\infty[}\int_{]0,y]} px^{p-1} dx d\mu_F(y) \\
&= \int_{]0,\infty[}\int_{[x,\infty[} px^{p-1} d\mu_F dx = \int_{]0,\infty[} px^{p-1}\mu_F([x,\infty[)dx \\
&\leq \int_{]0,\infty[}\int_{\{x\leq F\}} Gpx^{p-2} d\mu dx \quad \text{(by (1))} \\
&= \int_\Lambda \int_{]0,F(X)]} px^{p-2} dx G(X) d\mu(X) \\
&= \int_\Lambda \frac{p}{p-1} F^{p-1}\cdot G d\mu \leq \frac{p}{p-1}\left(\int_\Lambda F^p d\mu\right)^{\frac{p-1}{p}}\cdot\left(\int_\Lambda G^p d\mu\right)^{\frac{1}{p}}.
\end{aligned}$$

Dividing both ends of this computation by $\left(\int_\Lambda F^p d\mu\right)^{\frac{p-1}{p}}$, we obtain the result. Now assume that F is unbounded. It will be seen that for $n \in \mathbb{N}$

$$c\cdot\mu\{c \leq F\wedge n\} \leq \int_{\{c\leq F\wedge n\}} G d\mu \quad \text{for each positive } c \in \mathbb{R}. \tag{2}$$

If $n < c$, then $\{c \leq F\wedge n\} = \emptyset$. If $c \leq n$, then $\{c \leq F\wedge n\} = \{c \leq F\}$ and (2) follows from (1). We obtain, by Lebesgue's theorem,

$$\|F\|_p = \lim_{n\to\infty}\left(\int_\Lambda (F\wedge n)^p d\mu\right)^{\frac{1}{p}} \leq \frac{p}{p-1}\|G\|_p.$$

□

Theorem 2.4.2 (Doob [33]) *Fix a submartingale $M : \Lambda\times I \to \mathbb{R}_0^+$ and $p > 1$ and suppose that $M_H \in L^p$. Then*

$$\left\|\max_{t\in I} M_t\right\|_p \leq \frac{p}{p-1}\|M_H\|_p.$$

Proof We set

$$F := \max_{t\in I} M_t \text{ and } G := M_H.$$

Note that F is $\mathcal{C}$-measurable. By Proposition 2.3.1, for each $c \geq 0$,

$$c\cdot\mu(\{c \leq F\} \leq \int_{\{c\leq F\}} G d\mu.$$

By Lemma 2.4.1,

$$\left\| \max_{t \in I} M_t \right\|_p \leq \frac{p}{p-1} \|M_H\|_p . \qquad \square$$

2.5 The σ-algebra over the past of a stopping time

Let τ be a stopping time. Then we define for each process $F : \Lambda \times I \to \mathbb{R}$ the random variable F_τ by setting

$$F_\tau(X) := F(X, \tau(X) \wedge H).$$

Notice the difference between F_τ and F^τ, $F_\tau(X) = F^\tau(X, H)$. The σ-algebra

$$\mathcal{C}_\tau := \{A \in \mathcal{C} \mid A \cap \{\tau \leq t\} \in \mathcal{C}_t \text{ for all } t \in I\}$$

is called the **τ-past**. Note that

$$\mathcal{C}_\tau := \{A \in \mathcal{C} \mid A \cap \{\tau = t\} \in \mathcal{C}_t \text{ for all } t \in I\}.$$

Since the range of a stopping time is finite, the proof of the following lemma is simple and is left to the reader.

Lemma 2.5.1 *Fix stopping times τ and σ.*

(a) *τ is $\mathcal{C}_\tau$-measurable.*
(b) *If $\tau = s$ is constant, then $\mathcal{C}_\tau = \mathcal{C}_s$.*
(c) *If a process F is $(\mathcal{C}_t)_{t \in I}$-adapted, then F_τ is $\mathcal{C}_\tau$-measurable.*
(d) *If $\tau \leq \sigma$, then $\mathcal{C}_\tau \subseteq \mathcal{C}_\sigma$.*

The following result shows that the martingale property for the timeline I can be extended to the martingale property for stopping times:

Proposition 2.5.2 *Fix stopping times $\tau \leq \sigma$ and a martingale M. Then $\mathbb{E}^{\mathcal{C}_\tau} M_\sigma = M_\tau$ μ-a.s. It follows that, if $J \subset \mathbb{N}$ and $(\tau_n)_{n \in J}$ is a sequence of stopping times with $\tau_n \leq \tau_{n+1}$, then $\left(M_{\tau_n}\right)_{n \in J}$ is a $\left(\mathcal{C}_{\tau_n}\right)_{n \in J}$-martingale.*

Proof First we will show that $\mathbb{E}^{\mathcal{C}_\tau} M_H = M_\tau$ μ-a.s. Fix $A \in \mathcal{C}_\tau$. Then

$$\begin{aligned}
\mathbb{E}\mathbf{1}_A M_H &= \sum_{i=\square}^{H^+} \mathbb{E}\mathbf{1}_{A \cap \{\tau = i\}} M_H = \sum_{i=\square}^{H^+} \mathbb{E}\mathbf{1}_{A \cap \{\tau = i\}} \mathbb{E}^{\mathcal{C}_i} M_H \\
&= \sum_{i=\square}^{H^+} \mathbb{E}\mathbf{1}_{A \cap \{\tau = i\}} M_{i \wedge H} = \mathbb{E}\mathbf{1}_A M_\tau .
\end{aligned}$$

The general assertion follows from

$$\mathbb{E}^{C_\tau} M_\sigma = \mathbb{E}^{C_\tau} \mathbb{E}^{C_\sigma} M_H = \mathbb{E}^{C_\tau} M_H = M_\tau \ \mu\text{-a.s.}$$

□

2.6 L^p-spaces of martingales and the quadratic variation norm

Let us assume that martingales are defined on an ordered finite timeline I with $H := \max I$. Since a martingale M is μ-a.s. determined by M_H, we may identify the p-integrable martingales with the elements in L^p for $p \in [1, \infty[$. Let us identify two martingales M, N, provided that $M_H = N_H$ μ-a.s. We will use the **normalized counting measure** ν on I, setting $\nu(A) := \frac{|A|}{|I|}$ for each subset A of I, where $|A|$ denotes the number of elements of A.

Let $L^p_{[}$ be the set of martingales M such that $[M]_H^{\frac{1}{2}} \in L^p$. Recall that $[M]_H^{\frac{1}{2}} = \left(\sum_{s \in I} |\Delta M_s|^2\right)^{\frac{1}{2}}$. For each $M \in L^p_{[}$ set

$$\|M\|_{[,p} := \left\| [M]_H^{\frac{1}{2}} \right\|_p = \left(\mathbb{E}[M]_H^{\frac{p}{2}}\right)^{\frac{1}{p}}.$$

The spaces $L^p_{[}$ and L^p can be identified in the following sense.

Lemma 2.6.1

(a) *Fix $F \in L^p$ and set $M := \left(\mathbb{E}^{C_t} F\right)_{t \in I}$. Then $\|M\|_{[,p} \leq 2|I| \, \|F\|_p$.*
(b) *Fix $M \in L^p_{[}$. Then $\|M_H\|_p \leq \sqrt{|I|} \, \|M\|_{[,p}$.*

Proof Let ν be the normalized counting probability measure on I, i.e., for all subsets $A \subseteq I$,

$$\nu(A) := \frac{|A|}{|I|},$$

where $|A|$ is the number of elements of A.

(a) Since M is a martingale, we obtain

$$\|M\|_{[\cdot,p} = \left(\mathbb{E}\left(\sum_{s \in I} \Delta M_s^2\right)^{\frac{p}{2}}\right)^{\frac{1}{p}} \leq \left(\mathbb{E}\left(\sum_{s \in I} |\Delta M_s|\right)^p\right)^{\frac{1}{p}}$$

$$\leq 2\left(\mathbb{E}\left(\sum_{s \in I} |M_s|\right)^p\right)^{\frac{1}{p}} = 2\left(\mathbb{E}\left(\sum_{s \in I} |\mathbb{E}^{C_s} F|\right)^p\right)^{\frac{1}{p}}$$

$$= 2|I|\left(\mathbb{E}\left(\int_I |\mathbb{E}^{\mathcal{C}_s}F|\,d\nu(s)\right)^p\right)^{\frac{1}{p}} \leq 2|I|\left(\int_I \mathbb{E}|\mathbb{E}^{\mathcal{C}_s}F|^p\,d\nu(s)\right)^{\frac{1}{p}}$$

$$\leq 2|I|\left(\int_I \mathbb{E}\mathbb{E}^{\mathcal{C}_s}(|F|)^p\,d\nu(s)\right)^{\frac{1}{p}} = 2|I|\left(\mathbb{E}\int_I |F|^p\,d\nu\right)^{\frac{1}{p}}$$

$$= 2|I|\,\|F\|_p.$$

(b)

$$\|M_H\|_p = \left\|\sum_{s\in I}\Delta M_s\right\|_p \leq \left\|\sum_{s\in I}|\Delta M_s|\right\|_p = |I|\left\|\int_I |\Delta M_s|\,d\nu(s)\right\|_p$$

$$\leq |I|\left\|\left(\int_I |\Delta M_s|^2\,d\nu(s)\right)^{\frac{1}{2}}\right\|_p = \frac{|I|}{\sqrt{|I|}}\left\|\left(\sum_{s\in I}|\Delta M_s|^2\right)^{\frac{1}{2}}\right\|_p$$

$$= \sqrt{|I|}\,\|M\|_{[,p}.$$

□

Corollary 2.6.2

(a) $\left(L^p_{[}, \|\cdot\|_{[,p}\right)$ *is a Banach space.*

(b) $\left(L^2_{[}, \|\cdot\|_{[,2}\right)$ *is a Hilbert space with scalar product* $\langle M, N\rangle := \mathbb{E}M_H N_H$.

(c) $\left(L^2_{[}, \|\cdot\|_{[,2}\right)$ *is dense in* $\left(L^1_{[}, \|\cdot\|_{[,1}\right)$.

Proof (a) Using the triangle equalities for the Euclidean norm in $\mathbb{R}^{|I|}$ and for the norm $\|\cdot\|_p$ in L^p, we see that $\|\cdot\|_{[,p}$ is a norm. The completeness follows from Lemma 2.6.1.

(b) By Lemma 2.1.3 (c), $\|M\|_{[,2} = \sqrt{\langle M_H, M_H\rangle}$.

(c) Fix $M \in L^1_{[}$. By Lemma 2.6.1 (b), $M_H \in L^1$. For each $k \in \mathbb{N}$ set

$$N^k(X) := \begin{cases} M_H(X) & \text{if } |M_H|(X) \leq k \\ 0 & \text{otherwise.} \end{cases}$$

Set $M^k := \left(\mathbb{E}^{\mathcal{C}_t}N^k\right)_{t\in I}$. Then $M^k \in L^2_{[}$. Since N^k converges to M_H in L^1, by Lemma 2.6.1 (a), M^k converges to M in $L^1_{[}$. □

2.7 The supremum norm

For $p \in [1, \infty[$ let $L^p_\sim$ be the space of all martingales M such that $\max_{t\in I} |M_t| \in L^p$. For each $M \in L^p_\sim$ set

$$\|M\|_{\sim,p} := \left(\mathbb{E} \max_{t\in I} |M_t|^p\right)^{\frac{1}{p}}.$$

In analogy to Lemma 2.6.1, we shall now see that $L^p_\sim$ and L^p can be identified.

Lemma 2.7.1 *Assume that M is a martingale with $M_H \in L^p$. For $p > 1$*

$$\|M_H\|_p \le \|M\|_{\sim,p} \le \frac{p}{p-1} \|M_H\|_p,$$

and for $p = 1$

$$\|M_H\|_1 \le \|M\|_{\sim,1} \le |I| \, \|M_H\|_1.$$

Proof The first assertion is Doob's inequality; the second one follows from

$$\|M\|_{\sim,1} \le \sum_{s\in I} \mathbb{E} |M_s| = \sum_{s\in I} \mathbb{E} \left|\mathbb{E}^{C_s} M_H\right| \le \sum_{s\in I} \mathbb{E}\mathbb{E}^{C_s} |M_H| = |I| \, \|M_H\|_1.$$

□

Corollary 2.7.2 $(L^p_\sim, \|\cdot\|_{\sim,p})$ *is a Banach space and* $(L^2_\sim, \|\cdot\|_{\sim,2})$ *is dense in* $(L^1_\sim, \|\cdot\|_{\sim,1})$.

Proof The first assertion follows from Lemma 2.7.1. the proof of the second assertion is similar to the proof of Corollary 2.6.2 (c). □

We end this section with the following important remark.

Remark 2.7.3 By Lemmas 2.7.1 and 2.6.1, we may identify the spaces $L^p_\sim$ and $L^p_{[}$. However, in the estimates in these lemmas, the constants depend on the length $|I|$ of I. It is important to have these constants independent of the length of I, which is the topic of the following sections.

2.8 Martingales of bounded mean oscillation

Now other important norms on spaces of martingales are presented. For each $p \in [1, \infty[$ let BMO_p be the linear space of martingales M such that $\mathbb{E}^{C_s} |M_H - M_{s^-}|^p$ is bounded μ-a.s. for all $s \in I$. For all $M \in BMO_p$ set

$$\|M\|_{B,p} := \inf\left\{c \ge 0 \mid \mathbb{E}^{C_s} |M_H - M_{s^-}|^p \le c^p \ \mu\text{-a.s. for all } s \in I\right\}.$$

Note that $BMO_p \subseteq BMO_q$ if $1 \leq q \leq p$, thus $BMO_p \subseteq BMO_1$ for all $p \geq 1$. By the following lemma, the martingales in BMO_1 are bounded μ-a.s., thus, they belong to $L^p_{[}$ and $L^p_{\sim}$ (see Remark 2.7.3).

Lemma 2.8.1 *Fix $\Phi \in BMO_1$ and $s \in I$. Then*

(a) $|\Phi_s - \Phi_{s^-}| \leq \|\Phi\|_{B,1}$ *μ-a.s.*
(b) $|\Phi_s| \leq |I| \cdot \|\Phi\|_{B,1}$ *μ-a.s., thus Φ is bounded μ-a.s.*

Proof We obtain μ-a.s.

(a) $|\Phi_s - \Phi_{s^-}| = \left|\mathbb{E}^{C_s}(\Phi_H - \Phi_{s^-})\right| \leq \mathbb{E}^{C_s}|(\Phi_H - \Phi_{s^-})| \leq \|\Phi\|_{B,1}$.
(b) $|\Phi_s| = \left|\sum_{\rho \leq s}(\Phi_s - \Phi_{s^-})\right| \leq \sum_{\rho \leq s}|\Phi_s - \Phi_{s^-}| \leq |I| \cdot \|\Phi\|_{B,1}$. □

It follows that $\|\cdot\|_{B,p}$ defines a norm on BMO_p. In the following two sections we shall see that BMO_2 is the dual of $L^1_{[}$ and BMO_1 is the dual of $L^1_{\sim}$ in the following sense:

Theorem 2.8.2 *There exist linear bijective mappings*

$$\alpha : (L^1_{[})' \to BMO_2, \quad \beta : (L^1_{\sim})' \to BMO_1$$

such that, for $\varphi \in \left(L^1_{[}\right)'$ and $\Phi \in BMO_2$,

$$\frac{1}{3}\|\alpha(\varphi)\|_{B,2} \leq \|\varphi\|_{[,1} := \sup_{M \in L^1_{[}, \|M\|_{[,1} \leq 1} |\varphi(M)|,$$

$$\left\|\alpha^{-1}(\Phi)\right\|_{[,1} \leq \sqrt{2}\|\Phi\|_{B,2};$$

and, for $\varphi \in (L^1_{\sim})'$ and $\Phi \in BMO_1$,

$$\frac{1}{4}\|\beta(\varphi)\|_{B,1} \leq \|\varphi\|_{\sim,1} := \sup_{M \in L_{\sim,1}, \|M\|_{\sim,1} \leq 1} |\varphi(M)|,$$

$$\left\|\beta^{-1}(\Phi)\right\|_{\sim,1} \leq 8\|\Phi\|_{B,1}.$$

It follows that $\alpha, \beta, \alpha^{-1}, \beta^{-1}$ are continuous mappings. We shall present the construction of α and β and prove the preceding four inequalities. Note that the constants there are independent of the length of I!

2.9 $\left(L^1_{[}\right)'$ is $\mathbf{BMO_2}$

In the following sense we have $\left(L^1_{[}\right)' \subseteq BMO_2$:

Proposition 2.9.1 *Fix* $\varphi \in \left(L^1_{[}\right)'$. *Then there exists a* $\Phi \in BMO_2$ *with*

$$\varphi(M) = \mathbb{E}M_H \cdot \Phi_H \text{ for all } M \in L^1_{[} \quad \text{and} \quad \|\Phi\|_{B,2} \leq 3\,\|\varphi\|_{[,1}.$$

Note that Φ *is* μ*-a.s. uniquely determined by* φ. *Therefore we can define*

$$\alpha(\varphi) := \Phi.$$

Proof Since $\varphi \restriction L^2_{[} \in \left(L^2_{[}\right)'$, there exists a $\Phi \in L^2_{[}$ such that for all $M \in L^2_{[}$

(i) $\varphi(M) = \mathbb{E}M_H \cdot \Phi_H \; (= \langle \Phi, M \rangle)$.

We will now prove

(ii) $|\Phi_s - \Phi_{s^-}| \leq 2\,\|\varphi\|_{[,1}$ μ-a.s. for all $s \in I$.

Fix $A \in \mathcal{C}_s$ and set $A^+ := \{\Phi_{s^-} \leq \Phi_s\} \cap A$ and $A^- := \{\Phi_s < \Phi_{s^-}\} \cap A$. Note that it suffices to show that (3) and (4) are true with

$$\mathbb{E}\mathbf{1}_{A^+}(\Phi_s - \Phi_{s^-}) \leq 2\mu(A^+)\,\|\varphi\|_{[,1}, \tag{3}$$

$$\mathbb{E}\mathbf{1}_{A^-}(\Phi_s - \Phi_{s^-}) \leq 2\mu(A^-)\,\|\varphi\|_{[,1}. \tag{4}$$

We only prove (3); the proof of (4) is similar. If $s = \square$, set for all $t \in I$, $N_t := \mathbf{1}_{A^+}$ and, if $\square < s$, set

$$N_t := \mathbf{1}_{[s,H]}(t)\left(1_{A^+} - \mathbb{E}^{\mathcal{C}_{s^-}}\mathbf{1}_{A^+}\right).$$

By Example 2.1.1 (iv), $N := (N_t)_{t \in I}$ is a martingale in both cases. By Lemma 2.6.1 (a), $N \in L^2_{[}$. Now,

$$\begin{aligned}
\mathbb{E}\mathbf{1}_{A^+}(\Phi_s - \Phi_{s^-}) &= \mathbb{E}\sum_{t \in I}(N_t - N_{t^-})(\Phi_t - \Phi_{t^-}) \\
&= \mathbb{E}N_H \Phi_H \quad \text{(by 2.1.3 (b))} \\
&= \varphi(N) \leq \|\varphi\|_{[,1} \cdot \|N\|_{[,1} = \|\varphi\|_{[,1}\,\mathbb{E}\left(\left(\sum_{t \in I}(N_t - N_{t^-})^2\right)^{\frac{1}{2}}\right) \\
&= \|\varphi\|_{[,1}\,\mathbb{E}\left(\left(N_s^2\right)^{\frac{1}{2}}\right) = \|\varphi\|_{[,1}\,\mathbb{E}|N_s| \leq 2\,\|\varphi\|_{[,1}\,\mu(A^+).
\end{aligned}$$

This proves (ii). It follows that Φ_H is bounded μ-a.s., because

$$|\Phi_H| \leq \sum_{s\in I} |\Phi_s - \Phi_{s^-}| \leq 2\,|I|\,\|\varphi\|_{[,1}\ \mu\text{-a.s.}$$

By Corollary 2.6.2 (c), (i) is true for all $M \in L^1_{[}$. Now we will prove

(iii) $\mathbb{E}^{\mathcal{C}_s}(\Phi_H - \Phi_s)^2 \leq \left(\|\varphi\|_{[,1}\right)^2$ μ-a.s. for all $s \in I$.

Fix $s \in I$ and $A \in \mathcal{C}_s$ and set $\Psi := \mathbf{1}_A \cdot (\Phi_H - \Phi_s)$. It suffices to prove that $\mathbb{E}\Psi^2 \leq \mu(A) \cdot \left(\|\varphi\|_{[,1}\right)^2$. We may assume that $0 \neq \Psi$ in L^2. By the Hahn–Banach theorem, there exists an $f \in \left(L^2\right)' = L^2$ such that $\|f\|_2 \leq 1$ and $f(\Psi) = \mathbb{E}f \cdot \Psi = \|\Psi\|_2$. Therefore, it suffices to prove:

(iv) $\mathbb{E}f \cdot \Psi \leq \sqrt{\mu(A)} \cdot \|\varphi\|_{[,1}$.

To this end set $L_t := \mathbf{1}_{[s,H]}(t) \cdot \mathbf{1}_A \cdot \left(\mathbb{E}^{\mathcal{C}_t} f - \mathbb{E}^{\mathcal{C}_s} f\right)$. By Example 2.1.1 (iii) and since $f \in L^2$, we have $L := (L_t)_{t\in I} \in L^2_{[}$ (see Lemma 2.6.1 (a)). We obtain:

$$\mathbb{E}f \cdot \Psi = \mathbb{E}f \cdot \mathbf{1}_A \cdot (\Phi_H - \Phi_s) = \mathbb{E}L_H(\Phi_H - \Phi_s) = \mathbb{E}(L_H\Phi_H - L_H\Phi_s),$$

where $\mathbb{E}L_H\Phi_s = \mathbb{E}\mathbb{E}^{\mathcal{C}_s}(L_H\Phi_s) = \mathbb{E}\Phi_s\mathbb{E}^{\mathcal{C}_s}L_H = \mathbb{E}\Phi_s L_s = 0$. Therefore,

$$\mathbb{E}f \cdot \Psi = \mathbb{E}L_H\Phi_H = \varphi(L) \leq \|\varphi\|_{[,1} \cdot \|L\|_{[,1}.$$

By Hölder's inequality and Lemma 2.1.3 (a), we obtain:

$$\begin{aligned}
\|L\|_{[,1} &= \mathbb{E}\left(\left(\sum_{t\in I}(L_t - L_{t^-})^2\right)^{\frac{1}{2}}\right) = \mathbb{E}\left(\left(\sum_{t>s}(L_t - L_{t^-})^2\right)^{\frac{1}{2}}\right) \\
&= \mathbb{E}\left(1_A\left(\sum_{t>s}\left(\mathbb{E}^{\mathcal{C}_t}f - \mathbb{E}^{\mathcal{C}_{t^-}}f\right)^2\right)^{\frac{1}{2}}\right) \leq \sqrt{\mathbb{E}1_A}\left(\mathbb{E}\sum_{t>s}\left(\mathbb{E}^{\mathcal{C}_t}f - \mathbb{E}^{\mathcal{C}_{t^-}}f\right)^2\right)^{\frac{1}{2}} \\
&= \sqrt{\mu(A)}\left(\mathbb{E}\sum_{t>s}\left(\left(\mathbb{E}^{\mathcal{C}_t}f\right)^2 - \left(\mathbb{E}^{\mathcal{C}_{t^-}}f\right)^2\right)\right)^{\frac{1}{2}} \\
&= \sqrt{\mu(A)}\left(\mathbb{E}\left(f^2 - \left(\mathbb{E}^{\mathcal{C}_s}f\right)^2\right)\right)^{\frac{1}{2}} \leq \sqrt{\mu(A)}\,\|f\|_2 \leq \sqrt{\mu(A)}.
\end{aligned}$$

This proves (iv) and thus (iii) is also true.

The second assertion follows from (iii) and (ii): for all $s \in I$,

$$
\begin{aligned}
\left(\mathbb{E}^{\mathcal{C}_s}(\Phi_H - \Phi_{s-})^2\right)^{\frac{1}{2}} &= \left(\mathbb{E}^{\mathcal{C}_s}(\Phi_H - \Phi_s + \Phi_s - \Phi_{s-})^2\right)^{\frac{1}{2}} \\
&\leq \left(\mathbb{E}^{\mathcal{C}_s}(\Phi_H - \Phi_s)^2\right)^{\frac{1}{2}} + \left(\mathbb{E}^{\mathcal{C}_s}(\Phi_s - \Phi_{s-})^2\right)^{\frac{1}{2}} \\
&\leq \|\varphi\|_{[,1} + 2\|\varphi\|_{[,1}.
\end{aligned}
$$

This proves that $\|\Phi\|_{B,2} \leq 3\|\varphi\|_{[,1}$. □

In the following sense we have $BMO_2 \subseteq \left(L^1_{[}\right)'$.

Proposition 2.9.2 *Fix $\Phi \in BMO_2$ and define $\varphi(M) := \mathbb{E}M_H \cdot \Phi_H$ for all $M \in L^1_{[}$. Then φ is linear and continuous with*

$$
|\varphi(M)| \leq \sqrt{2}\,\|\Phi\|_{B,2}\,\|M\|_{[,1}.
$$

Proof For each $t \in I$ set $a_t(X) := \left(\sqrt{[M]_t(X)} + \sqrt{[M]_{t-}(X)}\right)^{\frac{1}{2}}$. Note that $a_t(X) \neq 0$ if $|M_t(X) - M_{t-}(X)| \neq 0$. Now, by Lemma 2.1.3 (b),

$$
\begin{aligned}
|\mathbb{E}M_H \cdot \Phi_H| &= \left|\mathbb{E}\sum_{t\in I}(M_t - M_{t-})(\Phi_t - \Phi_{t-})\right| \\
&= \left|\sum_{t\in I}\int_{\{|M_t - M_{t-}| \neq 0\}} a_t^{-1}(M_t - M_{t-})\,a_t(\Phi_t - \Phi_{t-})\,d\mu\right| \\
&\leq \sqrt{A}\cdot\sqrt{B},
\end{aligned}
$$

where, by the Cauchy–Schwarz inequality,

$$
\begin{aligned}
A &= \sum_{t\in I}\int_{\{|M_t - M_{t-}| \neq 0\}} a_t^{-2}(M_t - M_{t-})^2\,d\mu \\
&= \sum_{t\in I}\int_{\{|M_t - M_{t-}| \neq 0\}} \left(\sqrt{[M]_t} + \sqrt{[M]_{t-}}\right)^{-1}([M]_t - [M]_{t-})\,d\mu \\
&\leq \mathbb{E}\sum_{t\in I}\left(\sqrt{[M]_t} - \sqrt{[M]_{t-}}\right) = \mathbb{E}\sqrt{[M]_H} = \|M\|_{[,1},
\end{aligned}
$$

and

$$B = \mathbb{E}\sum_{t\in I} a_t^2(\Phi_t - \Phi_{t^-})^2 = \mathbb{E}\sum_{t\in I}\left(\sqrt{[M]_t} + \sqrt{[M]_{t^-}}\right)(\Phi_t - \Phi_{t^-})^2$$

$$\le 2\mathbb{E}\sum_{t\in I}\sqrt{[M]_t}(\Phi_t - \Phi_{t^-})^2 = 2\mathbb{E}\sum_{t\in I, s\le t}\left(\sqrt{[M]_s} - \sqrt{[M]_{s^-}}\right)(\Phi_t - \Phi_{t^-})^2$$

$$= 2\mathbb{E}\sum_{s\in I, t\ge s}(\Phi_t - \Phi_{t^-})^2\left(\sqrt{[M]_s} - \sqrt{[M]_{s^-}}\right)$$

$$= 2\mathbb{E}\sum_{s\in I}\left(\mathbb{E}^{\mathcal{C}_s}\sum_{t\ge s}(\Phi_t - \Phi_{t^-})^2\right)\left(\sqrt{[M]_s} - \sqrt{[M]_{s^-}}\right) \quad \text{(by 2.1.3 (d))}$$

$$= 2\mathbb{E}\sum_{s\in I}\mathbb{E}^{\mathcal{C}_s}(\Phi_H - \Phi_{s^-})^2\left(\sqrt{[M]_s} - \sqrt{[M]_{s^-}}\right)$$

$$\le 2\,\|\Phi\|_{B,2}^2\,\mathbb{E}\sum_{s\in I}\left(\sqrt{[M]_s} - \sqrt{[M]_{s^-}}\right)$$

$$= 2\,\|\Phi\|_{B,2}^2\,\mathbb{E}\sqrt{[M]_H} = 2\,\|\Phi\|_{B,2}^2\cdot\|M\|_{[,1}.$$

It follows that $|\mathbb{E}M_H\cdot\Phi_H| \le \sqrt{2}\,\|\Phi\|_{B,2}\,\|M\|_{[,1}$. □

2.10 $(L^1_\sim)'$ is BMO$_1$

In the following sense we have $(L^1_\sim)' \subseteq BMO_1$.

Proposition 2.10.1 *For each $\varphi \in (L^1_\sim)'$ there exists a $\Phi \in BMO_1$ with*

$$\varphi(M) = \mathbb{E}\Phi_H M_H \textit{ for all } M \in L^1_\sim \quad \textit{and} \quad \|\Phi\|_{B,1} \le 4\,\|\varphi\|_{\sim,1}.$$

Note that Φ is μ-a.s. uniquely determined by φ. Therefore we can define

$$\beta(\varphi) := \Phi.$$

Proof Fix $\varphi \in (L^1_\sim)'$. For each $N \in L^1$, define $\widetilde{\varphi}(N) := \varphi((\mathbb{E}^{\mathcal{C}_t}N)_{t\in I})$. By Lemma 2.7.1, $\widetilde{\varphi}$ is well defined. Moreover, $\widetilde{\varphi}$ is linear and continuous, because for $N \in L^1$ and the martingale $M := (\mathbb{E}^{\mathcal{C}_t}N)_{t\in I}$, by Lemma 2.7.1,

$$|\widetilde{\varphi}(N)| = |\varphi(M)| \le \|\varphi\|_{\sim,1}\,\|M\|_{\sim,1} \le \|\varphi\|_{\sim,1}\cdot|I|\cdot\|N\|_1.$$

It follows that $\|\widetilde{\varphi}\|_1 \le |I|\cdot\|\varphi\|_{\sim,1}$. Since $\widetilde{\varphi}\restriction L^2$ is continuous in the norm $\|\cdot\|_2$, there is a martingale $\Phi \in L^2$ such that $\widetilde{\varphi}(N) = \mathbb{E}\Phi_H N$ for all $N \in L^2$, thus

(i) $\varphi(M) = \widetilde{\varphi}(M_H) = \mathbb{E}\Phi_H M_H$ for all $M \in L^2_\sim$.

By Lemma 2.7.1, $\Phi \in L^2_\sim$.

Now the proof is similar to the proof of Proposition 2.9.1, in particular, to the proof of

(ii) $|\Phi_s - \Phi_{s-}| \le 2\,\|\varphi\|_{\sim,1}$ μ-a.s. for all $s \in I$.

(See Proposition 2.9.1 part (ii) of proof.) It follows that Φ_H is bounded μ-a.s. Since $L^2_\sim$ is dense in $L^1_\sim$, the first assertion is true. Now we will prove

(iii) $\mathbb{E}^{\mathcal{C}_s}|\Phi_H - \Phi_s| \le 2\,\|\varphi\|_{\sim,1}$ μ-a.s. for all $s \in I$.

Fix $A \in \mathcal{C}_s$ and set $\Psi := \mathbf{1}_A \cdot (\Phi_H - \Phi_s)$. It suffices to prove that $\|\Psi\|_1 \le 2\mu(A) \cdot \|\varphi\|_{\sim,1}$. We may assume that $0 \ne \Psi$ in L^1. By the Hahn–Banach theorem, there exists an $f \in (L^1)' = L^\infty$ such that $\|f\|_1 \le 1$ and $f(\Psi) = \mathbb{E}f \cdot \Psi = \|\Psi\|_1$. Therefore, it suffices to show that

(iv) $\mathbb{E}f \cdot \Psi \le 2\mu(A) \cdot \|\varphi\|_{\sim,1}$.

Let L be defined as in the proof of Proposition 2.9.1. Then

$$\mathbb{E}f \cdot \Psi = \varphi(L) \le \|\varphi\|_{\sim,1} \cdot \|L\|_{\sim,1} \le 2\,\|\varphi\|_{\sim,1}\,\mu(A).$$

The second assertion follows from (iii) and (ii). □

To prove that $BMO_1 \subseteq (L^1_\sim)'$, we need the following simple lemma.

Lemma 2.10.2 *Fix stopping times $\tau \le \sigma$, $s \in I$ and $\Phi \in BMO_1$.*

(a) $\mathbb{E}^{\mathcal{C}_s}|\Phi_H - \Phi_s| \le \|\Phi\|_{B,1}$ *μ-a.s.*
(b) $\mathbb{E}^{\mathcal{C}_\tau}|\Phi_H - \Phi_\tau| \le \|\Phi\|_{B,1}$ *μ-a.s.*
(c) $\mathbb{E}^{\mathcal{C}_\tau}|\Phi_\sigma - \Phi_\tau| \le \|\Phi\|_{B,1}$ *μ-a.s.*
(d) *If F is a $(\mathcal{C}_t)_{t\in I}$-adapted process, then* $\max_{s \le \sigma \wedge H}|F_s|$ *is $\mathcal{C}_\sigma$-measurable.*

Proof Set $\mathcal{C}_{H^+} := \mathcal{C}_H$ and $\Phi_{H^+} := \Phi_H$. Fix $A \in \mathcal{C}_s$.

(a) Note that $\mathbb{E}\mathbf{1}_A \cdot |\Phi_H - \Phi_s| = \mathbb{E}\mathbf{1}_A \cdot \mathbb{E}^{\mathcal{C}_{s^+}}|\Phi_H - \Phi_s| \le \mathbb{E}\mathbf{1}_A \cdot \|\Phi\|_{B,1}$. This implies (a).

(b) Suppose that $A \in \mathcal{C}_\tau$. Then (b) follows from (a), because

$$\mathbb{E}\mathbf{1}_A \cdot |\Phi_H - \Phi_\tau| = \sum_{i=\square}^{H^+} \mathbb{E}\mathbf{1}_{A\cap\{\tau=i\}} \cdot |\Phi_H - \Phi_i| \le \sum_{i=\square}^{H^+} \mathbb{E}\mathbf{1}_{A\cap\{\tau=i\}} \cdot \|\Phi\|_{B,1}.$$

(c) follows from Proposition 2.5.2, Lemma 2.5.1 and (b):

$$\mathbb{E}^{\mathcal{C}_\tau}|\Phi_\sigma - \Phi_\tau| = \mathbb{E}^{\mathcal{C}_\tau}\left|\mathbb{E}^{\mathcal{C}_\sigma}\Phi_H - \mathbb{E}^{\mathcal{C}_\sigma}\Phi_\tau\right| \leq \mathbb{E}^{\mathcal{C}_\tau}\mathbb{E}^{\mathcal{C}_\sigma}|\Phi_H - \Phi_\tau|$$
$$= \mathbb{E}^{\mathcal{C}_\tau}|\Phi_H - \Phi_\tau| \leq \|\Phi\|_{B,1}.$$

(d) Fix $c \in \mathbb{R}$, $t \in I$ and set $W := \left\{\max_{s \leq \sigma \wedge H}|F_s| \leq c\right\} \cap \{\sigma \leq t\}$. Then

$$W = \bigcup_{i=\square}^{t}\left\{\sigma = i \text{ and} \max_{s \leq i}|F_s| \leq c\right\} \in \mathcal{C}_t. \qquad \square$$

To see that each $\Phi \in BMO_1$ defines an element $\varphi \in \left(L^1_\sim\right)'$ by setting $\varphi(M) := \mathbb{E}M_H \cdot \Phi_H$ for all $M \in L^1_\sim$, it suffices to prove the following result.

Proposition 2.10.3 *Fix $\Phi \in BMO_1$ and $M \in L^1_\sim$. Then*

$$|\mathbb{E}\Phi_H M_H| \leq 8\,\|\Phi\|_{B,1} \cdot \|M\|_{\sim,1}.$$

Proof We first assume that $\|\Phi\|_{B,1} \leq 1$. Fix $c > 1$. By recursion, we define a sequence $(\tau_k)_{k \in \mathbb{N}}$ of stopping times τ_k, where $\Phi_{H^+} := \Phi_H$: set $\tau_0 := \square^-$ with $\square^- < t$ for all $t \in I$ and set

$$\tau_1 := \inf\{t \in I \mid c \leq |\Phi_t|\} = \inf\left\{t \in I \mid c \leq \left|\Phi_t - \Phi_{\tau_0}\right|\right\},$$
$$\tau_{k+1} := \inf\left\{t \in I \mid \tau_k < t \text{ and } c \leq \left|\Phi_t - \Phi_{\tau_k}\right|\right\}.$$

We define for all $t \in I$

$$A_t := \sum_{k \in \mathbb{N}} \mathbf{1}_{\{\tau_k \leq t\}},\ U_t := \sum_{k \in \mathbb{N}} \mathbf{1}_{\{\tau_k \leq t\}}\left(\Phi_{\tau_k} - \Phi_{\tau_{k-1}}\right).$$

Since I is finite, these sums are finite and A_t is the number of $k \in \mathbb{N}$ with $\tau_k \leq t$. Note that A_t and U_t are $\mathcal{C}_t$-measurable. Set

$$K := \max\{\tau_k \mid \tau_k \leq H\}.$$

Note that $U_H = \sum_{k \in \mathbb{N}, \tau_k \leq H}\left(\Phi_{\tau_k} - \Phi_{\tau_{k-1}}\right) = \Phi_K - \Phi_{\tau_0} = \Phi_K$. Now

$$|\mathbb{E}M_H\Phi_H| \leq \mathbb{E}(|M_H| \cdot |\Phi_H - \Phi_K|) + |\mathbb{E}M_H \cdot U_H|$$
$$\leq \mathbb{E}|M_H| \cdot c + |\mathbb{E}M_H \cdot U_H| \leq \|M\|_{\sim,1} \cdot c + |\mathbb{E}M_H \cdot U_H|.$$

Moreover, we have, using $M^\sim_s := \max_{r\leq s}|M_r|$,

$$
\begin{aligned}
|\mathbb{E}M_H\cdot U_H| &= \left|\mathbb{E}\sum_{s\in I} M_H(U_s-U_{s-})\right| = \left|\mathbb{E}\sum_{s\in I}\mathbb{E}^{\mathcal{C}_s} M_H(U_s-U_{s-})\right| \\
&\leq \mathbb{E}\sum_{s\in I}|M_s|\,|U_s-U_{s-}| \leq \mathbb{E}\sum_{s\in I} M^\sim_s\,|U_s-U_{s-}| \\
&\leq \mathbb{E}\sum_{s\in I} M^\sim_s \sum_{k\in\mathbb{N},\tau_k=s}\left|\Phi_{\tau_k}-\Phi_{\tau_{k-1}}\right| \\
&\leq \mathbb{E}\sum_{s\in I} M^\sim_s \sum_{k\in\mathbb{N},\tau_k=s}|\Phi_s-\Phi_{s-}|+\left|\Phi_{s-}-\Phi_{\tau_{k-1}}\right| \\
&\leq \mathbb{E}\sum_{s\in I} M^\sim_s \sum_{k\in\mathbb{N},\tau_k=s} 1+c \quad \text{(by 2.8.1 (a))} \\
&= \mathbb{E}\sum_{s\in I} M^\sim_s \sum_{k\in\mathbb{N},\tau_k=s}(1+c)\,(A_s-A_{s-}) \\
&\leq (c+1)\mathbb{E}\sum_{n\in\mathbb{N},\tau_n\leq H} M^\sim_{\tau_n}\left(A_{\tau_n}-A_{\tau_n^-}\right) \\
&= (c+1)\mathbb{E}\sum_{n\in\mathbb{N},\tau_n\leq H}\sum_{k\in\mathbb{N},k\leq n}\left(M^\sim_{\tau_k}-M^\sim_{\tau_{k-1}}\right)\left(A_{\tau_n}-A_{\tau_n^-}\right) \\
&= (c+1)\mathbb{E}\sum_{k\in\mathbb{N}}\left(\sum_{n\geq k,\tau_n\leq H}\left(A_{\tau_n}-A_{\tau_n^-}\right)\right)\left(M^\sim_{\tau_k}-M^\sim_{\tau_{k-1}}\right) \\
&\leq (c+1)\mathbb{E}\sum_{k\in\mathbb{N}}\left(M^\sim_{\tau_k}-M^\sim_{\tau_{k-1}}\right)\sum_{n\in\mathbb{N}}\mathbf{1}_{\{\tau_k\leq\tau_n\leq H\}} \\
&= (c+1)\mathbb{E}\sum_{k\in\mathbb{N},\tau_k\leq H}\left(M^\sim_{\tau_k}-M^\sim_{\tau_{k-1}}\right)\sum_{n\in\mathbb{N}}\mathbb{E}^{\mathcal{C}_{\tau_k}}\mathbf{1}_{\{\tau_k\leq\tau_n\leq H\}}
\end{aligned}
$$

(by 2.10.2 (d)). We will prove by induction that $\mathbb{E}^{\mathcal{C}_{\tau_k}}\mathbf{1}_{\{\tau_k\leq\tau_{k+m}\leq H\}}\leq\frac{1}{c^m}$ μ-a.s. for all $m\in\mathbb{N}_0$:

$$
m=0\colon \mathbb{E}^{\mathcal{C}_{\tau_k}}\mathbf{1}_{\{\tau_k\leq\tau_{k+0}\leq H\}}\leq 1=\frac{1}{c^0}\ \mu\text{-a.s.}
$$

$$
\begin{aligned}
m>0\colon &\mathbb{E}^{\mathcal{C}_{\tau_k}}\mathbf{1}_{\{\tau_k\leq\tau_{k+m}\leq H\}} \\
&\leq \frac{1}{c}\mathbb{E}^{\mathcal{C}_{\tau_k}} c\mathbf{1}_{\{\tau_k\leq\tau_{k+m-1}\leq H\}} \\
&\leq \frac{1}{c}\mathbb{E}^{\mathcal{C}_{\tau_k}}\left(\mathbb{E}^{\mathcal{C}_{\tau_{k+m-1}}}\left|\Phi_{\tau_{k+m}}-\Phi_{\tau_{k+m-1}}\right|\cdot\mathbf{1}_{\{\tau_k\leq\tau_{k+m-1}\leq H\}}\right)
\end{aligned}
$$

$$\leq \frac{1}{c}\mathbb{E}^{\mathcal{C}_{\tau_k}}\left(\mathbf{1}_{\{\tau_k \leq \tau_{k+m-1} \leq H\}}\right) \quad \text{(by 2.10.2 (c))}$$
$$\leq \frac{1}{c^m},$$

by the induction hypothesis. Therefore, we obtain

$$|\mathbb{E}M_H U_H| \leq (c+1)\mathbb{E}\sum_{k\in\mathbb{N},\tau_k\leq H}\left(M^{\sim}_{\tau_k} - M^{\sim}_{\tau_{k-1}}\right)\sum_{n\in\mathbb{N}}\mathbb{E}^{\mathcal{C}_{\tau_k}}\mathbf{1}_{\{\tau_k\leq\tau_n\leq H\}}$$
$$\leq (c+1)\mathbb{E}\sum_{k\in\mathbb{N},\tau_k\leq H}\left(M^{\sim}_{\tau_k} - M^{\sim}_{\tau_{k-1}}\right)\sum_{n\in\mathbb{N}_0}\frac{1}{c^n} \leq (c+1)\frac{c}{c-1}\mathbb{E}M^{\sim}_H.$$

It follows that $|\mathbb{E}M_H\Phi_H| \leq \left(c+(c+1)\frac{c}{c-1}\right)\|M\|_{\sim,1}$. If we set $f(c) := c+(c+1)\frac{c}{c-1}$ for $c > 1$, then 2 is the minimum of f and $f(2) = 8$. Therefore, in the case $\|\Phi\|_{B,1} \leq 1$, we obtain $\mathbb{E}M_H\Phi_H \leq 8\mathbb{E}\|M\|_{\sim,1}$. Suppose now that $\|\Phi\|_{B,1} > 1$. Then we obtain for $Z := \frac{\Phi}{\|\Phi\|_{B,1}}$:

$$\mathbb{E}M_H\Phi_H = \|\Phi\|_{B,1}\mathbb{E}M_H Z_H \leq 8\|\Phi\|_{B,1}\|M\|_{\sim,1}. \qquad \square$$

2.11 B–D–G inequalities for $p = 1$

From Theorem 2.8.2 we obtain the following beautiful and useful result.

Theorem 2.11.1 (Burkholder, Davis and Gandy [19]) *There exist real constants c_1 and d_1 such that, for each finite set I and each martingale $M : \Lambda \times I \to \mathbb{R}$,*

$$c_1\|M\|_{\sim,1} \leq \|M\|_{[,1} \leq d_1\|M\|_{\sim,1}.$$

We may choose $c_1 := \frac{1}{8}$, $d_1 := 24$. It should be mentioned that these constants are not optimal.

Proof Fix a martingale M. By Remark 2.7.3, $M \in L^1_{[}$ and $M \in L^1_{\sim}$. First we prove the second inequality. By the Hahn–Banach theorem, there exists a $\varphi \in \left(L^1_{[}\right)'$ such that $\varphi(M) = \|M\|_{[,1}$ and $\|\varphi\|_{[,1} \leq 1$. By Proposition 2.9.1, there exists a $\Phi \in BMO_2$ such that $\|\Phi\|_{B,2} \leq 3$ and $\varphi(M) = \mathbb{E}\Phi_H M_H$. By Proposition 2.10.3 and since $\|\Phi\|_{B,1} \leq \|\Phi\|_{B,2}$, we obtain

$$\|M\|_{[,1} = \varphi(M) = \mathbb{E}\Phi_H M_H \leq 8\|\Phi\|_{B,2}\|M\|_{\sim,1} \leq 24\|M\|_{\sim,1}.$$

To prove the first inequality, set

$$K := \min\left\{s \in I \mid |M_s| = \max_{t\in I}|M_t|\right\}$$

and

$$B_t(X) := \begin{cases} 1 & \text{if } K(X) \leq t \text{ and } M_{K(X)}(X) > 0 \\ 0 & \text{if } t < K(X) \text{ or } M_{K(X)}(X) = 0 \\ -1 & \text{if } K(X) \leq t \text{ and } M_{K(X)}(X) < 0. \end{cases}$$

Note that $\max_{s \in I} |M_s| = \sum_{s \in I} M_s \Delta B_s$. Set

$$B_t^+ := B_t \vee 0 \quad \text{and} \quad B_t^- := (-B_t) \vee 0.$$

Then $\left(B_t^+\right)_{t \in I}$ and $\left(B_t^-\right)_{t \in I}$ are monotone increasing and

$$\Delta B_t^{+(-)}(X) = \begin{cases} 1 & \text{if } K(X) = t \text{ and } 0 < M_t(X)\ (M_t(X) < 0) \\ 0 & \text{otherwise.} \end{cases}$$

We define recursively $A_t^{+(-)}$ by setting: $A_t^{+(-)} := \mathbb{E}^{\mathcal{C}_t} \Delta B_t^{+(-)} + A_{t-}^{+(-)}$ μ-a.s. Set $A_t := A_t^+ - A_t^-$. Then $A_t = \mathbb{E}^{\mathcal{C}_t} \Delta B_t + A_{t-}$ μ-a.s., and A_t^+, A_t^- and A_t are $\mathcal{C}_t$-measurable. Now

$$\begin{aligned} \|M\|_{\sim,1} &= \mathbb{E} \sum_{s \in I} M_s \Delta B_s = \mathbb{E} \sum_{s \in I} M_s \mathbb{E}^{\mathcal{C}_s} \Delta B_s \\ &= \mathbb{E} \sum_{s \in I} \left(\sum_{\rho \leq s} \Delta M_\rho \right) \Delta A_s \\ &= \mathbb{E} \sum_{\rho \in I} \left(\sum_{s \geq \rho} \Delta A_s \right) \Delta M_\rho = \mathbb{E} \sum_{\rho \in I} (A_H - A_{\rho-}) \Delta M_\rho \\ &= \mathbb{E} \sum_{\rho \in I} A_H \Delta M_\rho = \mathbb{E} A_H M_H. \end{aligned}$$

We define the martingale $\Phi := \left(\mathbb{E}^{\mathcal{C}_t} A_H\right)_{t \in I}$ and estimate $\|\Phi\|_{B,2}$. To this end we first estimate $\mathbb{E}^{\mathcal{C}_t}\left(A_H^+ - A_{t-}^+\right)$ and $\mathbb{E}^{\mathcal{C}_t}\left(A_H^- - A_{t-}^-\right)$:

$$\begin{aligned} \mathbb{E}^{\mathcal{C}_t}\left(A_H^+ - A_{t-}^+\right) &= \mathbb{E}^{\mathcal{C}_t}\left(\sum_{t \leq s} \Delta A_s^+ \right) = \mathbb{E}^{\mathcal{C}_t}\left(\sum_{t \leq s} \mathbb{E}^{\mathcal{C}_s} \Delta B_s^+ \right) \\ &= \mathbb{E}^{\mathcal{C}_t}\left(\sum_{t \leq s} \Delta B_s^+ \right) \leq \mathbb{E}^{\mathcal{C}_t} \Delta B_{K(\cdot)}^+ \leq 1 \ \mu\text{-a.s.} \end{aligned}$$

In the same way we have $\mathbb{E}^{\mathcal{C}_t}\left(A_H^- - A_{t-}^-\right) \leq 1$ μ-a.s. Now we obtain for all $s \in I$:

$$\left(\mathbb{E}^{\mathcal{C}_s}(\Phi_H - \Phi_{s-})^2\right)^{\frac{1}{2}} = \left(\mathbb{E}^{\mathcal{C}_s}(A_H - A_{s-} + A_{s-} - \Phi_{s-})^2\right)^{\frac{1}{2}} \leq \sqrt{\alpha^+} + \sqrt{\alpha^-} + \beta,$$

where

$$\alpha^+ = \mathbb{E}^{\mathcal{C}_s}\left(A_H^+ - A_{s-}^+\right)^2 = \mathbb{E}^{\mathcal{C}_s}\left(\sum_{s\le r}\Delta A_r^+\right)^2 \le 2\mathbb{E}^{\mathcal{C}_s}\sum_{s\le r\le t}\Delta A_r^+\cdot\Delta A_t^+$$
$$= 2\mathbb{E}^{\mathcal{C}_s}\sum_{s\le r}\Delta A_r^+\sum_{r\le t}\Delta A_t^+ = 2\mathbb{E}^{\mathcal{C}_s}\sum_{s\le r}\Delta A_r^+(A_H^+ - A_{r-}^+)$$
$$= 2\mathbb{E}^{\mathcal{C}_s}\sum_{s\le r}\Delta A_r^+\mathbb{E}^{\mathcal{C}_r}(A_H^+ - A_{r-}^+) \le 2\mathbb{E}^{\mathcal{C}_s}\sum_{s\le r}\Delta A_r^+$$
$$= 2\mathbb{E}^{\mathcal{C}_s}(A_H^+ - A_{s-}^+) \le 2\ \mu\text{-a.s.}$$

In the same way we obtain $\alpha^- = \mathbb{E}^{\mathcal{C}_s}\left(A_H^- - A_{s-}^-\right)^2 \le 2\ \mu$-a.s. Moreover, since $\Phi_H = A_H$, and since, as we have seen, $\mathbb{E}^{\mathcal{C}_t}\left(A_H^{+(-)} - A_{t-}^{+(-)}\right) \le 1\ \mu$-a.s., we have

$$\beta = |A_{s-} - \Phi_{s-}| = \left|\mathbb{E}^{\mathcal{C}_{s-}}(A_H - A_{s-})\right| \le \mathbb{E}^{\mathcal{C}_{s-}}\left|\mathbb{E}^{\mathcal{C}_s}(A_H - A_{s-})\right| \le 2\ \mu\text{-a.s..}$$

This proves that $\|\Phi\|_{B,2} \le 2\sqrt{2}+2$. Therefore, by Proposition 2.9.2, we obtain

$$\frac{1}{8}\|M\|_{\sim,1} = \frac{1}{8}\mathbb{E}A_H M_H = \frac{1}{8}\mathbb{E}\Phi_H M_H \le \frac{\sqrt{2}}{8}\|\Phi\|_{B,2}\|M\|_{[,1} \le \|M\|_{[,1}.$$

□

2.12 The B–D–G inequalities for the conditional expectation for $p = 1$

Theorem 2.12.1 (Burkholder, Davis and Gandy [19]) *Fix a martingale M and $s \in I$. Then μ-a.s.*

(a) $\mathbb{E}^{\mathcal{C}_s}\left(M_H^\sim - M_{s-}^\sim\right) \le 8\mathbb{E}^{\mathcal{C}_s}\sqrt{[M]_H}$. *(Recall that $M_t^\sim = \max_{s\le t}|M_s|$.)*
(b) $\mathbb{E}^{\mathcal{C}_s}\left(\sqrt{[M]_H} - \sqrt{[M]_{s-}}\right) \le 48\mathbb{E}^{\mathcal{C}_s}M_H^\sim$.

Proof We first assume that $s = \Box$. We will prove (a) and (b) with 24 instead of 48. Let $A \in \mathcal{C}_\Box$. Since $\mathbf{1}_A \cdot M$ is a martingale (see Example 2.1.1 (ii)), we obtain from Theorem 2.11.1

$$\mathbb{E}\mathbf{1}_A\cdot\left(M_H^\sim - M_{\Box-}^\sim\right) = \mathbb{E}\left(\mathbf{1}_A\cdot M\right)_H^\sim \le 8\mathbb{E}\sqrt{[\mathbf{1}_A\cdot M]_H} = 8\mathbb{E}\mathbf{1}_A\cdot\sqrt{[M]_H}.$$

This proves (a). (b) follows from Theorem 2.11.1:

$$\mathbb{E}\mathbf{1}_A\left(\sqrt{[M]_H} - \sqrt{[M]_{\Box-}}\right) = \mathbb{E}\sqrt{[\mathbf{1}_A\cdot M]_H} \le 24\mathbb{E}\left(\mathbf{1}_A\cdot M\right)_H^\sim = 24\mathbb{E}\mathbf{1}_A\cdot\left(M_H^\sim\right).$$

Now we assume that $s > \Box$. For all $r \geq s^-$ set $N_r := M_r - M_{s^-}$. Then $N := (N_r)_{s \leq r \in I}$ is an $(\mathcal{C}_r)_{s \leq r \in I}$-martingale with $N_{s^-} = 0$. Since s plays the same role for N as $\Box$ has played for M, we obtain

$$\text{(i)} \quad \begin{cases} \mathbb{E}^{\mathcal{C}_s} N_H^{\sim} \leq 8\mathbb{E}^{\mathcal{C}_s} \sqrt{[N]_H} \quad \text{and} \\ \mathbb{E}^{\mathcal{C}_s} \sqrt{[N]_H} \leq 24\mathbb{E}^{\mathcal{C}_s} N_H^{\sim}. \end{cases}$$

Note that

$$\text{(ii)} \quad \begin{cases} [M]_H - [M]_{s^-} = [N]_H \text{ and } M_H^{\sim} \leq N_H^{\sim} + M_{s^-}^{\sim} \quad \text{and} \\ N_H^{\sim} = \max_{s \leq t} |M_t - M_{s^-}| \leq \max_{s \leq t} (|M_t| + |M_{s^-}|) \leq 2M_H^{\sim}. \end{cases}$$

Since $\sqrt{\alpha} - \sqrt{\beta} \leq \sqrt{\alpha - \beta}$ for $0 \leq \beta \leq \alpha$, we obtain from (i) and (ii):

$$\mathbb{E}^{\mathcal{C}_s} \left(M_H^{\sim} - M_{s^-}^{\sim} \right) \leq \mathbb{E}^{\mathcal{C}_s} N_H^{\sim} \leq 8\mathbb{E}^{\mathcal{C}_s} \sqrt{[N]_H} \leq 8\mathbb{E}^{\mathcal{C}_s} \sqrt{[M]_H},$$

$$\begin{aligned} \mathbb{E}^{\mathcal{C}_s} \left(\sqrt{[M]_H} - \sqrt{[M]_{s^-}} \right) &\leq \mathbb{E}^{\mathcal{C}_s} \left(\sqrt{[M]_H - [M]_{s^-}} \right) \\ &= \mathbb{E}^{\mathcal{C}_s} \left(\sqrt{[N]_H} \right) \leq 24\mathbb{E}^{\mathcal{C}_s} N_H^{\sim} \leq 48\mathbb{E}^{\mathcal{C}_s} M_H^{\sim}. \end{aligned}$$

□

2.13 The B–D–G inequalities

Lemma 2.13.1 (Burkholder, Davis and Gandy [19]) *Fix $p \in [1, \infty[$ and monotone increasing non-negative $(\mathcal{C}_t)_{t \in I}$-adapted processes $(A_t)_{t \in I}$ and $(B_t)_{t \in I}$ such that B_H and A_H belong to L^p and $\mathbb{E}^{\mathcal{C}_s}(A_H - A_{s^-}) \leq \mathbb{E}^{\mathcal{C}_s} B_H$ μ-a.s. for all $s \in I$. Then*

$$\|A_H\|_p \leq p \|B_H\|_p.$$

Proof Since, by the mean value theorem, $a^p - b^p \leq p a^{p-1}(a - b)$ for all $a, b \in \mathbb{R}$ with $0 \leq b \leq a$, we obtain

$$\begin{aligned} \mathbb{E}A_H^p &= \mathbb{E} \sum_{s \in I} \Delta A_s^p \leq p\mathbb{E} \sum_{s \in I} A_s^{p-1} \Delta A_s = p\mathbb{E} \sum_{s \in I} \left(\sum_{\rho \leq s} \Delta A_\rho^{p-1} \right) \Delta A_s \\ &= p\mathbb{E} \sum_{\rho \in I} \left(\sum_{s \geq \rho} \Delta A_s \right) \Delta A_\rho^{p-1} = p\mathbb{E} \sum_{\rho \in I} (A_H - A_{\rho^-}) \Delta A_\rho^{p-1} \\ &= p\mathbb{E} \sum_{\rho \in I} \mathbb{E}^{\mathcal{C}_\rho} (A_H - A_{\rho^-}) \Delta A_\rho^{p-1} \leq p\mathbb{E} \sum_{\rho \in I} \mathbb{E}^{\mathcal{C}_\rho} B_H \Delta A_\rho^{p-1} \\ &= p\mathbb{E} \sum_{\rho \in I} B_H \Delta A_\rho^{p-1} = p\mathbb{E} B_H A_H^{p-1} \leq p \|B_H\|_p \left(\mathbb{E}A_H^p \right)^{1 - \frac{1}{p}}. \end{aligned}$$

It follows that $\|A_H\|_p \leq p\,\|B_H\|_p$. □

We obtain:

Theorem 2.13.2 (Burkholder, Davis and Gandy [19]) *Fix $p \in [1,\infty[$. Then there exist real constants c_p and d_p such that, for each finite set I and each p-integrable martingale $M : \Lambda \times I \to \mathbb{R}$,*

$$c_p \|M\|_{\sim,p} \leq \|M\|_{[,p} \leq d_p \|M\|_{\sim,p}.$$

We may choose $c_p := \frac{1}{8p}$ and $d_p := 48p$. It should be mentioned that these constants are not optimal.

Proof In order to prove the first inequality, set $A_t := M_t^{\sim}$ and $B_t := 8\sqrt{[M]_t}$. Then, by Theorem 2.12.1 (a), $\mathbb{E}^{\mathcal{C}_s}(A_H - A_{s^-}) \leq \mathbb{E}^{\mathcal{C}_s} B_H$ for all $s \in I$ μ-a.s. By Lemma 2.13.1,

$$\|M\|_{\sim,p} = \|A_H\|_p \leq p\,\|B_H\|_p = 8p\,\|M\|_{[,p}.$$

To prove the second equality, set $A_t := \sqrt{[M]_t}$ and $B_t := 48M_t^{\sim}$. Then, by Theorem 2.12.1 (b), $\mathbb{E}^{\mathcal{C}_s}(A_H - A_{s^-}) \leq \mathbb{E}^{\mathcal{C}_s} B_H$ for all $s \in I$. Lemma 2.13.1 implies

$$\|M\|_{[,p} = \|A_H\|_p \leq p\,\|B_H\|_p = 48p\,\|M\|_{\sim,p}.$$

□

2.14 The B–D–G inequalities for special convex functions

For applications to integration theory on Loeb spaces in Chapter 10, we need the B–D–G inequalities for slightly more general processes. Fix $p \in [1,\infty[$ and a sequence $(a_n)_{n \in \mathbb{N}_0}$ in $[0,\infty[$ with

$$a_0 = 0, 1 \leq a_1 \quad \text{and} \quad 4a_{n-1} < a_n$$

for all $n \in \mathbb{N}$. For all $x \in [0,\infty[$ and all $n \in \mathbb{N}_0$, set

$$\Phi(x) := (n+1)x - (a_1 + \ldots + a_n) \text{ if } a_n \leq x < a_{n+1} \quad \text{and} \quad \Psi(x) := \Phi\left(x^p\right).$$

For all $t \in [0,\infty[$, set

$$\psi(t) := \begin{cases} (n+1) \cdot p \cdot t^{p-1} & \text{if } 0 < t^p \in [a_n, a_{n+1}[. \\ 0 & \text{if } t = 0. \end{cases}$$

Note that $\int_0^x \psi(t)dt = \Psi(x)$ for all $x \in [0,\infty[$.

Lemma 2.14.1 (Dellacherie and Meyer [34]) *Fix non-negative monotone increasing μ-integrable $(\mathcal{C}_t)_{t\in I}$-adapted processes $(A_t)_{t\in I}$ and $(B_t)_{t\in I}$, fulfilling the inequality $\mathbb{E}^{\mathcal{C}_s}(A_H - A_{s^-}) \leq \mathbb{E}^{\mathcal{C}_s} B_H$ for all $s \in I$. Then*

$$\mathbb{E}\Psi \circ A_H \leq \mathbb{E}\Psi \circ 3p \cdot B_H.$$

Proof We proceed in several steps.

Claim 1: Fix $t \in [0,\infty[$. Then $t \cdot \psi(t) \leq 3p\Psi(t)$.

Proof There exists an $n \in \mathbb{N}_0$ with $t^p \in [a_n, a_{n+1}[$. For $n = 0$, the result is obvious. Let $n \geq 1$. Since

$$3\left(1 - \frac{\sum_{i=1}^n a_i}{(n+1)\,t^p}\right) \geq 3\left(1 - \frac{\sum_{i=1}^n a_i}{(n+1)\,a_n}\right) \geq 3\left(1 - \frac{1}{2}\sum_{i=0}^{\infty}\frac{1}{4^i}\right) = 1,$$

we see that $t\psi(t) = p(n+1)t^p \leq 3p\left((n+1)t^p - \sum_{i=1}^n a_i\right) = 3p\Psi(t)$. □

Claim 2:

(a) For $0 \leq b \leq a$, we have $a^p - b^p \leq pa^{p-1}(a-b)$.
(b) For $x \in [0,\infty[$ we have $f(x) := px^{p-1} - (p-1)x^p \leq 1$.

Proof (a) follows from the mean value theorem. (b) For $x = 0$ or $p = 1$, the result is obvious. Thus, assume that $0 < x$ and $1 < p$. Then $f'(x) = 0$ if and only if $x = 1$. Since $f' > 0$ on $[0,1[$ and $f' < 0$ on $]1,\infty[$, $f(1)$ is the maximum of f and $f(1) = 1$. □

Claim 3: $\Psi \circ A_t \leq \sum_{s\leq t}(\psi \circ A_s)\cdot(A_s - A_{s^-})$.

Proof By induction on $t \in I$. Let $t = \square$. If $A_\square^p \in [a_n, a_{n+1}[$, then we have $\Psi \circ A_\square = (n+1)A_\square^p - \sum_{i=1}^n a_i \leq p(n+1)A_\square^p = (\psi \circ A_\square)A_\square$. For the induction step we have two cases:

Case 1: $a_n \leq A_t^p < a_{n+1} \leq a_k \leq A_{t^+}^p < a_{k+1}$. Then,

$$\Psi \circ A_{t^+} = \Psi \circ A_t + (\Psi \circ A_{t^+} - \Psi \circ A_t) \leq \alpha + \beta,$$

where, by the induction hypothesis,

$$\alpha = \Psi \circ A_t \le \sum_{s \le t} (\psi \circ A_s) \cdot (A_s - A_{s^-}),$$

$$\begin{aligned}
\beta = \Psi \circ A_{t^+} - \Psi \circ A_t &= (k+1)A_{t^+}^p - (n+1)A_t^p - \sum_{i=n+1}^{k} a_i \\
&= (k+1)\left(A_{t^+}^p - A_t^p\right) + (k-n)A_t^p - \sum_{i=n+1}^{k} a_i \\
&\le (k+1)\left(A_{t^+}^p - A_t^p\right) \\
&\le (k+1) \cdot p \cdot A_{t^+}^{p-1} \cdot (A_{t^+} - A_t) = (\psi \circ A_{t^+}) \cdot (A_{t^+} - A_t). \quad \text{(by (a))}
\end{aligned}$$

This proves that $\Psi \circ A_{t^+} \le \sum_{s \le t^+} (\psi \circ A_s) \cdot (A_s - A_{s^-})$.

Case 2: $a_n \le A_t^p \le A_{t+1}^p < a_{n+1}$. The proof is similar, even simpler. □

Claim 4: $\mathbb{E}\Psi \circ A_H \le \mathbb{E}B_H (\psi \circ A_H)$.

Proof By Claim 3 and the assumption of the lemma, we obtain

$$\begin{aligned}
\mathbb{E}\Psi \circ A_H &\le \mathbb{E} \sum_{s \in I} (\psi \circ A_s) \cdot (A_s - A_{s^-}) \\
&= \mathbb{E} \sum_{s \in I} \left(\sum_{\rho \le s} (\psi \circ A_\rho - \psi \circ A_{\rho^-}) \right) \cdot (A_s - A_{s^-}) \\
&= \mathbb{E} \sum_{\rho \in I} \left(\sum_{s \ge \rho} (A_s - A_{s^-}) \right) \cdot \left(\psi \circ A_\rho - \psi \circ A_{\rho^-}\right) \\
&= \mathbb{E} \sum_{\rho \in I} \mathbb{E}^{\mathcal{C}_\rho} \left(A_H - A_{\rho^-}\right) \cdot \left(\psi \circ A_\rho - \psi \circ A_{\rho^-}\right) \\
&\le \mathbb{E} \sum_{\rho \in I} \mathbb{E}^{\mathcal{C}_\rho} B_H \cdot \left(\psi \circ A_\rho - \psi \circ A_{\rho^-}\right) = \mathbb{E}B_H \cdot \psi \circ A_H.
\end{aligned}$$

□

Claim 5: Fix $\mathcal{C}$-measurable $F, G : \Lambda \to \mathbb{R}_0^+$ with

$$\mathbb{E}F \cdot (\psi \circ F) \le \mathbb{E}G \cdot (\psi \circ F).$$

Then $\mathbb{E}\Psi \circ F \le \mathbb{E}\Psi \circ G$.

Proof Set $K(t) := t \cdot \psi(t) - \Psi(t)$ for all $t \in [0, \infty[$. Note that $K \ge 0$. We will first prove:

(c) $s \cdot \psi(t) \leq \Psi(s) + K(t)$ for all $t, s \in [0, \infty[$.

For $s = 0$ or $t = 0$ the result is obvious. Thus, let $t \neq 0$ and $s \neq 0$. Assume that $t^p \in [a_n, a_{n+1}[$ and $s^p \in [a_k, a_{k+1}[$.

Case 1: $k \leq n$. Note that we have to prove

$$s \cdot (n+1) \cdot p \cdot t^{p-1} \leq (k+1) \cdot s^p + \sum_{i=k+1}^{n} a_i + (n+1) \cdot (p-1) t^p.$$

Therefore, we have to prove that for $x := \frac{t}{s}$:

$$(n+1) \cdot p \cdot x^{p-1} \leq (k+1) + \frac{\sum_{i=k+1}^{n} a_i}{s^p} + (n+1) \cdot (p-1) x^p.$$

Since, by (b),

$$(n+1) \cdot p \cdot x^{p-1} - (n+1) \cdot (p-1) x^p \leq n+1,$$

it suffices to prove

$$n - k \leq \frac{\sum_{i=k+1}^{n} a_i}{s^p}.$$

But this is true, because $s^p < a_{k+1} < a_{k+2} < \ldots$.

Case 2: $n < k$. The proof is similar. This proves assertion (c).

If we replace t by F and s by G, then we obtain from the definition of K, the assumption of Claim 5 and (c):

$$\mathbb{E}\Psi \circ F = \mathbb{E}(F \cdot (\psi \circ F) - K \circ F) \leq \mathbb{E}(G \cdot (\psi \circ F) - K \circ F) \leq \mathbb{E}\Psi \circ G.$$

This proves Claim 5. □

Now we are able to prove $\mathbb{E}\Psi \circ A_H \leq \mathbb{E}\Psi \circ 3p \cdot B_H$. By Claim 5, it suffices to prove that $\mathbb{E}A_H \cdot (\psi \circ A_H) \leq \mathbb{E}3pB_H \cdot (\psi \circ A_H)$. But, by Claim 1 and Claim 4,

$$\mathbb{E}A_H \cdot (\psi \circ A_H) \leq \mathbb{E}3p\,(\Psi \circ A_H) \leq \mathbb{E}3pB_H\,(\psi \circ A_H).$$

The proof of the lemma is finished. □

Corollary 2.14.2 (Burkholder, Davis and Gandy [19]) *Let M be a martingale. Then for all $p \in [1, \infty[$*

$$\mathbb{E}\Phi \circ \left(M_H^{\sim}\right)^p \leq \mathbb{E}\Phi \circ (24p)^p \cdot [M]_H^{\frac{p}{2}},$$

$$\mathbb{E}\Phi \circ [M]_H^{\frac{p}{2}} \leq \mathbb{E}\Phi \circ (144p)^p \left(M_H^{\sim}\right)^p.$$

Proof Set $A_t := M_t^{\sim}$ and $B_t := 8[M]_t^{\frac{1}{2}}$. By Theorem 2.12.1 (a), we have $\mathbb{E}^{\mathcal{C}_s}(A_H - A_{s^-}) \leq \mathbb{E}^{\mathcal{C}_s} B_H$ for all $s \in I$. By Lemma 2.14.1,

$$\mathbb{E}\Phi \circ \left(M_H^{\sim}\right)^p = \mathbb{E}\Psi \circ A_H \leq \mathbb{E}\Psi \circ 3p \cdot B_H = \mathbb{E}\Phi \circ (24p)^p \cdot [M]_H^{\frac{p}{2}}.$$

The proof of the second inequality is similar. □

Exercises

2.1 The processes defined in Examples 2.1.1 are martingales.
2.2 Prove Proposition 2.1.2.
2.3 Prove Lemma 2.5.1.
2.4 Prove that the set $\mathfrak{A}_0$ of strongly $(\mathfrak{f}_t)_{t\in[0,\infty[}$-adapted sets is a σ-algebra.
2.5 $\left(M_{\tau_n}\right)_{n\in J}$, defined in Proposition 2.5.2, is a $\left(\mathcal{C}_{\tau_n}\right)_{n\in J}$-martingale.
2.6 Prove that the BMO_p are Banach spaces.
2.7 Prove Theorem 2.8.2 in detail.
2.8 Prove assertion (ii) under Proposition 2.10.1.
2.9 Prove the second assertion in Corollary 2.14.2.

3

Fourier and Laplace transformations

Fourier and Laplace transformations of measures provide a powerful tool, for example to prove equality of measures. Moreover, they are used to characterize the normal distribution, the independence of random variables and to represent Brownian motion by a martingale. All results in this chapter are well known.

3.1 Transformations of measures

We define for all $\lambda = (\lambda_1, \ldots, \lambda_n)$ and $\rho = (\rho_1, \ldots, \rho_n) \in \mathbb{R}^n$

$$\langle \lambda, \rho \rangle := \lambda_1 \rho_1 + \ldots + \lambda_n \rho_n$$

and $\|\lambda\| := \sqrt{\langle \lambda, \lambda \rangle}$. Let μ be a finite Borel measure on $\mathbb{R}$. Then $x \mapsto e^{ix}$ is μ-integrable. Therefore, the μ-integrability of $x \mapsto e^{zx}$ with $z \in \mathbb{C}$ only depends on the real part $\mathrm{Re}(z)$ of z. If μ is a Borel measure on $\mathbb{R}^n$, then the functions $\widetilde{\mu}$ and $\widetilde{\mu}^i$ defined on $\mathbb{R}^n$ by setting

$$\widetilde{\mu}(\lambda) := \int_{\mathbb{R}^n} e^{\langle \lambda, x \rangle} d\mu(x) \in \mathbb{R}_0^+ \cup \{\infty\}, \quad \widetilde{\mu}^i(\lambda) := \int_{\mathbb{R}^n} e^{i\langle \lambda, x \rangle} d\mu(x) \in \mathbb{C}$$

are called the **Laplace transformation** and **Fourier transformation of** μ, respectively. The following result is well known (see Ash [5] Theorem 8.1.3):

Lemma 3.1.1 *Assume that μ_1 and μ_2 are two finite Borel measures on $\mathbb{R}^n$. Then $\mu_1 = \mu_2$ if $\widetilde{\mu_1}^i(\lambda) = \widetilde{\mu_2}^i(\lambda)$ for all $\lambda \in \mathbb{R}^n$.*

Proposition 3.1.2 *Assume that μ_1 and μ_2 are two finite Borel measures on $\mathbb{R}^n$. Then $\mu_1 = \mu_2$ if there exists an $\varepsilon > 0$ such that for all $\lambda \in \mathbb{R}^n$ with $\|\lambda\| < \varepsilon$*

$$\widetilde{\mu_1}(\lambda) = \widetilde{\mu_2}(\lambda) < \infty. \tag{5}$$

Proof By Lemma 3.1.1, we have to show that $\widetilde{\mu_1}^i(\lambda) = \widetilde{\mu_2}^i(\lambda)$ for all $\lambda \in \mathbb{R}^n$. If $\lambda = 0$, by (5), this equality is true. Therefore, we may assume that $\lambda \neq 0$. Set $\delta := \frac{\varepsilon}{\|\lambda\|}$ and fix a complex number $z = a + ib$ with $(a,b) \in D :=]-\delta, \delta[\times \mathbb{R}$. Then $\left\| e^{z\langle \lambda, \cdot \rangle} \right\| = e^{a\langle \lambda, \cdot \rangle}$. Since $\|a \cdot \lambda\| < \varepsilon$, $\left\| e^{z\langle \lambda, \cdot \rangle} \right\|$ is μ_i-integrable, $i = 1, 2$. Let $N := \{x \in \mathbb{R}^n \mid \langle \lambda, x \rangle = 0\}$. For $z \in D$ define $f_i(z) := \int_{\mathbb{R}^n \setminus N} e^{z\langle \lambda, \cdot \rangle} d\mu_i$, $i = 1, 2$. Then,

$$\lim_{h \in \mathbb{C}, h \to 0} \frac{f_i(z+h) - f_i(z)}{h}$$

$$= \lim_{h \in \mathbb{C}, h \to 0} \int_{\mathbb{R}^n \setminus N} e^{z\langle \lambda, \cdot \rangle} \langle \lambda, \cdot \rangle \frac{e^{h\langle \lambda, \cdot \rangle} - 1}{h \langle \lambda, \cdot \rangle} d\mu_i = \int_{\mathbb{R}^n \setminus N} e^{z\langle \lambda, \cdot \rangle} \langle \lambda, \cdot \rangle \, d\mu_i.$$

Since $g_i := f_i + \int_N e^{z\langle \lambda, \cdot \rangle} d\mu_i = f_i + \mu_i(N)$ is analytic and, by (5), $g_1(z) = g_2(z)$ for all $z \in]-\delta, \delta[$, by the identity theorem for analytic functions, $g_1(z) = g_2(z)$ for all $z \in D$, in particular, $\int_{\mathbb{R}^n} e^{i\langle \lambda, \cdot \rangle} d\mu_1 = \int_{\mathbb{R}^n} e^{i\langle \lambda, \cdot \rangle} d\mu_2$. □

In the remainder of this section we fix a complete probability space $(\Lambda, \mathcal{C}, \mu)$ and a real random variable f, defined on Λ.

3.2 Laplace characterization of $\mathcal{N}(0, \sigma)$-distribution

Fix $n \in \mathbb{N}$ and $\sigma > 0$. The **centred Gaussian measure** γ_σ^n with variance σ is a Borel measure on $\mathbb{R}^n$ defined by

$$\gamma_\sigma^n(B) := \int_B \exp\left[-\frac{1}{2\sigma} \sum_{i=1}^n x_i^2 \right] d(x_i)_{i \leq n} \left(\frac{1}{\sqrt{2\pi\sigma}} \right)^n.$$

By Fubini's Theorem, this measure γ_σ^n is the n-fold product of the one-dimensional measure γ_σ^1.

Lemma 3.2.1 $\int_{\mathbb{R}} e^{-\frac{1}{2}x^2} dx = \sqrt{2\pi}$.

Proof First note that $\int_{\mathbb{R}} e^{-\frac{1}{2}x^2} dx < \infty$. By Fubini's theorem and using polar coordinates we obtain

$$\left(\int_{\mathbb{R}} e^{-\frac{1}{2}x^2} dx \right)^2 = \int_{\mathbb{R}} \int_{\mathbb{R}} e^{-\frac{1}{2}(x^2+y^2)} dx dy = \int_{\mathbb{R}^2} e^{-\frac{1}{2}(x^2+y^2)} dxy$$

$$= \int_0^{2\pi} \int_0^{\infty} e^{-\frac{1}{2}r^2} r dr d\varphi = 2\pi \lim_{k \to \infty} \int_0^k e^{-\frac{1}{2}r^2} r dr$$

$$= 2\pi \lim_{k \to \infty} \left(-e^{-\frac{1}{2}k^2} + 1 \right) = 2\pi.$$

It follows that $\int_{\mathbb{R}} e^{-\frac{1}{2}x^2}\,dx = \sqrt{2\pi}$. □

Using the transformation rule we obtain:

Corollary 3.2.2 *γ_σ^1 is a probability measure, thus γ_σ^n is a probability measure for each $n \in \mathbb{N}$.*

The function f is called **$\mathcal{N}(0,\sigma)$-distributed** if f is **normally distributed** with **variance** σ and **mean** 0, i.e., γ_σ^1 is the image measure of μ by f.

Proposition 3.2.3 *The function f is $\mathcal{N}(0,\sigma)$-distributed if and only if for all $\lambda \in \mathbb{R}$*

$$\mathbb{E}_\mu e^{\lambda f} = e^{\frac{\sigma\lambda^2}{2}}.$$

Proof '$\Rightarrow$' Suppose that f is $\mathcal{N}(0,\sigma)$-distributed. Fix $\lambda \in \mathbb{R}$. Then

$$\begin{aligned}\mathbb{E}_\mu e^{\lambda f} &= \int_{\mathbb{R}} e^{\lambda \cdot}\,d\mu_f = \frac{1}{\sqrt{2\pi\sigma}}\int_{\mathbb{R}} e^{\lambda t} e^{-\frac{t^2}{2\sigma}}\,dt \\ &= \frac{1}{\sqrt{2\pi\sigma}} e^{\frac{\lambda^2\sigma}{2}} \int_{\mathbb{R}} e^{-\frac{(t-\lambda\sigma)^2}{2\sigma}}\,dt = e^{\frac{\lambda^2\sigma}{2}}.\end{aligned}$$

'$\Leftarrow$' There always exists an $\mathcal{N}(0,\sigma)$-distributed random variable g on a probability space with measure, say ρ. (See Bauer [6] Theorem 29.1.) By '$\Rightarrow$', and the hypothesis, $\mathbb{E}_\rho e^{\lambda g} = e^{\frac{\sigma\lambda^2}{2}}$ and $\mathbb{E}_\mu e^{\lambda f} = e^{\frac{\sigma\lambda^2}{2}}$ for all $\lambda \in \mathbb{R}$. It follows that for each $\lambda \in \mathbb{R}$

$$\int_{\mathbb{R}} e^{\lambda t}\,d\rho_g(t) = \mathbb{E}_\rho e^{\lambda g} = \mathbb{E}_\mu e^{\lambda f} = \int_{\mathbb{R}} e^{\lambda t}\,d\mu_f(t) < \infty.$$

By Proposition 3.1.2, the measures ρ_g and μ_f coincide, thus f is also $\mathcal{N}(0,\sigma)$-distributed. □

Corollary 3.2.4 *If f is $\mathcal{N}(0,\sigma)$-distributed and $\beta \neq 0$ in $\mathbb{R}$, then $\beta \cdot f$ is $\mathcal{N}(0,\beta^2\sigma)$-distributed.*

3.3 Fourier and Laplace characterization of independence

A k-tuple of real random variables $(g_1,\ldots,g_k)$ is called **independent** if for all Borel sets $B_1,\ldots,B_k \subseteq \mathbb{R}$

$$\mu(g_1^{-1}[B_1] \cap \ldots \cap g_k^{-1}[B_k]) = \mu(g_1^{-1}[B_1]) \cdot \ldots \cdot \mu(g_k^{-1}[B_k]).$$

Therefore, the independence of $(g_1,\ldots,g_k)$ is equivalent to the fact that the image measure $\mu_{(g_1,\ldots,g_k)}$ of μ by $(g_1,\ldots,g_k)$ on $\mathcal{B}(\mathbb{R}^k)$ and the product measure $\mu_{g_1}\otimes\ldots\otimes\mu_{g_k}$ on $\mathcal{B}(\mathbb{R}^k)$ are identical.

Let $\mathcal{S}\subseteq\mathcal{C}$ with $\Lambda\in\mathcal{S}$ and assume that $\mathcal{S}$ is closed under complements. Then $(f,\mathcal{S})$ is called **independent** if $(f,\mathbf{1}_C)$ is independent for all $C\in\mathcal{S}$. Note that $(f,\mathcal{S})$ is independent if and only if for all Borel sets $B\subseteq\mathbb{R}$ and all $S\in\mathcal{S}$

$$\mu(f^{-1}[B]\cap S])=\mu(f^{-1}[B])\cdot\mu(S).$$

Let us put together some simple results on independence, which we will apply again and again:

Proposition 3.3.1 *Fix k-tuples* $g:=(g_1,\ldots,g_k)$, $f:=(f_1,\ldots,f_k)$ *of random variables.*

(a) *Then g is independent if and only if for all* $\tau\in\mathbb{R}^k$

$$\mathbb{E}e^{i\cdot\langle\tau,g\rangle}=\prod_{j=1}^{k}\mathbb{E}e^{i\cdot\tau_j\cdot g_j}.$$

(b) *Fix* $\varepsilon\in\mathbb{R}^+\cup\{\infty\}$. *Suppose that* $\mathbb{E}_\mu e^{\lambda f}<\infty$ *for all* $\lambda\in]-\varepsilon,\varepsilon[$. *Then* $(f,\mathcal{S})$ *is independent if and only if for all* $\lambda\in]-\varepsilon,\varepsilon[$

$$\mathbb{E}^{\mathcal{S}}e^{\lambda f}=\mathbb{E}e^{\lambda f}\ \mu\text{-a.s.}$$

(c) *Assume that f and g are independent and that* f_j *has the same distribution as* g_j *for all* $j=1,\ldots,k$. *Then* $\sum_{j=1}^k f_j$ *has the same distribution as* $\sum_{j=1}^k g_j$.

(d) *Assume that* $(g_1,\ldots,g_k)$ *is independent and the* g_i *are integrable. Then* $g_1\cdot\ldots\cdot g_k$ *is integrable and*

$$\mathbb{E}(g_1\cdot\ldots\cdot g_k)=\mathbb{E}(g_1)\cdot\ldots\cdot\mathbb{E}(g_k).$$

(e) *Assume that there is an* $\varepsilon\in\mathbb{R}^+\cup\{\infty\}$ *such that* $\mathbb{E}e^{\langle\tau,g\rangle}<\infty$ *for all* $\tau\in\mathbb{R}^n$ *with* $\|\tau\|<\varepsilon$. *Then* $(g_1,\ldots,g_k)$ *is independent if and only if for all* $\tau\in\mathbb{R}^n$ *with* $\|\tau\|<\varepsilon$

$$\mathbb{E}e^{\langle\tau,g\rangle}=\prod_{j=1}^{k}\mathbb{E}e^{\tau_j\cdot g_j}.$$

Proof (a) By the transformation rule and Fubini's theorem, $\mathbb{E}e^{i\cdot\langle\tau,g\rangle}=\int_{\mathbb{R}^k}e^{i\cdot\langle\tau,x\rangle}d\mu_g(x)$ and $\prod_{j=1}^k\mathbb{E}e^{i\cdot\tau_j g_j}=\int_{\mathbb{R}^k}e^{i\cdot\langle\tau,x\rangle}d\left(\mu_{g_1}\otimes\ldots\otimes\mu_{g_k}\right)(x)$. Part (a) now follows from Lemma 3.1.1.

(b) '$\Rightarrow$' Fix $\lambda \in]-\varepsilon, \varepsilon[$, and assume that $(f, \mathcal{S})$ is independent. Note that $(e^{\lambda f}, \mathcal{S})$ is also independent. It follows that $\mathbb{E}^{\mathcal{S}} e^{\lambda f} = \mathbb{E} e^{\lambda f}$ μ-a.s. (See Bauer [6] 54.5.)

'$\Leftarrow$' Let $Q : \Lambda \times \mathcal{B}(\mathbb{R}) \to [0,1]$ be the **regular conditional probability** of f given $\mathcal{S}$ (see Ash [5] Theorem 6.6.4), i.e.,

(i) $Q(X, \cdot)$ is a probability measure on the Borel algebra $\mathcal{B}(\mathbb{R})$ of $\mathbb{R}$ for all $X \in \Lambda$.
(ii) $\int_{\mathbb{R}} \mathbf{1}_B(t) dQ(\cdot, t) = Q(\cdot, B) = \mathbb{E}^{\mathcal{S}}(\mathbf{1}_{f^{-1}[B]}) = \mathbb{E}^{\mathcal{S}}(\mathbf{1}_B \circ f)$ μ-a.s. for all $B \in \mathcal{B}(\mathbb{R})$.

It follows that for all non-negative Borel measurable functions $\varphi : \mathbb{R} \to \mathbb{R}$

$$\int_{\mathbb{R}} \varphi(t) dQ(\cdot, t) = \mathbb{E}^{\mathcal{S}}(\varphi \circ f), \text{ in particular } \int_{\mathbb{R}} e^{\lambda t} dQ(\cdot, t) = \mathbb{E}^{\mathcal{S}} e^{\lambda f} \ \mu\text{-a.s.}$$

Therefore, there exists a set $U \in \mathcal{C}$ with $\mu(U) = 1$ such that for all rationals $\lambda \in]-\varepsilon, \varepsilon[$ and all $X \in U$

$$\int_{\mathbb{R}} e^{\lambda t} dQ(X, t) = \mathbb{E} e^{\lambda f} = \int_{\mathbb{R}} e^{\lambda t} d\mu_f(t).$$

By the continuity of $e^{\cdot f}$ and the dominated convergence theorem, these equalities are true for all $\lambda \in]-\varepsilon, \varepsilon[$. By Proposition 3.1.2, $Q(X, B) = \mu_f(B)$ for all $X \in U$ and all Borel subsets $B \subset \mathbb{R}$. We obtain for all $C \in \mathcal{S}$

$$\begin{aligned} \mu(f^{-1}[B] \cap C) &= \int_C \mathbf{1}_{f^{-1}[B]} d\mu = \int_C Q(\cdot, B) d\mu \\ &= \int_C \mu_f(B) d\mu = \mu(f^{-1}[B]) \mu(C). \end{aligned}$$

(c) Lemma 3.1.1 and (a) tell us that $\mu_f = \mu_g$, whence f and g have the same distribution. Then, by the transformation rule, we obtain for all $\tau \in \mathbb{R}$

$$\begin{aligned} \int_{\mathbb{R}} e^{i \cdot \tau \cdot x} d\mu_{\sum_{j=1}^k f_j} &= \mathbb{E} e^{i \cdot \tau \cdot \sum_{j=1}^k f_j} = \int_{\mathbb{R}^k} e^{i \cdot \langle (\tau, \ldots, \tau), x \rangle} d\mu_f \\ &= \int_{\mathbb{R}^k} e^{i \cdot \langle (\tau, \ldots, \tau), x \rangle} d\mu_g = \int_{\mathbb{R}} e^{i \cdot \tau \cdot x} d\mu_{\sum_{j=1}^k g_j}. \end{aligned}$$

This proves (c).

(d) Since $\mu_{(g_1,\dots,g_k)} = \mu_{g_1} \otimes \ldots \otimes \mu_{g_k}$, we obtain

$$\begin{aligned}\mathbb{E}(g_1)\cdot\ldots\cdot\mathbb{E}(g_k) &= \int_{\mathbb{R}} x d\mu_{g_1} \cdot \ldots \cdot \int_{\mathbb{R}} x d\mu_{g_k} \\ &= \int_{\mathbb{R}^k} x_1 \cdot \ldots \cdot x_k d\mu_{g_1} \otimes \ldots \otimes \mu_{g_k} \\ &= \int_{\mathbb{R}^k} x_1 \cdot \ldots \cdot x_k d\mu_{(g_1,\dots,g_k)} = \mathbb{E}(g_1 \cdot \ldots \cdot g_k).\end{aligned}$$

(e) follows from Proposition 3.1.2. □

3.4 Discrete Lévy processes and their representation

Here is an application of the results of the preceding section, which are simple but important for the whole book. Recall the notation in Section 2.1.

What has been said about martingales at the beginning of Section 2.1 also holds for Lévy processes. First of all we study Lévy processes defined on standard finite timelines. Later the notion 'finite' is extended and the standard results are transferred to a finite timeline in the extended sense. Then it will be shown that all Lévy processes defined on the continuous timeline $[0,\infty[$ are infinitely close to Lévy processes defined on the finite timeline, finite in the extended sense. This result holds not only for finite-dimensional Lévy processes but also for infinite-dimensional Brownian motion.

For simplicity, let us study one-dimensional Lévy processes. In view of the infinite-dimensional Brownian motion we then have only two dimensions, $d = 1$ and $d = \infty$. Lévy processes in this section are one-dimensional.

A mapping $L : \Lambda \times I \to \mathbb{R}$ is called **$\mathcal{C}$-measurable** if L_t is $\mathcal{C}$-measurable for all $t \in I$. We always define $L_{\square^-} := 0$. Recall that $I_t := \{s \in T \mid s \le t\}$. If $s \le t$ in $I \cup \{\square^-\}$, then the number of all $r \in I$ with $s < r \le t$ is called the **distance from s to t**. If L is $\mathcal{C}$-measurable, then L is called a **discrete (one-dimensional) Lévy process**, provided that the following conditions (L 1), (L 2) and (L 3) hold.

(L 1) The $|I|$-tuple $(\Delta L_t)_{t\in I}$ is independent. Then, by Proposition 3.3.1 (a), $(L_{t_{2j}} - L_{t_{2j-1}})_{j\in\{1,\dots,n\}}$ is independent for all $n \in \mathbb{N}$, and each $2n$-tuple $(t_1,\dots,t_{2n})$ in $I \cup \{\square^-\}$ with $t_1 < t_2 \le t_3 < t_4 \le \ldots \le t_{2n-1} < t_{2n}$.

(L 2) For all $s,t \in I$, ΔL_t has the same distribution as ΔL_s. Then, by Proposition 3.3.1 (c) for all $s \le t, \widetilde{s} \le \widetilde{t}$ in I, $(L_t - L_s)$ has the same distribution as $L_{\widetilde{t}} - L_{\widetilde{s}}$, provided that the distance from s to t equals the distance from $\widetilde{s}$ to $\widetilde{t}$.

(L 3) $\mathbb{E}L_{\square}^2 < \infty$, in particular, $\mathbb{E}\,|L_{\square}| < \infty$.

Remark 3.4.1 It should be mentioned that Condition (L 3) is a quite weak assumption. We shall see in Theorem 15.8.1 later that any standard Lévy process can be represented by a discrete Lévy process defined on a finite timeline in the extended sense.

Lemma 3.4.2 *Assume that L is a discrete Lévy process. Then:*

(a) $\mathbb{E}(L_t - L_s)^j = \mathbb{E}(L_{\widetilde{t}} - L_{\widetilde{s}})^j$ *for* $j \in \{1,2\}$, *provided that the distance from s to t equals the distance from* $\widetilde{s}$ *to* $\widetilde{t}$.

(b) $\mathbb{E}L_t = \mathbb{E}\left(\sum_{s \le t} \Delta L_s\right) = |I_t| \cdot \mathbb{E}L_\square$ *for all* $t \in I$.

(c) $\mathbb{E}(\Delta L_s \cdot \Delta L_t) = \mathbb{E}\Delta L_s \cdot \mathbb{E}\Delta L_t$ *if* $s \neq t$ *in* I.

(d) $\Sigma_{r<s\le t \in I} \mathbb{E}(\Delta L_s \cdot \Delta L_r) = \frac{|I_t|\cdot(|I_t|-1)}{2} (\mathbb{E}L_\square)^2$ *for a fixed* $t \in I$.

(e) $\mathbb{E}L_t^2 = \mathbb{E}\left(\sum_{s \le t} \Delta L_s\right)^2 = |I_t| \cdot \mathbb{E}L_\square^2 + |I_t| \cdot (|I_t| - 1)(\mathbb{E}L_\square)^2$.

Proof (a) follows from the transformation rule and from the fact that the image measures of μ by $L_t - L_s$ and by $L_{\widetilde{t}} - L_{\widetilde{s}}$ are identical.

(b) follows from (a).

(c) follows from Proposition 3.3.1 (d).

(d) follows from (a) and (c) using that $\frac{|I_t|\cdot(|I_t|-1)}{2}$ is the number of pairs (r,s) with $r < s \le t$ in I.

(e) follows from (a) and (d). □

Often Lévy processes are modified by restricting the increments to certain Borel sets of $\mathbb{R}$. Fix a Borel set A in $\mathbb{R}$ and let L be a discrete Lévy process. Then

$$L^A : \Lambda \times T \to \mathbb{R}, (X,t) \mapsto \sum_{s \in I_t} \mathbf{1}_A(\Delta L_s(X)) \cdot \Delta L_s(X)$$

is again a discrete Lévy process (with $L^A_{\square^-} := 0$). We say that the increments of L are **restricted to** A.

If L is a discrete Lévy process, then we call the canonical martingale $M^L : \Lambda \times I \to \mathbb{R}$, $(X,t) \mapsto L_t - \mathbb{E}_\mu L_t$ the **martingale associated to** L. Recall that M^L is a $\left(\mathcal{B}_t^L\right)_{t \in I}$-martingale, where

$$\mathcal{B}_t^L := \left\{ (\Delta L_s)_{s \in I_t}^{-1} [D] \mid D \text{ is a Borel set in } \mathbb{R}^{I_t} \right\}.$$

Moreover, M^L is μ-square-integrable. Note that

$$M_t^L = \sum_{s \le t} \left(\Delta L_s - \mathbb{E}_\mu L_\square \right).$$

It is easy to see that M^L is again a discrete Lévy process.

Let us fix a Borel probability measure μ^1 on $\mathbb{R}$ with $\mathbb{E}_{\mu^1}x^2 < \infty$. Let μ be the product measure of μ^1 on $\Omega := \mathbb{R}^I$. Define $B : \Omega \times I \to \mathbb{R}$ by setting $B_t(X) := \sum_{s\leq t} X_s$. Of course, B_t is a smooth function. This fact will play an important role later on. Set $(\mathcal{B}_t)_{t\in I} := \left(\mathcal{B}_t^B\right)_{t\in I}$. Note that

$$\mathcal{B}_t = \left\{D \times \mathbb{R}^{I\setminus I_t} \mid D \text{ is a Borel set in } \mathbb{R}^{I_t}\right\}.$$

It is also easy to see that each discrete Lévy process can be represented by such a smooth function B: Fix a discrete Lévy process L, defined on our probability space $(\Lambda, \mathcal{C}, \mu)$. Let ρ^1 be the image measure of μ by $L_\square$. By Proposition 3.3.1 and Lemma 3.4.2, we obtain

Proposition 3.4.3 *Fix $n \in \mathbb{N}$ and a strictly increasing $n+1$-tuple $(t_0, \ldots, t_n)$ in $I \cup \{\square^-\}$. Then $(L_{t_j} - L_{t_{j-1}})_{j\in\{1,\ldots,n\}}$ has the same distribution by μ as $(B_{t_j} - B_{t_{j-1}})_{j\in\{1,\ldots,n\}}$ has by ρ. Moreover, if L is a $\left(\mathcal{B}_t^L\right)_{t\in T}$-$\mu$-martingale, then B is a $(\mathcal{B}_t)_{t\in T}$-ρ-martingale.*

The process B is called the **representation of** L **by** (ρ, μ). By Proposition 3.4.3, we may identify L and B; in particular, we have

Corollary 3.4.4 *For all $t \in I$, $(L_s)_{s\in I_t}$ has the same distribution by μ as $(B_s)_{s\in I_t}$ has by ρ.*

The process B is fixed, but the Borel measure ρ^1 on $\mathbb{R}$ characterizes the special Lévy process. The increment of $B(X, \cdot)$ to t is the tth-component of the vector X. Note that $M^B : (X,t) \mapsto \sum_{s\leq t}\left(X_s - \mathbb{E}_{\rho^1}x\right)$ is the canonical ρ-martingale associated to B. In order to indicate that M^B depends only on ρ, indeed only on ρ^1, let us set

$$M^\rho := M^B.$$

By the triangle and Doob's inequality, we obtain

Lemma 3.4.5 *Fix $t \in I$. Set $\beta_t^\rho := |I_t|\mathbb{E}_{\rho^1}x$. Then*

$$\left\|\max_{s\leq t}|B_s|\right\|_2 \leq \left\|\max_{s\leq t}|M_s^\rho|\right\|_2 + |\beta_t^\rho| \leq 2\,\|M_t^\rho\|_2 + |\beta_t^\rho|,$$

$$\left\|\max_{s\leq t}|M_s^\rho|\right\|_2 \leq 2\,\|M_t^\rho\|_2 \leq 2\left(\|B_t\|_2 + |\beta_t^\rho|\right) \leq 2\left(\left\|\max_{s\leq t}|B_s|\right\|_2 + |\beta_t^\rho|\right).$$

Remark 3.4.6 My Bachelor student Benedikt Rehle has applied results along the lines in this section to Lindstrøm's approach [67] to Lévy processes. In particular, he used the representation result of discrete Lévy processes.

3.5 Martingale characterization of Brownian motion

Now we use Laplace transformations to prove a well-known martingale characterization of Brownian motion. A process $d : \Lambda \times [0,\infty[\to \mathbb{R}$ is called a **one-dimensional Brownian motion under a filtration** $(\mathfrak{f}_t)_{t\in[0,\infty[}$ on $\mathcal{C}$ if the following conditions hold:

(i) d_t is $\mathfrak{f}_t$-measurable for all $t \in [0,\infty[$.
(ii) $d(X,\cdot)$ is continuous for μ-almost all $X \in \Lambda$.
(iii) $d_t - d_s$ is $\mathcal{N}(0,t-s)$-distributed for all $s,t \in [0,\infty[$, $s < t$.
(iv) $d_t - d_s$ is independent of $\mathfrak{f}_s$ for all $s,t \in [0,\infty[$, $s < t$.
(v) $d(X,0) = 0$ for μ-almost all $X \in \Lambda$.

Proposition 3.5.1 *Let $d : \Lambda \times [0,\infty[\to \mathbb{R}$ be a Brownian motion under $(\mathfrak{f}_t)_{t\in[0,\infty[}$ on $\mathcal{C}$. Fix $t_0,\dots,t_k \in [0,\infty[$ with $0 \le t_0 < \dots < t_k$. Then $\left(d_{t_i} - d_{t_{i-1}}\right)_{i\le k}$ is independent.*

Proof Fix $\lambda_1,\dots,\lambda_k \in \mathbb{R}$. By Proposition 3.2.3 and Hölder's inequality, $\mathbb{E}e^{\sum_{i=1}^k \lambda_i\left(d_{t_i}-d_{t_{i-1}}\right)} < \infty$. We obtain:

$$
\begin{aligned}
&\mathbb{E}\exp\left[\sum_{i=1}^{k}\lambda_i\left(d_{t_i}-d_{t_{i-1}}\right)\right]\\
&=\mathbb{E}\left(\exp\left[\sum_{i=1}^{k-1}\lambda_i\left(d_{t_i}-d_{t_{i-1}}\right)\right]\cdot\mathbb{E}^{\mathfrak{f}_{t_{k-1}}}\exp\left[\lambda_k\left(d_{t_k}-d_{t_{k-1}}\right)\right]\right)\\
&=\left(\mathbb{E}\exp\left[\sum_{i=1}^{k-1}\lambda_i\left(d_{t_i}-d_{t_{i-1}}\right)\right]\right)\cdot\mathbb{E}\exp\left[\lambda_k\left(d_{t_k}-d_{t_{k-1}}\right)\right] \quad \text{(by 3.3.1 (b))}\\
&=\dots=\prod_{i=1}^{k}\mathbb{E}\exp\left[\lambda_i\left(d_{t_i}-d_{t_{i-1}}\right)\right]<\infty.
\end{aligned}
$$

By Proposition 3.3.1 (e), the proof is finished. □

Theorem 3.5.2 (Lévy [63]) *Let $d : \Lambda \times [0,\infty[\to \mathbb{R}$ be a μ-a.s. continuous process with $d(X,0) = 0$ for μ-almost all $X \in \Lambda$. Then d is a Brownian motion under $(\mathfrak{f}_t)_{t\in[0,\infty[}$ if and only if for all $\lambda \in \mathbb{R}$ the process*

$$\Lambda \times [0,\infty[\ni (X,t) \mapsto \exp\left[\lambda d(X,t) - \frac{1}{2}\lambda^2 t\right]$$

is an $(\mathfrak{f}_t)_{t\in[0,\infty[}$-martingale.

Proof '$\Rightarrow$' Assume that d is a Brownian motion under $(\mathfrak{f}_t)_{t\in[0,\infty[}$. Since d_t is $\mathfrak{f}_t$-measurable, $\exp\left[\lambda d_t - \frac{1}{2}\lambda^2 t\right]$ is also $\mathfrak{f}_t$-measurable. Since $d_t - d_s$ is $\mathcal{N}(0, t-s)$-distributed for $s < t$, by Proposition 3.2.3,

$$\mathbb{E}\exp[\lambda(d_t - d_s)] = \exp\left[\frac{\lambda^2}{2}(t-s)\right] < \infty \text{ for all } \lambda \in \mathbb{R}.$$

Since $d_t - d_s$ is independent of $\mathfrak{f}_s$, by Proposition 3.3.1 (b), for all $\lambda \in \mathbb{R}$,

$$\mathbb{E}^{\mathfrak{f}_s}\exp[\lambda(d_t - d_s)] = \mathbb{E}\exp[\lambda(d_t - d_s)] = \exp\left[\frac{\lambda^2}{2}(t-s)\right] \mu\text{-a.s.}$$

Thus we obtain for all $\lambda \in \mathbb{R}$ and for $s < t$ μ-a.s., where $\alpha := \exp\left[\lambda d_s - \frac{1}{2}\lambda^2 t\right]$,

$$\begin{aligned}
\mathbb{E}^{\mathfrak{f}_s}\exp\left[\lambda d_t - \frac{1}{2}\lambda^2 t\right] &= \exp\left[-\frac{1}{2}\lambda^2 t\right]\mathbb{E}^{\mathfrak{f}_s}\exp[\lambda d_s + \lambda(d_t - d_s)] \\
&= \alpha\mathbb{E}^{\mathfrak{f}_s}\exp[\lambda(d_t - d_s)] = \alpha\exp\left[\frac{\lambda^2}{2}(t-s)\right] \\
&= \exp\left[\lambda d_s - \frac{\lambda^2}{2}s\right].
\end{aligned}$$

'$\Leftarrow$' Suppose that $\left(\exp\left[\lambda d_t - \frac{\lambda^2}{2}t\right]\right)_{t\in[0,\infty[}$ is an $(\mathfrak{f}_t)_{t\in[0,\infty[}$-martingale for all $\lambda \in \mathbb{R}$. Then d_t is $\mathfrak{f}_t$-measurable for all $t \in [0,\infty[$. Fix $\lambda \in \mathbb{R}$ and $s < t$ in $[0,\infty[$. Then μ-a.s.:

$$\begin{aligned}
\mathbb{E}^{\mathfrak{f}_s}\exp[\lambda(d_t - d_s)] &= \exp\left[-\lambda d_s + \frac{\lambda^2}{2}t\right]\mathbb{E}^{\mathfrak{f}_s}\exp\left[\lambda d_t - \frac{\lambda^2}{2}t\right] \\
&= \exp\left[-\lambda d_s + \frac{\lambda^2}{2}t\right]\exp\left[\lambda d_s - \frac{\lambda^2}{2}s\right] = \exp\left[\frac{\lambda^2}{2}(t-s)\right] \\
&= \mathbb{E}\exp\left[\frac{\lambda^2}{2}(t-s)\right] = \mathbb{E}\mathbb{E}^{\mathfrak{f}_s}\exp[\lambda(d_t - d_s)] \\
&= \mathbb{E}\exp[\lambda(d_t - d_s)].
\end{aligned}$$

By Proposition 3.3.1 (b), $d_t - d_s$ is independent of $\mathfrak{f}_s$ and, by Proposition 3.2.3, $d_t - d_s$ is $\mathcal{N}(0, t-s)$-distributed. □

Exercises

3.1 The set of finitely many disjoint unions of right rectangles $I_1 \times \ldots \times I_k$, where $I_1, \ldots, I_k$ are right intervals in $\mathbb{R}$, is an algebra on $\mathbb{R}^k$.

3.2 Assume that $\mathcal{A} \subseteq \mathcal{P}(\Lambda)$ is $\cap$-stable (**intersection stable**), i.e., $A \cap B \in \mathcal{A}$ for all $A, B \in \mathcal{A}$. Prove that $\mathcal{A}$ is an algebra (σ-algebra) on Λ if and only if
   (i) $\Lambda \in \mathcal{A}$,
   (ii) $A \setminus B \in \mathcal{A}$ for all $A, B \in \mathcal{A}$ with $B \subseteq A$,
   (iii) $A \cup B \in \mathcal{A}$ for all $A, B \in \mathcal{A}$ with $B \cap A = \emptyset$,
   (iv) ($\bigcup_{n \in \mathbb{N}} A_n \in \mathcal{A}$ for pairwise disjoint $A_n \in \mathcal{A}$).

3.3 Let $\mathcal{G}$ be a linear space of real $\mathcal{C}$-measurable functions, defined on Λ. Let $\sigma(\mathcal{G})$ be the σ-algebra, generated by finite intersections of sets of the form $f^{-1}[B]$, where $f \in \mathcal{G}$ and B is a Borel set in $\mathbb{R}$.
   (i) Prove that $\sigma(\mathcal{G})$ is the smallest σ-algebra such that each $f \in \mathcal{G}$ is Borel measurable.
   (ii) Let μ_1 and μ_2 two finite measures on $\sigma(\mathcal{G})$. Prove that $\mu_1 = \mu_2$ if $\int_\Lambda e^{if} \, d\mu_1 = \int_\Lambda e^{if} \, d\mu_2$.

3.4 Prove that $\int_{\mathbb{R}} e^{-\frac{1}{2}x^2} < \infty$.

3.5 Prove Corollary 3.2.2.

3.6 Prove Corollary 3.2.4.

3.7 Recall the notation in Section 3.3. Prove that $(f, \mathcal{S})$ is independent if and only if $\mu(f^{-1}[B] \cap S]) = \mu(f^{-1}[B]) \cdot \mu(S)$ for all Borel sets $B \subseteq \mathbb{R}$ and all $S \in \mathcal{S}$.

3.8 Prove Corollary 3.4.4.

3.9 Fix a discrete Lévy process L on $(\Lambda, \mathcal{C}, \rho)$ and let μ^1 be the image measure of ρ by $L_\square$. Moreover, fix a Borel probability measure κ on $\mathbb{R}$. Prove the following.
   (i) Let A be a Borel set in $\mathbb{R}$. Then κ is the image measure of ρ by $L_\square^A$ if and only if, for all internal Borel sets C in $\mathbb{R}$,

$$\kappa(C) = \begin{cases} \mu^1(A \cap C) & \text{if } 0 \notin C \\ \mu^1((\mathbb{R} \setminus A) \cup C) & \text{if } 0 \in C. \end{cases}$$

   (ii) Let M^L be the martingale associated to L. Then κ is the image measure of ρ by $M_\square^L$ if and only if, for all Borel sets C in $\mathbb{R}$,

$$\kappa(C) = \mu^1(C + \mathbb{E}_\rho(L_\square)).$$

Define by recursion for all $n \in \mathbb{N}_0$:

$$H_n(x) = \frac{(-1)^n}{n!} e^{\frac{1}{2}x^2} \frac{d^n}{dx^n} e^{-\frac{1}{2}x^2}, \quad H_{-1} := 0.$$

H_n is called the nth **Hermite polynomial**, $n \geq 0$. Set $f(x) := e^{-\frac{1}{2}x^2}$ and $f^{(n)}(x) := \frac{d^n}{dx^n} e^{-\frac{1}{2}x^2}$ with $f^{(-1)}(x) := 0$.

3.10 Prove that for all $x\in\mathbb{R}$ the following hold.

(i) $f^{(n)}(x)=-f^{(n-1)}(x)\cdot x-(n-1)f^{(n-2)}(x)$ for $n\geq 1$.

(ii) $\lim_{x\to\infty}f^{(n)}(x)\cdot x^k=0$ for all $k,n\in\mathbb{N}_0$. It follows that

$$\lim_{x\to\infty}f^{(n)}(x)\cdot p(x)=0 \text{ for each polynomial } p.$$

(iii) $e^{x\cdot t-\frac{1}{2}t^2}=\sum_{n=0}^{\infty}t^nH_n(x)$.

(iv) $H_n'=H_{n-1}$ for each $n\in\mathbb{N}_0$ with $H_{-1}:=0$.

(v) $(n+1)H_{n+1}+H_{n-1}=H_n\cdot id$ for all $n\in\mathbb{N}$.

(vi) For each $n\in\mathbb{N}_0$, H_n is a polynomial of degree n and each polynomial of degree n is a linear combination of $H_0,\ldots,H_n$.

(vii) The sequence $(H_n)_{n\in\mathbb{N}_0}$ is an orthogonal basis of $L^2(\gamma_1^1)$ with

$$\int_{\mathbb{R}}H_n^2d\gamma_1^1=\frac{1}{n!}.$$

We may define the derivative on $L^2(\gamma_1^1)$ as a densely defined operator. Since $\left(\sqrt{n!}H_n\right)_{n\in\mathbb{N}_0}$ is an ONB of $L^2(\gamma_1^1)$, each $\varphi\in L^2(\gamma_1^1)$ has the decomposition

$$\varphi=\sum_{n=0}^{\infty}\left\langle\varphi,\sqrt{n!}H_n\right\rangle_{\gamma_1^1}\sqrt{n!}H_n=\sum_{n=0}^{\infty}n!\left\langle\varphi,H_n\right\rangle_{\gamma_1^1}H_n.$$

Moreover, $\sum_{n=0}^{\infty}\langle\varphi,H_n\rangle_{\gamma_1^1}^2\cdot n!<\infty$. We define the operator d on $L^2(\gamma_1^1)$ by setting

$$d\varphi=\sum_{n=0}^{\infty}\langle\varphi,H_n\rangle_{\gamma_1^1}n!H_n'=\sum_{n=1}^{\infty}\langle\varphi,H_n\rangle_{\gamma_1^1}n!H_{n-1}$$

if this series converges in $L^2(\gamma_1^1)$, which is equivalent to

$$\sum_{n=0}^{\infty}\left(\langle\varphi,H_n\rangle_{\gamma_1^1}n!\right)^2\mathbb{E}_{\gamma_1^1}(H_{n-1})^2=\sum_{n=0}^{\infty}\langle\varphi,H_n\rangle_{\gamma_1^1}^2n!n<\infty.$$

It follows that $d\varphi$ converges in $L^2(\gamma_1^1)$ if and only if $\sum_{n=0}^{\infty}\sqrt{n}\langle\varphi,H_n\rangle_{\gamma_1^1}n!H_n$ converges in $L^2(\gamma_1^1)$. It will be seen later that a similar result holds for the Malliavin derivative.

Moreover, if $\varphi=\sum_{n=0}^{k}\langle\varphi,H_n\rangle_{\gamma_1^1}n!H_n$ is a finite sum, then $d\varphi$ exists in $L^2(\gamma_1^1)$. It follows that the domain of d is a dense subspace of $L^2(\gamma_1^1)$; d is called the **derivative on** $L^2(\gamma_1^1)$.

Let $\varphi:\mathbb{R}\to\mathbb{R}\in L^2(\gamma_1^1)$ be continuously differentiable with $\varphi'\in L^2(\gamma_1^1)$, and fix $n\in\mathbb{N}_0$.

3.11 Prove the following.

(i) For each polynomial p

$$\lim_{x\to\infty}\varphi(x)p(x)f^{(n)}(x)=0 \text{ and } \lim_{x\to-\infty}\varphi(x)p(x)f^{(n)}(x)=0.$$

(ii) $\varphi\cdot H_n$ is γ_1^1-integrable and $\langle\varphi',H_{n-1}\rangle_{\gamma_1^1}=n\cdot\langle\varphi,H_n\rangle_{\gamma_1^1}$.

(iii) The domain of d is a dense subspace of $L^2(\gamma_1^1)$.

(iv) $d\varphi=\varphi'$.

4

Abstract Wiener–Fréchet spaces

Abstract Wiener spaces were introduced by Leonhard Gross [41] to construct Gaussian measures on infinite-dimensional spaces. The fact that a Hilbert space can be densely embedded into each Fréchet space is crucial. We refer to the books of Bogachev [14] and Kuo [56].

4.1 Projective systems of measures and their limit

It is well known that it is impossible to define a translation invariant measure on the Borel σ-algebra of an infinite-dimensional Hilbert space such that non-empty open sets have measure greater than 0 (see Kuo [56]). Let us now try to introduce Gaussian measures on infinite-dimensional Hilbert spaces $\mathbb{H}$: in a first step we introduce Gaussian measures γ_σ^E of a fixed variance σ on the Borel σ-algebra of any $E \in \mathcal{E}(\mathbb{H})$, where $\mathcal{E}(\mathbb{H})$ denotes the set of finite-dimensional subspaces of $\mathbb{H}$. We shall see that these measures can be used to define a finitely additive measure γ_σ on the algebra of cylinder sets of $\mathbb{H}$. However, as we shall also see, this measure is not σ-additive. We get stuck.

A possible solution to the problem of finding σ-additive Gaussian measures in infinite-dimensional spaces is based on the concept of abstract Wiener spaces as introduced by Gross [41]. He used a weaker topology on $\mathbb{H}$ such that γ_σ restricted to the cylinder sets with respect to the new topology can be extended to a σ-additive measure on the new Borel σ-algebra. In this chapter we present the details in a slightly more general setting.

A family $(\mu_E)_{E\in\mathcal{E}(\mathbb{H})}$ of Borel probability measures μ_E on $E \in \mathcal{E}(\mathbb{H})$ is called a **projective system** (of measures) if, for each $E, F \in \mathcal{E}(\mathbb{H})$ with $E \subseteq F$, μ_E is the image measure of μ_F under the orthogonal projection π_{FE}, i.e., $\mu_E(B) = \mu_F(\pi_{FE}^{-1}[B])$ for all $B \in \mathcal{B}(E)$. The proofs of the following lemmas are left to the reader.

Lemma 4.1.1 *Fix* $E \subseteq F$ *in* $\mathcal{E}(\mathbb{H})$. *Then* $\pi_{FE}^{-1}[B] = B + (E^{\perp} \cap F) \in \mathcal{B}(F)$ *for all* $B \in \mathcal{B}(E)$.

Lemma 4.1.2 *For all* $E, F \in \mathcal{E}(\mathbb{H})$

$$E^{\perp} = \left(E^{\perp} \cap (E+F)\right) + (E+F)^{\perp}.$$

Lemma 4.1.3 *Fix* $E, F \in \mathcal{E}(\mathbb{H})$ *and* $A \subseteq E$. *Then*

(a) $A \in \mathcal{B}(E)$ *if and only if* $A + E^{\perp} \in \mathcal{B}(\mathbb{H})$,
(b) *for all* $B \in \mathcal{B}(E)$ *and* $D \in \mathcal{B}(F)$ *we have the implication*

$$B + E^{\perp} = D + F^{\perp} \Rightarrow \mu_E(B) = \mu_F(D).$$

A subset $Z \subseteq \mathbb{H}$ is called a **cylinder set in** $\mathbb{H}$ if there exist $E \in \mathcal{E}(\mathbb{H})$ and $B \in \mathcal{B}(E)$ such that $Z = B + E^{\perp}$. Let $\mathcal{Z}$ be the set of all cylinder sets in $\mathbb{H}$. The following result characterizes cylinder sets in $\mathbb{H}$ by means of functionals in $\mathbb{H}'$. This characterization can be used to extend the notion of cylinder sets to locally convex spaces. We need the following notation: fix $E \in \mathcal{E}(\mathbb{H})$, $B \in \mathcal{B}(E)$ and an ONB $\mathfrak{E} := (\mathfrak{e}_1, \ldots, \mathfrak{e}_n)$ of E. Set

$$B^{\mathfrak{E}} := \left\{(x_1, \ldots, x_n) \in \mathbb{R}^n \mid x_1\mathfrak{e}_1 + \ldots + x_n\mathfrak{e}_n \in B\right\}.$$

Since $(x_1, \ldots, x_n) \mapsto x_1\mathfrak{e}_1 + \ldots + x_n\mathfrak{e}_n$ is continuous, $B^{\mathfrak{E}}$ is a Borel set in $\mathbb{R}^n$.

Proposition 4.1.4 *A subset* $Z \subseteq \mathbb{H}$ *is a cylinder set in* $\mathbb{H}$ *if and only if there exist an* $n \in \mathbb{N}$, *a Borel set* D *in* $\mathbb{R}^n$ *and* $\varphi_1, \ldots, \varphi_n \in \mathbb{H}'$ *such that*

$$Z = \left\{x \in \mathbb{H} \mid (\varphi_i(x))_{i \leq n} \in D\right\}.$$

Proof Assume that Z is a cylinder set in $\mathbb{H}$ with $Z = B + E^{\perp}$ where $E \in \mathcal{E}(\mathbb{H})$ and $B \in \mathcal{B}(E)$. Set $\varphi_i := \langle \mathfrak{e}_i, \cdot \rangle$, where $(\mathfrak{e}_i)_{i \leq n}$ is an ONB of E. Then $\varphi_i \in \mathbb{H}'$ and, by the projection theorem,

$$Z = \left\{ a + b \in \mathbb{H} \mid \sum_{i=1}^{n} \langle \mathfrak{e}_i, a \rangle\, \mathfrak{e}_i \in B \text{ and } b \in E^{\perp} \right\}$$

$$= \left\{ x \in \mathbb{H} \mid \sum_{i=1}^{n} \langle \mathfrak{e}_i, x \rangle\, \mathfrak{e}_i \in B \right\} = \left\{x \in \mathbb{H} \mid (\langle \mathfrak{e}_i, x \rangle)_{i \leq n} \in B^{\mathfrak{E}}\right\}.$$

Conversely, assume that $Z = \left\{x \in \mathbb{H} \mid (\varphi_i(x))_{i \leq n} \in D\right\}$, where D is a Borel set in $\mathbb{R}^n$ and $\varphi_i \in \mathbb{H}' = \mathbb{H}$. Set $E := \operatorname{span}\{\varphi_1, \ldots, \varphi_n\} \in \mathcal{E}(\mathbb{H})$. Then $Z =$

$\{x \in E \mid (\varphi_i(x))_{i\leq n} \in D\} + E^{\perp}$ and $\{x \in E \mid (\varphi_i(x))_{i\leq n} \in D\}$ is a Borel set in E, because $(\varphi_1,\ldots,\varphi_n) : E \to \mathbb{R}^n$ is continuous. □

By Lemma 4.1.3 (b), we may define the **limit μ of a projective system** $(\mu_E)_{E\in\mathcal{E}(\mathbb{H})}$ by setting

$$\mu(B+E^{\perp}) := \mu_E(B) \quad \text{for each cylinder set } B+E^{\perp} \text{ in } \mathbb{H}.$$

Proposition 4.1.5

(a) *$\mathcal{Z}$ is an algebra.*
(b) *μ is finitely additive.*

Proof Fix $Y, Z \in \mathcal{Z}$. By Lemma 4.1.3 (a) and 4.1.2, we may assume that $Y = B+E^{\perp}$ and $Z = D+E^{\perp}$, where $E \in \mathcal{E}(\mathbb{H})$ and $B, D \in \mathcal{B}(E)$.

(a) $\mathbb{H} \in \mathcal{Z}$, because $\mathbb{H} = \{0\} + \{0\}^{\perp}$. Note that $Y \cup Z = (B \cup D) + E^{\perp} \in \mathcal{Z}$ and $\mathbb{H} \setminus Z = (E \setminus D) + E^{\perp} \in \mathcal{Z}$. This proves that $\mathcal{Z}$ is an algebra.

(b) Suppose that $Y \cap Z = \emptyset$. Then $B \cap D = \emptyset$. Therefore,

$$\mu(Y \cup Z) = \mu_E(B \cup D) = \mu_E(B) + \mu_E(D) = \mu(Y) + \mu(Z).$$

□

In the following section we shall prove that μ is not σ-additive in general.

4.2 Gaussian measures in Hilbert spaces

Our aim now is to extend γ_σ^n to a measure γ_σ^E on the Borel σ-algebra of any n-dimensional subspace E of $\mathbb{H}$, using an ONB $\mathfrak{E}$ of E. In order to show that γ_σ^E does not depend on $\mathfrak{E}$, we need the following lemma.

Lemma 4.2.1 *Let $\mathfrak{E} := (\mathfrak{e}_1,\ldots,\mathfrak{e}_n)$, $\mathfrak{B} := (\mathfrak{b}_1,\ldots,\mathfrak{b}_n)$ be ONBs of $E \in \mathcal{E}(\mathbb{H})$. Then for all $B \in \mathcal{B}(E)$*

$$\gamma_\sigma^n(B^{\mathfrak{E}}) = \gamma_\sigma^n(B^{\mathfrak{B}}).$$

Proof Let $\mathfrak{e}_i = \sum_{j=1}^n \alpha_{ji}\mathfrak{b}_j$. Set for all $(\beta_1,\ldots,\beta_n) \in \mathbb{R}^n$

$$\Phi(\beta_1,\ldots,\beta_n) := \left(\alpha_{j1}\beta_1 + \ldots + \alpha_{jn}\beta_n\right)_{j\leq n}.$$

Note that $B^{\mathfrak{E}} = \Phi^{-1}\left[B^{\mathfrak{B}}\right]$ and $f \circ \Phi = f : (x_i)_{i\leq n} \mapsto e^{-\frac{1}{2\sigma}\sum_{i=1}^n x_i^2}$. Moreover, the determinant of the Jacobian $D\Phi$ of Φ is 1 or -1. By the transformation rule, we obtain

$$\gamma_\sigma^n(B^{\mathfrak{B}}) = \frac{\int_{B^{\mathfrak{B}}} f d\lambda^n}{\left(\sqrt{2\pi\sigma}\right)^n} = \frac{\int_{\Phi^{-1}[B^{\mathfrak{B}}]} f \circ \Phi \, |\det D\Phi| \, d\lambda^n}{\left(\sqrt{2\pi\sigma}\right)^n} = \frac{\int_{B^{\mathfrak{E}}} f d\lambda^n}{\left(\sqrt{2\pi\sigma}\right)^n} = \gamma_\sigma^n(B^{\mathfrak{E}}),$$

which proves the lemma. □

We define for each $E \in \mathcal{E}(\mathbb{H})$ with $\dim E = n$ and each Borel set $B \in \mathcal{B}(E)$ and for each ONB $\mathfrak{E}$ of E

$$\gamma_\sigma^E(B) := \gamma_\sigma^n(B^{\mathfrak{E}}).$$

By Lemma 4.2.1, γ_σ is well defined, and we have:

Corollary 4.2.2 *For all $\mathcal{B}(E)$-measurable functions f defined on E, and any orthonormal basis $(\mathfrak{e}_1, \ldots, \mathfrak{e}_n)$ of E,*

$$\int_E f d\gamma_\sigma^E = \int_{\mathbb{R}^n} f\left(\sum_{i \leq n} \alpha_i \mathfrak{e}_i\right) d\gamma_\sigma^n(\alpha_i)_{i \leq n},$$

provided that at least one integral exists. Note that

$$\int_E f d\gamma_\sigma^E = \int_{\mathbb{R}^n} f\left(\sum_{i \leq n} \alpha_i \mathfrak{e}_i\right) e^{-\frac{1}{2\sigma}\sum_{i \leq n} \alpha_i^2} d(\alpha_i)_{i \leq n} \left(\frac{1}{\sqrt{2\pi\sigma}}\right)^n.$$

Proposition 4.2.3 $\left(\gamma_\sigma^E\right)_{E \in \mathcal{E}(\mathbb{H})}$ *is a projective system.*

Proof Suppose that $E, F \in \mathcal{E}(\mathbb{H})$ with $E \subseteq F$ and $\dim(E) = n$ and $\dim(F) = m$. Let $\mathfrak{E}$ be an ONB of E and let $\mathfrak{F}$ be an extension of $\mathfrak{E}$ to an ONB of F. Fix $B \in \mathcal{B}(E)$. By Lemma 4.1.1, we have to show that $\gamma_\sigma^F\left(B + \left(E^\perp \cap F\right)\right) = \gamma_\sigma^E(B)$. Therefore, it suffices to show that the equality $\gamma_\sigma^m((B + (E^\perp \cap F))^{\mathfrak{F}}) = \gamma_\sigma^n(B^{\mathfrak{E}})$ holds. But this is true, because

$$\gamma_\sigma^m\left(\left(B + \left(E^\perp \cap F\right)\right)^{\mathfrak{F}}\right) = \gamma_\sigma^m\left(B^{\mathfrak{E}} \times \mathbb{R}^{m-n}\right) = \gamma_\sigma^n\left(B^{\mathfrak{E}}\right).$$

□

The limit of $\left(\gamma_\sigma^E(B)\right)_{E \in \mathcal{E}(\mathbb{H})}$ is denoted by γ_σ. By Proposition 4.1.5, γ_σ is a finitely additive measure on $\mathcal{Z}$.

Proposition 4.2.4 (Kuo [56]) γ_σ *is not σ-additive.*

Proof Fix $n \in \mathbb{N}$. Since $x \mapsto e^{-\frac{1}{2\sigma}x^2}$ is continuous and $\gamma_\sigma^1(\mathbb{R}) = 1$, $\gamma_\sigma^1([-n,n]) < 1$. Therefore, there exists a $k_n \in \mathbb{N}$ with $\gamma_\sigma^{k_n}\left([-n,n]^{k_n}\right) < \frac{1}{2^{n+1}}$. Fix an ONB $\mathfrak{E} := (\mathfrak{e}_i)_{i\in\mathbb{N}}$ of $\mathbb{H}$ and set

$$Z_n := \left\{x \in \mathbb{H} \mid \left(\langle \mathfrak{e}_1, x\rangle, \ldots, \langle \mathfrak{e}_{k_n}, x\rangle\right) \in [-n,n]^{k_n}\right\}.$$

By Proposition 4.1.4, $Z_n \in \mathcal{Z}$. By the Cauchy–Schwarz inequality, $\mathbb{H} = \bigcup_{n\in\mathbb{N}} Z_n$. Now assume that γ_σ is σ-additive. Then

$$1 = \gamma_\sigma(\mathbb{H}) \leq \sum_{n=1}^{\infty} \gamma_\sigma(Z_n) = \sum_{n=1}^{\infty} \gamma_\sigma^{k_n}\left([-n,n]^{k_n}\right) \leq \sum_{n=1}^{\infty} \frac{1}{2^{n+1}} = \frac{1}{2},$$

which is a contradiction. □

4.3 Abstract Wiener spaces

As we have already mentioned, it is possible to obtain σ-additive Gaussian measures on infinite-dimensional spaces using measurable semi-metrics: a semi-metric d on $\mathbb{H}$ is called **measurable for** $\sigma \in]0,\infty[$ if for each $m \in \mathbb{N}$ there exists an $E_{\sigma,m} \in \mathcal{E}(\mathbb{H})$ such that for any $E \in \mathcal{E}(\mathbb{H})$ with $E \perp E_{\sigma,m}$:

$$\gamma_\sigma^E \left\{x \in E \mid d(x,0) \geq \frac{1}{2^m}\right\} \leq \frac{1}{2^{m+1}}.$$

Intuitively, d being measurable means that the larger the finite subspaces of $\mathbb{H}$ are to which a vector x is orthogonal, the smaller is the probability that $d(x,0)$ is different from 0 (see Proposition 10.2.1 below). If d is measurable for each $\sigma \in]0,\infty[$, then d is called **measurable**. Note that d is measurable if and only if there exists a function $g :]0,\infty[\times \mathbb{N} \to \mathcal{E}(\mathbb{H})$ with $g(\sigma,m) \subseteq g(\sigma,n)$ for $m \leq n$ such that for each $(\sigma,m) \in]0,\infty[\times \mathbb{N}$ and each $E \perp g(\sigma,m)$

$$\gamma_\sigma^E \left\{x \in E \mid d(x,0) \geq \frac{1}{2^m}\right\} \leq \frac{1}{2^{m+1}}. \tag{6}$$

The function g is called a **witness for the measurability** of d. Later we will give an elegant proof of the following result.

Proposition 4.3.1 *Let d be a metric on $\mathbb{H}$, generated by a separating sequence $(|\cdot|_j) := (|\cdot|_j)_{j\in\mathbb{N}}$ of semi-norms on $\mathbb{H}$. Then d is measurable for $\sigma \in]0,\infty[$ if and only if, for all $j \in \mathbb{N}$, $|\cdot|_j$ is measurable for σ.*

Corollary 4.3.2 *Under the assumptions of Proposition 4.3.1, d is measurable if and only if d is measurable for* 1.

Examples 4.3.3 (i) Set $\mathbb{H} = l^2$. Define

$$\left|(a_i)_{i\in\mathbb{N}}\right|_j := \sqrt{\sum_{i=1}^{j} a_i^2}$$

for $j \in \mathbb{N}$. Then each $|\cdot|_j$ is measurable.

(ii) Let S be a **Hilbert–Schmidt operator** on $\mathbb{H}$, i.e., S is continuous and for some ONB $(\mathfrak{e}_i)_{i\in\mathbb{N}}$ of $\mathbb{H}$ we have $\sum_{i=1}^{\infty} \|S(\mathfrak{e}_i)\|^2 < \infty$. It follows that $\sum_{i=1}^{\infty} \|S(\mathfrak{b}_i)\|^2 = \sum_{i=1}^{\infty} \|S(\mathfrak{e}_i)\|^2$ for each *ONB* $(\mathfrak{b}_i)_{i\in\mathbb{N}}$ of $\mathbb{H}$. Define a new semi-norm $|\cdot|$ on $\mathbb{H}$ by setting

$$|a| := \|S(a)\|.$$

Then $|\cdot|$ is a measurable semi-norm on $\mathbb{H}$.

(iii) Let $\mathbb{B}$ be a finite-dimensional space. Then each semi-norm on $\mathbb{B}$ is measurable.

It can be seen that measurable semi-norms $|\cdot|$ are considerably weaker than the original norm $\|\cdot\|$ for infinite-dimensional Hilbert spaces. We only need the following result.

Proposition 4.3.4 (Kuo [56]) *Fix a measurable semi-norm* $|\cdot|$ *on* $\mathbb{H}$ *for variance* 1. *There exists a constant* $c \in \mathbb{R}$ *such that* $|a| \le c\,\|a\|$ *for each* $a \in \mathbb{H}$.

Proof By Corollary 3.2.2 and the continuity of the integral, there exists an $\alpha > 0$ such that $\frac{1}{8} = \int_\alpha^\infty e^{-\frac{1}{2}x^2}dx\frac{1}{\sqrt{2\pi}}$. Fix $y \in E_{1,1}^{\perp}$. If $|y| = 0$, then $|y| \le \frac{\|y\|}{2\alpha}$. Now assume that $|y| > 0$. Then $y \neq 0$ and $\left\{\frac{1}{\|y\|}y\right\}$ is an ONB of $E := \operatorname{span}\{y\}$. We obtain

$$\begin{aligned}\frac{1}{4} \ge \gamma_1^E\left\{x \in E \mid |x| \ge \frac{1}{2}\right\} &= \gamma_1^1\left\{\beta \in \mathbb{R} \,\middle|\, \left|\frac{\beta}{\|y\|}y\right| \ge \frac{1}{2}\right\}\\ &= 2\cdot\gamma_1^1\left\{\beta \in \mathbb{R} \mid \beta \ge \frac{\|y\|}{2\,|y|}\right\} = 2\int_{\frac{\|y\|}{2|y|}}^\infty e^{-\frac{1}{2}x^2}dx\frac{1}{\sqrt{2\pi}}.\end{aligned}$$

It follows that $\frac{1}{8} \ge \int_{\frac{\|y\|}{2|y|}}^\infty e^{-\frac{1}{2}x^2}dx\frac{1}{\sqrt{2\pi}}$, thus $\alpha \le \frac{\|y\|}{2|y|}$, i.e., $|y| \le \frac{\|y\|}{2\alpha}$ for $y \in E_{1,1}^{\perp}$. To estimate $|y|$ for $y \in E_{1,1}$, fix an ONB $(e_1,\ldots,e_n)$ of $E_{1,1}$ and define for $y = \Sigma\alpha_i e_i \in E_{1,1}$, $\|y\|_1 := \Sigma_{i=1}^n |\alpha_i|$. Then $\|\cdot\|_1$ is a norm on $E_{1,1}$ and $|y| \le \gamma\,\|y\|_1$ with $\gamma = \max\{|e_1|,\ldots,|e_n|\}$. Since $\|\cdot\|_1$ and $\|\cdot\|$ are equivalent on $E_{1,1}$ (see Heuser

[45] 109.6), there exists a $k \in \mathbb{R}^+$ such that $|y| \leq k \|y\|$ for all $y \in E_{1,1}$. Now fix $a \in \mathbb{H}$. By the projection theorem there exist $x \in E_{1,1}$ and $y \in E_{1,1}^{\perp}$ with $a = x + y$. Since $2|x|\,|y| \leq |x|^2 + |y|^2$ and $\|a\|^2 = \|x\|^2 + \|y\|^2$, we obtain

$$|a|^2 \leq (|x| + |y|)^2 \leq 2\left(|x|^2 + |y|^2\right) \leq 2\left(k^2 \|x\|^2 + \frac{\|y\|^2}{4\alpha^2}\right)$$

$$\leq 2\left(k^2 + \frac{1}{4\alpha^2}\right)\left(\|x\|^2 + \|y\|^2\right) = 2\left(k^2 + \frac{1}{4\alpha^2}\right)\|a\|^2.$$

This proves the result. □

Corollary 4.3.5 *Let the metric d be generated by a separating sequence of measurable semi-norms on $\mathbb{H}$. If $\mathbb{H}$ is the domain of a continuous φ with respect to d, then φ is continuous with respect to $\|\cdot\|$.*

Now let d be a metric, generated by a separating sequence of measurable semi-norms on $\mathbb{H}$. Let $(\mathbb{B}, d)$ be the completion of $(\mathbb{H}, d)$. By Corollary 4.3.5, the restriction $\varphi \upharpoonright \mathbb{H}$ of $\varphi \in \mathbb{B}'$ is also continuous with respect to $\|\cdot\|$ and $\iota : \mathbb{B}' \ni \varphi \mapsto \varphi \upharpoonright \mathbb{H} \in \mathbb{H}' = \mathbb{H}$ is injective. Moreover, $\left\{\iota(\varphi) \mid \varphi \in \mathbb{B}'\right\}$ is a dense subspace of $(\mathbb{H}, \|\cdot\|)$, because $a = 0$, provided $a \perp i(\varphi)$ for each $\varphi \in \mathbb{B}'$. This proves the following important result.

Proposition 4.3.6 *By identification of φ with $\iota(\varphi)$, we obtain*

(a) $\left(\mathbb{B}', \|\cdot\|\right)$ *is a dense subspace of* $\left(\mathbb{H}', \|\cdot\|\right) = (\mathbb{H}, \|\cdot\|)$ *and*
(b) $(\mathbb{H}, d)$ *is a dense subspace of* $(\mathbb{B}, d)$.

Now $(\mathbb{B}, d)$ is called an **abstract Wiener (Fréchet) space over** $(\mathbb{H}, \|\cdot\|)$.

Example 4.3.7 Let $\mathbb{B}$ be the space of real-valued convergent sequences endowed with the topology of pointwise convergence. By Example 4.3.3 (i), $\mathbb{B}$ is an abstract Wiener–Fréchet space over l^2.

According to Proposition 4.1.4, a subset $Z \subseteq \mathbb{B}$ is called a **cylinder set in** $\mathbb{B}$ if there exist an $n \in \mathbb{N}$, a Borel set $B \subseteq \mathbb{R}^n$ and $\varphi_1, \ldots, \varphi_n \in \mathbb{B}'$ such that

$$Z = \{a \in \mathbb{B} \mid (\varphi_1(a), \ldots, \varphi_n(a)) \in B\}.$$

Since $Z \cap \mathbb{H}$ is a cylinder set in $\mathbb{H}$ for each cylinder set $Z \subseteq \mathbb{B}$, we may define for each $\sigma \in]0, \infty[$

$$\gamma_\sigma(Z) := \gamma_\sigma(Z \cap \mathbb{H})$$

for each cylinder set Z in $\mathbb{B}$.

Theorem 4.3.8 (Gross [41]) *The measure γ_σ on the cylinder sets in $\mathbb{B}$ can be extended to a σ-additive measure on the Borel σ-algebra of $\mathbb{B}$.*

In Corollary 11.8.5 we obtain this result by using a $\mathbb{B}$-valued Brownian motion b on a suitable probability space: since $\varphi \circ b_\sigma$ is normally distributed with variance σ for each $\varphi \in \mathbb{B}'$ with $\|\varphi \restriction \mathbb{H}\| = 1$, the image measure under b_σ is the required measure.

4.4 Cylinder sets in Fréchet spaces generate the Borel sets

Let $\mathcal{Z}_\mathbb{B}$ be the σ-algebra generated by the family of cylinder sets of a Fréchet space $\mathbb{B}$, constructed in Section 4.3. Our aim is to prove the well-known result that $\mathcal{Z}_\mathbb{B}$ is identical to the Borel σ-algebra $\mathcal{B}(\mathbb{B})$ of $\mathbb{B}$. By Proposition 4.3.4, dense subsets of $\mathbb{H}$ are dense in $\mathbb{B}$. Since $\mathbb{H}$ is separable, we have:

Lemma 4.4.1 *$\mathbb{B}$ is separable.*

Lemma 4.4.2 *Fix $Z \in \mathcal{Z}_\mathbb{B}$. Then $a + Z \in \mathcal{Z}_\mathbb{B}$ for each $a \in \mathbb{B}$ and $\alpha \cdot Z \in \mathcal{Z}_\mathbb{B}$ for each $\alpha \in \mathbb{R}$.*

Proof Suppose first that $Z := (\varphi_1, \ldots, \varphi_k)^{-1}[B]$ is a cylinder set in $\mathbb{B}$, where $k \in \mathbb{N}$, $B \in \mathcal{B}(\mathbb{R}^k)$ and $\varphi_1, \ldots, \varphi_k \in \mathbb{B}'$. Then

$$\begin{aligned} a + Z &= \{a + x \in \mathbb{B} \mid (\varphi_1, \ldots, \varphi_k)(x) \in B\} \\ &= \{a + x \in \mathbb{B} \mid (\varphi_1, \ldots, \varphi_k)(a + x) \in (B + (\varphi_1, \ldots, \varphi_k)(a))\} \\ &= \{y \in \mathbb{B} \mid (\varphi_1, \ldots, \varphi_k)(y) \in (B + (\varphi_1, \ldots, \varphi_k)(a))\} \end{aligned}$$

is a cylinder set in $\mathbb{B}$. It is easy to prove that $\mathcal{S} := \{D \subseteq \mathbb{B} \mid a + D \in \mathcal{Z}_\mathbb{B}\}$ is a σ-algebra. Since each cylinder set belongs to $\mathcal{S}$, $\mathcal{Z}_\mathbb{B} \subseteq \mathcal{S}$. In the same way one can prove the second assertion for $\alpha \neq 0$. The proof for $\alpha = 0$ is left to the reader. □

Theorem 4.4.3 (Kuo [56] Theorem 4.2) $\mathcal{Z}_\mathbb{B} = \mathcal{B}(\mathbb{B})$.

Proof Since $(\varphi_1, \ldots, \varphi_k) : \mathbb{B} \to \mathbb{R}^k$ is continuous for any $\varphi_1, \ldots, \varphi_k \in \mathbb{B}'$, $(\varphi_1, \ldots, \varphi_k)^{-1}[B] \in \mathcal{B}(\mathbb{B})$ for each $B \in \mathcal{B}(\mathbb{R}^k)$, thus $\mathcal{Z}_\mathbb{B} \subseteq \mathcal{B}(\mathbb{B})$. To prove the reverse inclusion, we first prove that, for all $i \in \mathbb{N}$, all $a \in \mathbb{B}$ and $\varepsilon > 0$,

$$B_{\varepsilon,i}(a) := \{x \in \mathbb{B} \mid |x - a|_i \leq \varepsilon\} \in \mathcal{Z}_\mathbb{B}.$$

Since $B_{\varepsilon,i}(a) = a + \varepsilon B_{1,i}(0)$, it suffices to prove that $B_{1,i}(0) \in \mathcal{Z}_\mathbb{B}$ (see Lemma 4.4.2). Fix a countable dense subset $\mathbb{D} \subseteq \mathbb{B}$. For each $a \in \mathbb{D}$ and each $\lambda \in \mathbb{R}$

set $\varphi_a(\lambda \cdot a) := \lambda |a|_i$. Then $\varphi_a : \text{span}\{a\} \to \mathbb{R}$ is linear. By the Hahn–Banach theorem, there exists a linear extension $\varphi_a : \mathbb{B} \to \mathbb{R}$ with $|\varphi_a| \leq |\cdot|_i$. We prove that

$$B_{1,i}(0) = D := \bigcap_{a \in \mathbb{D}} \{x \in \mathbb{B} \mid |\varphi_a(x)| \leq 1\}.$$

Since $x \in B_{1,i}(0)$ implies $|\varphi_a(x)| \leq |x|_i \leq 1$, $B_{1,i}(0) \subseteq D$. Now suppose that $x \notin B_{1,i}(0)$. Then there exists an $a \in \mathbb{D}$ with $|x - a|_i < \frac{|x|_i - 1}{2}$. Therefore, $|a|_i > \frac{|x|_i + 1}{2}$, because otherwise $|x|_i \leq |x - a|_i + |a|_i < |x|_i$. We obtain

$$|\varphi_a(x)| \geq |\varphi_a(a)| - |\varphi_a(a - x)| \geq |a|_i - |x - a|_i > \frac{|x|_i + 1}{2} - \frac{|x|_i - 1}{2} = 1,$$

thus $x \notin D$. It follows that, for each $m \in \mathbb{N}$, $B_{\frac{1}{m}}(a) := \bigcap_{j \leq m} B_{\frac{1}{m},j}(a) \in \mathcal{Z}_{\mathbb{B}}$. Since $\mathbb{B}$ is separable, each open set in $\mathbb{B}$ is a countable union of sets of the form $B_{\frac{1}{m}}(a)$. This proves that each open set in $\mathbb{B}$ belongs to $\mathcal{Z}_{\mathbb{B}}$, thus $\mathcal{B}(\mathbb{B}) \subseteq \mathcal{Z}_{\mathbb{B}}$. □

Let $\mathcal{Z}_{\mathbb{B}}^s$ be the σ-algebra generated by the family of all **simple cylinder sets**; these are sets of the form $\{a \in \mathbb{B} \mid \varphi(a) < c\}$ with $\varphi \in \mathbb{B}'$ and $c \in \mathbb{R}$. Since the family of Borel sets in $\mathbb{R}^k$ is generated by the family of rectangles of the form $B_1 \times \ldots \times B_k$, where $B_i \in \mathcal{B}(\mathbb{R})$, and $\mathcal{B}(\mathbb{R})$ is generated by the family of sets of the form $]-\infty, c[$ with $c \in \mathbb{R}$, we obtain:

Corollary 4.4.4 $\mathcal{Z}_{\mathbb{B}}^s = \mathcal{Z}_{\mathbb{B}} = \mathcal{B}(\mathbb{B})$.

Finally, we shall see that each Fréchet space given by a separating sequence of semi-norms appears as an abstract Wiener space. In the finite-dimensional case this result follows from Example 4.3.3 (iii). So we may assume that the Fréchet space is infinite-dimensional.

The proofs of the following lemma and theorem are slight modifications of the proof of Theorem 4.4 in Kuo [56].

Lemma 4.4.5 *Let F be a linear space with countable basis $(\mathfrak{b}_i)_{i \in \mathbb{N}}$ and let $(|\cdot|_j)_{j \in \mathbb{N}}$ be a separating sequence of semi-norms on F with $|\cdot|_{j+1} \leq |\cdot|_j$ on $F_j := \text{span}\{\mathfrak{b}_1, \ldots, \mathfrak{b}_j\}$ for all $j \in \mathbb{N}$. Then there exists a sequence $(\alpha_j)_{j \in \mathbb{N}}$ of positive reals such that for all sequences $(\beta_i)_{i \in \mathbb{N}}$ in $\mathbb{R}$ and all $n, j \in \mathbb{N}$ with $j \leq n$ the following implication holds:*

$$\sum_{i=1}^{n} \beta_i^2 \leq 1 \Rightarrow \left| \sum_{i=1}^{n} \beta_i \alpha_i \mathfrak{b}_i \right|_j < 1. \tag{7}$$

Proof The sequence (α_j) is constructed by recursion. For $n = 1$, choose $\alpha_1 > 0$ small enough such that $|b_1|_1 < \frac{1}{\alpha_1}$. Assume that $(\alpha_i)_{i \leq n}$ is already defined and implication (7) is true for all $j \leq n$. Set

$$\Phi : \mathbb{R}^{n+1} \to F_{n+1}, (\beta_1, \ldots, \beta_{n+1}) \mapsto \sum_{i=1}^{n} \beta_i \alpha_i \mathfrak{b}_i + \beta_{n+1} \mathfrak{b}_{n+1},$$

and

$$S := \{a \in F_{n+1} \mid |a|_1, \ldots, |a|_{n+1} < 1\}.$$

Then $\Phi^{-1}[S]$ is open and $A := \left\{ ((\beta_i)_{i \leq n}, 0) \mid \sum_{i=1}^{n} \beta_i^2 \leq 1 \right\}$ is a closed subset of $\Phi^{-1}[S]$. It follows that there exists an $\alpha_{n+1} > 0$ such that the thin ellipsoid

$$E := \left\{ (\beta_i)_{i \leq n+1} \mid \sum_{i=1}^{n} \beta_i^2 + \frac{\beta_{n+1}^2}{\alpha_{n+1}^2} \leq 1 \right\}$$

lies between A and $\Phi^{-1}[S]$. Obviously, implication (7) is true for $n+1$ instead of n and for all $j \leq n+1$. □

Theorem 4.4.6 *Let $\mathbb{B}$ be an infinite-dimensional Fréchet space given by a separating sequence $(|\cdot|_j)_{j \in \mathbb{N}}$ of semi-norms. Then there exists a Hilbert space $\mathbb{H} \subseteq \mathbb{B}$ such that $(|\cdot|_j)_{j \in \mathbb{N}}$ is a sequence of measurable semi-norms on $\mathbb{H}$ and $\mathbb{B}$ is the completion of $\left(\mathbb{H}, (|\cdot|_j)_{j \in \mathbb{N}}\right)$.*

Proof By the separability of $\mathbb{B}$, there exists a countable sequence $(\mathfrak{b}_n)$ of linear independent vectors $\mathfrak{b}_n$ in $\mathbb{B}$ such that $F := \bigcup_{n \in \mathbb{N}} F_n$ is dense in $\mathbb{B}$, where $F_n :=$ span $\{\mathfrak{b}_1, \ldots, \mathfrak{b}_n\}$. The first aim is to find a Hilbert space norm $\|\cdot\|_0$ on F such that each Cauchy sequence with respect to $\|\cdot\|_0$ is a Cauchy sequence in $(F, (|\cdot|_j))$. To this end fix a single semi-norm $|\cdot|_j$. By Lemma 4.4.5, there exists a sequence $\left(\alpha_n^j\right)$ of positive real numbers such that for all $n \in \mathbb{N}$ the following implication holds:

$$\sum_{i=1}^{n} \beta_i^2 \leq 1 \Rightarrow \left| \sum_{i=1}^{n} \beta_i \alpha_i^j \mathfrak{b}_i \right|_j < 1.$$

Set $\mathfrak{e}_n^j := \alpha_n^j \mathfrak{b}_n$ and define an inner product $[\cdot,\cdot]_j$ on F such that $\left(\mathfrak{e}_n^j\right)$ is an orthonormal sequence. The associated norm is denoted by $[\cdot]_j$. Then for all $x \in F$,

$$[x]_j \leq 1 \Rightarrow |x|_j < 1, \text{ thus } |x|_j \leq [x]_j.$$

In order to apply Lemma 4.4.5 again, we define new norms $\|\cdot\|_j$ on F such that $\|\cdot\|_{j+1} \leq \|\cdot\|_j$ on F_j and such that $(\|\cdot\|_j)_{j \in \mathbb{N}}$ defines the same topology on F as

$([\cdot]_j)_{j\in\mathbb{N}}$. Since all norms on finite-dimensional spaces are equivalent, for each $n\in\mathbb{N}$ there exists a constant $c_n \geq 1$ such that $[a]_{n+1} \leq c_n [a]_n$ for all $a\in F_n$. Set $\|\cdot\|_1 := [\cdot]_1$ and for each $n\in\mathbb{N}$ define

$$\varepsilon_n := \frac{1}{c_n \cdot c_{n-1} \cdot \ldots \cdot c_1} \quad \text{and} \quad \|\cdot\|_{n+1} := \varepsilon_n [\cdot]_{n+1}.$$

Note that, for all $n\in\mathbb{N}$ and all $a\in F_n$,

$$\|a\|_{n+1} \leq \|a\|_n,$$

and $(\|\cdot\|_j)$ defines the same topology on F as the sequence $([\cdot]_j)$.

By Lemma 4.4.5 again, there exists a sequence (α_n) of positive real numbers such that for all $n\in\mathbb{N}$ and all $j\leq n$ the following implication holds:

$$\sum_{i=1}^{n} \beta_i^2 \leq 1 \Rightarrow \left\| \sum_{i=1}^{n} \beta_i \alpha_i \mathfrak{b}_i \right\|_j < 1.$$

Set $\mathfrak{e}_n := \alpha_n \mathfrak{b}_n$ and define an inner product $\langle\cdot,\cdot\rangle_0$ on F such that $(\mathfrak{e}_n)$ is an orthonormal sequence. The associated norm is denoted by $\|\cdot\|_0$. Obviously, for all $j\in\mathbb{N}$ and all $x\in F$,

$$\|x\|_0 \leq 1 \Rightarrow \|x\|_j < 1, \quad \text{thus } \|x\|_j \leq \|x\|_0.$$

It follows that each Cauchy sequence with respect to $\|\cdot\|_0$ is a Cauchy sequence with respect to $(\|\cdot\|_j)$, and therefore it is also a Cauchy sequence with respect to $(|\cdot|_j)$. This was our first goal. To continue, let $\mathbb{H}_0$ be the completion of $(F, \langle\cdot,\cdot\rangle_0)$. Then we may identify $\mathbb{H}_0$ with a dense subspace of $\mathbb{B}$ under $(|\cdot|_j)$. Let T be an injective Hilbert–Schmidt operator on $\mathbb{H}_0$, given, for example, by

$$T(a) := \left(\sum_{i=1}^{\infty} \frac{1}{2^i} \langle a, \mathfrak{e}_i \rangle_0 \, \mathfrak{e}_i \right).$$

Define $\langle x,y\rangle := \left\langle T^{-1}(x), T^{-1}(y)\right\rangle_0$ for all $x,y\in T[\mathbb{H}_0] := \mathbb{H}$. The norm on $\mathbb{H}$ is denoted by $\|\cdot\|$. Since $F\subseteq\mathbb{H}$, $\mathbb{B}$ is the completion of $\mathbb{H}$. Since $\|a\|_0 = \|T(a)\|$, $\|\cdot\|_0$ is a measurable norm on the space $\|\mathbb{H}, \langle\cdot,\cdot\rangle\|$ by Example 4.3.3 (ii). It follows that $(|\cdot|_j)$ is a separating sequence of measurable semi-norms on $\mathbb{H}$. □

4.5 Cylinder sets in Fréchet space valued continuous functions

Our final aim is to develop Malliavin calculus on the space $C_{\mathbb{B}}$ of continuous functions on $[0,\infty[$ with values in $\mathbb{B}$. The space $C_{\mathbb{B}}$ is a Fréchet space, endowed with the separating sequence $(\lfloor\cdot\rfloor_m)_{m\in\mathbb{N}}$ of **supremum semi-norms** $\lfloor\cdot\rfloor_m$, i.e., $\lfloor f\rfloor_m := \sup_{t\in[0,m]} \max_{j\leq m} |f(t)|_j$. A neighbourhood base of a function $f:[0,\infty[\to\mathbb{B}$ in $C_{\mathbb{B}}$ are sets of the form

$$U_{\frac{1}{m}}(f) := \left\{ g\in C_{\mathbb{B}} \mid \lfloor g-f\rfloor_m < \frac{1}{m} \right\},$$

where $m\in\mathbb{N}$.

We take $C_{\mathbb{B}}$ instead of an arbitrary space $\mathbb{B}$ in order to have the notions 'time-anticipating' and 'non-time-anticipating' according to the classical Wiener space $C_{\mathbb{R}}$. In this section we shall prove that the Borel σ-algebra $\mathcal{B}(C_{\mathbb{B}})$ of $C_{\mathbb{B}}$ is generated by the cylinder sets of $C_{\mathbb{B}}$.

Cylinder sets in $C_{\mathbb{B}}$ are sets $\{f\in C_{\mathbb{B}} \mid f(t)\in B\}$, where $t\in[0,\infty[$ and B is a Borel set in $\mathbb{B}$. Let $\mathcal{Z}_{C_{\mathbb{B}}}$ be the σ-algebra generated by the family of cylinder sets in $C_{\mathbb{B}}$. Let $\mathcal{Z}^s_{C_{\mathbb{B}}}$ be the σ-algebra generated by the family of **simple cylinder sets in** $C_{\mathbb{B}}$; these are sets of the form $\{f\in C_{\mathbb{B}} \mid f(t)\in Z\}$ where Z is a simple cylinder set in $\mathbb{B}$ and $t\in[0,\infty[$. Therefore, simple cylinder sets in $C_{\mathbb{B}}$ are sets of the form $\{f\in C_{\mathbb{B}} \mid \varphi\circ f(t) < c\}$ with $\varphi\in\mathbb{B}'$, $t\in[0,\infty[$ and $c\in\mathbb{R}$.

Proposition 4.5.1 *$\mathcal{Z}^s_{C_{\mathbb{B}}} = \mathcal{Z}_{C_{\mathbb{B}}} = \mathcal{B}(C_{\mathbb{B}})$, where $\mathcal{B}(C_{\mathbb{B}})$ is the Borel σ-algebra on $C_{\mathbb{B}}$ with respect to the separating sequence $(\lfloor\cdot\rfloor_\sigma)_{\sigma\in\mathbb{N}}$ of supremum semi-norms.*

Proof The first equality is obtained by arguments similar to those by which we obtained Corollary 4.4.4.

To prove the second equality, note that each simple cylinder set $G := \{f\in C_{\mathbb{B}} \mid \varphi\circ f(t) < c\}$ is open: if $f\in G$, then there exists an $m\in\mathbb{N}$ with $m>t$ such that $U_{\frac{1}{m}}(f(t)) \subseteq \varphi^{-1}[]-\infty,c[]$ (see Chapter 1). It follows that $U_{\frac{1}{m}}(f)\subseteq G$. Therefore, $\mathcal{Z}_{C_{\mathbb{B}}} \subseteq \mathcal{B}(C_{\mathbb{B}})$. To prove the reverse inclusion, first note that $C_{\mathbb{B}}$ is separable: fix a countable dense subset $\mathbb{D}$ of $\mathbb{H}$. Then $\mathbb{D}$ is dense in $\mathbb{B}$ by Proposition 4.3.6 (b). Fix $k,m\in\mathbb{N}$ and $a_0,\ldots a_M\in\mathbb{D}$, where $\frac{M-1}{k} < m \leq \frac{M}{k}$. Note that the set $\mathbb{D}_{C_{\mathbb{B}}}$ of functions $g_m^{k,a_0,\ldots,a_M}$ builds a countable dense subset of $C_{\mathbb{B}}$, where $g_m^{k,a_0,\ldots,a_M}$ results from the function $\frac{i}{k}\mapsto a_i$, $i=0,\ldots,M$, by linear interpolation (see the proof of Lemma 11.8.1 for details). Set $g_m^{k,a_0,\ldots,a_M}(x) = a_M$ for $x>\frac{M}{k}$. It follows that each open set in $C_{\mathbb{B}}$ is a countable union of sets of the form $U_{\frac{1}{n}}(g)$

with $g \in \mathbb{D}_{C_{\mathbb{B}}}$. Note that

$$U_{\frac{1}{n}}(g) = \bigcup_{k\in\mathbb{N}} \bigcap_{j\leq n} \bigcap_{t\in[0,n]\cap\mathbb{Q}} \left\{ f \in C_{\mathbb{B}} \mid |f(t)-g(t))|_j < \frac{1}{n} - \frac{1}{k} \right\},$$

where $\left\{f \in C_{\mathbb{B}} \mid |f(t)-g(t))|_j < \frac{1}{n} - \frac{1}{k}\right\}$ is a cylinder set in $C_{\mathbb{B}}$, because $\left\{x \in \mathbb{B} \mid |x-a|_j < c\right\}$ is an open set in $\mathbb{B}$. This proves that $\mathcal{B}(C_{\mathbb{B}}) \subseteq \mathcal{Z}_{C_{\mathbb{B}}}$. □

4.6 Tensor products

In this section we establish some basic results on tensor products of separable real Hilbert spaces $\mathbb{H}$ with scalar product $\langle \cdot, \cdot \rangle$ and norm $\|\cdot\|$. For each $d \in \mathbb{N}_0$ let $\mathbb{H}^{\otimes d}$ denote the **d-fold tensor product** of $\mathbb{H}$, i.e., $\mathbb{H}^{\otimes d}$ is the space of all real-valued continuous **multilinear forms** f, i.e., f is linear in each argument defined on $\mathbb{H}^d$ such that

$$f_{\mathfrak{E}} := \sum_{\mathfrak{e}\in\mathfrak{E}^d} f^2(\mathfrak{e}) < \infty,$$

where $\mathfrak{E}$ is an ONB of $\mathbb{H}$. We don't assume that the elements of $\mathbb{H}^{\otimes d}$ are symmetric. Since $\mathbb{H}^0 = \{\emptyset\}$, we may identify $f \in \mathbb{H}^{\otimes 0}$ with $f(\emptyset) \in \mathbb{R}$. In case $d = 1$, $\mathbb{H}^{\otimes 1} = \mathbb{H}'$. Since we identify $f \in \mathbb{H}'$ with $a_f \in \mathbb{H}$, where $\langle a_f, h\rangle = f(h)$ for each $h \in \mathbb{H}$, we see that $\mathbb{H}^{\otimes 1} = \mathbb{H}$ and $f_{\mathfrak{E}} = \|f\|^2$. This proves that $f_{\mathfrak{E}}$ does not depend on the chosen ONB. By induction, this result can be extended to higher dimensions. On $\mathbb{H}^{\otimes d}$ we use the **Hilbert–Schmidt norm** $\|\cdot\|_{\mathbb{H}^d}$, i.e.,

$$\|f\|_{\mathbb{H}^d} := \left(\sum_{\mathfrak{e}\in\mathfrak{E}^d} f^2(\mathfrak{e})\right)^{\frac{1}{2}} = (f_{\mathfrak{E}})^{\frac{1}{2}},$$

where $\mathfrak{E}$ is an ONB of $\mathbb{H}$. Note that $\mathbb{H}^{\otimes d}$ is a Hilbert space with scalar product

$$\langle f, g\rangle_{\mathbb{H}^d} := \sum_{\mathfrak{e}\in\mathfrak{E}^d} f(\mathfrak{e}) \cdot g(\mathfrak{e}).$$

Since a scalar product can be defined in terms of the associated norm, $\langle f, g\rangle_{\mathbb{H}^d}$ does not depend on the ONB.

Now we will show that $\mathbb{H}^{\otimes d}$ is separable: for $(a_1, \ldots, a_d) \in \mathbb{H}^d$ define $[a_1, \ldots, a_n] := a_1 \otimes \ldots \otimes a_d : \mathbb{H}^d \to \mathbb{R}$ by

$$a_1 \otimes \ldots \otimes a_d(x_1,\ldots,x_d) := \langle a_1, x_1\rangle \cdot \ldots \cdot \langle a_d, x_d\rangle .$$

Obviously, $a_1 \otimes \ldots \otimes a_d$ is multilinear and continuous. This multilinear form is called the **tensor product** or simply the **product** of $(a_1,\ldots,a_d)$.

Proposition 4.6.1 *The set* $B := \{[\mathfrak{e}] \mid \mathfrak{e} \in \mathfrak{E}^d\}$ *is an ONB of* $\mathbb{H}^{\otimes d}$. *For* $f \in \mathbb{H}^{\otimes d}$ *we have*

$$f = \sum_{\mathfrak{e} \in \mathfrak{E}^d} f(\mathfrak{e}) \cdot [\mathfrak{e}].$$

Proof We only prove that B is a basis for $\mathbb{H}^{\otimes 2}$. Fix $f \in \mathbb{H}^{\otimes 2}$ such that $f \perp B$. Then, for each $(\mathfrak{e}_i, \mathfrak{e}_j) \in \mathfrak{E}^2$,

$$0 = \sum_{(\alpha,\beta)\in\mathbb{N}^2} f(\mathfrak{e}_\alpha, \mathfrak{e}_\beta) \cdot \left([\mathfrak{e}_i, \mathfrak{e}_j](\mathfrak{e}_\alpha, \mathfrak{e}_\beta)\right) = f(\mathfrak{e}_i, \mathfrak{e}_j).$$

Since f is continuous and multilinear, $f(a,b) = 0$ for all $a, b \in \mathbb{H}$. This proves that $f = 0$. □

A set Z is called a **cylinder set in** $\mathbb{H}^{\otimes d}$ if there exist a $c \in \mathbb{R}$ and an $\mathfrak{s} := (\mathfrak{e}_1,\ldots,\mathfrak{e}_d) \in \mathbb{H}^d$, where the $\mathfrak{e}_i$ are pairwise orthonormal, such that

$$Z = \{f \in \mathbb{H}^{\otimes d} \mid f(\mathfrak{s}) < c\}.$$

Let $\mathcal{Z}$ be the σ-algebra generated by the cylinder sets in $\mathbb{H}^{\otimes d}$. Our aim is to show that $\mathcal{Z} = \mathcal{B}(\mathbb{H}^{\otimes d})$, the Borel σ-algebra on $\mathbb{H}^{\otimes d}$.

Lemma 4.6.2 *Fix* $g \in \mathbb{H}^{\otimes d}$, *a Borel set* B *in* $\mathbb{R}$ *and* $\mathfrak{s}_1,\ldots,\mathfrak{s}_k$ *in* $\mathbb{H}^d$ *such that the components of each* $\mathfrak{s}_i$ *are orthonormal. Then*

(a) $\{f \in \mathbb{H}^{\otimes d} \mid (f-g)(\mathfrak{s}_1) \in B\} \in \mathcal{Z}$,
(b) *if* $h : \mathbb{R}^k \to \mathbb{R}$ *is continuous, then*

$$\{f \in \mathbb{H}^{\otimes d} \mid h\left(((f-g)(\mathfrak{s}_i))_{i\le k}\right) \in B\} \in \mathcal{Z}.$$

Proof Part (a) is obvious. It suffices to prove part (b) for all open sets B in $\mathbb{R}$. Since $h^{-1}[B]$ is open, there are k sequences $(I^1_m)_{m\in\mathbb{N}},\ldots,(I^k_m)_{m\in\mathbb{N}}$ of open and bounded intervals I^j_m in $\mathbb{R}$ with $h^{-1}[B] = \bigcup_{m\in\mathbb{N}} (I^1_m \times \ldots \times I^k_m)$. We obtain

$$\begin{aligned}&\{f \in \mathbb{H}^{\otimes d} \mid h\left(((f-g)(\mathfrak{s}_i))_{i\le k}\right) \in B\}\\ &= \bigcup_{m\in\mathbb{N}} \{f \in \mathbb{H}^{\otimes d} \mid (f-g)(\mathfrak{s}_i) \in I^i_m \text{ for all } i = 1,\ldots,k\} \in \mathcal{Z}.\end{aligned}$$

□

Proposition 4.6.3 $\mathcal{Z} = \mathcal{B}\left(\mathbb{H}^{\otimes d}\right)$.

Proof '$\subseteq$' Let $Z := \left\{f \in \mathbb{H}^{\otimes d} \mid f(\mathfrak{s}) < c\right\}$ be a cylinder set in $\mathbb{H}^{\otimes d}$. To show that Z is open, fix $g \in Z$. Set $\varepsilon := c - g(\mathfrak{s})$ and extend $\mathfrak{s}$ to an ONB $\mathfrak{E}$ of $\mathbb{H}$. Then

$$U_\varepsilon(g) = \left\{f \in \mathbb{H}^{\otimes d} \mid \sum_{\mathfrak{e} \in \mathfrak{E}^d} (f-g)^2(\mathfrak{e}) < \varepsilon^2\right\}$$
$$\subseteq \left\{f \mid (f-g)^2(\mathfrak{s}) < \varepsilon^2\right\} = \{f \mid |f-g|(\mathfrak{s}) < \varepsilon\} \subseteq Z.$$

'$\supseteq$' Since $\mathbb{H}^{\otimes d}$ is separable, it suffices to prove that $U_\varepsilon(g) \in \mathcal{Z}$ for each $g \in \mathbb{H}^{\otimes d}$ and each $\varepsilon > 0$. By Lemma 4.6.2 (b), we obtain

$$U_\varepsilon(g) = \{f \mid \|f-g\|_{\mathbb{H}^d} < \varepsilon\} = \bigcup_{m \in \mathbb{N}} \left\{f \mid \|f-g\|_{\mathbb{H}^d} \leq \varepsilon - \frac{1}{m}\right\}$$
$$= \bigcup_{m \in \mathbb{N}} \bigcap_{k \in \mathbb{N}} \left\{f \in \mathbb{H}^{\otimes d} \mid \left(\sum_{i_1,\ldots,i_d \leq k} (f-g)^2\left(\mathfrak{e}_{i_1},\ldots,\mathfrak{e}_{i_d}\right)\right)^{\frac{1}{2}} \leq \varepsilon - \frac{1}{m}\right\} \in \mathcal{Z}.$$

□

4.7 Bochner integrable functions

Later we use Bochner integrable functions with values $\mathbb{H}^{\otimes d}$. Here we prepare the basis for this concept. Fix a σ-finite measure space $(\Lambda, \mathcal{C}, \mu)$ and $p \in [1, \infty[$. A function $f : \Lambda \to \mathbb{H}^{\otimes d}$ is called $\mathcal{C}$**-simple** if f is $\mathcal{C}$-measurable and the range of f is finite, i.e., f has the form

$$f = \sum_{i=1}^{n} \alpha_i \mathbf{1}_{B_i} \quad \text{with } n \in \mathbb{N}, \alpha_i \in \mathbb{H}^{\otimes d} \text{ and } B_i \in L_\mu(\mathcal{C}),$$

where the B_i are pairwise disjoint and $\mu(B_i) < \infty$. The integral of f is defined by

$$\int_\Lambda f d\mu := \sum_{i=1}^{n} \alpha_i \mu(B_i) \in \mathbb{H}^{\otimes d}.$$

A $\mathcal{C}$-measurable function $f : \Lambda \to \mathbb{H}^{\otimes d}$ is called p**-times Bochner** μ**-integrable** if there exists a sequence $(f_n)_{n \in \mathbb{N}}$ of $\mathcal{C}$-simple functions, a so-called **witness for the integrability** of f, such that

(i) $(f_n)_{n\in\mathbb{N}}$ **converges to** f **in measure,** i.e., for each $\varepsilon > 0$,

$$\lim_{n\to\infty} \mu\left\{x\in\Lambda \mid \|f_n(x)-f(x)\|_{\mathbb{H}^d} \geq \varepsilon\right\} = 0$$

(ii) $\lim_{n,m\to\infty}\int_\Lambda \|f_n(x)-f_m(x)\|^p_{\mathbb{H}^d}\, d\mu = 0.$

The proof that $\lim_{n\to\infty}\int_\Lambda \|f_n(x)-g(x)\|^p_{\mathbb{H}^d}\, d\mu = 0$ for some $\mathcal{C}$-measurable function g is very similar to the proof that real-valued L^p-spaces are complete (see Heuser [45]). By (i), $g = f$ μ-a.e. In case $p = 1$, we set

$$\int_\Lambda f d\mu := \lim_{n\to\infty}\int_\Lambda f_n d\mu.$$

We denote the space of p-times Bochner integrable functions by $L^p(\mu, \mathbb{H}^{\otimes d})$, equipped with the norm

$$\|f\|_{p,\mu,\mathbb{H}^d} := \left(\int_\Lambda \|f\|^p_{\mathbb{H}^d}\, d\mu\right)^{\frac{1}{p}}.$$

We identify two functions $f, g : \Lambda \to \mathbb{B}$ if $f = g$ μ-a.e. The space $L^2(\mu, \mathbb{H}^{\otimes d})$ is a Hilbert space with respect to the scalar product

$$\langle f, g\rangle_{\mu,\mathbb{H}^d} := \int_\Lambda \langle f(x), g(x)\rangle_{\mathbb{H}^d}\, d\mu(x).$$

We drop the index $\mathbb{H}^d$ if it is clear which space we mean. Recall that, in the case $n = 1$, $\mathbb{H}^{\otimes 1} = \mathbb{H}$. If $n = 0$, then $\mathbb{H}^{\otimes 0} = \mathbb{R}$ and $L^p(\mu, \mathbb{H}^{\otimes 0}) = L^p(\mu)$.

For the remainder of this section fix a separable Hilbert space $\mathbb{H}$ with ONB $\mathfrak{E}$ and assume that $L^2(\mu)$ is separable with ONB $(f_i)_{i\in\mathbb{N}}$. Recall that $([\mathfrak{e}])_{\mathfrak{e}\in\mathfrak{E}^d}$ is an ONB of $\mathbb{H}^d$. Define $f_i \otimes [\mathfrak{e}] : \Lambda \to \mathbb{H}^{\otimes d}$ by

$$(f_i \otimes [\mathfrak{e}])\,(x)(a) := f_i(x)\cdot \langle \mathfrak{e}, a\rangle_{\mathbb{H}^d}\,.$$

Proposition 4.7.1 *The double sequence* $(f_i \otimes [\mathfrak{e}])_{i\in\mathbb{N},\mathfrak{e}\in\mathfrak{E}^d}$ *is an ONB of* $L^2(\mu, \mathbb{H}^{\otimes d})$.

Proof To prove that $(f_i \otimes [\mathfrak{e}])_{i\in\mathbb{N},\mathfrak{e}\in\mathfrak{E}^d}$ is an ONS in $L^2(\mu, \mathbb{H}^{\otimes d})$, note that

$$\left\langle f_\alpha \otimes [\mathfrak{e}_\beta], f_i \otimes [\mathfrak{e}_j]\right\rangle_\mu = \begin{cases} 0 & \text{if } (i,j) \neq (\alpha,\beta) \\ 1 & \text{if } (i,j) = (\alpha,\beta). \end{cases}$$

To prove that $(f_i \otimes [\mathfrak{e}])_{i\in\mathbb{N},\mathfrak{e}\in\mathfrak{E}^d}$ is maximal in $L^2(\mu,\mathbb{H}^{\otimes d})$, fix an $f \in L^2(\mu,\mathbb{H}^{\otimes d})$ orthogonal to each $(f_i \otimes [\mathfrak{e}])$. Then for all $i \in \mathbb{N}$, $\mathfrak{e} \in \mathfrak{E}^d$

$$0 = \int_\Lambda \langle f(x), f_i(x) \cdot [\mathfrak{e}]\rangle_{\mathbb{H}^d}\, d\mu(x) = \int_\Lambda \langle f(x), [e]\rangle_{\mathbb{H}^d} \cdot f_i(x) d\mu(x).$$

Since $(f_i)_{i\in\mathbb{N}}$ is maximal in $L^2(\mu)$, $\langle f, [e]\rangle_{\mathbb{H}^d} = 0$ μ-a.s. It follows that $f = 0$ in $L^2(\mu,\mathbb{H}^{\otimes d})$. □

A function $g : \Lambda \to \mathbb{H}^{\otimes d}$ is called **very simple** if there exists a $B \in \mathcal{C}$ with $\mu(B) < \infty$ and an $\mathfrak{e} \in \mathfrak{E}^d$ such that $g = \mathbf{1}_B \otimes [\mathfrak{e}]$.

Proposition 4.7.2 *The space of linear combinations of very simple functions is a dense subspace of* $L^2(\mu,\mathbb{H}^{\otimes d})$.

Proof Suppose that $f \perp g$ in $L^2(\mu,\mathbb{H}^{\otimes d})$ for each very simple function g, i.e., for all $\mathfrak{e} \in \mathfrak{E}^d$ and all $B \in \mathcal{C}$ with $\mu(B) < \infty$ we have

$$\int_B \langle f, \mathfrak{e}\rangle_{\mathbb{H}^d}\, d\mu = 0.$$

It follows that $\langle f, \mathfrak{e}\rangle_{\mathbb{H}^d} = 0 \mu$-a.e. for all $\mathfrak{e} \in \mathfrak{E}^d$. Therefore, $f = 0$ in $\mathbb{H}^{\otimes d}$ μ-a.s., thus $f = 0$ in $L^2(\mu,\mathbb{H}^{\otimes d})$. □

4.8 The Wiener measure on $C_\mathbb{B}$ is the centred Gaussian measure of variance 1

In this final section about abstract Wiener spaces we present an interesting result, due to J. Kuelbs and R. Lepage [55]: The Wiener measure on $C_\mathbb{B}$ is identical to the Gaussian measure of variance 1, constructed in Section 4.3 by Gross' method.

Fix a separable Hilbert space $\mathbb{H}$ with scalar product $\langle\cdot,\cdot\rangle$ and norm $\|\cdot\|$. In Chapter 11 we shall construct a probability space Ω with measure $\widehat{\Gamma}$, a filtration $(\mathfrak{b}_t)_{t\in[0,\infty[}$ on Ω and for each abstract Wiener–Fréchet space $\mathbb{B}$ over $\mathbb{H}$ a Brownian motion $b_\mathbb{B} : \Omega \times [0,\infty] \to \mathbb{B}$ under $(\mathfrak{b}_t)_{t\in[0,\infty[}$. The probability space Ω is rich in the following sense: for each continuous function $f : [0,\infty] \to \mathbb{B}$ there exists an $X \in \Omega$ with $b_\mathbb{B}(X,\cdot) = f$. Moreover, Ω, $\widehat{\Gamma}$ and $(\mathfrak{b}_t)_{t\in[0,\infty[}$ depend only on $\mathbb{H}$. A mapping $b_\mathbb{B} : \Omega \times]0,\infty] \to \mathbb{B}$ is called a **Brownian motion in** $(\mathbb{H},\mathbb{B})$ **under** $(\mathfrak{b}_t)_{t\in[0,\infty[}$ if the following two conditions are fulfilled:

(A) Each component of $b_\mathbb{B}$ is a one-dimensional Brownian motion, i.e., $\varphi \circ b_\mathbb{B}$ is a one-dimensional Brownian motion under $(\mathfrak{b}_t)_{t\in[0,\infty[}$ for each $\varphi \in \mathbb{B}'$ with $\|\varphi\| = 1$. Recall from Proposition 4.3.6 that $\mathbb{B}' \subseteq \mathbb{H}$.

(B) The components of $b_{\mathbb{B}}$ are running independently on orthogonal axes, i.e., $(\varphi \circ b_{\mathbb{B}}(\cdot,t), \psi \circ b_{\mathbb{B}}(\cdot,t))$ is independent for all $\varphi, \psi \in \mathbb{B}'$ with $\varphi \perp \psi$ and all $t \in [0,\infty]$.

From Proposition 3.3.1 (a) it follows that $(\varphi_i \circ b_{\mathbb{B}}(\cdot,t))_{i \leq k}$ is independent for all $k \in \mathbb{N}$, for all pairwise orthogonal $\varphi_1, \ldots, \varphi_k$ in $\mathbb{B}'$ and all $t \in [0,\infty]$.

It was already mentioned that the random variable $\kappa : \Omega \to C_{\mathbb{B}}, X \mapsto b_{\mathbb{B}}(X,\cdot)$ is surjective. The **Wiener measure** $W := W_{C_{\mathbb{B}}}$ on $\mathcal{B}(C_{\mathbb{B}})$ is the image measure of the probability measure $\widehat{\Gamma}$ on Ω by κ. We shall see later that $W(C_{\mathbb{B}} \setminus C_{\mathbb{B}}^0) = 0$, where

$$C_{\mathbb{B}}^0 = \{f \in C_{\mathbb{B}} \mid f(0) = 0\}.$$

Fix an abstract Wiener space $\mathbb{B}$ over $\mathbb{H}$. Our aim now is to prove that the Wiener measure coincides with the Gaussian measure γ_1 constructed by using the Hilbert space of absolutely continuous functions in $C_{\mathbb{B}}^0$. To this end fix a cylinder set Z in $C_{\mathbb{B}}$. Recall Proposition 4.5.1 and recall that Z has the form

$$Z = \{f \in C_{\mathbb{B}} \mid (\varphi_1 \circ f(t_1), \ldots, \varphi_n \circ f(t_n)) \in B\},$$

where $\varphi_1, \ldots, \varphi_n \in \mathbb{B}'$, $t_1, \ldots, t_n \in [0,\infty[$ and $B \subseteq \mathbb{R}^n$ is a Borel set. Then

$$W(Z) = \widehat{\Gamma}(\{X \mid (\varphi_1 \circ b_{\mathbb{B}}(X,t_1), \ldots, \varphi_n \circ b_{\mathbb{B}}(X,t_n)) \in B\}).$$

Choose $\sigma \in \mathbb{N}$ with $t_1, \ldots, t_n \leq \sigma$. Then we may assume that Z is a cylinder set in $C_{\mathbb{B},\sigma} := \{f : [0,\sigma] \to \mathbb{B} \mid f \text{ is continuous}\}$. Our aim is to find a separable Hilbert space $\mathbb{H}_\sigma \subseteq C_{\mathbb{B},\sigma}^0 := \{f \in C_{\mathbb{B},\sigma} \mid f(0) = 0\}$, such that $Z \cap \mathbb{H}_\sigma$ is a cylinder set in $\mathbb{H}_\sigma$ and

$$\gamma_1(Z) = \gamma_1(Z \cap \mathbb{H}_\sigma) = W(Z) \quad \text{(see Theorem 4.3.8).}$$

A square-integrable function $g : [0,\sigma] \to \mathbb{H}$ is called **absolutely continuous** if there exists a square-integrable $g' : [0,\sigma] \to \mathbb{H}$ such that for all $a \in \mathbb{H}$ and $s \in [0,\sigma]$

$$\langle a, g(s) \rangle = \int_0^s \langle a, g'(r) \rangle d\lambda(r),$$

i.e., $g(s) = \int_0^s g' d\lambda$ in the Pettis' integral sense. Let us call g' the **derivative** of g. Note that the space $\mathbb{H}_\sigma$ of absolutely continuous functions is a separable Hilbert space with scalar product $\langle g, f \rangle_\sigma$, given by

$$\langle g, f \rangle_\sigma := \int_0^\sigma \langle g', f' \rangle d\lambda.$$

Proposition 4.8.1 (Kuelbs and Lepage [55]) $W = \gamma_1$.

Proof We have already seen that it suffices to prove that $\gamma_1(Z \cap \mathbb{H}_\sigma) = W(Z)$. Choose $i \in \{1,\ldots,n\}$ and set $\varphi := \varphi_i$ and $t := t_i$ and define $\Phi := \Phi_{\varphi,t} : \mathbb{H}_\sigma \to \mathbb{R}$ by $\Phi(f) := \varphi \circ f(t)$ for all $f \in \mathbb{H}_\sigma$. Obviously, Φ is linear. We will prove that Φ is continuous: Fix $f \in \mathbb{H}_\sigma$. Then

$$|\Phi(f)| \leq \|\phi\| \cdot \|f(t)\| = \|\phi\| \cdot \left\| \int_0^t f' d\lambda \right\|.$$

Now

$$\begin{aligned}\left\| \int_0^t f' d\lambda \right\|^2 &= \left\langle \int_0^t f' d\lambda, \int_0^t f' d\lambda \right\rangle = \int_{[0,t]^2} \langle f'(r), f'(s) \rangle d\lambda^2(r,s) \\ &\leq \int_{[0,t]^2} \|f'(r)\| \cdot \|f'(s)\| \, d\lambda^2(r,s) = \left(\int_0^t \|f'\| \, d\lambda \right)^2 \\ &\leq t \cdot \int_0^\sigma \|f'\|^2 d\lambda = t \cdot \langle f, f \rangle_\sigma = t \|f\|_\sigma^2 .\end{aligned}$$

It follows that $|\Phi(f)| \leq \|\phi\| \cdot \sqrt{t} \cdot \|f\|_\sigma$, which proves that Φ is continuous, thus $\Phi \in \mathbb{H}_\sigma$. Note that $\Phi' : s \mapsto \mathbf{1}_{[0,t]}(s) \cdot \varphi$ is the derivative of Φ, because for all $f \in \mathbb{H}_\sigma$

$$\int_0^\sigma \langle \Phi', f' \rangle d\lambda = \varphi \circ f(t).$$

Let E be the finite-dimensional subspace of $\mathbb{H}_\sigma$, generated by the Φ_{φ_i,t_j}, $i,j \in \{1,\ldots,n\}$. Then

$$\gamma_1(Z \cap \mathbb{H}_\sigma) = \gamma_1^E \left\{ a \in E \mid \left(\langle \Phi_{\varphi_1,t_1}, a \rangle_\sigma, \ldots, \langle \Phi_{\varphi_n,t_n}, a \rangle_\sigma \right) \in B \right\}.$$

Let u, v be the image measures of the measures γ_1^E and $\widehat{\Gamma}$ by

$$\begin{aligned} E \ni a &\mapsto \left(\langle \Phi_{\varphi_1,t_1}, a \rangle_\sigma, \ldots, \langle \Phi_{\varphi_n,t_n}, a \rangle_\sigma \right) \in \mathbb{R}^n, \\ \Omega \ni X &\mapsto (\varphi_1 \circ b_{\mathbb{B}}(X, t_1), \ldots, \varphi_n \circ b_{\mathbb{B}}(X, t_n)) \in \mathbb{R}^n, \end{aligned}$$

respectively. We may assume that $t_0 := 0 \leq t_1 \leq \ldots \leq t_n$.

In order to prove that $\gamma_1(Z \cap \mathbb{H}_\sigma) = W(Z)$, it suffices to prove that u and v are the same measures. By Proposition 3.1.2, it suffices to prove that for all $\lambda_1, \ldots, \lambda_n \in \mathbb{R}$

$$\int_{\mathbb{R}^n} e^{\lambda_1 x_1 + \ldots + \lambda_n x_n} du(x_1, \ldots, x_n) = \int_{\mathbb{R}^n} e^{\lambda_1 x_1 + \ldots + \lambda_n x_n} dv(x_1, \ldots, x_n) < \infty. \quad (8)$$

Let $(\mathfrak{e}_i)_{i\in k}$ be an orthonormal basis of span $\{\varphi_1,\ldots,\varphi_n\}\subseteq\mathbb{B}'\subseteq\mathbb{H}$. Recall that we identify k with $\{1,\ldots,k\}$.

Let us first compute the left-hand side of equality (8). Assume that $\{r_1,\ldots,r_l\}=\{t_0,t_1,\ldots,t_n\}\setminus\{0\}$ with $r_0:=0<r_1<\ldots<r_l$. Define for each $i=1,\ldots,k, j=1,\ldots,l$,

$$\mathfrak{e}_{i,j}:=\frac{1}{\sqrt{r_j-r_{j-1}}}\left(\Phi_{\mathfrak{e}_i,r_j}-\Phi_{\mathfrak{e}_i,r_{j-1}}\right).$$

Note that $\left(\mathfrak{e}_{i,j}\right)_{i\in k,j\in l}$ is an ONB of E. We will write $\sum_{i,j}$ for $\sum_{i\in k,j\in l}$ and set $\beta_{i,j}:=\sum_{m\in n}\lambda_m\left\langle\Phi_{\varphi_m,t_m},\mathfrak{e}_{i,j}\right\rangle_\sigma$. If $r_j\leq t_m$, then $\left\langle\Phi_{\varphi_m,t_m},\mathfrak{e}_{i,j}\right\rangle_\sigma=\sqrt{r_j-r_{j-1}}\langle\varphi_m,\mathfrak{e}_i\rangle$. If $\left\langle\Phi_{\varphi_m,t_m},\mathfrak{e}_{i,j}\right\rangle_\sigma\neq 0$, then $r_j\leq t_m$. Therefore, $\beta_{i,j}:=\sum_{m\in n,r_j\leq t_m}\lambda_m\left\langle\Phi_{\varphi_m,t_m},\mathfrak{e}_{i,j}\right\rangle_\sigma$. Now we have, using Corollary 4.2.2 and the transformation rule,

$$\begin{aligned}
&\int_{\mathbb{R}^n}\exp\left(\lambda_1x_1+\ldots+\lambda_nx_n\right)du(x_1,\ldots,x_n)\\
&=\int_E\exp\left(\sum_{m\leq n}\lambda_m\left\langle\Phi_{\varphi_m,t_m},a\right\rangle_\sigma\right)d\gamma_1^E(a)\\
&=\int_{\mathbb{R}^{k\cdot l}}\exp\left(\sum_{i,j}x_{i,j}\sum_{m\in n}\lambda_m\left\langle\Phi_{\varphi_m,t_m},\mathfrak{e}_{i,j}\right\rangle_\sigma\right)d\gamma_1^{k\cdot l}(x_{i,j})_{i\in k,j\in l}\\
&=\int_{\mathbb{R}^{k\cdot l}}\exp\left(\sum_{i,j}x_{i,j}\beta_{i,j}-\frac{1}{2}\sum_{i,j}x_{i,j}^2\right)d\left(x_{i,j}\right)_{i\in k,j\in l}\frac{1}{\sqrt{2\pi}^{k\cdot l}}\\
&=\int_{\mathbb{R}^{k\cdot l}}\exp\left(\sum_{i,j}-\frac{1}{2}\left(x_{i,j}-\beta_{i,j}\right)^2+\frac{1}{2}\sum_{i,j}\beta_{i,j}^2\right)d\left(x_{i,j}\right)_{i\in k,j\in l}\frac{1}{\sqrt{2\pi}^{k\cdot l}}\\
&=\int_{\mathbb{R}^{k\cdot l}}\exp\left(\sum_{i,j}-\frac{1}{2}x_{i,j}^2+\frac{1}{2}\sum_{i,j}\beta_{i,j}^2\right)d\left(x_{i,j}\right)_{i\in k,j\in l}\frac{1}{\sqrt{2\pi}^{k\cdot l}}\\
&=\exp\left(\frac{1}{2}\sum_{i,j}\beta_{i,j}^2\right)=\exp\left(\frac{1}{2}\sum_{i,j}\left(\sum_{m\in n}\lambda_m\left\langle\Phi_{\varphi_m,t_m},\mathfrak{e}_{i,j}\right\rangle_\sigma\right)^2\right)\\
&=\exp\left(\frac{1}{2}\sum_{i,j}\left(\sum_{m\in n,r_j\leq t_m}\lambda_m\langle\varphi_m,\mathfrak{e}_i\rangle\right)^2(r_j-r_{j-1})\right).\\
&=\prod_{i\in k,j\in l}\exp\left(\frac{1}{2}\left(\sum_{m\in n,r_j\leq t_m}\lambda_m\langle\varphi_m,\mathfrak{e}_i\rangle\right)^2(r_j-r_{j-1})\right).
\end{aligned}$$

Here are preparations to compute the right-hand side of (8). Set $b:=b_{\mathbb{B}}$. Let $a=\sum_{i=1}^k\alpha_i\mathfrak{e}_i\in$ span $\{\varphi_1,\ldots,\varphi_n\}$. Since $\left(\alpha_i\mathfrak{e}_i(b_{r_j}-b_{r_{j-1}})\right)_{i\leq k}$ is independent and

$\alpha_i \mathfrak{e}_i(b_{r_j} - b_{r_{j-1}})$ is $\mathcal{N}\left(0, \alpha_i^2 \cdot (r_j - r_{j-1})\right)$-distributed, we see that $a(b_{r_j} - b_{r_{j-1}})$ is $\mathcal{N}\left(0, \|a\|^2 \cdot (r_j - r_{j-1})\right)$-distributed, by Corollary 3.2.4. Therefore,

$$\begin{aligned}
&\int_{\mathbb{R}^n} \exp\left(\lambda_1 x_1 + \ldots + \lambda_n x_n\right) dv(x_1, \ldots, x_n) \\
&= \mathbb{E}\exp\left(\sum_{m \in n} \lambda_m \varphi_m(b_{t_m})\right) \\
&= \mathbb{E}\exp\left(\sum_{j \in l}\left(\sum_{m \in n, r_j \leq t_m} \lambda_m \varphi_m\right)\left(b_{r_j} - b_{r_{j-1}}\right)\right) \\
&= \prod_{j \in l} \mathbb{E}\exp\left(\left(\sum_{m \in n, r_j \leq t_m} \lambda_m \varphi_m\right)\left(b_{r_j} - b_{r_{j-1}}\right)\right) \quad \text{(by 3.5.1 and 3.3.1 (e))} \\
&= \prod_{j \in l} \exp\left(\frac{1}{2}\left\|\sum_{m \in n, r_j \leq t_m} \lambda_m \varphi_m\right\|^2 (r_j - r_{j-1})\right) \quad \text{(by 3.2.3)} \\
&= \prod_{j \in l} \exp\left(\frac{1}{2}\sum_{i \in k}\left(\sum_{m \leq n, r_j \leq t_m} \lambda_m \langle \varphi_m, \mathfrak{e}_i \rangle\right)^2 (r_j - r_{j-1})\right) \\
&= \prod_{i \in k, j \in l} \exp\left(\frac{1}{2}\left(\sum_{m \leq n, r_j \leq t_m} \lambda_m \langle \varphi_m, \mathfrak{e}_i \rangle\right)^2 (r_j - r_{j-1})\right).
\end{aligned}$$

This proves equation (8) and the theorem. □

The results in this section have shown that $C^0_{\mathbb{B},\sigma}$ is an abstract Wiener space over $\mathbb{H}_\sigma$ for each $\sigma \in \mathbb{N}$.

Exercises

4.1 Prove Lemmas 4.1.1, 4.1.2 and 4.1.3.

4.2 Prove Corollary 4.3.2.

4.3 Prove the assertions in Examples 4.3.3.

4.4 Prove Lemma 4.4.2 for $\alpha = 0$.

4.5 The definition of $\|f\|$ for $f \in \mathbb{H}^{\otimes n}$ does not depend on the chosen ONB.

4.6 $\mathbb{H}^{\otimes n}$ is a Hilbert space.

4.7 Let $\mathbb{H}$ be a separable Hilbert space and let $S : \mathbb{H} \to \mathbb{H}$ be a Hilbert–Schmidt operator. Prove that $\sum_{i=1}^{\infty} \|S(\mathfrak{b}_i)\|^2 = \sum_{i=1}^{\infty} \|S(\mathfrak{e}_i)\|^2$ for all ONBs $(\mathfrak{e}_i)_{i \in \mathbb{N}}$, $(\mathfrak{b}_i)_{i \in \mathbb{N}}$ of $\mathbb{H}$.

4.8 Prove that $(\mathbb{H}_\sigma, \langle \cdot, \cdot \rangle_\sigma)$ is a Hilbert space.

5

Two concepts of no-anticipation in time

In this chapter we will briefly discuss some basic concepts of $L^p(\mu)$-spaces over a complete σ-finite measure space $(\Lambda, \mathcal{C}, \mu)$.

Moreover, fix a complete probability space $(\Omega, \mathcal{F}, \rho)$ and a filtration $(\mathfrak{f}_r)_{r\in[0,\infty[}$ on $\mathcal{F}$. We assume that this filtration fulfils the Doob–Meyer Conditions. Using continuous approximations of the Dirac δ-function, we will first prove the equivalence of two notions of 'no-anticipation in time', namely adaptedness and predictability, following ideas in the book by Albeverio *et al.* [2].

5.1 Predictability and adaptedness

Fix r, s in $[0,\infty[$ with $r \leq s$. Sets of the form $B\times]r,s]$, $B\times]r,\infty[$ with $B \in \mathfrak{f}_r$ or sets $B \times [0,s], B \times [0,\infty[$ with $B \in \mathfrak{f}_0$ are called $(\mathfrak{f}_t)_{t\in[0,\infty[}$-**predictable rectangles**. The proof of the following result is an application of Exercise 3.2 and is left to the reader.

Proposition 5.1.1 *The set* $\widetilde{\mathfrak{P}}$ *of finite-disjoint unions of* $(\mathfrak{f}_t)_{t\in[0,\infty[}$*-predictable rectangles is an algebra and* $\rho \otimes \lambda$ *is* σ*-finite on* $\Omega \times [0,\infty[$.

Let $\mathfrak{P}_0$ be the σ-algebra generated by $\widetilde{\mathfrak{P}}$ and set $\mathfrak{P} := \mathfrak{P}_0 \vee \mathcal{N}_{\rho\otimes\lambda}$. A function $f : \Omega \times [0,\infty[\to \mathbb{R}$ is called $(\mathfrak{f}_t)_{t\in[0,\infty[}$-**predictable** (**strongly** $(\mathfrak{f}_t)_{t\in[0,\infty[}$-**predictable**) if f is $\mathfrak{P}$-($\mathfrak{P}_0$-)measurable.

From the following result it follows that each predictable (adapted) process is equivalent to a strongly predictable (strongly adapted) process. In the following two results fix a separable metric space $\mathbb{B}$ and a σ-subalgebra $\mathcal{S}$ of $\mathcal{C}$.

Proposition 5.1.2 *A random variable* $f : \Lambda \to \mathbb{B}$ *is* $\mathcal{S} \vee \mathcal{N}_\mu$*-measurable if and only if there exists an* $\mathcal{S}$*-measurable function* $g : \Lambda \to \mathbb{B}$ *such that* $\mu\{g \neq f\} = 0$.

Proof Obviously, if $\mu\{g \neq f\} = 0$ and g is $\mathcal{S}$-measurable, then f is $\mathcal{S} \vee \mathcal{N}_\mu$-measurable. To prove the converse, first note that $\mathcal{S} \cup \mathcal{N}_\mu$ is a subset of the σ-algebra $\{B \in \mathcal{C} \mid \exists A \in \mathcal{S}(A \,\Delta\, B \in \mathcal{N}_\mu)\}$. Therefore,

$$\mathcal{S} \vee \mathcal{N}_\mu \subseteq \{B \in \mathcal{C} \mid \exists A \in \mathcal{S}(A \,\Delta\, B \in \mathcal{N}_\mu)\}.$$

It follows that, if f is an $\mathcal{S} \vee \mathcal{N}_\mu$-simple function, then there exists an $\mathcal{S}$-simple function g with $\mu\{g \neq f\} = 0$. Now we obtain the assertion, by approximating an arbitrary $\mathcal{S} \vee \mathcal{N}_\mu$-measurable f by $\mathcal{S} \vee \mathcal{N}_\mu$-simple functions and their $\mathcal{S}$-simple versions. □

Corollary 5.1.3 *The set $\mathcal{S}$ has the* ***approximation property w.r.t.*** *$\mathcal{S} \vee \mathcal{N}_\mu$, i.e., $B \in \mathcal{S} \vee \mathcal{N}_\mu$ if and only if there is an $A \in \mathcal{S}$ such that $A \,\Delta\, B$ is a μ-nullset.*

Here are partial results for the equivalence of predictability and adaptedness.

Proposition 5.1.4 *Each $(\mathfrak{f}_r)_{r\in[0,\infty[}$-predictable process is $(\mathfrak{f}_r)_{r\in[0,\infty[}$-adapted, and 'strongly predictable' implies 'strongly adapted'.*

Proof By Proposition 5.1.2, it suffices to show that $(\mathfrak{f}_t)$-predictable rectangles are strongly $(\mathfrak{f}_t)$-adapted. Let $C = B \times]s,t]$ with $B \in \mathfrak{f}_s$ and $s \leq t$ in $[0,\infty[$ and let $r \in [0,\infty[$. If $r \notin]s,t]$, then $C(\cdot,r) = \emptyset \in \mathfrak{f}_r$. If $r \in]s,t]$, then $C(\cdot,r) = B \in \mathfrak{f}_s \subseteq \mathfrak{f}_r$. For the other predictable rectangles the proof is similar. □

Proposition 5.1.5 *Let $f : \Omega \times [0,\infty[\to \mathbb{R}$ be a continuous μ-a.s. and strongly $(\mathfrak{f}_r)_{r\in[0,\infty[}$-adapted process. Then f is $(\mathfrak{f}_r)_{r\in[0,\infty[}$-predictable.*

Proof For each $n \in \mathbb{N}$ we define

$$f_n(x,t) := \begin{cases} f(x,0) & \text{if } 0 \leq t \leq \frac{1}{n} \\ f(x, \frac{i-1}{n}) & \text{if } \frac{i-1}{n} < t \leq \frac{i}{n},\, i \in \mathbb{N},\, i \geq 2. \end{cases}$$

Then for all $c \in \mathbb{R}$

$$\{(x,t) \in \Omega \times [0,\infty[\mid f_n(x,t) \leq c\}$$
$$= \left(\{x \mid f(x,0) \leq c\} \times \left[0, \frac{1}{n}\right]\right) \cup \bigcup_{i=2}^{\infty} \left(\left\{x \mid f\left(x, \frac{i-1}{n}\right) \leq c\right\} \times \left]\frac{i-1}{n}, \frac{i}{n}\right]\right)$$

is a countable union of $(\mathfrak{f}_r)_{r\in[0,\infty[}$-predictable rectangles, thus, f_n is strongly $(\mathfrak{f}_r)_{r\in[0,\infty[}$-predictable. Let $f(x,\cdot)$ be continuous on $[0,\infty[$. We prove that $\lim_{n\to\infty} f_n(x,\cdot) = f(x,\cdot)$. If $0 < t \in \mathbb{R}$, then there exists a sequence (a_n) in $\mathbb{N}$ such that $\frac{a_n-1}{n} < t \leq \frac{a_n}{n}$ for all $n \in \mathbb{N}$. It follows that $\lim_{n\to\infty} \frac{a_n-1}{n} = t$. Now

$$\lim_{n\to\infty} f_n(x,t) = \lim_{n\to\infty} f\left(x, \frac{a_n-1}{n}\right) = f(x,t).$$

If $t=0$ then $\lim_{n\to\infty} f_n(x,0) = f(x,0)$. Therefore, $\lim_{n\to\infty} f_n = f$ $\mu\otimes\lambda$-a.s., thus f is $(\mathfrak{f}_r)_{r\in[0,\infty[}$-predictable. □

5.2 Approximations of the Dirac δ-function

A sequence $(\delta_\varepsilon)_{\varepsilon\in\mathbb{R}^+}$ of continuous functions $\delta_\varepsilon : [0,\infty[\to \mathbb{R}_0^+$ is called an **approximation of the Dirac δ-function** if

(i) $[0,\varepsilon[$ is the **support** of δ_ε, i.e., $\delta_\varepsilon = 0$ on $[\varepsilon,\infty[$
(ii) $\int_0^\varepsilon \delta_\varepsilon d\lambda = 1$,
(iii) $\lim_{\varepsilon\downarrow 0} \delta_\varepsilon(0) = \infty$.

In L. Schwartz' distribution theory there are many types of approximations of the Dirac δ-function considered. This section uses a very special and simple one. The functions δ_ε with

$$\delta_\varepsilon(x) := \begin{cases} \frac{2(\varepsilon - x)}{\varepsilon^2} & \text{if } 0 \leq x < \varepsilon \\ 0, & \text{otherwise} \end{cases}$$

provide an approximation of the Dirac δ-function. Fix $\sigma \in \mathbb{N}$. For each bounded Lebesgue measurable function $f : [0,\sigma] \to \mathbb{R}$ we define the **convolution** $f_\varepsilon : [0,\sigma] \to \mathbb{R}$ of f and δ_ε by setting

$$f_\varepsilon(t) := \int_0^{t\wedge\varepsilon} f(t-s)\delta_\varepsilon(s)ds.$$

Proposition 5.2.1 *Suppose that* $f : [0,\sigma] \to \mathbb{R}$ *is bounded and Lebesgue measurable. Then*

(a) $\lim_{\varepsilon\to 0} f_\varepsilon = f$ *in* L^1_σ, *the space of real Lebesgue integrable functions g, defined on* $[0,\sigma]$
(b) f_ε *is continuous.*

Proof (a) Fix $\eta > 0$. By Theorem 2.4.14 in Ash [5], there exists a sequence (f_n) of continuous functions $f_n : [0,\sigma] \to \mathbb{R}$ converging to f in L^1_σ. We can assume that $|f_n|$ and $|f|$ are bounded by an $M \in \mathbb{R}$. Note that for each $\varepsilon > 0$

$$|f_\varepsilon| \leq M.$$

Moreover, note that for each $\varepsilon > 0$ and for each $s \in [0,\varepsilon]$

$$\int_s^{\sigma+s} |f_n(t-s) - f(t-s)|\, dt = \int_0^\sigma |f_n(t) - f(t)|\, dt = \|f_n - f\|_1 .$$

It follows that there exists an $m \in \mathbb{N}$ such that for each $\varepsilon > 0$ and for each $s \in [0, \varepsilon]$

$$\int_s^{\sigma+s} |f_m(t-s) - f(t-s)|\, dt, \quad \int_0^{\sigma} |f_m(t) - f(t)|\, dt < \frac{\eta}{4}.$$

Since f_m is uniformly continuous, there exists a $\delta > 0$ such that for all $s, t \in [0, \sigma]$ with $s \leq t \wedge \delta$

$$|f_m(t-s) - f_m(t)| < \frac{\eta}{4 \cdot \sigma}.$$

We can choose δ such that $\delta \cdot 2M < \frac{\eta}{4}$. Then for each ε with $0 < \varepsilon < \delta$

$$\|f_\varepsilon - f\|_1 = \int_0^{\sigma} \left| \int_0^{\varepsilon \wedge t} f(t-s)\delta_\varepsilon(s)ds - \int_0^{\varepsilon} f(t)\delta_\varepsilon(s)ds \right| dt \leq A + B + C$$

where

$$A = \int_0^{\varepsilon} |f_\varepsilon - f|\, d\lambda \leq \int_0^{\varepsilon} 2M d\lambda < \frac{\eta}{4},$$

$$B = \int_\varepsilon^{\sigma} \int_0^{\varepsilon} |f(t-s) - f(t) - (f_m(t-s) - f_m(t))|\, \delta_\varepsilon(s)ds\, dt$$

$$\leq \int_0^{\varepsilon} \int_\varepsilon^{\sigma} |f(t-s) - f_m(t-s)|\, dt \delta_\varepsilon(s)ds + \int_0^{\varepsilon} \int_\varepsilon^{\sigma} |f - f_m|\, dt\ \delta_\varepsilon(s)ds < \frac{\eta}{2}$$

and

$$C = \int_0^{\varepsilon} \int_\varepsilon^{\sigma} |f_m(t-s) - f_m(t)|\, dt\ \delta_\varepsilon(s)ds < \frac{\eta}{4}.$$

It follows that for each ε with $0 < \varepsilon < \delta$

$$\|f_\varepsilon - f\|_1 < \eta.$$

(b) Assume that $t \in [0, \sigma]$ and $h \neq 0$ with $t + h \in [0, \sigma]$ and $|h| < \varepsilon$. Let f be bounded by M. By the substitution rule,

$$\begin{aligned} &|f_\varepsilon(t+h) - f_\varepsilon(t)| \\ &= \left| \int_0^{(t+h)\wedge\varepsilon} f(t-(s-h))\delta_\varepsilon(s)ds - \int_0^{t\wedge\varepsilon} f(t-s)\delta_\varepsilon(s)ds \right| \\ &= \left| \int_{-h}^{t\wedge(\varepsilon-h)} f(t-s)\delta_\varepsilon(s+h)ds - \int_0^{t\wedge\varepsilon} f(t-s)\delta_\varepsilon(s)ds \right|. \end{aligned}$$

Now we can easily compare both summands, because they have the same factor $f(t-s)$. Assume first that $h > 0$. Then

$$|f_\varepsilon(t+h) - f_\varepsilon(t)| \leq \rho_1(h) + \rho_2(h) + \rho_3(h),$$

where

$$\rho_1(h) = \int_{-h}^{0} |f(t-s)|\,\delta_\varepsilon(s+h)ds \le M \cdot \int_{-h}^{0} \delta_\varepsilon(s+h)ds \to_{h\to 0} 0,$$

$$\rho_2(h) = \int_0^{t\wedge(\varepsilon-h)} |f(t-s)|\,|\delta_\varepsilon(s+h)-\delta_\varepsilon(s)|\,ds$$

$$\le M \int_0^t |\delta_\varepsilon(s+h)-\delta_\varepsilon(s)|\,ds \to_{h\to 0} 0,$$

$$\rho_3(h) = \int_{t\wedge(\varepsilon-h)}^{t\wedge\varepsilon} |f(t-s)|\,\delta_\varepsilon(s)ds \le M \int_{t\wedge(\varepsilon-h)}^{t\wedge\varepsilon} \delta_\varepsilon(s)ds \to_{h\to 0} 0.$$

Now assume that $h<0$. Since $t+h\in[0,\sigma]$, $0<t$. Let $-h<t\wedge\varepsilon$. We obtain in a similar way to before,

$$|f_\varepsilon(t+h)-f_\varepsilon(t)| \le \sigma_1(h)+\sigma_2(h)+\sigma_3(h)$$

with

$$\sigma_1(h) = \int_0^{-h} |f(t-s)|\,\delta_\varepsilon(s)ds \le M \int_0^{-h} \delta_\varepsilon(s)ds \to_{h\to 0} 0,$$

$$\sigma_2(h) = \int_{-h}^{t\wedge\varepsilon} |f(t-s)|\,|\delta_\varepsilon(s+h)-\delta_\varepsilon(s)|\,ds$$

$$\le M \int_0^t |\delta_\varepsilon(s+h)-\delta_\varepsilon(s)|\,ds \to_{h\to 0} 0,$$

$$\sigma_3(h) = \int_{t\wedge\varepsilon}^{(\varepsilon-h)\wedge t} |f(t-s)|\,\delta_\varepsilon(s+h)ds$$

$$\le M \int_{t\wedge\varepsilon}^{(\varepsilon-h)\wedge t} \delta_\varepsilon(s+h)ds \to_{h\to 0} 0.$$

This proves the continuity of f_ε. □

5.3 Convolutions of adapted functions are adapted

In order to obtain this result we need the following lemma.

Lemma 5.3.1 *Let $g:\Omega\times[0,\sigma]\to\mathbb{R}$ be a bounded process and let $\mathcal{S}$ be a σ-subalgebra of $\mathcal{F}$, containing all ρ-nullsets. Suppose that $g(\cdot,s)$ is $\mathcal{S}$-measurable* *for all* $s \in [0,\sigma]$. *Then*

$$\int_0^\sigma g(\cdot,s)d\lambda(s) \text{ is } \mathcal{S}\text{-measurable.}$$

Proof Let $\mathcal{E}$ be the set of all rectangles of the form $A \times B$ with $A \in \mathcal{F}$ and $B \in \text{Leb}[0,\sigma]$. There exists a sequence (φ_n) of $\mathcal{E}$-simple functions converging to g in L^1_σ. We obtain

$$\int_{\Omega\times[0,\sigma]} \left|\mathbb{E}^{\mathcal{S}}(\varphi_n(\cdot,s)) - g(\cdot,s)\right| d\rho\otimes\lambda(\cdot,s)$$

$$= \int_0^\sigma \mathbb{E}_\rho \left|\mathbb{E}^{\mathcal{S}}(\varphi_n(\cdot,s)) - \mathbb{E}^{\mathcal{S}} g(\cdot,s)\right| d\lambda(s) \leq \int_0^\sigma \mathbb{E}_\rho \mathbb{E}^{\mathcal{S}} \left|(\varphi_n(\cdot,s) - g(\cdot,s)\right| d\lambda(s)$$

$$= \int_0^\sigma \mathbb{E}_\rho \left|(\varphi_n(\cdot,s) - g(\cdot,s)\right| d\lambda(s) \to_{n\to\infty} 0.$$

It follows that $\mathbb{E}_\rho \left|\int_0^\sigma \mathbb{E}^{\mathcal{S}}(\varphi_n(\cdot,s)) - g(\cdot,s)d\lambda(s)\right| \to_{n\to\infty} 0$. Therefore, there exists a subsequence (φ_{k_n}) of (φ_n) such that

$$\lim_{k\to\infty} \int_0^\sigma \mathbb{E}^{\mathcal{S}}(\varphi_{k_n}(\cdot,s)d\lambda(s) = \int_0^\sigma g(\cdot,s)d\lambda(s) \ \rho\text{-a.s.}$$

Since $\int_0^\sigma \mathbb{E}^{\mathcal{S}}(\varphi_{k_n}(\cdot,s)d\lambda(s)$ is $\mathcal{S}$-measurable and $\mathcal{S}$ is complete, $\int_0^\sigma g(\cdot,s)d\lambda(s)$ is $\mathcal{S}$-measurable. □

Proposition 5.3.2 *Assume that* $f : \Omega \times [0,\sigma] \to \mathbb{R}$ *is bounded and strongly* $(\mathfrak{f}_r)_{r\in[0,\sigma]}$*-adapted (see Section 2.1). Then for each* $\varepsilon > 0$

$$f_\varepsilon : \Omega \times [0,\sigma] \ni (x,t) \mapsto f(x)_\varepsilon(t) \text{ is strongly } (\mathfrak{f}_r)_{r\in[0,\sigma]}\text{-adapted.}$$

Proof Fix $r \in [0,\sigma]$. Define $g(x,s) := \mathbf{1}_{[0,r\wedge\varepsilon]}(s) f(x,r-s)\delta_\varepsilon(s)$ for all $(x,s) \in \Omega \times [0,\sigma]$. Since $g(\cdot,s)$ is $\mathfrak{f}_r$-measurable for all $s \in [0,\sigma]$, by Lemma 5.3.1,

$$f_\varepsilon(\cdot,r) = \int_0^\sigma g(\cdot,s)d\lambda(s) = \int_0^{r\wedge\varepsilon} f(\cdot,r-s)\delta_\varepsilon(s)d\lambda(s) \text{ is } \mathfrak{f}_r\text{-measurable.}$$

□

5.4 Adaptedness is equivalent to predictability

Now we can prove one of the main results in this chapter.

Theorem 5.4.1 *A process* $g : \Omega \times [0,\infty[\to \mathbb{R}$ *is* $(\mathfrak{f}_t)$*-adapted if and only if* f *is* $(\mathfrak{f}_t)$*-predictable.*

Proof We may assume that the support of g is a subset of $\Omega \times [0,\sigma]$. By Proposition 5.1.4, each predictable process is adapted. Now assume that the process $g : \Omega \times [0,\sigma] \to \mathbb{R}$ is $(\mathfrak{f}_r)_{r\in[0,\sigma]}$-adapted. Since each process is the limit of a sequence of bounded processes, we may assume that g is bounded. By Proposition 5.1.2, there exists a strongly $(\mathfrak{f}_r)_{r\in[0,\sigma]}$-adapted bounded process f with $f = g$ $\rho \otimes \lambda$-a.e. By Proposition 5.2.1 (a) and the dominated convergence theorem,

$$\lim_{\varepsilon \to 0} f_\varepsilon = f \text{ in } \mathcal{L}^1(\rho \otimes \lambda).$$

It follows that there exists a sequence (a_n) in $]0,1]$ converging to 0 such that $\lim_{n\to\infty} f_{a_n} = f$ $\rho \otimes \lambda$-a.e. By Proposition 5.3.2, f_{a_n} is strongly $(\mathfrak{f}_t)$-adapted. Since $f(x,\cdot)$ is Lebesgue measurable for ρ-almost all $x \in \Omega$, by Proposition 5.2.1 (b), $f_{a_n}(x,\cdot)$ is continuous for ρ-almost all $x \in \Omega$. Since, by Proposition 5.1.5, each f_{a_n} is $(\mathfrak{f}_r)_{r\in[0,\sigma]}$-predictable, f is $(\mathfrak{f}_r)_{r\in[0,\sigma]}$-predictable and therefore so is g. □

Since the cylinder sets of a separable Hilbert space $\mathbb{H}$ generate the Borel-σ-algebra of $\mathbb{H}$ (see Theorem 4.4.3), we obtain

Corollary 5.4.2 *The following statements are equivalent for processes $f : \Omega \times [0,\infty[\to \mathbb{H}$.*

(i) *f is $(\mathfrak{f}_t)$-adapted.*
(ii) *$\langle a,f \rangle$ is $(\mathfrak{f}_t)$-adapted for all $a \in \mathbb{H}$.*
(iii) *$\langle a,f \rangle$ is $(\mathfrak{f}_t)$-predictable for all $a \in \mathbb{H}$.*
(iv) *f is $(\mathfrak{f}_t)$-predictable.*

5.5 The weak approximation property

The approximation property, used in the preceding sections, will also play an important role in connection with Loeb spaces. We also need the so-called weak approximation property. Let $\mathcal{S}$ be a subset of $\mathcal{C}$. We say that $\mathcal{S}$ has the **weak approximation property w.r.t.** $\mathcal{C}$ if for each $\varepsilon > 0$ and each $C \in \mathcal{C}$ with $\mu(C) < \infty$ there exists a finite union E of disjoint elements of $\mathcal{S}$ such that $\mu(E \,\Delta\, C) < \varepsilon$. In the next result we give some important examples, which follow from Theorem 11.4 in Billingsley [13].

Examples 5.5.1 (i) If $\mathcal{D} \subseteq \mathcal{C}$ is an algebra generating $\mathcal{C}$ and μ is σ-finite on $\mathcal{D}$, then $\mathcal{D}$ has the weak approximation property w.r.t. $\mathcal{C}$.

(ii) If $\mathcal{C}$ is a product σ-algebra of two complete σ-finite measure spaces, then the set of measurable rectangles has the weak approximation property w.r.t. $\mathcal{C}$.

(iii) The set of predictable rectangles has the weak approximation property w.r.t. $\mathfrak{P}$ (see Proposition 5.1.1 and Example 5.5.1 (i)).

(iv) The set of right rectangles in $\mathbb{R}^n$ has the weak approximation property w.r.t. the Borel algebra $\mathcal{B}(\mathbb{R}^n)$ on $\mathbb{R}^n$.

5.6 Elementary facts about L^p-spaces

Now we will show that apparently very weak closure properties of a linear subspace M of an L^p-space with $p \in [1, \infty[$ force the equality of M and the whole space. Fix a separable Hilbert space $\mathbb{H}$ and $d \in \mathbb{N}_0$. Recall the notation of Section 4.6.

Proposition 5.6.1 *Suppose that $\mathcal{S}$ has the weak approximation property w.r.t. $\mathcal{C}$. Let M be a subset of $L^p(\mu, \mathbb{H}^{\otimes d})$ with the following properties:*

(A 1) *M is a linear space.*

(A 2) *$\mathbf{1}_A \cdot a \in M$ for each $A \in \mathcal{S}$ with $\mu(A) < \infty$ and each $a \in \mathbb{H}^{\otimes d}$.*

(A 3) *M is closed.*

Then $M = L^p(\mu, \mathbb{H}^{\otimes d})$.

Proof Let $f \in L^p(\mu, \mathbb{H}^{\otimes d})$. First assume that $f = \mathbf{1}_B \cdot a$ with $a \in \mathbb{H}^{\otimes d}$ and $B \in \mathcal{C}$ with $\mu(B) < \infty$. There exists a sequence (A_k) of finite unions A_k of pairwise disjoint elements in $\mathcal{S}$ such that $\mu(A_k \triangle B) < \frac{1}{2^k}$. By the Borel–Cantelli lemma,

$$N := \bigcap_{k_o \in \mathbb{N}} \bigcup_{k_o \le k} B \triangle A_k$$

is a μ-nullset and $\lim_{k\to\infty} \mathbf{1}_{A_k}(x) \cdot a = \mathbf{1}_B(x) \cdot a$ for all $x \in \Lambda \setminus N$. It follows that

$$\lim_{k\to\infty} \mathbf{1}_{A_k} \cdot a = \mathbf{1}_B \cdot a \text{ in } L^p(\mu, \mathbb{H}^{\otimes d}).$$

Therefore, there exists a sequence of $\mathcal{S}$-simple functions converging to $\mathbf{1}_B \cdot a$ in $L^p(\mu, \mathbb{H}^{\otimes d})$. By (A 1), ..., (A 3), $\mathbf{1}_B \cdot a \in M$. By (A 1), each $\mathcal{C}$-simple function belongs to M. Now, by approximating $f \in L^p(\mu, \mathbb{H}^{\otimes d})$ by $\mathcal{C}$-simple functions, we obtain the result. □

Here are first applications of the preceding result.

Corollary 5.6.2 *Assume that $\mathcal{S}$ has the weak approximation property w.r.t. $\mathcal{C}$. Let $f \in L^p(\mu, \mathbb{H}^{\otimes d})$. Then there exists a sequence $(\varphi_n)_{n\in\mathbb{N}}$ of $\mathcal{S}$-simple functions φ_n converging to f in $L^p(\mu, \mathbb{H}^{\otimes d})$.*

Proof Let M be the set of all $f \in L^p(\mu, \mathbb{H}^{\otimes d})$ such that the assertion is true for f. Obviously, $\mathbf{1}_A \cdot a \in M$ for each $a \in \mathbb{H}^{\otimes d}$ and each $A \in \mathcal{S}$ with $\mu(A) < \infty$, and M is a linear space. Therefore, in order to prove that $M = L^p(\mu, \mathbb{H}^{\otimes d})$, it suffices to show that M is complete. Let (f_n) be a Cauchy sequence in M with limit f in $L^p(\mu, \mathbb{H}^{\otimes d})$. For each $n \in \mathbb{N}$ there exists a sequence $(\varphi_k^n)_{k\in\mathbb{N}}$ of $\mathcal{S}$-simple functions φ_k^n converging to f_n in $L^p(\mu, \mathbb{H}^{\otimes d})$. Therefore, there exists a sequence $(g_k)_{k\in\mathbb{N}}$ of $\mathcal{S}$-simple functions g_k, converging to f in $L^p(\mu, \mathbb{H}^{\otimes d})$. □

In case μ is the Lebesgue measure we obtain

Corollary 5.6.3 *Let $M \subseteq L^p(\lambda^n, \mathbb{H}^{\otimes d})$ be linear and complete. Then $M = L^p(\lambda^n, \mathbb{H}^{\otimes d})$ if, for all products $[h_1, \ldots, h_d] \in \mathbb{H}^{\otimes d}$ and for all bounded right rectangles $B_1 \times \ldots \times B_n \subseteq [0, \infty[^n$,*

$$\mathbf{1}_{B_1 \times \ldots \times B_n} \otimes [h_1, \ldots, h_d] \in M.$$

Two L^p-spaces are called **equivalent** if there exists a canonical isomorphic isometry between them; **canonical** means that the isometry does not depend on the choice of a possible basis.

Corollary 5.6.4 *Suppose that there exists a measure-preserving mapping κ from $(\Lambda, \mathcal{C}, \mu)$ onto $(\Lambda', \mathcal{C}', \mu')$, i.e., $\kappa : \Lambda \to \Lambda'$ is surjective, $\mathcal{C} = \{\kappa^{-1}[B] \mid B \in \mathcal{C}'\}$, augmented by the μ-nullsets, and $\mu'(B) = \mu(\kappa^{-1}[B])$ for all $B \in \mathcal{C}'$. Then*

$$J : L^p(\mu', \mathbb{H}^{\otimes d}) \to L^p(\mu, \mathbb{H}^{\otimes d}), \quad J(\varphi) := \varphi \circ \kappa$$

is a canonical isomorphic isometry from $L^p(\mu', \mathbb{H}^{\otimes d})$ onto $L^p(\mu, \mathbb{H}^{\otimes d})$.

Proof To prove that J is surjective, let M be the space of all $\psi \in L^p(\mu, \mathbb{H}^{\otimes d})$ such that there exists a $\varphi \in L^p(\mu', \mathbb{H}^{\otimes d})$ with $J(\varphi) = \psi$. Apply Proposition 5.6.1 to prove that $M = L^p(\mu, \mathbb{H}^{\otimes d})$. □

Finally, we are concerned with a useful result on convergence in L^2-spaces.

Proposition 5.6.5 *Fix a Cauchy sequence $(\sum_{n=0}^{\infty} \varphi_n^k)_{k\in\mathbb{N}}$ in $L^2(\mu, \mathbb{H}^{\otimes d})$ such that $\varphi_n^k \perp \varphi_m^l$ for all $m \neq n$ and all $k, l \in \mathbb{N}$. Moreover, assume that $\lim_{k\to\infty} \varphi_n^k = \varphi_n$ in $L^2(\mu, \mathbb{H}^{\otimes d})$. Then:*

$$\sum_{n=0}^{\infty} \varphi_n \in L^2(\mu, \mathbb{H}^{\otimes d})$$

and

$$\lim_{k\to\infty}\left(\sum_{n=0}^{\infty}\varphi_n^k\right)=\sum_{n=0}^{\infty}\varphi_n \quad \text{in } L^2(\mu,\mathbb{H}^{\otimes d}).$$

Proof Since $\left\{\|\sum_{n=0}^{\infty}\varphi_n^k\|_{\mathbb{H}^d} \mid k\in\mathbb{N}\right\}$ is bounded, there exists an $s\in\mathbb{R}$ such that for all $k\in\mathbb{N}$

$$s\geq\int_\Lambda\left\|\sum_{n=0}^{\infty}\varphi_n^k\right\|_{\mathbb{H}^d}^2 d\mu=\sum_{n=0}^{\infty}\int_\Lambda\|\varphi_n^k\|_{\mathbb{H}^d}^2\,d\mu.$$

We obtain for all $m\in\mathbb{N}$

$$s\geq\lim_{k\to\infty}\sum_{n=0}^{m}\int_\Lambda\|\varphi_n^k\|_{\mathbb{H}^d}^2\,d\mu=\sum_{n=0}^{m}\lim_{k\to\infty}\int_\Lambda\|\varphi_n^k\|_{\mathbb{H}^d}^2\,d\mu=\sum_{n=0}^{m}\int_\Lambda\|\varphi_n\|_{\mathbb{H}^d}^2\,d\mu.$$

It follows that $\sum_{n=0}^{\infty}\mathbb{E}\,\|\varphi_n\|_{\mathbb{H}^d}^2\leq s<\infty$, thus $\sum_{n=0}^{\infty}\varphi_n$ exists in $L^2(\mu,\mathbb{H}^{\otimes d})$.

Now we will show that $\lim_{k\to\infty}\int_\Lambda\|\sum_{n=0}^{\infty}(\varphi_n^k-\varphi_n)\|_{\mathbb{H}^d}^2\,d\mu=0$. We have, for all $k\in\mathbb{N}$, $n,m\in\mathbb{N}_0, m\neq n$,

$$\int_\Lambda\langle\varphi_n^k-\varphi_n,\varphi_m^k-\varphi_m\rangle_{\mathbb{H}^d}\,d\mu=\lim_{l\to\infty}\int_\Lambda\langle\varphi_n^k-\varphi_n^l,\varphi_m^k-\varphi_m^l\rangle_{\mathbb{H}^d}\,d\mu=0.$$

Therefore, it suffices to show that $\lim_{k\to\infty}\sum_{n=0}^{\infty}\int_\Lambda\|\varphi_n^k-\varphi_n\|_{\mathbb{H}^d}^2\,d\mu=0$. Fix $\varepsilon>0$. Then there exists a $k_0\in\mathbb{N}$ such that for all $k,l\geq k_0$

$$\begin{aligned}\varepsilon>\int_\Lambda\left\|\sum_{n=0}^{\infty}\varphi_n^k-\sum_{n=0}^{\infty}\varphi_n^l\right\|_{\mathbb{H}^d}^2 d\mu&=\int_\Lambda\left\|\sum_{n=0}^{\infty}(\varphi_n^k-\varphi_n^l)\right\|_{\mathbb{H}^d}^2 d\mu\\&=\sum_{n=0}^{\infty}\int_\Lambda\|\varphi_n^k-\varphi_n^l\|_{\mathbb{H}^d}^2\,d\mu.\end{aligned}$$

We obtain for all $m\in\mathbb{N}_0$ and all $k\geq k_0$

$$\begin{aligned}\varepsilon\geq\lim_{l\to\infty}\sum_{n=0}^{m}\int_\Lambda\|\varphi_n^k-\varphi_n^l\|_{\mathbb{H}^d}^2\,d\mu&=\sum_{n=0}^{m}\lim_{l\to\infty}\int_\Lambda\|\varphi_n^k-\varphi_n^l\|_{\mathbb{H}^d}^2\,d\mu\\&=\sum_{n=0}^{m}\int_\Lambda\|\varphi_n^k-\varphi_n\|_{\mathbb{H}^d}^2\,d\mu.\end{aligned}$$

Therefore, for all $k\geq k_0$, $\sum_{n=0}^{\infty}\int_\Lambda\|\varphi_n^k-\varphi_n\|_{\mathbb{H}^d}^2\,d\mu\leq\varepsilon$. □

Exercises

5.1 Prove Proposition 5.1.1. (Use the notation in Proposition 5.1.2.)

5.2 Prove that $\{B \in \mathcal{C} \mid \exists A \in \mathcal{S}(A \,\Delta\, B \in \mathcal{N}_\mu)\}$ is a σ-algebra.

5.3 Assume that $\mu\{f \neq g\} = 0$. If g is $\mathcal{S}$-measurable, then f is $\mathcal{S} \vee \mathcal{N}_\mu$-measurable.

5.4 Prove Corollary 5.4.2.

5.5 Prove the assertion in Corollary 5.6.4.

6

†Malliavin calculus on real sequences

Although the techniques in this chapter can be used to establish the general Malliavin calculus later on, it is possible to jump to the next chapter. Only the techniques in this chapter, not the results, are used later.

Following [88], we deal with calculus for discrete Lévy processes. In an application we obtain Malliavin calculus for Poisson processes and for Brownian motion with values in abstract Wiener spaces over 'little' l^2. To obtain similar results for Lévy processes defined on the continuous timeline $[0, \infty[$, and for Brownian motion with values in abstract Wiener spaces over any separable Hilbert space, the space $\mathbb{R}^{\mathbb{N}}$ is replaced by an extension ${}^*\left(\mathbb{R}^{\mathbb{N}}\right)$ of $\mathbb{R}^{\mathbb{N}}$ and $\mathbb{N}$ is replaced by $[0, \infty[$. We will identify two separable Hilbert spaces only if there exists a canonical, i.e., basis independent, isomorphic isometry between them.

The seminal paper of Malliavin [75] was designed to study smoothness of solutions to stochastic differential equations. Here the Itô integral and Malliavin derivative are used to obtain the Clark–Ocone formula. This formula plays an important role in mathematics of finance (cf. Aase *et al.* [1] and Di Nunno *et al.* [31]).

Smolyanov and von Weizsäcker [108] use differentiability to study measures on $\mathbb{R}^{\mathbb{N}}$. They admit products of different measures. In contrast to their work, our approach is based on chaos decomposition, and measures are included which are not necessarily smooth. However, each measure has to be the product of a single fixed Borel measure on $\mathbb{R}$.

6.1 Orthogonal polynomials

In this section we fix a Borel probability measure μ^1 on $\mathbb{R}$. Let μ denote the product of μ^1 on $\mathcal{B}$, the product σ-algebra on $\mathbb{R}^{\mathbb{N}}$. We assume that $\mathbb{E}_{\mu^1} x^n < \infty$

for all $n \in \mathbb{N}$ and construct from $1, x, x^2, \ldots$ orthogonal polynomials $p_0, p_1, p_2, \ldots$ in $L^2(\mu^1)$, using the Gram–Schmidt orthonormalization procedure.

Using Proposition 6.1.1 below, it is possible to study probability measures μ^1 for which polynomials are not necessarily integrable, for example, the Cauchy distribution, i.e., the measure μ^1 with density $\frac{1}{\pi}\frac{1}{x^2+1}$. Then one can apply the orthonormalization procedure to $\mathfrak{E}^0, \mathfrak{E}^1, \mathfrak{E}^2, \ldots$ instead of $1, x, x^2, \ldots$.

Proposition 6.1.1 ([88]) *There exists a constant c such that for each Borel probability measure μ^1 on $\mathbb{R}$ there is a Borel measurable bijection $\mathfrak{E}$ on $\mathbb{R}$ such that $\mathfrak{E}^{-1}$ is Borel measurable and $\mathbb{E}_{\mu^1} e^{|\mathfrak{E}|} \leq c$.*

Moreover, non-smooth measures are admitted, for example the binomial distribution $\mu^1(A) := \sum_{i \in A \cap \{0,\ldots,n\}} \binom{n}{i} p^i (1-p)^{n-i}$ or the geometric distribution $\mu^1(A) := \sum_{i \in A \cap \mathbb{N}} p(1-p)^{i-1}, 0 < p < 1$.

Set $p_0 := 1$ and call 0 an **uncritical exponent**. Assume that p_{n-1} is already defined and assume that $u_0 < \ldots < u_k$ are the uncritical exponents smaller than n. Define

$$p_n(x) := x^n - \sum_{i=0}^{k} \frac{\mathbb{E}_{\mu^1}\left(x^n \cdot p_{u_i}\right)}{\mathbb{E}_{\mu^1} p_{u_i}^2} p_{u_i}(x) \in L^2\left(\mu^1\right).$$

If $\mathbb{E}_{\mu^1} p_n^2 = 0$, then n is called **critical**; otherwise n is called **uncritical**. Let $\mathbb{N}_\mu$ be the set of uncritical exponents in $\mathbb{N}$. Note that $0 \notin \mathbb{N}_\mu$. If $i \in \mathbb{N}_\mu$, then we may assume that $\mathbb{E}_{\mu^1} p_i^2 = 1$ by normalization.

Proposition 6.1.2

(a) $(p_n)_{n \in \mathbb{N}_0}$ *is an orthonormal sequence in* $L^2(\mu^1)$.
(b) $(p_i)_{i \in \mathbb{N}_\mu \cup \{0\}}$ *is an orthonormal basis of* $L^2(\mu^1)$.

Proof Part (a) is obvious. To prove part (b), assume that $\varphi \in L^2(\mu^1)$ and $\varphi \perp p_i$ for all $i \in \mathbb{N}_\mu \cup \{0\}$. Note that each polynomial p belongs to the linear hull of $\{p_i \mid i \in \mathbb{N}_\mu \cup \{0\}\}$ in $L^2(\mu^1)$. It follows that $\varphi \perp \alpha \cdot x^n$ for all $n \in \mathbb{N}_0$, $\alpha \in \mathbb{R}$. We have to prove that $\varphi = 0$ in $L^2(\mu^1)$. Let μ^+, μ^- be the finite Borel measures on $\mathbb{R}$ with densities $\varphi^+ := \varphi \vee 0$ and $\varphi^- := (-\varphi) \vee 0$, respectively, with respect to μ^1. To prove that $\mu^+ = \mu^-$, fix $\lambda \in \mathbb{R}$. Then

$$\int_{\mathbb{R}} e^{i \cdot \lambda \cdot x} d\mu^+ = \sum_{n=0}^{\infty} \mathbb{E}_{\mu^1} \frac{i^n \lambda^n x^n}{n!} \varphi^+ = \sum_{n=0}^{\infty} \mathbb{E}_{\mu^1} \frac{i^n \lambda^n x^n}{n!} \varphi^- = \int_{\mathbb{R}} e^{i \cdot \lambda \cdot x} d\mu^-.$$

By Lemma 3.1.1, $\mu^+ = \mu^-$, thus $\varphi = 0$ in $L^2(\mu^1)$. □

6.2 Integration

In this section we assume that $\mathbb{N}_\mu \neq \emptyset$. It follows that $1 \in \mathbb{N}_\mu$ and $\mu^1\{a\} \neq 1$ for all $a \in \mathbb{R}$. Let $(\mathcal{B}_n)_{n\in\mathbb{N}}$ be the **natural filtration** on $\mathbb{R}^\mathbb{N}$, i.e.,

$$\mathcal{B}_n := \left\{ A \times \mathbb{R}^{\{k\in\mathbb{N} \mid n<k\}} \mid A \text{ is a Borel set in } \mathbb{R}^n \right\},$$

augmented by the set $\mathcal{N}_\mu$ of all μ-nullsets. Often we use the filtration $(\mathcal{B}_{n^-})_{n\in\mathbb{N}}$ with $n^- := n-1$ and $\mathcal{B}_0 := \{\mathbb{R}^\mathbb{N}, \emptyset\} \vee \mathcal{N}_\mu$.

In order to establish the integral, we introduce the following L^2-spaces. Set

$$L^2(\mu \otimes c) := \left\{ f : \mathbb{R}^\mathbb{N} \times \mathbb{N} \to \mathbb{R} \mid \sum_{n=1}^{\infty} \mathbb{E}_\mu \left| f(\cdot, n) \right|^2 < \infty \right\},$$

$$L^2(c \otimes \mu \otimes c) := \left\{ f : \mathbb{N}_\mu \times \mathbb{R}^\mathbb{N} \times \mathbb{N} \to \mathbb{R} \mid \sum_{k\in\mathbb{N}_\mu, n\in\mathbb{N}} \mathbb{E}_\mu \left| f_k(\cdot, n) \right|^2 < \infty \right\}.$$

Note that c is the counting measure on $\mathbb{N}$ and on $\mathbb{N}_\mu$. The spaces $L^2(\mu \otimes c)$ and $L^2(c \otimes \mu \otimes c)$ are Hilbert spaces. Before using these spaces, let us establish the following simple but important lemma.

Lemma 6.2.1 *Fix* $k, l \in \mathbb{N}_\mu$. *Then*

(a) $\mathbb{E}^{\mathcal{B}_{m^-}} p_k(X_n) p_l(X_m) = 0$ *if* $n < m$,
(b) $\mathbb{E}^{\mathcal{B}_{m^-}} p_k(X_m) p_l(X_m) = 0$ *if* $k \neq l$,
(c) $\mathbb{E}^{\mathcal{B}_{m^-}} p_k^2(X_m) = 1$.

Fix $k \in \mathbb{N}_\mu$. For each $(\mathcal{B}_{n^-})_{n\in\mathbb{N}}$-adapted $\varphi \in L^2(\mu \otimes c)$ we define the ***k*th integral of** φ, setting

$$\int^V \varphi p_k : \mathbb{R}^\mathbb{N} \to \mathbb{R}, \quad X \mapsto \sum_{n\in\mathbb{N}} \varphi(X, n) \cdot p_k(X_n).$$

Thus, we integrate with respect to the martingale $X \mapsto \left(\sum_{i=1}^n p_k(X_i) \right)_{n\in\mathbb{N}}$. Note that the increments of this discrete martingale are orthogonal polynomials. We shall see later on that it is possible to orthonormalize the increments of much more general Lévy processes, which is the key to Malliavin calculus in our setting.

Let $\varphi \in L^2(c \otimes \mu \otimes c)$ and assume that φ_k is $(\mathcal{B}_{n^-})_{n\in\mathbb{N}}$-adapted for all $k \in \mathbb{N}_\mu$. The **integral** $\int^V \varphi p : \mathbb{R}^\mathbb{N} \to \mathbb{R}$ is defined by setting

$$\int^V \varphi p := \sum_{k\in\mathbb{N}_\mu} \int^V \varphi_k p_k.$$

The following result shows, in particular, that this integral is well defined.

Proposition 6.2.2 *Fix $k, m \in \mathbb{N}_\mu$.*

(a) *Fix $(\mathcal{B}_{n^-})_{n\in\mathbb{N}}$-adapted $\varphi, \psi \in L^2(\mu\otimes c)$. Then*

$$\mathbb{E}_\mu\left(\int^V \varphi p_k \cdot \int^V \psi p_m\right) = \begin{cases} \sum_{n\in\mathbb{N}} \mathbb{E}_\mu(\varphi(\cdot,n)\cdot\psi(\cdot,n)) & \text{if } m=k \\ 0 & \text{if } m\neq k. \end{cases}$$

(b) *Fix $\varphi, \psi \in L^2(c\otimes\mu\otimes c)$ and assume that all the φ_k, ψ_k are $(\mathcal{B}_{n^-})_{n\in\mathbb{N}}$-adapted. Then*

$$\mathbb{E}_\mu\left(\int^V \varphi p \cdot \int^V \psi p\right) = \sum_{k\in\mathbb{N}_\mu, n\in\mathbb{N}} \mathbb{E}_\mu(\varphi_k(\cdot,n)\cdot\psi_k(\cdot,n)).$$

From Proposition 6.2.2 it follows that all the kth Itô integrals are continuous and therefore the integral is also continuous.

Theorem 6.2.3 *Fix $k \in \mathbb{N}_\mu$.*

(a) *If $(\varphi^m)_{m\in\mathbb{N}}$ is a sequence of $(\mathcal{B}_{n^-})_{n\in\mathbb{N}}$-adapted $\varphi^m \in L^2(\mu\otimes c)$, converging to φ in $L^2(\mu\otimes c)$, then φ is $(\mathcal{B}_{n^-})_{n\in\mathbb{N}}$-adapted and $\int^V \varphi^m p_k$ converges to $\int^V \varphi p_k$ in $L^2(\mu)$.*

(b) *Let $(\varphi^m)_{m\in\mathbb{N}}$ be a sequence in $L^2(c\otimes\mu\otimes c)$, where each φ^m_k is $(\mathcal{B}_{n^-})_{n\in\mathbb{N}}$-adapted. Suppose that (φ^m) converges to φ in $L^2(c\otimes\mu\otimes c)$, i.e., $\lim_{m\to\infty}\sum_{k\in\mathbb{N}_\mu, n\in\mathbb{N}} \mathbb{E}_\mu\left(\varphi^m_k - \varphi_k\right)^2(\cdot,n) = 0$. Then each φ_k is $(\mathcal{B}_{n^-})_{n\in\mathbb{N}}$-adapted and $\int^V \varphi^m p$ converges to $\int^V \varphi p$ in $L^2(\mu)$.*

6.3 Iterated integrals

We use the following notation: Set

$$\mathbb{N}^n_< := \{a\in\mathbb{N}^n \mid a_1 < \ldots < a_n\}, \quad \mathbb{N}^n_{\neq} := \{a\in\mathbb{N}^n \mid a_i \neq a_j \text{ for } i\neq j\}.$$

The iterated integral is defined for elements in the Hilbert space $l^2_{n,<}$ of functions $f:\mathbb{N}^n_< \to \mathbb{R}$, $n\in\mathbb{N}_0$, such that $\sum_{i\in\mathbb{N}^n_<} f^2(i) < \infty$. If $n=0$, then $f\in\mathbb{R}$. Fix $f\in l^2_{n,<}$. For each $k=(k_1,\ldots,k_n)\in\mathbb{N}^n_\mu$ we define the **iterated integral** $I_k(f):\mathbb{R}^\mathbb{N}\to\mathbb{R}$ **of f of order** k by

$$I_k(f): X \mapsto \sum_{t\in\mathbb{N}^n_<} f(t)\cdot p_{k_1}(X_{t_1})\cdot\ldots\cdot p_{k_n}(X_{t_n}).$$

The function f is called the **kernel** of $I_k(f)$. For $f \in \mathbb{R}$, $I_\emptyset(f) = f$. Instead of $I_{(1)}(f)$ we write $I(f)$; $I(f)$ is called the **μ-integral** of f. Because of the orthonormality of the p_k, we obtain:

Proposition 6.3.1 *Fix $f \in l^2_{n,<}, g \in l^2_{m,<}$. Then*

$$\mathbb{E}_\mu \left(I_{(\cdot)}(f) \cdot I_{(\circ)}(g) \right) = \begin{cases} \sum_{t \in \mathbb{N}^n_<} (f \cdot g)(t) & \text{if } (\cdot) = (\circ) \\ 0 & \text{if } (\cdot) \neq (\circ). \end{cases}$$

The continuity of the iterated Itô integral follows immediately:

Corollary 6.3.2 *If $(f_m)_{m \in \mathbb{N}}$ converges to f in $l^2_{n,<}$, then $\left(I_{(\circ)}(f_m)\right)_{m \in \mathbb{N}}$ converges to $I_{(\circ)}(f)$ in $L^2(\mu)$.*

6.4 Chaos decomposition

Our aim now is to prove a chaos decomposition theorem for all functionals in $L^2(\mu)$, using Fourier transformations of measures. It shows that functionals in $L^2(\mu)$ are uniquely determined by deterministic functions. This result will be used to establish Malliavin calculus. The key is the following simple but important lemma. The iterated integrals were defined in order to obtain this lemma. The proof uses induction over the degree of polynomials and is left to the reader.

Lemma 6.4.1 *Let $f : \mathbb{N} \to \mathbb{R}$ be a function with finite support. Then, for each polynomial $Q : \mathbb{R} \to \mathbb{R}$, $Q(I(f))$ is a finite linear combination of iterated integrals with kernels having finite support and which are tensor products of 1-ary functions.*

Proposition 6.4.2 *Fix $\varphi \in L^2(\mu)$.*

(a) *Then for each $n \in \mathbb{N}_0$ there exists a sequence $(f_k)_{k \in \mathbb{N}^n_\mu}$ of functions $f_k \in l^2_{n,<}$ such that*

$$\varphi = \sum_{n=0}^{\infty} \sum_{k \in \mathbb{N}^n_\mu} I_k(f_k).$$

Note that $\mathbb{E}\varphi = I_\emptyset(f_\emptyset) = f_\emptyset$.

(b) *If $\varphi = \sum_{n=0}^{\infty} \sum_{k \in \mathbb{N}^n_\mu} I_k(g_k)$, where $g_k \in l^2_{n,<}$, then for all $n \in \mathbb{N}_0$,*

$$\sum_{t \in \mathbb{N}^n_<} (g_k - f_k)^2(t) = 0.$$

Proof (a) Let M be the set of all $\varphi \in L^2(\mu)$ having a decomposition $\varphi = \sum_{n=0}^{\infty} \sum_{k \in \mathbb{N}_\mu^n} I_k(f_k)$ with $f_k \in l^2_{n,<}$. Then M is a closed linear subspace of $L^2(\mu)$. In order to prove that $M = L^2(\mu)$, we have to prove that, if $\varphi \in L^2(\mu)$ and $\varphi \perp M$, then $\varphi = 0$ in $L^2(\mu)$. Assume that $\varphi \perp M$. Then $\varphi \perp I_k(f)$ for all $n \in \mathbb{N}_0$, $k \in \mathbb{N}_\mu^n$ and all $f \in l^2_{n,<}$. By Lemma 6.4.1, $\varphi \perp Q(I(f))$ for all functions $f : \mathbb{N} \to \mathbb{R}$ with finite support and all polynomials Q. Let φ^+, φ^- be the positive, respectively negative, parts of φ. Then $\mathbb{E}_\mu(Q(I(f)) \cdot \varphi^+) = \mathbb{E}_\mu(Q(I(f)) \cdot \varphi^-)$. Let $\mu^{+(-)}$ be the measures with density $\varphi^{+(-)}$. To prove that $\mu^+ = \mu^-$, fix $\lambda \in \mathbb{R}$ and $f : \mathbb{N} \to \mathbb{R}$ with finite support. Then,

$$\begin{aligned}\int_{\mathbb{R}^{\mathbb{N}}} e^{i \cdot \lambda I(f)} d\mu^+ &= \sum_{n=0}^{\infty} \mathbb{E}_\mu \frac{i^n (\lambda I(f))^n}{n!} \varphi^+ \\ &= \sum_{n=0}^{\infty} \mathbb{E}_\mu \frac{i^n (\lambda I(f))^n}{n!} \varphi^- = \int_{\mathbb{R}^{\mathbb{N}}} e^{i \cdot \lambda I(f)} d\mu^- .\end{aligned}$$

Since $\{I(f) \mid f$ has finite support$\}$ is a linear space, and $\mu^+_{I(f)} = \mu^-_{I(f)}$ for all $I(f)$ in this set, $\varphi = 0$ in $L^2(\mu)$.

(b) The uniqueness follows from Proposition 6.3.1. □

In order to obtain the Malliavin derivative and the Skorokhod integral, we use a slight modification of Proposition 6.4.2: fix $n \in \mathbb{N}$ and $f \in l^2_{n,n,\neq}$, i.e., $f : \mathbb{N}_\mu^n \times \mathbb{N}_{\neq}^n \to \mathbb{R}$ and $\sum_{t \in \mathbb{N}_\mu^n \times \mathbb{N}_{\neq}^n} f^2(t) < \infty$. The random variable

$$I_n(f) := \sum_{k \in \mathbb{N}_\mu^n} I_k(f_k)$$

is called the ***n*-fold multiple integral of *f***. The function f is called the **kernel** of $I_n(f)$. If $n = 0$, then $l^2_{n,n,\neq} = \mathbb{R}$. We call f **symmetric** if

$$f_{(k_{\sigma_1}, \ldots, k_{\sigma_n})}(t_{\sigma_1}, \ldots, t_{\sigma_n}) = f_{(k_1, \ldots, k_n)}(t_1, \ldots, t_n)$$

for all permutations σ of $\{1, \ldots, n\}$. Now we obtain the following result from Proposition 6.4.2.

Theorem 6.4.3 *Fix* $\varphi \in L^2(\mu)$.

(a) *There exists a sequence* $(f_n)_{n \in \mathbb{N}_0}$ *of symmetric functions* $f_n \in l^2_{n,n,\neq}$ *with* $\varphi = \sum_{n=0}^{\infty} I_n(f_n)$.

(b) *If* $g_n \in l^2_{n,n,\neq}$ *is symmetric for all* $n \in \mathbb{N}$ *and* $\varphi = \sum_{n=0}^{\infty} I_n(g_n)$, *then* $\sum_{k \in \mathbb{N}_\mu^n, t \in \mathbb{N}_{\neq}^n} (g_{n,k}(t) - f_{n,k}(t))^2 = 0$.

(c) $f_{n,k}(t) = \mathbb{E}\left(\varphi \cdot p_{k_1}(X_{t_1}) \cdot \ldots \cdot p_{k_n}(X_{t_n})\right)$

Proof (a) Let $\varphi = \sum_{n=0}^{\infty} \sum_{k \in \mathbb{N}_\mu^n} I_k(f_k)$ be the decomposition according to Proposition 6.4.2. For all $n \in \mathbb{N}$, $k \in \mathbb{N}_\mu^n$ and all $t \in \mathbb{N}_{\neq}^n$ define

$$f_n : \mathbb{N}_\mu^n \times \mathbb{N}_{\neq}^n \to \mathbb{R} : (k,t) \mapsto f_{n,k}(t) := f_{(k_{\sigma_1},\ldots,k_{\sigma_n})}(t_{\sigma_1},\ldots,t_{\sigma_n}),$$

where σ is the permutation of $\{1,\ldots,n\}$ with $t_{\sigma_1} < \ldots < t_{\sigma_n}$. Note that f_n is symmetric and $f_n \in l^2_{n,n,\neq}$. By Proposition 6.4.2, $\varphi = \sum_{n=0}^{\infty} I_n(f_n)$.

The proofs of (b) and (c) are left to the reader. □

Part (c) of Theorem 6.4.3 presents a method for the computations of the kernels of the chaos decomposition. From Theorem 6.4.3 we obtain a decomposition result for stochastic processes:

Corollary 6.4.4 *Fix a sequence $\varphi \in L^2(c \otimes \mu \otimes c)$. Then $\varphi_k(\cdot,t)$ has the expansion*

$$\varphi_k(\cdot,t) = \sum_{n=0}^{\infty} I_n(f_n(\cdot,k,\cdot,t)),$$

where $f_n : \mathbb{N}_\mu^{n+1} \times \left(\mathbb{N}_{\neq}^n \times \mathbb{N}\right) \to \mathbb{R} \in l^2_{n+1,n+1}$ and f_n is symmetric in the first n variables.

6.5 Malliavin derivative and Skorokhod integral

Fix $\varphi \in L^2(\mu)$ with decomposition $\varphi = \sum_{n=0}^{\infty} I_n(f_n)$, according to Theorem 6.4.3. Then the **Malliavin derivative** $D\varphi$ belongs to $L^2(c \otimes \mu \otimes c)$, where for all $k \in \mathbb{N}_\mu, t \in \mathbb{N}$

$$D\varphi(k,\cdot,t) = \sum_{n=1}^{\infty} I_{n-1}(f_n(\cdot,k,\cdot,t))$$

if $D\varphi$ converges in $L^2(c \otimes \mu \otimes c)$. The Malliavin derivative D is defined on a dense subspace of $L^2(\mu)$, because $D\varphi$ exists if and only if $\sum_{n=1}^{\infty} \sqrt{n} I_n(f_n)$ converges in $L^2(\mu)$ as in the situation of the classical Wiener space. In this case φ is called **Malliavin differentiable.**

In order to define the Skorokhod integral, we use the following common notation. For each $g : \mathbb{N}_\mu^{n+1} \times \mathbb{N}^{n+1} \to \mathbb{R}$ we define $\widetilde{g} : \mathbb{N}_\mu^{n+1} \times \mathbb{N}_{\neq}^{n+1} \to \mathbb{R}$, by setting

$$\widetilde{g}_{(k_1,\ldots,k_{n+1})}(t_1,\ldots,t_{n+1})$$
$$:= \sum_{i=1}^{n+1} g_{(k_1,\ldots,k_{i-1},k_{i+1},\ldots,k_{n+1},k_i)}(t_1,\ldots,t_{i-1},t_{i+1},\ldots,t_{n+1},t_i).$$

Note that, if g is symmetric in the first n components, then $\widetilde{g}$ is symmetric. If $g \in l^2_{n+1,n+1}$, so is $\widetilde{g}$. Let $\varphi \in L^2(c \otimes \mu \otimes c)$ with decomposition

$$\varphi(k,\cdot,t) = \sum_{n=0}^{\infty} I_n(f_n(\cdot,k,\cdot,t)),$$

according to Corollary 6.4.4. The **Skorokhod integral** $\delta\varphi : \mathbb{R}^{\mathbb{N}} \to \mathbb{R}$ is defined by

$$\delta\varphi := I_{n+1}(\widetilde{f_n})$$

if this series converges in $L^2(\mu)$, in which case φ is called **Skorokhod integrable**.

It is straightforward to see that δ is the adjoint operator of D. In fact we obtain:

Theorem 6.5.1 *Let φ be Skorokhod integrable and ψ be Malliavin differentiable. Then*

$$\langle \delta\varphi, \psi \rangle_{L^2(\mu)} = \langle \varphi, D\psi \rangle_{L^2(c\otimes\mu\otimes c)},$$

i.e., $\mathbb{E}_\mu(\delta\varphi \cdot \psi) = \sum_{k\in\mathbb{N}_\mu, t\in\mathbb{N}} \mathbb{E}_\mu\left(\varphi_k(\cdot,t) \cdot (D\psi)_k(\cdot,t)\right).$

6.6 The integral as a special case of the Skorokhod integral

In this section, we will show that the integral introduced in this chapter is a special case of the Skorokhod integral. Let us call a function $f : \mathbb{N}^{n+1} \to \mathbb{R}$ **non-anticipating** if $f(t_1,\ldots,t_{n+1}) \neq 0$ implies $t_1,\ldots,t_n < t_{n+1}$.

Fix $\varphi \in L^2(c \otimes \mu \otimes c)$ such that each φ_k is $(\mathcal{B}_{t^-})_{t\in\mathbb{N}}$-adapted. Choose the decomposition of $\varphi_k(\cdot,t)$, according to Corollary 6.4.4. Then

$$\varphi_k(\cdot,t) = \mathbb{E}^{\mathcal{B}_{t^-}} \varphi_k(\cdot,t) = \sum_{n=0}^{\infty} \mathbb{E}^{\mathcal{B}_{t^-}} I_n(f_n(\cdot,k,\cdot,t)).$$

Note that

$$\mathbb{E}^{\mathcal{B}_{t^-}} I_n(f_n(\cdot,k,\cdot,t)) = \sum_{(k_1,\ldots,k_n)\in\mathbb{N}_\mu^n} \sum_{t_1<\ldots<t_n<t} f_{n,(k_1,\ldots,k_n,k)}(t_1,\ldots,t_n,t) \cdot p_{k_1}(X_{t_1})\cdot\ldots\cdot p_{k_n}(X_{t_n}).$$

Therefore, we may assume that each $f_{n,(k_1,\ldots,k_n,k)}$ is non-anticipating. We obtain:

$$\begin{aligned}\mathbb{E}_\mu\left(\sum_{n=0}^{\infty} I_{n+1}(\widetilde{f}_n)\right)^2 &= \mathbb{E}_\mu\left(\sum_{n=0}^{\infty} I_{n+1}(f_n)\right)^2 \\ &= \sum_{n\in\mathbb{N}_0} \sum_{(k_1,\ldots,k_{n+1})\in\mathbb{N}_\mu^{n+1}} \sum_{t\in\mathbb{N}_<^{n+1}} f^2_{n,(k_1,\ldots,k_{n+1})}(t) \\ &\le \sum_{k\in\mathbb{N}_\mu, t\in\mathbb{N}} \mathbb{E}_\mu \varphi_k^2(\cdot,t) < \infty.\end{aligned}$$

It follows that φ is Skorokhod integrable. Moreover,

$$\sum_{k\in\mathbb{N}_\mu} \int^V \varphi_k p_k = \delta\varphi.$$

This proves:

Theorem 6.6.1 *If $\varphi \in L^2(c\otimes\mu\otimes c)$ and each φ_k is $(\mathcal{B}_{t^-})_{t\in\mathbb{N}}$-adapted, then φ is Skorokhod integrable and the Skorokhod integral of φ is identical to the integral of φ.*

6.7 The Clark–Ocone formula

This formula tells us that each $\varphi \in L^2(\mu)$ is the sum of a constant and the integral of a sequence of adapted processes. It plays an important role in mathematics of finance (see for example [1]).

Theorem 6.7.1 *Fix $\varphi \in L^2(\mu)$ with decomposition*

$$\varphi = \mathbb{E}\varphi + \sum_{n=1}^{\infty} I_n(f_n),$$

according to Theorem 6.4.3. Then

$$\varphi = \mathbb{E}\varphi + \int^V (k,X,t) \mapsto \left(\sum_{n\in\mathbb{N}} \mathbb{E}^{\mathcal{B}_{t-}} I_{n-1}(f_n(\cdot,k,\cdot,t)(X)) \right) p.$$

If φ is Malliavin differentiable, then this equality has the form

$$\varphi = \mathbb{E}\varphi + \int^V t \mapsto \left(\mathbb{E}^{\mathcal{B}_{t-}} D\varphi_k(\cdot,t)\right) p.$$

Proof From the chaos decomposition of φ it follows that φ has the form $\varphi - \mathbb{E}\varphi = \sum_{n=1}^{\infty} I_n(f_n)$, where

$$\begin{aligned} I_n(f_n) &= \sum_{k\in\mathbb{N}_\mu, t\in\mathbb{N}} \sum_{l\in\mathbb{N}_\mu^{n-1}} \sum_{s\in\mathbb{N}_<^{n-1}, s<t} f_n(l,k,s,t) \prod_{i=1}^{n-1} p_{l_i}(X_{s_i}) \cdot p_k(X_t) \\ &= \sum_{k\in\mathbb{N}_\mu, t\in\mathbb{N}} \left(\mathbb{E}^{\mathcal{B}_{t-}} I_{n-1}(f_n(\cdot,k,\cdot,t))\right) p_k(X_t) \\ &= \int^V \left((k,\cdot,t) \mapsto \left(\mathbb{E}^{\mathcal{B}_{t-}} I_{n-1}(f_n(\cdot,k,\cdot,t))\right)\right) p. \end{aligned}$$

Now we apply Theorem 6.2.3 part (b) and obtain

$$\sum_{n=1}^{\infty} I_n(f_n) = \int^V \left((k,\cdot,t) \mapsto \mathbb{E}^{\mathcal{B}_{t-}} \sum_{n=1}^{\infty} I_{n-1}(f_n(\cdot,k,\cdot,t)) \right) p,$$

which in the case when φ is Malliavin differentiable is equal to

$$\int^V \left((k,\cdot,t) \mapsto \mathbb{E}^{\mathcal{B}_{t-}} D\varphi_k(\cdot,t)\right) p.$$

□

Note that $(k,\cdot,t) \mapsto \sum_{n\in\mathbb{N}} \mathbb{E}^{\mathcal{B}_{t-}} I_{n-1}(f_n(\cdot,k,\cdot,t)) \in L^2(c\otimes\mu\otimes c)$, although $(k,\cdot,t) \mapsto \sum_{n\in\mathbb{N}} I_{n-1}(f_n(\cdot,k,\cdot,t)) \notin L^2(c\otimes\mu\otimes c)$ if φ is not Malliavin differentiable. Therefore, one can say that the conditional expectation heals the loss of convergence.

6.8 Examples

Example I: The classical Poisson space

Let π_β^1 be the exponential distribution on $\mathbb{R}$ with rate $\beta > 0$, i.e., for all Borel sets B in $\mathbb{R}$, $\pi_\beta^1(B) = \int_{B\cap[0,\infty[} \beta e^{-\beta x} dx$. We may apply the preceding results to the infinite product π_β of π^1_β on $\mathbb{R}^{\mathbb{N}}$. Since the sequence $(X_i)_{i\in\mathbb{N}}$ of projections $X \mapsto X_i$ is a family of independent identically distributed exponential random variables each having mean $\frac{1}{\beta}$, the process

$$N:\left(\mathbb{R}_0^+\right)^{\mathbb{N}}\times\mathbb{R}_0^+\to\mathbb{N}_0,\ (X,t)\mapsto\max\left\{n\in\mathbb{N}_0\mid\sum_{i=1}^{n}X_i\leq t\right\}$$

is π_β-a.s. well defined and a Poisson process with parameter β. Note that $N\in D$ π_β-a.s., where

$$D:=\{f:[0,\infty[\to\mathbb{N}_0\mid f \text{ is monotone increasing and càdlàg}\}.$$

The symbol D has also been used to denote the Malliavin derivative, but this should not cause any confusion.

Proposition 6.8.1 *For π_β-almost all X,*

(a) $N_t(X)-N_{t^-}(X)\leq 1$ *for all* $t\in\mathbb{R}_0^+$,
(b) $N(X)$ *has infinite range,*
(c) *for each* $f\in D$ *there exists an* $X\in\left(\mathbb{R}_0^+\right)^{\mathbb{N}}$ *such that* $N(X)=f$.

Proof (a) Since $\left\{\exists t\in\mathbb{R}_0^+\,(N_t-N_{t^-})\geq 2\right\}\subseteq\left\{\exists k\forall m\in\mathbb{N}\left(X_k\leq\frac{1}{m}\right)\right\}$ and $\pi_\beta\left\{\forall m\in\mathbb{N}\ \left(X_k\leq\frac{1}{m}\right)\right\}=\lim_{m\to\infty}\int_0^{\frac{1}{m}}1d\pi_\beta=0$, (a) is true.

(b) Note that $\lim_{m\to\infty}\pi_\beta\left\{\sum_{i=1}^{k}X_i\leq m\right\}=1$ for all $k\in\mathbb{N}$. This implies that the range of N is infinite π_β-a.s.

(c) Fix $f\in D$. Let us assume that the range $k_0<\ldots<k_n<\ldots$ of f is infinite. Since f is monotone increasing and right continuous, $f^{-1}[\{k_i\}]$ is an interval such that $m_i:=\min f^{-1}[\{k_i\}]$ exists. Then $f=\sum_{i=1}^{\infty}k_{i-1}\mathbf{1}_{[m_{i-1},m_i[}$ and $m_0=0$. Now the sequence $(X_i)_{i\in\mathbb{N}}\in\left(\mathbb{R}_0^+\right)^{\mathbb{N}}$ is defined by

$$X_j:=\begin{cases}(m_i-m_{i-1}) & \text{if } j=k_{i-1}+1\\ 0 & \text{otherwise.}\end{cases}$$

Note that $N(X)=f$. If the range of f is finite, the proof is similar. □

It follows that the mapping $\kappa:\left(\mathbb{R}_0^+\right)^{\mathbb{N}}\to D,\ X\mapsto N(X)$ is surjective. Let $\mathfrak{P}$ be the σ-algebra on D generated by the **cylinder sets**

$$Z_{c,m}:=\{f\in D\mid f(c)=m\},\quad c\in\mathbb{R}_0^+,m\in\mathbb{N}_0.$$

Note that $\kappa^{-1}[B]\in\mathcal{B}$ for each $B\in\mathfrak{P}$. Let P_β be the image measure of π_β by κ on $\mathfrak{P}$. Then P_β is called the **Poisson measure** on $\mathfrak{P}$ and $(D,\mathfrak{P},P_\beta)$ is called the **classical Poisson space**.

Proposition 6.8.1 implies that the set of functions in D with jumps larger than 1 or with a finite range is a P_β-nullset.

Proposition 6.8.2 *Let $\mathcal{P}$, $\mathcal{P}_I$ be generated by N and by the π_β-integrals, respectively, augmented by the π_β-nullsets. Then*

$$\mathcal{B} \subseteq \mathcal{P} \subseteq \mathcal{P}_I \subseteq \mathcal{B}.$$

Proof The result follows from the facts that $\mathcal{B}$ is generated by the projections, and that, for all $c \in \mathbb{R}_0^+$ and all $n \in \mathbb{N}$, $\left\{\sum_{i \leq n} X_i \leq c\right\} = \{N(\cdot, c) \geq n\}$. □

It follows that the spaces $L^2(P_\beta)$ on D and $L^2(\pi_\beta)$ on $\left(\mathbb{R}_0^+\right)^{\mathbb{N}}$ can be identified: the mapping $\iota : L^2(P_\beta) \to L^2(\pi_\beta), \iota(\varphi)(X) := \varphi(N(X))$ defines a canonical (i.e., basis-independent) isometric isomorphism from $L^2(P_\beta)$ onto $L^2(\pi_\beta)$. Therefore all results on $L^2(\pi_\beta), L^2(\pi_\beta \otimes c)$ and $L^2(c \otimes \pi_\beta \otimes c)$, established in the previous sections, can be transferred to $L^2(P_\beta), L^2(P_\beta \otimes c)$ and $L^2(c \otimes P_\beta \otimes c)$ via the mapping $X \mapsto N(X)$.

Malliavin calculus for jump processes has been initiated in the work of Itô [50] and Bichteler, Gravereaux and Jacod [11]. The authors use integration with respect to the Poisson martingale $(N_t - \beta t)_{t \in \mathbb{R}_0^+}$. In this special case, we may use integration and multiple integration with respect to Laguerre polynomials p_k^β, the orthonormal polynomials with respect to π_β^1. In his work [96], Privault uses multiple stochastic integrals $I_n(f)$, $n \in \mathbb{N}$, where the increments of the integrators are also Laguerre polynomials. The kernels f belong to the Fock space over l^2 and $I_n(f)$ is an element of the L^2-space over the classical Wiener space, endowed with a suitable σ-algebra.

In their work [31], Di Nunno, Meyer-Brandis, Øksendal and Proske use Laguerre polynomials and orthogonal polynomials for Lévy measure to develop Malliavin calculus for Lévy processes.

Example II: Abstract Wiener–Fréchet spaces over l^2

Let γ^1 be the centred Gaussian distribution on $\mathbb{R}$ with variance 1. We may apply the results in the preceding sections to the product γ of γ^1 on $\mathbb{R}^{\mathbb{N}}$.

We have seen that the space $\left(\mathbb{R}^{\mathbb{N}}, \mathcal{B}, \gamma\right)$ is a perfect example of an abstract Wiener–Fréchet space over l^2. Following [82] and in more details [83], we shall see in Chapter 11 that there exist a rich probability space Ω with measure $\widehat{\Gamma}$, depending only on l^2, and a σ-isomorphism $\kappa_{\mathbb{B}}$ from Ω onto any abstract Wiener–Fréchet space $\mathbb{B}$ over l^2. Here we use a certain σ-algebra $\mathcal{W}_{l^2}$ on Ω. Let $\gamma_{\mathbb{B}}$ be the Gaussian measure on the Borel sets on $\mathbb{B}$ induced by the Gaussian measure γ^1 on the cylinder sets of l^2. In case $\mathbb{B} = \mathbb{R}^{\mathbb{N}}$ the Gaussian measure $\gamma_{\mathbb{B}}$ is the infinite product γ of γ^1.

It follows that the L^2-spaces $L^2_{W_{l^2}}(\widehat{\Gamma})$ and $L^2(\gamma_{\mathbb{B}})$ can be identified, because there exists a canonical (i.e., basis independent) isometric isomorphism ι from $L^2(\gamma_{\mathbb{B}})$ onto $L^2(\widehat{\Gamma})$, given by $\iota(\varphi)(X) := \varphi(\kappa_{\mathbb{B}}(X))$. Thus, since $L^2(\gamma_{\mathbb{B}})$ can therefore be identified with $L^2(\gamma)$, our results can be transferred from $L^2(\gamma), L^2(\gamma \otimes c), L^2(c \otimes \gamma \otimes c)$ to $L^2(\gamma_{\mathbb{B}}), L^2(\gamma_{\mathbb{B}} \otimes c), L^2(c \otimes \gamma_{\mathbb{B}} \otimes c)$ for any abstract Wiener–Fréchet space $\mathbb{B}$ over l^2.

It is interesting to study the relationship between the chaos decomposition of functionals φ in $L^2(\gamma)$, introduced here, and the chaos decomposition of the associated functionals $\varphi \circ \kappa_{\mathbb{R}^{\mathbb{N}}}$ in $L^2(\widehat{\Gamma})$ later. It may also be interesting to study the relationship between the kernels of the chaos decomposition of functionals φ in $L^2(P_\beta)$ and the kernels of their associated functionals $\varphi \circ N$ in $L^2(\pi_\beta)$.

Exercises

6.1 Prove the results in this chapter for which proofs are omitted.

The symmetrization operator $\widetilde{\cdot}$ on $L^2(c \otimes \mu \otimes c)$ is defined as follows: Fix φ in $L^2(c \otimes \mu \otimes c)$ with decomposition, according to Corollary 6.4.4, given by

$$\varphi_k(\cdot, t) = \sum_{n=0}^{\infty} I_n(f_n(\cdot, k, \cdot, t)).$$

We define for each $k \in \mathbb{N}_\mu, t \in \mathbb{N}$

$$\widetilde{\varphi}_k(\cdot, t) := \sum_{n=0}^{\infty} I_n(\widetilde{f}_n(\cdot, k, \cdot, t))$$

if this series converges in $L^2(\mu)$.

6.2 Prove the **fundamental theorem of calculus**: if φ is Skorokhod integrable and $\delta\varphi$ is Malliavin differentiable, then

$$\sum_{k \in \mathbb{N}_\mu, t \in \mathbb{N}} \mathbb{E}_\mu \widetilde{\varphi}_k(\cdot, t)^2 < \infty$$

and

$$D(\delta\varphi)_k(t) = \widetilde{\varphi}_k(\cdot, t).$$

7

Introduction to poly-saturated models of mathematics

As has already been mentioned, one aim of this book is to extend the results in Chapter 6 to more general Lévy processes, using similar techniques. In particular, we want to orthonormalize the increments of a Lévy process rather than the whole process itself. To this end we use an extension ${}^*\left(\mathbb{R}^{\mathbb{N}}\right)$ of $\mathbb{R}^{\mathbb{N}}$ in a poly-saturated model of mathematics. Instead of the timeline $\mathbb{N}$ we take the timeline $\frac{{}^*\mathbb{N}}{H}$, where H is an infinitely large positive integer. Then $\frac{{}^*\mathbb{N}}{H}$ is an infinitely fine partition of $[0,\infty[$ and both timelines $\frac{{}^*\mathbb{N}}{H}$ and $[0,\infty[$ can be identified in a certain sense. As a matter of fact, it will be made mathematically precise that time can be seen as discrete and continuous as well.

We have to extend the notion of finiteness. Let us write *finite to indicate that we mean 'finite' in the extended sense. Although *finite sets may be infinite, they have the same properties as the standard finite sets in a certain sense, as we shall see. We will also show that all objects studied so far take their values in *finite-dimensional Euclidean spaces. This is true for Bochner integrable functions with values in tensor products of infinite-dimensional Hilbert spaces. Moreover, a *finite-dimensional representation of the infinite-dimensional Brownian motion exists. It thus turns out that Malliavin calculus reduces to elementary analysis in *finite-dimensional Euclidean spaces. Moreover, we will represent finite-dimensional Lévy functionals and Brownian functionals in the infinite-dimensional case by smooth functions in the new sense.

To carry out this project, first of all we give a quite informal introduction to poly-saturated models of mathematics, avoiding technical details as far it is possible. In Appendix A the reader can find the full details. I trust that she or he does not need these details and, although not familiar with model theory, will be able to understand the mathematical methods provided by poly-saturated models. Mathematicians can work in these models in the usual way. The advantage however of these models is the fact that any set within such a model is at least countably compact.

7.1 Models of mathematics

Let us start with the notion of **standard model** of mathematics. What should we demand from such a model, let us call it $\mathfrak{V}$?

$\mathfrak{V}$ is a set containing sets and individuals as elements. If a is an element of $\mathfrak{V}$, then, of course, we write $a \in \mathfrak{V}$. One of the main characteristics of individuals is the fact that we are not interested in their elements (see Sections A.1 and A.2 in Appendix A). For example, if $\mathbb{H}$ is a general Hilbert space, then each $a \in \mathbb{H}$ is an individual. The elements of the dual space of $\mathbb{H}$ are real-valued mappings, whence they are sets. If $\mathbb{H} = l^2$, then its elements are sequences of real numbers, thus sets. The real numbers can be seen as individuals or as sets as well. If we identify the real numbers with their associated equivalence classes of Cauchy sequences of rationals, then real numbers become sets. It depends on the mathematical background we have in mind, which entities we take either as individuals or as sets. Sometimes, individuals are also called **urelements**. Since individuals and sets must be treated differently (see Proposition B.1.1), we have to assume and can assume (see Proposition A.2.1) that individuals are different from sets.

A standard model $\mathfrak{V}$ should contain at least the complex numbers as elements, and the sets $\mathbb{R}$ and $\mathbb{R}^n$, $n \in \mathbb{N}$, should also be elements. Let S be a set with $S \in \mathfrak{V}$. If $a \in S$, then $a \in \mathfrak{V}$ (a logician would say that the model is **transitive**). Moreover, each subset of S, the powerset $\mathcal{P}(S)$ of S, $\mathcal{PP}(S)$, and so on, should be elements of $\mathfrak{V}$. It is also required that the functions $+, \cdot$ and the relation $<$ are elements of each standard model. Here we identify $<$ with the set of all ordered pairs $(a,b) := \{\{a\}, \{a,b\}\}$ of real numbers a, b such that $a < b$. Analogously, we identify $+$ with the set of all $((a,b),c)$ such that $a+b=c$. More generally, we identify each relation R with the set $\{(a,b) \mid a\,R\,b\}$. A standard model should also contain more complex functions as elements, for example the function $\int_0^1$, which assigns to every integrable function f, defined on $[0,1]$, its integral $\int_0^1 f(x)dx \in \mathbb{R}$. Note that $\int_0^1$ is a set and the integrable functions are also sets and elements of a standard model. Moreover, the derivative operator, which assigns to every differentiable function its derivative, is supposed to be an element of a standard model. It follows that standard models are very rich. We may also assume that the abstract Wiener spaces we are working with are elements of each standard model.

Along with the notion 'standard model' we have the notion 'model'. Standard models belong to the class of models, and each model, let us call it $\mathfrak{W}$, is also transitive and contains many sets and individuals as elements. However, in contrast to a standard model, we do not assume that, if A is a set and an element of $\mathfrak{W}$, then each subset of A is necessarily an element of $\mathfrak{W}$. Sets which are subsets of a set $A \in \mathfrak{W}$, but which are not elements of $\mathfrak{W}$, are called **external** (in $\mathfrak{W}$). Otherwise, if they are elements of $\mathfrak{W}$, they are called **internal** (in $\mathfrak{W}$).

Here is the reason why we need external sets: we want to construct extensions of mathematical entities; in particular, we use a strict extension $({}^*\mathbb{R}, {}^*{+}, {}^*{\cdot}, 0, 1, {}^*{<})$ of the ordered field $(\mathbb{R}, +, \cdot, 0, 1, <)$ of real numbers, which is again an ordered field, as we shall see. It is well known that $({}^*\mathbb{R}, {}^*{+}, {}^*{\cdot}, 0, 1, {}^*{<})$ cannot be complete. However, we shall see that ${}^*\mathbb{R}$ is complete for the internal subsets of ${}^*\mathbb{R}$. To give up the full completeness of ${}^*\mathbb{R}$ is not at all a serious restriction, because the internal subsets have very nice closure properties, which are even better than the closure properties of measurable sets.

It turns out that all finite subsets of a set which is an element of a model are internal. Elements of internal sets and the individuals of a model are internal. Moreover, each set that can be defined by using only internal objects of the model is internal. To explain this phrase 'sets defined by using only internal objects' in more detail, we need sentences of the language $L_{\mathfrak{W}}$, associated to a fixed model $\mathfrak{W}$:

(i) To obtain the simplest sentences of $L_{\mathfrak{W}}$, fix $a, b \in \mathfrak{W}$. Then $(a \dot{=} b)$ is a sentence of $L_{\mathfrak{W}}$ and $(a \dot{\in} b)$ is a sentence of $L_{\mathfrak{W}}$ if b is a set in $\mathfrak{W}$. There are reasons for using these dotted syntactic symbols $\dot{\in}$ and $\dot{=}$ instead of the associated semantic symbols $\in$ and $=$; see below.

(ii) To obtain more complex sentences, fix sentences A, B of $L_{\mathfrak{W}}$. Then $(A \vee B), (\neg A)$ are sentences of $L_{\mathfrak{W}}$, where '$\vee$', '$\neg$' are shorthand for 'or', 'not', respectively. If $a \in \mathfrak{W}$ is a set in $\mathfrak{W}$, then $(\exists x \dot{\in} a A_c(x))$ is a sentence of $L_{\mathfrak{W}}$, where $A_c(x)$ results from A by replacing some $c \in \mathfrak{W}$ with a variable x at all places where c occurs in A. We assume that the chosen variable does not occur in A. Instead of x we may take any variable $y, z, u, x_1, \ldots, x_n, \ldots$, not occurring in A. We define $(A \Rightarrow B)$, $(A \wedge B)$, $(A \Leftrightarrow B)$, $(\forall x \dot{\in} a A_b(x))$ in the usual way (see Section A.4 of Appendix A). The elements of $\mathfrak{W}$ occurring in a sentence A of $L_{\mathfrak{W}}$, are called the **parameters** of A. It is presupposed that the logical signs $\dot{\in}, \dot{=}, \wedge, \vee, \neg, \Rightarrow, \Leftrightarrow, \forall, \exists, (,)$ and the variables are different from the parameters. We will drop parentheses in sentences if there is no risk of confusion (see Section A.3 of Appendix A). For example, instead of $(\neg (a \dot{\in} b))$ we write $\neg a \dot{\in} b$.

Note that $(\exists x \dot{\in} a\,(x \dot{=} b))$ is a sentence of $\mathfrak{W}$ if $a, b \in \mathfrak{W}$ and a is a set in $\mathfrak{W}$. Here $(x \dot{=} b)$ equals $((c \dot{=} b)_c\,(x))$ if $c \neq b$. A further example: $\forall y \dot{\in} b \exists x \dot{\in} a\,(\neg x \dot{=} y)$ is also a sentence of $\mathfrak{W}$ if a, b are sets in $\mathfrak{W}$.

The **truth** of a sentence A of $L_{\mathfrak{W}}$ **in** $\mathfrak{W}$ is equivalent to the statement $\widetilde{A}$, where $\widetilde{A}$ results from A by replacing $\dot{\in}$ by $\in$ and $\dot{=}$ by $=$. A quick example: let $\mathbb{N}, 5, \pi \in \mathfrak{W}$. Then $5 \dot{\in} \mathbb{N}$ is true in $\mathfrak{W}$, because $5 \in \mathbb{N}$ $(= \widetilde{5 \dot{\in} \mathbb{N}})$, and $\pi \dot{\in} \mathbb{N}$ is not true in $\mathfrak{W}$, because $\neg(\pi \in \mathbb{N})$ $(= \widetilde{\neg(\pi \dot{\in} \mathbb{N})})$. Instead of $\neg(a \in b)$ we write $a \notin b$ and instead of $\neg(a = b)$ we write $a \neq b$.

Examples

Suppose that $\mathfrak{W}$ is a model with $\mathbb{R}, \mathbb{N} \in \mathfrak{W}$. Then $\mathbb{R}$ is a subset of $\mathfrak{W}$, because models are transitive.

(i) $\frac{1}{2} \dot{\in} \mathbb{N}$ is a sentence of $L_{\mathfrak{W}}$. This sentence is not true in $\mathfrak{W}$, because $\frac{1}{2}$ is not a positive integer. Therefore, $\neg \frac{1}{2} \dot{\in} \mathbb{N}$ is true in $\mathfrak{W}$.

(ii) Here we consider a fixed standard model $\mathfrak{V}$. Then $\exists x \dot{\in} \mathbb{N}\,(x > 5)$ is a sentence of $L_{\mathfrak{V}}$, where we use the following abbreviations:
 $x > 5$ is short for $(x, 5) \dot{\in} >$,
 $(x, 5) \dot{\in} >$ is short for $\exists y \dot{\in} > (y \dot{=} (x, 5))$,
 $y \dot{=} (x, 5)$ is short for $\forall z \dot{\in} \mathcal{P}(\mathbb{N})\,(z \dot{\in} y \Leftrightarrow z \dot{=} \{x\} \vee z \dot{=} \{x, 5\})$,
 $z \dot{=} \{x\}$ is short for $\forall u \dot{\in} \mathbb{N}\,(u \dot{\in} z \Leftrightarrow u \dot{=} x)$,
 $z \dot{=} \{x, 5\}$ is short for $\forall u \dot{\in} \mathbb{N}\,(u \dot{\in} z \Leftrightarrow (u \dot{=} x \vee u \dot{=} 5))$.

The sentence $\exists x \dot{\in} \mathbb{N}\,(x > 5)$ is true in $\mathfrak{V}$.

(iii) Every mathematical statement about elements of a standard model $\mathfrak{V}$ can be expressed by sentences of $L_{\mathfrak{V}}$. However, as we have seen in Example (ii), these sentences may be very long.

Before we continue, we have to make a remark on the notion ‘parameter’.

Remark 7.1.1 The symbols $\dot{\in}$ and $\dot{=}$ do not belong to the set of parameters. On the other hand, $\in$ or $=$ may be parameters in a sentence, where we again identify $\in, =$ with the set of all pairs (a, b) with $a \in b, a = b$, respectively. For example $(0, 1) \dot{\in} =$ may be a sentence, also $(0, \{1\}) \dot{\in} \in$. Both sentences are not true if $0 \neq 1$. Later on, we shall see why we use the notation $\dot{\in}$ and $\dot{=}$ (see Remark 7.1.3 below).

Now we can define the following: a model $\mathfrak{W}$ is **closed under definition** if each set that is defined by using only internal objects of $\mathfrak{W}$ is an internal set in $\mathfrak{W}$. More precisely, $\mathfrak{W}$ is closed under definition if for all sets $S \in \mathfrak{W}$ and for all sentences A of $L_{\mathfrak{W}}$

$$\{a \in S \mid A_b(a) \text{ is true in } \mathfrak{W}\} \in \mathfrak{W},$$

where $A_b(a)$ results from A by replacing some parameter b by a at all places, where b occurs in A. Of course, a standard model $\mathfrak{V}$ is closed under definition, because each subset of a set in $\mathfrak{V}$ is an element of $\mathfrak{V}$. We may, and will, assume that all models, including those having external sets, are closed under definition. Here are some further examples.

Examples

Let $\mathfrak{W}$ be closed under definition. Fix a set $S \in \mathfrak{W}$.

(iv) Suppose that $\{a_1, \ldots, a_k\}$ is a finite subset of S. Then

$$\{a_1, \ldots, a_k\} = \{a \in S \mid a \dot{=} a_1 \vee \ldots \vee a \dot{=} a_k \text{ is true in } \mathfrak{W}\} \in \mathfrak{W},$$

thus finite subsets of S are internal. Recall that

$$a \dot{=} a_1 \vee \ldots \vee a \dot{=} a_k \text{ is true in } \mathfrak{W} \Leftrightarrow a = a_1 \vee \ldots \vee a = a_k.$$

(v) $\emptyset = \{a \in S \mid a \neq a\} \in \mathfrak{W}$, where $a \neq a$ is equivalent to $\neg(a \dot{=} a)$ is true in $\mathfrak{W}$.

(vi) If $A, B \in \mathfrak{W}$ and $A, B \subseteq S$, then $A \cup B, A \cap B, A \setminus B \in \mathfrak{W}$, because, for example, $A \setminus B = \{a \in S \mid a \in A \wedge a \notin B\} \in \mathfrak{W}$.

(vii) Fix $f : A \to B$, where A, B and f are sets in $\mathfrak{W}$. Then the range $f[A]$ of f is an internal set in $\mathfrak{W}$, because

$$f[A] = \{b \in B \mid \exists x \in A\, ((x, b) \in f)\} \in \mathfrak{W}.$$

Our aim now is to define the notion of a poly-saturated model of a fixed standard model $\mathfrak{V}$. Let us call an arbitrary set M **small by** $\mathfrak{V}$ if there exists an injective mapping from M into $\mathfrak{V}$. (Thus, the cardinality of small sets by $\mathfrak{V}$ is not larger than the cardinality of $\mathfrak{V}$.) Let us call a set $\mathcal{B}$ of sets **deep** if $\mathcal{B}$ has the **finite intersection property**, i.e., $A_1 \cap \ldots \cap A_k \neq \emptyset$ for any finite collection $A_1, \ldots, A_k \in \mathcal{B}$. A model $\mathfrak{W}$ which is closed under definition is called a **poly-saturated model of** $\mathfrak{V}$ if for each small by $\mathfrak{V}$ and deep set $\mathcal{B}$ of (internal) sets in $\mathfrak{W}$ the whole intersection of all sets in $\mathcal{B}$ is non-empty, i.e.,

$$\bigcap \mathcal{B} := \{x \mid x \in B \text{ for all } B \in \mathcal{B}\} \neq \emptyset.$$

Note that poly-saturation depends on the size of the fixed standard model of mathematics. (A logician would say that a poly-saturated model of $\mathfrak{V}$ is κ^+-**saturated**, where κ is the cardinality of $\mathfrak{V}$.)

Examples

(viii) The set $\mathcal{B} := \{]0, \frac{1}{n}] \mid n \in \mathbb{N}\}$ is deep and small, but $\bigcap \mathcal{B} = \emptyset$. It follows that standard models are never poly-saturated.
Assume now that $\mathfrak{W}$ is a poly-saturated model of some standard model $\mathfrak{V}$. Recall that $\mathbb{N}$ is small by $\mathfrak{V}$.

(ix) We can work with any (internal) set of $\mathfrak{W}$ as though it were **compact** in the following sense: let I be a small set by $\mathfrak{V}$ and assume that A and A_i are internal sets in $\mathfrak{W}$ for all $i \in I$ with

$$A \subseteq \bigcup_{i \in I} A_i. \tag{9}$$

Then there exist finitely many $A_{i_1}, \ldots, A_{i_k}$ with $A \subseteq A_{i_1} \cup \ldots \cup A_{i_k}$.

Proof Assume the assertion fails. Then $\mathcal{B} := \{A \setminus A_i \mid i \in I\}$ is deep. Since $\mathcal{B}$ is small by $\mathfrak{V}$, there is an $a \in \bigcap_{i \in I} A \setminus A_i$, contradicting (9). □

The following examples are immediate consequences of (ix).

(x) Let $(A_i)_{i \in \mathbb{N}}$ be a sequence of internal sets $A_i \in \mathfrak{W}$ with $A_i \subseteq A_{i+1}$. If $\bigcup_{i \in \mathbb{N}} A_i$ is internal in $\mathfrak{W}$, then there is a $k \in \mathbb{N}$ with $A_k = \bigcup_{i \in \mathbb{N}} A_i$.

(xi) Suppose that A is small by $\mathfrak{V}$ and internal in $\mathfrak{W}$. Then A is finite.

Proof Since $A \subseteq \bigcup_{a \in A} \{a\}$, there exists a $k \in \mathbb{N}$ and $a_1, \ldots, a_k \in A$ with

$$A \subseteq \{a_1\} \cup \ldots \cup \{a_k\} = \{a_1, \ldots, a_k\}.$$

Therefore, A is finite. □

Remark 7.1.2 It follows that any internal set A in $\mathfrak{W}$ is either finite or very large, larger than the cardinality of $\mathfrak{V}$. The reader may think that this fact is too drastic a restriction, because we do not have enough internal sets, for example, all countably infinite sets are external.

However, this supposed 'defect' has been used first by Loeb [69] and later by other authors (see, for example, Aldaz and Render [3], Landers and Rogge [59]) to construct extensions of measures. One reason is that internal countable unions of internal sets reduce to finite unions (see Example (x)). Therefore, each finitely additive measure on an internal algebra $\mathcal{C}$ of sets is already countably additive. Therefore, by Caratheodory's extension method, it can be extended to a measure, defined on the σ-algebra $\sigma(\mathcal{C})$, generated by $\mathcal{C}$. It turns out that, by saturation, $\mathcal{C}$ is very rich: we shall see that each $B \in \sigma(\mathcal{C})$ is **equivalent to** an element $A \in \mathcal{C}$, i.e., the symmetric difference of B and A is a nullset.

One of the main results for poly-saturated models states that for each standard model $\mathfrak{V}$ there exists a poly-saturated model $\mathfrak{W}$ with the same properties as $\mathfrak{V}$. What does this mean? Fix two models $\mathfrak{U}$ and $\mathfrak{W}$. Then we say that $\mathfrak{W}$ has the **same properties as** $\mathfrak{U}$ if there exists a mapping $* : \mathfrak{U} \to \mathfrak{W}$ such that for each sentence φ of $L_{\mathfrak{U}}$

$$\varphi \text{ is true in } \mathfrak{U} \Leftrightarrow {}^*\varphi \text{ is true in } \mathfrak{W},$$

where ${}^*\varphi$ results from φ by replacing each parameter a in φ by $*(a)$. A logician would say the $*$ is an **elementary embedding from $\mathfrak{U}$ into $\mathfrak{W}$**.

Attention: The relation '$\mathfrak{W}$ has the same properties as $\mathfrak{U}$' does not define an equivalence relation on the class of models. This relation is reflexive and transitive but not symmetric: by Theorem 7.2.1 below, for each standard model $\mathfrak{V}$ there exists a poly-saturated model $\mathfrak{W}$ with the same properties as $\mathfrak{V}$. Since $\mathfrak{W}$ is much larger than $\mathfrak{V}$ and elementary embeddings are injective (see Example (xii)), there does not exist an elementary embedding from $\mathfrak{W}$ into $\mathfrak{V}$, thus $\mathfrak{V}$ has not the same properties as $\mathfrak{W}$.

In other words, elementary embeddings from $\mathfrak{U}$ into $\mathfrak{W}$ are not surjective, in general. There may exist sentences in $L_{\mathfrak{W}}$ which are not $*$-images of sentences in $L_{\mathfrak{U}}$.

Let us write ${}^*a$ instead of $*(a)$.

Attention: If $A \in \mathfrak{U}$ and a set, then ${}^*A$ is the $*$-image $*(A)$ of A and not the set $*[A]$, the set of all $*(a)$ with $a \in A$. However, $*[A] \subseteq {}^*A$, as we see under Example (xiii).

Examples

We refer to Appendix A, where the reader can find more examples and their proofs. Suppose that $*$ is an elementary embedding from $\mathfrak{U}$ into $\mathfrak{W}$.

(xii) $*$ is injective, because, if $a \neq b$ with $a, b \in \mathfrak{U}$, then $\neg(a \doteq b)$ is true in $\mathfrak{U}$, thus $\neg({}^*a \doteq {}^*b)$ is true in $\mathfrak{W}$, which implies ${}^*a \neq {}^*b$.

(xiii) In the same way we obtain the following. Fix a set $A \in \mathfrak{U}$ and $a \in A$. Then ${}^*a \in {}^*A$. If we identify a and ${}^*a$, then we see that ${}^*A$ is an extension of A. It follows that an elementary embedding is an injective homomorphism with respect to the $\in$-relation. However, elementary embeddings are much stronger than $\in$-homomorphisms.

(xiv) Let S be a set in $\mathfrak{U}$. Then for all sentences A of L,

$$^*\{a \in S \mid A_b(a) \text{ holds in } \mathfrak{U}\} = \left\{a \in {}^*S \mid \left({}^*A\right)_{*b}(a) \text{ holds in } \mathfrak{W}\right\}.$$

(See Proposition A.6.1 (g).)

(xv) Since $\mathfrak{U}$ is closed under definition, $\mathfrak{W}$ is closed under definition as well. (See Proposition A.6.1 (f).)

Now assume that $\mathbb{R}, +, \cdot, <$ are sets in $\mathfrak{U}$. Then

(xvi) $({}^*\mathbb{R},\ {}^*+,\ {}^*\cdot,\ {}^*0,\ {}^*1,\ {}^*<)$ is an ordered field and ${}^*+,\ {}^*\cdot,\ {}^*<$ are extensions of $+,\ \cdot,\ <$, respectively (see (xiii)).

Now assume that $*$ is an elementary embedding from a standard model into some model. What can we say about the completeness of ${}^*\mathbb{R}$? If A, B are subsets of $\mathbb{R}$, then we write $A \leq B$ if each element of A is less than or equal to any element of B. The real numbers fulfil the **Dedekind Cut Axiom**, i.e.,

$$\forall x, y \dot{\in} \mathcal{P}(\mathbb{R})\,(x, y \neq \emptyset \wedge x \leq y \Rightarrow \exists z \dot{\in} \mathbb{R}\,(x \leq z \leq y)),$$

where $y \neq \emptyset$ is a shorthand for $\exists z \dot{\in} y(z \dot{=} z)$ and $x \leq y$ is a shorthand for $\forall z \dot{\in} x \forall u \dot{\in} y\,((z, u) \dot{\in} \leq)$. Since * is an elementary embedding, the Dedekind Cut Axiom is true for all pairs of non-empty subset of ${}^*\mathbb{R}$ which belong to ${}^*\mathcal{P}(\mathbb{R})$. Thus we have

(xvii) $({}^*\mathbb{R},\ {}^*+,\ {}^*\cdot,\ 0, 1,\ {}^*<)$ is an ordered field and is complete for all cuts of ${}^*\mathbb{R}$, where the left and right class are elements of ${}^*\mathcal{P}(\mathbb{R})$.

Remark 7.1.3 Here is the reason why we use $\dot{\in}$ and $\dot{=}$ instead of $\in$ and $=$: since $\in$ and $=$ are sets, they belong to the set of parameters. Therefore, they have to be replaced in the *-image of a sentence by ${}^*\in$ and ${}^*=$. These new relations may be different from the usual $\in$- and $=$-relations, but we always interpret the signs $\dot{\in}$ and $\dot{=}$ by the usual $\in$- and $=$-relations.

7.2 The main theorem

Following [86], we will give an elementary proof of Theorem 7.2.1 below in Appendix B. 'Elementary' here means that the existence of κ-good incomplete ultrafilters is not used in the proof. The proof in Appendix B slightly modifies Sacks' corresponding proof for first order logic.

Theorem 7.2.1 (Sacks [102]) *Fix a standard model $\mathfrak{V}$. Then there exists a model $\mathfrak{W}$ such that the following three principles hold, where Principle (D) actually follows from Principle (T) (see Example (xv)).*

(T) *(transfer principle) There exists an elementary embedding * from $\mathfrak{V}$ into $\mathfrak{W}$; thus $\mathfrak{W}$ has the same properties as $\mathfrak{V}$.*
(D) *(internal definition principle) $\mathfrak{W}$ is closed under definition.*
(S) *(saturation principle) $\mathfrak{W}$ is a poly-saturated model of $\mathfrak{V}$.*

Throughout this book, we fix a poly-saturated model $\mathfrak{W}$ of a standard model $\mathfrak{V}$ and an elementary embedding $*$ from $\mathfrak{V}$ into $\mathfrak{W}$. Let us write **small** instead of 'small by $\mathfrak{V}$'.

The idea of the proof of Theorem 7.2.1 is similar to a proof of the following result in algebra:

Theorem 7.2.2 *Fix a field $\mathfrak{V}$. Then there exists an algebraically closed extension $\mathfrak{W}$ of $\mathfrak{V}$.*

In the proof of Theorem 7.2.2 one has to find roots of polynomials; in the proof of Theorem 7.2.1 we have to find elements in the intersection of deep and small families of internal sets. In both cases one may proceed in two steps.

Proof of Theorem 7.2.2
In the first step one enumerates all irreducible polynomials over $\mathfrak{V}$ by say $(p_\alpha)_{\alpha<\gamma}$, where γ is the cardinality of the set of all these polynomials. Then an increasing chain $(\mathfrak{V}_\alpha)_{\alpha<\gamma}$ of fields $\mathfrak{V}_\alpha$ is constructed with $\mathfrak{V}_0=\mathfrak{V}$ such that, if α is a successor ordinal, then $\mathfrak{V}_\alpha$ contains at least one root of $p_{\alpha-1}$. If α is a limit number, then $\mathfrak{V}_\alpha$ is the union of the preceding fields. It turns out that $\mathfrak{V}_\gamma$ is an extension of $\mathfrak{V}$ such that each irreducible polynomial over $\mathfrak{V}$ has a root in $\mathfrak{V}_\gamma$.

In a second step, one takes a countable increasing chain $(\mathfrak{W}_n)_{n\in\mathbb{N}_0}$ of fields $\mathfrak{W}_n$ such that $\mathfrak{W}_0=\mathfrak{V}$ and so that $\mathfrak{W}_{n+1}$ contains a root of each irreducible polynomial over $\mathfrak{W}_n$. The union $\mathfrak{W}$ of all the $\mathfrak{W}_n$ provides an algebraically closed extension of $\mathfrak{V}$. □

Proof of Theorem 7.2.1
Instead of an increasing chain $(\mathfrak{V}_\alpha)_{\alpha<\gamma}$ of fields $\mathfrak{V}_\alpha$ we take an **elementary chain** $\left((\mathfrak{V}_\alpha)_{\alpha<\gamma},(*_{\alpha,\beta})_{\alpha\le\beta<\gamma}\right)$ of models $\mathfrak{V}_\alpha$. Here $*_{\alpha,\beta}$ is an elementary embedding from $\mathfrak{V}_\alpha$ into $\mathfrak{V}_\beta$ where $*_{\alpha,\alpha}$ is the identity map and $*_{\beta,\delta}\circ *_{\alpha,\beta}=*_{\alpha,\delta}$ for $\alpha\le\beta\le\delta<\gamma$.

The **elementary limit** $\mathfrak{V}_\gamma$ of $\left((\mathfrak{V}_\alpha)_{\alpha<\gamma},(*_{\alpha,\beta})_{\alpha\le\beta<\gamma}\right)$ is the smallest model into which all the models $\mathfrak{V}_\beta$ with $\beta<\gamma$ can be elementarily embedded, by say $*_{\beta,\gamma}$.

In analogy to the proof in algebra we enumerate all deep families of sets in $\mathfrak{V}$ by say $(p_\alpha)_{\alpha<\gamma}$. Then an elementary chain $\left((\mathfrak{V}_\alpha)_{\alpha<\gamma},(*_{\alpha,\beta})_{\alpha\le\beta<\gamma}\right)$ of models $\mathfrak{V}_\alpha$ is constructed with $\mathfrak{V}_0 = \mathfrak{V}$ such that, if α is a successor ordinal, then $\mathfrak{V}_\alpha$ contains at least one 'root' in the whole intersection of all the sets in ${}^{*0,\alpha-1}[p_{\alpha-1}] = \{{}^{*0,\alpha-1}A \mid A \in p_{\alpha-1}\}$. More precisely, we use the compactness theorem (see Theorem B.2.4) to construct $\mathfrak{V}_\alpha$ from $\mathfrak{V}_{\alpha-1}$. This theorem tells us that for $\mathfrak{V}_{\alpha-1}$ there exist a model $\mathfrak{V}_\alpha$ and an elementary embedding $*_{\alpha-1,\alpha}$ from $\mathfrak{V}_{\alpha-1}$ into $\mathfrak{V}_\alpha$ such that $\bigcap\{{}^{*\alpha-1,\alpha}A \mid A \in {}^{*0,\alpha-1}[p_{\alpha-1}]\} \neq \emptyset$ in $\mathfrak{V}_\alpha$. If α is a limit number, then $\mathfrak{V}_\alpha$ is the elementary limit of the preceding models. Finally, there exists a model $\mathfrak{V}_\gamma$, the elementary limit of the $\mathfrak{V}_\alpha$ with $\alpha < \gamma$, and an elementary embedding $*_{0,\gamma}$ from $\mathfrak{V}_0$ into $\mathfrak{V}_\gamma$ such that $\bigcap\{{}^{*0,\gamma}A \mid A \in \mathcal{B}\} \neq \emptyset$ for each deep family $\mathcal{B}$ of sets in $\mathfrak{V}$.

To finish the proof, first recall from set theory that κ^+, the successor cardinal of $\kappa :=$ cardinality($\mathfrak{V}$), has similar properties with respect to smaller cardinals as $\mathbb{N}$ has with respect to finite numbers. In particular, $\sup_{\alpha<\delta} o_\alpha$ is strictly smaller than κ^+ for each small sequence $(o_\alpha)_{\alpha<\delta}$ of ordinals below κ^+. By the preceding paragraph, there exists an elementary chain $\left((\mathfrak{W}_\alpha)_{\alpha<\gamma}, (*_{\alpha,\beta})_{\alpha\leq\beta<\gamma}\right)$ of models $\mathfrak{W}_\alpha$ such that $\bigcap\{{}^{*\alpha,\beta}A \mid A \in \mathcal{B}\} \neq \emptyset$ for each deep family $\mathcal{B}$ of sets in $\mathfrak{W}_\alpha$, provided that $\alpha < \beta$. Of course, $\mathfrak{W}_0$ has to be the model $\mathfrak{V}$. Now the elementary limit $\mathfrak{W} := \mathfrak{W}_{\kappa^+}$ of all the $\mathfrak{W}_\alpha$ with $\alpha < \kappa^+$ is a desired model and $* := *_{0,\kappa^+}$ is a desired elementary embedding from $\mathfrak{V}$ into $\mathfrak{W}$. □

We have seen that the internal sets of $\mathfrak{W}$ (and of $\mathfrak{V}$) have nice closure properties, while the external sets do not enjoy good closure properties in general. For example, the union of two external sets may be internal: since $\mathbb{N} \in \mathfrak{V}$ we have ${}^*\mathbb{N} \in \mathfrak{W}$, thus ${}^*\mathbb{N}$ is internal. We may assume that $\mathbb{N} \subseteq {}^*\mathbb{N}$, by identification of $a \in \mathbb{N}$ with ${}^*a$. Then $\mathbb{N}$, as a small infinite subset of ${}^*\mathbb{N}$, is external and ${}^*\mathbb{N} \setminus \mathbb{N}$ is also external (otherwise $\mathbb{N}$ could be defined by means of the internal sets ${}^*\mathbb{N} \setminus \mathbb{N}$ and ${}^*\mathbb{N}$ and thus would be internal). We see that the internal set ${}^*\mathbb{N}$ is the union of the external sets $\mathbb{N}$ and ${}^*\mathbb{N} \setminus \mathbb{N}$. We may characterize external sets as those subsets E of some set $S \in \mathfrak{W}$ such that the assumption that E is internal leads to a contradiction.

Since $*$ is a structure-conserving mapping, we may identify $A \in \mathfrak{V}$ with ${}^*A \in \mathfrak{W}$, provided that we are not interested in the elements of A and ${}^*A$. For instance, the real numbers r can be identified with ${}^*r \in {}^*\mathbb{R}$. The 'star', even on sets, will often be omitted if this does not cause confusion: for example, if f is a function in $\mathfrak{V}$ and $a_1, \ldots, a_k, a$ are elements of $\mathfrak{W}$ then $f(a_1, \ldots, a_k) = a$ of course means that ${}^*f(a_1, \ldots, a_n) = a$, because, otherwise, $f(a_1, \ldots, a_n)$ may not be defined.

Since ${}^*\mathbb{N}_0$ is a strict extension of $\mathbb{N}_0$, we have an extension of the standard notion of finiteness in the following sense: let us call a set F *finite if there exists an $n \in {}^*\mathbb{N}_0$ and an internal bijection $f : F \leftrightarrow \{k \in {}^*\mathbb{N}_0 | k < n\}$. Then n is the number of elements in F. For more details see Section 8.3. In particular, we shall see that F is infinite if $n \in {}^*\mathbb{N}$ and n is not a standard number. So, intentionally, F has two cardinalities: F is finite in the sense of the model $\mathfrak{W}$, but infinite in the sense of everything surrounding naive set theory, in which we have constructed the models $\mathfrak{V}$ and $\mathfrak{W}$.

Since $\mathfrak{W}$ has the same properties as $\mathfrak{V}$, *finite sets have the same properties as standard finite sets.

Remark 7.2.3 Finally, let us give an answer to the following natural question. Is it possible to obtain new results or at least new insights into standard mathematics, using the new model $\mathfrak{W}$?

At a first glance the answer is: NO!, because the model $\mathfrak{W}$ has the same properties as the standard model $\mathfrak{V}$ from the internal point of view. However, since external sets in $\mathfrak{W}$ will be extensively used, we leave our home of elementary equivalence.

For example, Loeb measure spaces, which will play an important role later on, contain highly external sets of $\mathfrak{W}$. Therefore, these spaces belong neither to $\mathfrak{V}$ nor to $\mathfrak{W}$, although they are complete measure spaces in the common sense. Loeb spaces exist in (naive) set theory in which models have been constructed. We shall see and use that Loeb spaces are very rich, richer than many standard spaces, in particular, richer than the Lebesgue space. Sun [113] used Loeb measures to prove that in all relevant situations new product spaces arise. In these spaces it is possible, as Sun showed, to study a jointly measurable continuum of independent random variables which has applications in mathematical economics. Important is that the Fubini theorem is true for these product spaces (a result of Keisler [53]).

In this book we use the smooth and *finite-dimensional representation of Lévy functionals, mentioned in the first paragraph of this chapter, to obtain new insights into the theory of Lévy processes.

Another example of new standard structures, arising from external sets of $\mathfrak{W}$, are the nonstandard hulls of topological vector spaces and vector lattices. They were introduced by Luxemburg [74] and have been systematically studied by Henson and Moore (see for example [44]) and in [71].

Exercises

7.1 Prove that $\{\{a\},\{a,b\}\}=\{\{\alpha\},\{\alpha,\beta\}\}$ if and only if $a=\alpha$ and $b=\beta$.

Let $\mathfrak{V}$ be a standard model.

7.2 Find a sentence A in $L_{\mathfrak{V}}$ with parameters $n, m, \mathbb{N}, \cdot$ such that A is true in $\mathfrak{V}$ if and only if n divides m.

7.3 Find a sentence A in $L_{\mathfrak{V}}$ with parameters $1, n, \mathbb{N}, \cdot$ such that A is true in $\mathfrak{V}$ if and only if n is a prime number.

7.4 Find a sentence A in $L_{\mathfrak{V}}$ with parameters $1, +, \cdot, n, \mathbb{N}$ such that A is true in $\mathfrak{V}$ if and only if n and $n+2$ are prime numbers; n and $n+2$ are called **prime-twins**.

7.5 Find a sentence A in $L_{\mathfrak{V}}$ with parameters $1, +, \cdot, \mathbb{N}$ such that A is true in $\mathfrak{V}$ if and only if $\mathbb{N}$ contains infinitely many prime-twins.

7.6 Find a sentence A in $L_{\mathfrak{V}}$ with parameters $+, \cdot, 0, 1, <, \mathbb{R}$ such that A is true in $\mathfrak{V}$ if and only if $(\mathbb{R}, +, \cdot, 0, 1, <)$ is an ordered field.
Hint: Use shorthands, as in Example (ii), for $x+y=z$ and $x \cdot y = z$.

7.7 Suppose that $*$ is an elementary embedding from $\mathfrak{U}$ into $\mathfrak{W}$. Let $\{a_1, \ldots, a_n\}$ be a finite set in $\mathfrak{U}$. Prove that

$$^*\{a_1, \ldots, a_n\} = \{^*a_1, \ldots, ^*a_n\}.$$

8

Extension of the real numbers and properties

In this chapter the reader will find details of some elementary applications of the transfer and saturation principles to the *extension of the real numbers.

8.1 *ℝ as an ordered field

By transfer (T), we have the following result:

Proposition 8.1.1 $({}^*\mathbb{R}, {}^*+, {}^*\cdot, 0, 1, {}^*<)$ *is an ordered field which is complete for the internal subsets of* ${}^*\mathbb{R}$.

The absolute value $|\cdot|$ on $\mathbb{R}$ is a set in $\mathfrak{V}$. Therefore, ${}^*|\cdot|$ is a set in $\mathfrak{W}$. Again by the transfer principle, we obtain, omitting the $*$ on $\geq, 0, \cdot, +, \leq$, the following:

Proposition 8.1.2 ${}^*|\cdot|$ *is a mapping from* ${}^*\mathbb{R}$ *into* ${}^*\left(\mathbb{R}_0^+\right)$ *such that for all* $a, b \in {}^*\mathbb{R}$*:*

(a) ${}^*|a| \geq 0$ *and* ${}^*|a| = 0$ *if and only if* $a = 0$,
(b) ${}^*|a \cdot b| = {}^*|a| \cdot {}^*|b|$,
(c) ${}^*|a + b| \leq {}^*|a| + {}^*|b|$.

8.2 The *extension of the positive integers

We have already seen that $\mathbb{N}$ is a strict subset of ${}^*\mathbb{N}$. By the transfer principle, we obtain, omitting * on $<, +, 1$:

Proposition 8.2.1 *Fix* $n \in {}^*\mathbb{N}$.

(a) *The number* 1 *is the smallest element of* ${}^*\mathbb{N}$ *with respect to* ${}^*<$.
(b) *There is no* $m \in {}^*\mathbb{N}$ *such that* $n < m < n + 1$.

(c) *If $n \neq 1$, then there exists an $m \in {}^*\mathbb{N}$ with $m+1=n$.*

(d) *Each internal non-empty set $B \subseteq {}^*\mathbb{N}$ has a smallest element with respect to ${}^*<$.*

(e) *(Internal transfinite induction) For each internal $B \subseteq {}^*\mathbb{N}$ we have the following implication:*

$$1 \in B \quad \textit{and} \quad \forall x(x \in B \Rightarrow x+1 \in B) \Rightarrow B = {}^*\mathbb{N}.$$

(f) *Let $A \subseteq {}^*\mathbb{N} \setminus \mathbb{N}$ be a small set. Then there exists an $N_\infty \in {}^*\mathbb{N} \setminus \mathbb{N}$ such that $N_\infty \leq A$, i.e., $N_\infty \leq n$ for all $n \in A$.*

(g) *There exists an $H \in {}^*\mathbb{N}$ such that each $n \in \mathbb{N}$* ***divides*** *H, i.e., there exists a $k \in {}^*\mathbb{N}$ with $k \cdot n = H$, in which case we shall write $n \mid H$.*

Proof Parts (a)–(e) follow from the transfer principle (T).

(f) For each $a \in A$ and $i \in \mathbb{N}$ set $A_{a,i} := \{\beta \in {}^*\mathbb{N} \mid i \leq \beta \leq a\}$. Then, by Principle (D), $A_{a,i}$ is internal and $\mathcal{B} := \{A_{a,i} \mid a \in A, i \in \mathbb{N}\}$ is small and deep. By saturation, there exists an $N_\infty \in \bigcap \mathcal{B}$, i.e., $N_\infty \in {}^*\mathbb{N} \setminus \mathbb{N}$ and $N_\infty \leq A$.

(g) For $n \in \mathbb{N}$ set $B_n := \{m \in {}^*\mathbb{N} \mid n \mid m\}$. Then B_n is internal and $\mathcal{B} := \{B_n \mid n \in \mathbb{N}\}$ is small and deep, thus there exists an $H \in \bigcap \mathcal{B}$. □

From Proposition 8.2.1 (a) and (b) it follows that for each $n \in {}^*\mathbb{N}$

$$n \in {}^*\mathbb{N} \setminus \mathbb{N} \text{ if and only if } m < n \text{ for all } m \in \mathbb{N},$$

i.e., ${}^*\mathbb{N}$ is an **end extension** of $\mathbb{N}$. We call the elements of ${}^*\mathbb{N} \setminus \mathbb{N}$ **unlimited**.

Sequences $(c_k)_{k \in \mathbb{N}}$ can be defined as mappings f, defined on $\mathbb{N}$, with $f(k) = c_k$. Since $\{(k, c_k) \mid k \in \mathbb{N}\}$ is an infinite small set, each sequence $(c_k)_{k \in \mathbb{N}}$ in $\mathfrak{W}$ is external. However, by saturation, this sequence can be extended to an internal sequence, more precisely:

Proposition 8.2.2 *Fix an internal set A in $\mathfrak{W}$. Assume that $(a_n)_{n \in \mathbb{N}}$ is a sequence in A. Then $(a_n)_{n \in \mathbb{N}}$ can be extended to an internal sequence $(a_n)_{n \in {}^*\mathbb{N}}$ in A.*

Proof Set for all $k \in \mathbb{N}$

$$B_k := \{F : {}^*\mathbb{N} \to A \text{ internal} \mid F(1) = a_1 \text{ and } \dots \text{ and } F(k) = a_k\}$$

for each $k \in \mathbb{N}$. Since $\mathfrak{W}$ is closed under definition, each B_k is an internal set and $\mathcal{B} := \{B_k \mid k \in \mathbb{N}\}$ is small. Moreover, $\mathcal{B}$ is deep, because for a fixed $a \in A$ and each standard $k \in \mathbb{N}$

$$F_k : {}^*\mathbb{N} \ni i \mapsto \begin{cases} a_i & \text{if } i \in \{1,\ldots,k\} \\ a & \text{otherwise,} \end{cases}$$

belongs to B_k. By saturation, $\bigcap \mathcal{B} \neq \emptyset$. Each element in this intersection is an internal extension of $(a_k)_{k\in\mathbb{N}}$. □

8.3 Hyperfinite sets and summation in *ℝ

Using the extension $^*\mathbb{N}$ of $\mathbb{N}$ we obtain an extension of finiteness, which plays an important role. There exist infinite sets in $\mathfrak{W}$ which behave like finite sets: Fix a set $J \in \mathfrak{W}$. A subset $E \subseteq J$ is called ***finite**, if there exists an $n \in {}^*\mathbb{N}_0$ and an internal bijection $f : E \leftrightarrow \{k \in {}^*\mathbb{N}_0 | k < n\}$. In this case, n is called the **internal cardinality** of E which is denoted by $|E|$. By the internal definition principle (D), E is an internal set in $\mathfrak{W}$.

We have already seen (Section 7.2) that the set $F := \{k \in {}^*\mathbb{N}_0 | k < n\}$ has two different cardinalities if $n \in {}^*\mathbb{N}$ is unlimited: the internal *finite cardinality n and the external large cardinality. From the internal point of view, F can be handled as a finite set.

Finite summation $\sum$ in $\mathbb{R}$ assigns to each $k \in \mathbb{N}_0$ and each k-tuple $(a_1,\ldots,a_k)$ in $\mathbb{R}$ the sum $a_1 + \ldots + a_k$. Since $\mathfrak{V}$ is a standard model, $\sum$ is a set in $\mathfrak{V}$. By Transfer, $^*\sum$ is a set in $\mathfrak{W}$ and assigns to each internal k-tuple $(a_1,\ldots,a_k)$ in $^*\mathbb{R}$, $k \in {}^*\mathbb{N}_0$, an element $a \in {}^*\mathbb{R}$. This element a, which is uniquely determined by $(a_1,\ldots,a_k)$, is denoted by $a_1 + \ldots + a_k$. By transfer, the function $^*\sum$ has the same formal properties as $\sum$, for example,

$$^*\sum \emptyset = 0, \quad ^*\sum (a_1,\ldots,a_{k+1}) = {}^*\sum (a_1,\ldots,a_k) + a_{k+1}$$

for each internal $k+1$-tuple $(a_1,\ldots,a_{k+1})$ in $^*\mathbb{R}$. We always drop * on $\sum$.

We see that $\sum$ is defined for certain infinite families without using any topology.

Finally, we will show that there exist k-tuples of internal entities in $\mathfrak{W}$, which are not internal. Let $k \in {}^*\mathbb{N} \setminus \mathbb{N}$. Define $f : \{i \in {}^*\mathbb{N} \mid i \leq k\} \to \{0,1\}$, by setting

$$f(i) := \begin{cases} 0 & \text{if } i \in \mathbb{N} \\ 1 & \text{if } i \in {}^*\mathbb{N} \setminus \mathbb{N}. \end{cases}$$

Then f is not internal, because otherwise $\mathbb{N} = f^{-1}[\{0\}]$ is internal, which is not true.

An internal set E in $\mathfrak{W}$ is called **hyperfinite** if E is *finite, but not finite.

8.4 The underspill and overspill principles

The underspill principle tells us that an internal property holds for at least one standard natural number if it holds for arbitrarily small unlimited natural numbers. The overspill principle tells us that an internal property holds for arbitrarily small unlimited natural number if it holds for arbitrarily large standard (limited) natural numbers. In other words, 'potential infinity' is equivalent to 'actual infinity'. More precisely:

Proposition 8.4.1 *Suppose that A is an internal subset of* ${}^*\mathbb{N}$.

(a) *(The underspill principle) Assume that for each unlimited $G \in {}^*\mathbb{N}$ there exists an unlimited $K \in A$ with $K \leq G$. Then A contains arbitrarily large standard positive integers.*
(b) *(The overspill principle) Assume that $A \cap \mathbb{N}$ is infinite. Then for each unlimited number $M \in {}^*\mathbb{N}$ there is an unlimited $K \in A$ with $K < M$.*

Proof (a) Fix $\sigma \in \mathbb{N}$. Since $A_\sigma := \{a \in A \mid \sigma < a\}$ is internal and non-empty, by Proposition 8.2.1 (d) A_σ has a smallest element m. Assume that m is unlimited. Then $m-1$ is unlimited and there exists an unlimited $K \in A$ with $K \leq m-1$, thus $\sigma < K$, which contradicts the minimality of m. It follows that $m \in \mathbb{N}$.

(b) Fix an $M \in {}^*\mathbb{N} \setminus \mathbb{N}$. Set $B_k := \{n \in A \mid k \leq n < M\}$ for each $k \in \mathbb{N}$. Then B_k is internal and $B_k \neq \emptyset$ for all $k \in \mathbb{N}$. It follows that $\mathcal{B} := \{B_k \mid k \in \mathbb{N}\}$ is deep. Since $\mathcal{B}$ is small, by saturation, $\bigcap \mathcal{B} \neq \emptyset$. Each $K \in \bigcap \mathcal{B}$ is unlimited and $K < M$. □

Example 8.4.2 We have the following characterization of an unsolved problem: there exist infinitely many prime numbers n such that $n+2$ is prime if and only if there exists at least one unlimited prime number $n \in^* \mathbb{N}$ such that $n+2$ is prime. So, one example may prove the existence of infinitely many examples. However, this single example is infinitely large, which implies the existence of infinitely many prime-twins.

8.5 The infinitesimals

An element $i \in {}^*\mathbb{R}$ is called **infinitesimal** if, for all standard $n \in \mathbb{N}$,

$$|i| < \frac{1}{n},$$

which is equivalent to $|i| < r$ for each positive standard number $r \in \mathbb{R}$. For example, 0 is infinitesimal, and, if K is an unlimited natural number, then $\frac{1}{K}$ is infinitesimal. We see that, in contrast to $\mathbb{R}$, in ${}^*\mathbb{R}$ there exist infinitesimals different from 0. By $\widetilde{0}$ we denote the set of infinitesimals in $^*\mathbb{R}$. Two elements $a,b \in {}^*\mathbb{R}$ are called **infinitely close** if $a-b \in \widetilde{0}$. Then we will write $a \approx b$.

Proposition 8.5.1 (Robinson [100])

(a) *$\widetilde{0}$ is a subring of $^*\mathbb{R}$.*
(b) *$\sup_{^*<} \widetilde{0}$ and $\inf_{^*<} \widetilde{0}$ do not exist.*
(c) *$\widetilde{0}$ is external.*
(d) *The relation $\approx$ is an equivalence relation on $^*\mathbb{R}$.*

Proof We prove only (b) and (c):

(b) Assume that $i := \sup \widetilde{0}$ exists. Then $0 < i$. Case 1: $i \in \widetilde{0}$. Then, by (a), $i < i+i \in \widetilde{0}$, which contradicts the fact that i is an upper bound of $\widetilde{0}$. Case 2: $i \notin \widetilde{0}$. Then there exists an $n \in \mathbb{N}$ with $\frac{1}{n} < i$. Since $\frac{1}{n}$ is an upper bound of $\widetilde{0}$, we obtain a contradiction to the fact that i is the smallest upper bound of $\widetilde{0}$. In the same way one can show that $\inf \widetilde{0}$ does not exist.

(c) Assume that $\widetilde{0}$ is internal. Since $\widetilde{0}$ is not empty and bounded from above by the number 1 for example, by Proposition 8.1.1, $\widetilde{0}$ has a least upper bound, which contradicts (b). □

We have the following technical, but very useful, corollary to Proposition 8.4.1.

Corollary 8.5.2 *Suppose that $0 < \varepsilon \approx 0$ and $H \in {}^*\mathbb{N}$ with $n \mid H$ for all $n \in \mathbb{N}$ (see Proposition 8.2.1 (g)). Let $(a_{n,m})_{n,m \in {}^*\mathbb{N}}$ be an internal sequence. If $a_{n,m} \approx 0$ for all $n \in \mathbb{N}$ and $m \in {}^*\mathbb{N}$ with $\varepsilon < \frac{m}{H} \approx 0$, then for all $n \in \mathbb{N}$ there exists an $m \in {}^*\mathbb{N}$ such that $m \mid H$ and $\frac{m}{H}$ is not infinitesimal and $\left|a_{n,m}\right| < \frac{1}{2^n}$.*

Proof Assume that the assertion is not true. Then there exists an $n \in \mathbb{N}$ with $\mathbb{N} \subseteq \left\{k \in {}^*\mathbb{N} \mid k \mid H \wedge \varepsilon < \frac{1}{k} \Rightarrow \left|a_{n,\frac{H}{k}}\right| \geq \frac{1}{2^n}\right\}$. By Proposition 8.4.1 (b), this set contains an unlimited number $K \in {}^*\mathbb{N}$. Let $m \cdot K = H$. Then $\varepsilon < \frac{m}{H} = \frac{1}{K} \approx 0$ and $\left|a_{n,\frac{H}{K}}\right| = \left|a_{n,m}\right| \geq \frac{1}{2^n}$, which is a contradiction. □

Fix an $n \in \mathbb{N}$. Then a vector $(i_1,\ldots,i_n) \in {}^*\mathbb{R}^n$ is called **infinitesimal** if each i_k, $k = 1,\ldots,n$, is infinitesimal.

8.6 Limited and unlimited numbers in $^*\mathbb{R}$

An element $a \in {}^*\mathbb{R}$ is called **limited** if there exists a standard $n \in \mathbb{N}$ such that $|a| < n$; otherwise, a is called **unlimited**. For example, infinitesimals are limited and all standard real numbers are limited. We have seen in Section 8.2 that there exist unlimited numbers in ${}^*\mathbb{R}$. An element $a \in {}^*\mathbb{R}$, $a \neq 0$, is unlimited if and only if $\frac{1}{a}$ is infinitesimal. The set of limited numbers is denoted by $\widetilde{\mathbb{R}}$.

Proposition 8.6.1 (Robinson [100])

(a) $\widetilde{\mathbb{R}}$ *is a subring of* ${}^*\mathbb{R}$.
(b) $\widetilde{0}$ *is a maximal ideal in* $\widetilde{\mathbb{R}}$. *It follows that* $\widetilde{\mathbb{R}}$ *factorized by* $\widetilde{0}$ *is* $\mathbb{R}$.
(c) $\widetilde{\mathbb{R}}$ *is external.*
(d) *Given an internal subset A of* $\widetilde{\mathbb{R}}$, *then there exists an* $n \in \mathbb{N}$ *such that*

$$A \subseteq \left\{a \in {}^*\mathbb{R} \mid |a| \leq n\right\}.$$

Proof (a) The proof is left to the reader.

(b) Proposition 8.5.1 (a) implies that $\widetilde{0}$ is a subring of $\widetilde{\mathbb{R}}$. In order to show that $\widetilde{0}$ is an ideal in $\widetilde{\mathbb{R}}$, fix $b \in \widetilde{\mathbb{R}}$, $i \in \widetilde{0}$ and $n \in \mathbb{N}$. There exists an $m \in \mathbb{N}$ with $|b| \leq m$. We obtain

$$|b \cdot i| = |b| \cdot |i| < m \cdot \frac{1}{n \cdot m} = \frac{1}{n}.$$

It follows that $b \cdot i \in \widetilde{0}$. To prove maximality, assume that I is an ideal in $\widetilde{\mathbb{R}}$ with $\widetilde{0} \subsetneqq I$. Choose $b \in I \setminus \widetilde{0}$. Then there exists an $n \in \mathbb{N}$ with $\frac{1}{n} < |b|$, thus $\left|\frac{1}{b}\right| < n$. Therefore $\frac{1}{b}$ is limited. Fix $c \in \widetilde{\mathbb{R}}$. Then

$$c = b \cdot \left(\frac{1}{b} \cdot c\right),$$

where $\frac{1}{b} \cdot c \in \widetilde{\mathbb{R}}$. Since I is an ideal in $\widetilde{\mathbb{R}}$ and $b \in I$, we have $c \in I$. It follows that $\widetilde{\mathbb{R}} = I$, which proves that $\widetilde{0}$ is maximal.

(c) Since $\widetilde{\mathbb{R}} = \bigcup_{n \in \mathbb{N}} A_n$ with $A_n := \{a \in {}^*\mathbb{R} \mid |a| \leq n\}$ and each A_n is internal and $(A_n)_{n \in \mathbb{N}}$ is strictly increasing, the result follows from Example (x) in Section 7.1.

(d) follows from the underspill principle. □

8.7 The standard part map on limited numbers

The next result shows that only the limited elements in ${}^*\mathbb{R}$ are infinitely close to a uniquely determined standard real number. In the proof we use a standard subset A of the real numbers, which is defined by means of a nonstandard element. However, A is internal in $\mathfrak{V}$, because $\mathfrak{V}$ has no external sets.

Proposition 8.7.1 (Robinson [100]) *An element* $\alpha \in {}^*\mathbb{R}$ *is limited if and only if there exists an* $a \in \mathbb{R}$ *with* $\alpha \approx a$. *This* a *is uniquely determined by* α.

Proof Assume that α is limited. Then there exists an $n \in \mathbb{N}$ with $|\alpha| \leq n$. We set $A := \{b \in \mathbb{R} \,|\, b \leq \alpha\}$. Then A is a non-empty set, because $-n \leq \alpha$, and A is bounded from above by n. Since $\mathbb{R}$ is complete, A has a least upper bound a with respect to the standard $<$-relation. We will show that $\alpha \approx a$: Assume that $\alpha \not\approx a$. Then there exists an $m \in \mathbb{N}$ with $\frac{1}{m} < |\alpha - a|$. If $\alpha < a$, then $\alpha < a - \frac{1}{m}$, which contradicts the fact that a is the least upper bound of A. If $a < \alpha$, then $a + \frac{1}{m} < \alpha$, which contradicts the fact that a is an upper bound of A.

Conversely, assume that $a \in \mathbb{R}$ with $\alpha \approx a$. Then $|\alpha| \approx |a|$, because

$$||\alpha| - |a|| \leq |\alpha - a| < r \quad \text{for each } r \in \mathbb{R}, r > 0.$$

Choose $n \in \mathbb{N}$ with $|a| < n$. It follows that $|\alpha| < n$. Therefore, α is limited.

To prove the uniqueness, let $a, a' \in \mathbb{R}$ with $a \approx \alpha$ and $a' \approx \alpha$. By Proposition 8.5.1 (d), $a \approx a'$. Since a, a' are real numbers, this is only possible if $a = a'$. □

If $\alpha \in \widetilde{\mathbb{R}}$, we set

$${}^\circ\alpha := a$$

if a is the uniquely determined element in $\mathbb{R}$ with $\alpha \approx a$. An element $\alpha \in {}^*\mathbb{R}$ is called **nearstandard** if there exists an $a \in \mathbb{R}$ with $\alpha \approx a$, in which case a is called the **standard part of** α. The **standard part map** $\mathrm{st} : \widetilde{\mathbb{R}} \to \mathbb{R}$ is given by

$$\mathrm{st} : \widetilde{\mathbb{R}} \ni \alpha \mapsto {}^\circ\alpha \in \mathbb{R}.$$

We have seen that the set $\widetilde{\mathbb{R}}$ of limited numbers is identical to the set of nearstandard numbers in ${}^*\mathbb{R}$.

Proposition 8.7.2 (Robinson [100]) *The standard part map* $\mathrm{st} : \widetilde{\mathbb{R}} \to \mathbb{R}$ *is a homomorphism with respect to addition and multiplication, i.e., for all* $\alpha, \beta \in \widetilde{\mathbb{R}}$

$${}^\circ(\alpha + \beta) = {}^\circ\alpha + {}^\circ\beta \quad \text{and} \quad {}^\circ(\alpha \cdot \beta) = {}^\circ\alpha \cdot {}^\circ\beta.$$

Proof Fix $\alpha \approx a \in \mathbb{R}$ and $\beta \approx b \in \mathbb{R}$. By the triangle equality, $\alpha + \beta \approx a + b$. It follows that ${}^\circ(\alpha + \beta) = a + b = {}^\circ\alpha + {}^\circ\beta$. Since

$$|\alpha\beta - ab| = |\alpha\beta - a\beta + a\beta - ab| \leq |\beta|\,|\alpha - a| + |a|\,|\beta - b|,$$

by Proposition 8.6.1, $\alpha\beta \approx ab$. It follows that ${}^\circ(\alpha\beta) = ab = {}^\circ\alpha{}^\circ\beta$. □

Exercises

8.1 Fix $a \in {}^*\mathbb{R}$ with $a \neq 0$. Then a is infinitesimal if and only if $\frac{1}{a}$ is unlimited.

8.2 Fix $a \in {}^*\mathbb{R}$ with $a \geq 0$. Then $\frac{a}{1+a} \approx 0$ if and only if $a \approx 0$.
8.3 Prove Proposition 8.6.1 (a) and (d).
8.4 Prove Proposition 8.5.1 (a) and (d).
8.5 Prove the assertion in Example 8.4.2.

9

Topology

In this chapter we apply the saturation principle to elementary topology as far as it is necessary for our purposes.

9.1 Monads

Let $(X, \mathcal{T})$ be a topological space. We always assume that X is a set in our standard model $\mathfrak{V}$. Then $\mathcal{T}$ is a set in $\mathfrak{V}$ too and ${}^*X$ and ${}^*\mathcal{T}$ are sets in our model $\mathfrak{W}$. The elements of $\mathcal{T}$ are called **open in** X. A subset $U \subseteq X$ is called a **neighbourhood of** $a \in X$ **in** $\mathcal{T}$ if there exists a $G \in \mathcal{T}$ such that $a \in G \subseteq U$. We define

$$\widetilde{a}^{\mathcal{T}} := \bigcap \{ {}^*U \mid U \text{ is a neighborhood of } a \text{ in } \mathcal{T} \},$$

and call $\widetilde{a}^{\mathcal{T}}$ the **monad of** a **in** $\mathcal{T}$. Note that ${}^*a \in \widetilde{a}^{\mathcal{T}}$. If $\alpha \in \widetilde{a}^{\mathcal{T}}$, then we say that α is **infinitely close to** a **or to** ${}^*a$ **in** $\mathcal{T}$ and write $\alpha \approx_{\mathcal{T}} a$ or $\alpha \approx_{\mathcal{T}} {}^*a$. We drop the index '$\mathcal{T}$' if it is clear which topology we mean.

A point $\alpha \in {}^*X$ is called **nearstandard** if there exists an $a \in X$ with $\alpha \approx_{\mathcal{T}} {}^*a$. Let $\widetilde{X}$ denote the **set of nearstandard points** of X.

In $\mathbb{R}^n$ these notions of 'infinitely close' and 'nearstandard' coincide with the definitions in Sections 8.5 and 8.7.

It should be mentioned here that the notion of 'infinitely close' in arbitrary topological spaces is just a relation between elements of X and ${}^*X$. If X is a uniform space (for example, X is a metric space or a topological vector space), then this relation can be extended to pairs of elements of ${}^*X$.

Since Fréchet spaces are important here, we now fix a Fréchet space $(\mathbb{B}, (|\cdot|_j)) = (\mathbb{B}, d)$ within the standard model. For example, $\mathbb{B} = \mathbb{H}^{\otimes d}$ or $\mathbb{B} = \mathbb{R}$. Since $|\cdot|_j$ is a function defined on $\mathbb{B}$ with values in $[0, \infty[$, by the transfer principle, ${}^*|\cdot|_j$ is a mapping defined on ${}^*\mathbb{B}$ with values in ${}^*[0, \infty[$ with the same formal properties. We drop the star on $|\cdot|_j$. The 'infinitely close' relation in the topology, induced by d, is denoted by $\approx$. For $\alpha, \beta \in {}^*\mathbb{B}$ we will write $\alpha \approx_j \beta$ if and only if $|\alpha - \beta|_j \approx 0$. We obtain $\alpha \approx \beta$ if and only if $\alpha \approx_j \beta$ for all $j \in \mathbb{N}$. For each $\varepsilon > 0$ and each $a \in \mathbb{B}$ set

$$U_\varepsilon(a) := \{x \in X \mid d(x,a) < \varepsilon\} \quad \text{and} \quad \overline{U_\varepsilon(a)} := \{x \in X \mid d(x,a) \leq \varepsilon\}.$$

The next lemma is important for what follows. It says that the monad of a point a can be sandwiched between all *images of neighbourhoods of a and a nonstandard neighbourhood of ${}^*a$:

Lemma 9.1.1 *For each $a \in X$ there exists a $G \in {}^*\mathcal{T}$ with ${}^*a \in G \subseteq \widetilde{a}$.*

Proof For each neighbourhood U of a fixed $a \in X$ let

$$B_U := \left\{O \in {}^*\mathcal{T} \mid {}^*a \in O \subseteq {}^*U\right\}.$$

Then B_U is an internal set and $\mathcal{B} := \{B_U \mid U \text{ is neighbourhood of } a\}$ is small and deep. By saturation, there exists a $G \in \bigcap \mathcal{B}$. Therefore, ${}^*a \in G \subseteq \widetilde{a}$ and $G \in {}^*\mathcal{T}$. □

Proposition 9.1.2 (Robinson [100]) *Fix $G \subseteq X$. Then G is open if and only if $\widetilde{a} \subseteq {}^*G$ for each $a \in G$.*

Proof Let G be open and $a \in G$. Since G is a neighbourhood of a, $\widetilde{a} \subseteq {}^*G$.

Conversely, suppose that $\widetilde{a} \subseteq {}^*G$ for each $a \in G$. Fix $b \in G$. By Lemma 9.1.1, there exists an $A \in {}^*\mathcal{T}$ with ${}^*b \in A \subseteq {}^*G$. By transfer, there exists an $A \in \mathcal{T}$ with $b \in A \subseteq G$, thus G is open. □

Since closed sets are complements of open sets we obtain the following corollary:

Corollary 9.1.3 (Robinson [100]) *A subset $F \subseteq X$ is closed if and only if $a \in F$ for all $a \in X$ with $\widetilde{a} \cap {}^*F \neq \emptyset$.*

9.2 Hausdorff spaces

Proposition 9.2.1 (Robinson [100]) *A topological space $(X, \mathcal{T})$ is a Hausdorff space if and only if $\widetilde{a} \cap \widetilde{b} = \emptyset$ for each $a, b \in X$ with $a \neq b$.*

Proof Assume that $(X, \mathcal{T})$ is a Hausdorff space and $a \neq b$, $a, b \in X$. Then there exist neighbourhoods U and V of a and b, respectively, such that $U \cap V = \emptyset$. By the transfer principle, ${}^*U \cap {}^*V = \emptyset$. Since $\widetilde{a} \subseteq {}^*U$ and $\widetilde{b} \subseteq {}^*V$, $\widetilde{a} \cap \widetilde{b} = \emptyset$.

Now assume that $\widetilde{a} \cap \widetilde{b} = \emptyset$. By Lemma 9.1.1, there exist $G_a, G_b \in {}^*\mathcal{T}$ such that ${}^*a \in G_a \subseteq \widetilde{a}$ and ${}^*b \in G_b \subseteq \widetilde{b}$, thus $G_a \cap G_b = \emptyset$. By transfer, there are $G_a, G_b \in \mathcal{T}$ such that $a \in G_a$, $b \in G_b$, $G_a \cap G_b = \emptyset$. □

In Hausdorff spaces $(X, \mathcal{T})$ for each $\alpha \in \widetilde{X}$ there exists a unique $a \in X$ with $\alpha \in \widetilde{a}$. In analogy to the real numbers, this a is called the **standard part** of α, denoted by $°\alpha$, and the mapping $\mathrm{st} : \widetilde{X} \to X$, $\alpha \mapsto °\alpha$ is called the **standard part map on** X.

9.3 Continuity

Proposition 9.3.1 (Robinson [100]) *Fix two topological spaces $(X, \mathcal{T})$ and $(Y, \mathcal{S})$ in the standard model and fix $a \in X$. Then $f : X \to Y$ is continuous in a if and only if $({}^*f)(\alpha) \approx f(a)$ for all $\alpha \in {}^*X$ with $\alpha \approx a$.*

Proof Assume that f is continuous in a and $\alpha \in \widetilde{a}$. In order to show that ${}^*f(\alpha) \approx f(a)$, fix a neighbourhood $U_{f(a)}$ of $f(a)$. Then there exists a neighbourhood U_a of a such that $f[U_a] \subseteq U_{f(a)}$. By transfer, ${}^*f[{}^*U_a] \subseteq {}^*U_{f(a)}$. Since $\alpha \in {}^*U_a$, we have $({}^*f)(\alpha) \in {}^*U_{f(a)}$. Since $U_{f(a)}$ was an arbitrary neighbourhood of $f(a)$, $({}^*f)(\alpha) \approx f(a)$.

Conversely, assume that $({}^*f)(\alpha) \approx f(a)$ for all $\alpha \in \widetilde{a}$. In order to show that f is continuous in a, fix a neighbourhood $U_{f(a)}$ of $f(a)$. By Lemma 9.1.1, there exists $G \in {}^*\mathcal{T}$ with ${}^*a \in G \subseteq \widetilde{a}$, thus $({}^*f)[G] \subseteq {}^*U_{f(a)}$. By transfer, there is a $G \in \mathcal{T}$ with $a \in G$ and $f[G] \subseteq U_{f(a)}$, which proves the continuity of f in a. □

9.4 Compactness

Now we will give a nice characterization of compactness due to Robinson [100].

Proposition 9.4.1 (Robinson [100]) *Let $(X, \mathcal{T})$ be a topological space in the standard model. A subset $K \subseteq X$ is compact if and only if for each $\alpha \in {}^*K$ there exists an $a \in K$ with $\alpha \approx {}^*a$.*

Proof Let K be compact and $\alpha \in {}^*K$. Assume that $\alpha \not\approx {}^*a$ for each $a \in K$. Then for each $a \in K$ there exists an open neighbourhood $U_a(a)$ of a such that $\alpha \notin {}^*U_a(a)$. Since $\{U_a(a) \mid a \in K\}$ is an open cover of K and K is compact, there exist finitely many $a_1, \ldots, a_k \in K$ with

$$K \subseteq U_{a_1}(a_1) \cup \ldots \cup U_{a_k}(a_k),$$

thus

$$^*K \subseteq {}^*U_{a_1}(a_1) \cup \ldots \cup {}^*U_{a_k}(a_k).$$

Since $\alpha \in {}^*K$, there exists an $i \in \{1,\ldots,k\}$ such that $\alpha \in {}^*U_{a_i}(a_i)$, which is a contradiction.

Conversely, to prove that K is compact, fix an open cover $\mathcal{G}$ of K. Set $A_G := {}^*K \setminus {}^*G$ for each $G \in \mathcal{G}$ and set $\mathcal{B} := \{A_G \mid G \in \mathcal{G}\}$. Assume that there does not exist a finite subset $\{G_1,\ldots,G_k\} \subseteq \mathcal{G}$ with $K \subseteq G_1 \cup \ldots \cup G_k$. Since $K \setminus G \neq \emptyset$ implies ${}^*K \setminus {}^*G \neq \emptyset$, $\mathcal{B}$ is deep. Since $\mathcal{B}$ is small, by saturation, there exists an $\alpha \in {}^*K \setminus {}^*G$ for each $G \in \mathcal{G}$. Let $a \in K$ with $\alpha \approx a$. Since there exists a $G \in \mathcal{G}$ with $a \in G$ and since G is open and $\alpha \approx a$, we have $\alpha \in {}^*G$, which is a contradiction. □

The previous characterization shows how closely linked compactness and finiteness are: Note that a subset $K \subseteq X$ is finite if and only if ${}^*[K] := \{{}^*a \mid a \in K\} = {}^*K$, i.e.,

$$\forall \alpha \in {}^*K \exists a \in K (\alpha = {}^*a). \tag{10}$$

Now K is compact if and only if we replace '$=$' in (10) by '$\approx$', i.e., K is compact if and only if

$$\forall \alpha \in {}^*K \exists a \in K (\alpha \approx {}^*a).$$

9.5 Convergence

The following technical result is a useful tool. In particular, part (b) opens the possibility to obtain the limit of a Cauchy sequence $(b_n)_{n \in \mathbb{N}}$ by taking the standard part of some b_M where $M \in {}^*\mathbb{N}$ is an unlimited number.

Proposition 9.5.1 (Robinson [100]) *Let $(a_n)_{n \in \mathbb{N}}$ and $(b_{n,m})_{n,m \in \mathbb{N}}$ be sequences of nearstandard points in ${}^*\mathbb{B}$ and let $a \in \mathbb{B}$. (By Proposition 8.2.2, there exist internal extensions $(a_n)_{n \in {}^*\mathbb{N}}$, $(b_{n,m})_{n,m \in {}^*\mathbb{N}}$ in ${}^*\mathbb{B}$ of $(a_n)_{n \in \mathbb{N}}$, $(b_{n,m})_{n,m \in \mathbb{N}}$, respectively.)*

(a) *Suppose that $\lim_{n,m \to \infty} {}^\circ b_{n,m} = 0$. Then there exists a strictly monotone increasing function $g : \mathbb{N} \to \mathbb{N}$ and an unlimited $M \in {}^*\mathbb{N}$ such that for all unlimited $K \in {}^*\mathbb{N}$ with $K \leq M$ and all $m \in \mathbb{N}$*

$$d(b_{K,g(m)}, 0) < \frac{1}{m}.$$

(b) *$\lim_{n \to \infty} ({}^\circ a_n) = a$ if and only if there exists an unlimited $K \in {}^*\mathbb{N}$ such that $a_M \approx a$ for each unlimited $M \in {}^*\mathbb{N}$ with $M \leq K$.*

(c) *Let* $f : \mathbb{N} \to \mathbb{B}$ *be a sequence in* $\mathbb{B}$ *and* $b \in \mathbb{B}$. *Then* $\lim_{k\to\infty} f_k = b$ *if and only if* ${}^*f(K) \approx {}^*b$ *for all unlimited* $K \in {}^*\mathbb{N}$.

(d) *Let* p *be a limited number in* ${}^*\mathbb{R}$. *Then* $\left(1+\frac{p}{H}\right)^H \approx e^p$ *for all unlimited* $H \in {}^*\mathbb{N}$.

Proof (a) It is obvious that there exists a strictly monotone increasing function $g : \mathbb{N} \to \mathbb{N}$ such that for all $m \in \mathbb{N}$

$$\mathbb{N} \subseteq \left\{ n \in {}^*\mathbb{N} \mid \forall k \in {}^*\mathbb{N}(g(m) \leq k \leq n \to \left(d(b_{k,g(m)}, 0) < \frac{1}{m} \right) \right\}$$

where we have dropped the stars on g and d. By Proposition 8.4.1 (b), for each $m \in \mathbb{N}$ there exists an unlimited $N_m \in {}^*\mathbb{N}$ such that $d(b_{K,g(m)}, 0) < \frac{1}{m}$ for all unlimited $K \in {}^*\mathbb{N}$ with $K \leq N_m$. By Proposition 8.2.1 (f), there exists an unlimited $M \in {}^*\mathbb{N}$ with $M \leq N_m$ for all $m \in \mathbb{N}$. It follows that $d(b_{K,g(m)}, 0) < \frac{1}{m}$ for all $m \in \mathbb{N}$ and all unlimited $K \in {}^*\mathbb{N}$ with $K \leq M$.

(b) Suppose that $\lim_{n\to\infty}({}^\circ a_n) = a$. Then there exists a strictly increasing function $g : \mathbb{N} \to \mathbb{N}$ such that $d({}^\circ a_n, a) < \frac{1}{m}$, thus $d(a_n, a) < \frac{1}{m}$, for all $m, n \in \mathbb{N}$ with $n \geq g(m)$. The proof is now similar to the proof of (a).

Conversely, assume that $\lim_{n\to\infty}({}^\circ a_n) = a$ fails. Then there exists a neighbourhood U of a such that the set of all $n \in \mathbb{N}$ with ${}^\circ a_n \notin U$ is infinite. Since $\mathbb{B}$ is a Fréchet space, we may assume that U is closed. By Proposition 9.1.2, the set of all $n \in \mathbb{N}$ with $a_n \notin {}^*U$ is also infinite. By Proposition 8.4.1 (b), for each unlimited $K \in {}^*\mathbb{N}$ there exists an unlimited $M \in {}^*\mathbb{N}$ with $M \leq K$ and $a_M \notin {}^*U$, thus $a_M \not\approx a$.

(c) The proof is similar to the proof of Proposition 9.3.1 and is left to the reader. The proof of (d) is also left to the reader. □

9.6 The standard part of an internal set of nearstandard points is compact

The result in the title of this section is due to Luxemburg [74], who proved it for arbitrary regular topological spaces. An element $\alpha \in {}^*\mathbb{B}$ is called **limited** if $|\alpha|_j$ is limited for all $j \in \mathbb{N}$. As we have seen in Section 8.7, in the case of the real numbers the notion 'limited' is equivalent to 'nearstandard'. However, in general, nearstandard is much stronger than limited. The reason is that nearstandardness has to do with compactness, which follows from the next proposition. The proof shows that the result can be generalized to regular spaces. Recall that a topological space is called **regular** if each point has a neighbourhood base of closed sets. For example, pre-Fréchet spaces are regular.

Proposition 9.6.1 (Luxemburg [74]) *Let A be an internal set of nearstandard points in ${}^*\mathbb{B}$. Then*

$$\mathrm{st}\,[A] = \{{}^\circ\alpha \mid \alpha \in A\} \text{ is compact.}$$

Proof Let $\{G_i \mid i \in I\}$ be a cover of $\mathrm{st}\,[A]$ by open sets G_i. Because the set of open sets in $\mathbb{B}$ is small, we may assume that I is small. For each $a \in \mathrm{st}\,[A]$ and each $i \in I$ such that $a \in G_i$ there exists a standard $\varepsilon_{a,i} > 0$ such that

$$a \in U_{\varepsilon_{a,i}}(a) \subseteq \overline{U_{\varepsilon_{a,i}}(a)} \subseteq G_i,$$

where $\overline{U_{\varepsilon_{a,i}}(a)}$ is the closed hull of $U_{\varepsilon_{a,i}}(a)$. By Proposition 9.1.2

$$A \subseteq \bigcup \left\{ {}^*U_{\varepsilon_{a,i}}(a) \mid i \in I \text{ and } a \in G_i \cap \mathrm{st}[A] \right\}.$$

By saturation, there exist finitely many ${}^*U_{\varepsilon_{a_1,i_1}}(a_1), \ldots, {}^*U_{\varepsilon_{a_k,i_k}}(a_k)$ such that $A \subseteq {}^*U_{\varepsilon_{a_1,i_1}}(a_1) \cup \ldots \cup {}^*U_{\varepsilon_{a_k,i_k}}(a_k)$. We shall see now that

$$\mathrm{st}[A] \subseteq F := \overline{U_{\varepsilon_{a_1,i_1}}(a_1)} \cup \ldots \cup \overline{U_{\varepsilon_{a_k,i_k}}(a_k)}. \tag{11}$$

Fix $a \in \mathrm{st}[A]$. Then there exists an $\alpha \in A$ with $\alpha \approx a$. Assume that $a \notin F$. By Proposition 9.1.2, $\alpha \notin {}^*F \supseteq A$, which is a contradiction. This proves that (11) is true. It follows that $\{G_i \mid i \in I\}$ contains a finite subset covering $\mathrm{st}[A]$. □

9.7 From S-continuous to continuous functions

In order to construct Brownian motion, the Itô integral and the iterated Itô integral in abstract Wiener spaces we will use the fact that it is possible to construct standard continuous functions from internal S-continuous functions, defined on a *finite set.

Here we fix an unlimited $H \in {}^*\mathbb{N}$ such that, for technical reasons, each $n \in \mathbb{N}$ divides H and we fix $T := \left\{ \frac{1}{H}, \ldots, \frac{H^2}{H} \right\}$ (see Proposition 8.2.1 (g)) and set

$$\widetilde{T} := \{t \in T \mid t \text{ is limited}\}.$$

It is easy to see that ${}^\circ T := \left\{ {}^\circ t \mid t \in \widetilde{T} \right\} = [0, \infty[$.

Let $\mathbb{B}$ now be a metric space with metric d. Then $\alpha \approx \beta$ in $\mathbb{B}$ if and only if ${}^*d(\alpha, \beta) \approx 0$. An internal function $F : T \to {}^*\mathbb{B}$ is called S-**continuous** if

(a) for all $t \in \widetilde{T}$, $F(t)$ is nearstandard in $\mathbb{B}$ and
(b) for all $s, t \in \widetilde{T} (s \approx t \Rightarrow F(s) \approx F(s) \approx F(t))$.

If condition (b) is true, we call F **pre-S-continuous.**

Let F be S-continuous. We may define for all $a \in [0,\infty[$ and all $t \in \widetilde{T}$ with $t \approx a$

$$f(a) := {}^\circ F(t).$$

Because of (a) and (b), $f : [0,\infty[\to \mathbb{B}$ is well defined. This function is called the **continuous standard part** of F, which is justified by

Proposition 9.7.1 (Albeverio *et al.* [2]) *$f : [0,\infty[\to \mathbb{B}$ is continuous.*

Proof Assume that f is not continuous in $a \in [0,\infty[$. Then there exists a standard $\varepsilon > 0$ such that for each $k \in \mathbb{N}$ there is $b \in [0,\infty[$ with $|b-a| < \frac{1}{k}$ and $d(f(a), f(b)) > \varepsilon$. Let $s,t \in \widetilde{T}$ with $s \approx a$ and $t \approx b$. Then $|s-t| < \frac{1}{k}$ and, since $F(s) \approx f(a)$ and $F(t) \approx f(b)$, we have ${}^*d(F(s), F(t)) > \varepsilon$ (see Proposition 9.1.2). We obtain for all $k \in \mathbb{N}$

$$\exists s,t \in T \left(a - \frac{1}{k} < s,t < a + \frac{1}{k} \wedge {}^*d(F(s), F(t)) > \varepsilon \right). \tag{12}$$

By Proposition 8.4.1 (b), there exists an unlimited $K \in {}^*\mathbb{N}$ such that (12) is true with K instead of k. It follows that there exist $s,t \in T$ with $s \approx t \approx a$ and ${}^*d(F(s), F(t)) > \varepsilon$, contradicting the S-continuity of F. □

9.8 Hyperfinite representation of the tensor product

Real separable Hilbert spaces play an important role in the theory of abstract Wiener spaces. Instead of an infinite-dimensional Hilbert space $\mathbb{H}$ (inside the standard model $\mathfrak{V}$) we would like to use a suitable *finite-dimensional linear space $\mathbb{F}$ over ${}^*\mathbb{R}$ between $\mathbb{H}$ and ${}^*\mathbb{H}$, which exists, by saturation. Here we identify each element in $\mathbb{H}$ with its *-image in the poly-saturated model.

If $E \in \mathcal{E}(\mathbb{H})$, then ${}^*E$ is a finite-dimensional space over ${}^*\mathbb{R}$: $({}^*\mathfrak{e}_1, \ldots, {}^*\mathfrak{e}_n)$ is an (orthonormal) basis of ${}^*E$ if $(\mathfrak{e}_1, \ldots, \mathfrak{e}_n)$ is an (orthonormal) basis of E. Let $\mathbb{F} \in {}^*\mathcal{E}(\mathbb{H})$ with ONB $\mathfrak{E} := (\mathfrak{e}_1, \ldots, \mathfrak{e}_k)$. Then $k \in {}^*\mathbb{N}$. For each $d \in \mathbb{N}$ and each $F \in {}^*\mathbb{H}^{\otimes d}$ define

$$\|F\|_{\mathbb{F}^d} := \left(\sum_{\mathfrak{e} \in \mathfrak{E}^d} F^2(\mathfrak{e}) \right)^{\frac{1}{2}}.$$

By Lemma 4.2.1 and the transfer principle, the value does not depend on the ONB.

Theorem 9.8.1 *There exists an* $\mathbb{F} \in {}^*\mathcal{E}(\mathbb{H})$ *such that*

(a) ${}^*E \subseteq \mathbb{F}$ *for all* $E \in \mathcal{E}(\mathbb{H})$,

(b) $\|f\|_{\mathbb{F}^d} := \|{}^*f\|_{\mathbb{F}^d} \approx \|f\|_{\mathbb{H}^d}$ *for all* $d \in \mathbb{N}$ *and all* $f \in \mathbb{H}^{\otimes d}$,

(c) *each ONB* $(\mathfrak{e}_i)_{i\in\mathbb{N}}$ *of* $\mathbb{H}$ *can be extended to an internal ONB* $(\mathfrak{b}_i)_{i\leq\omega}$ *of* $\mathbb{F}$, *i.e.,* $(\mathfrak{b}_i)_{i\leq\omega}$ *is an ONB of* $\mathbb{F}$ *and* $\mathfrak{b}_i = {}^*\mathfrak{e}_i$ *for all* $i \in \mathbb{N}$.

Proof We will first show that there exists an $\mathbb{F} \in {}^*\mathcal{E}(\mathbb{H})$ such that (a) and (13) are true, where

$$\|{}^*f\|^2_{\mathbb{F}^d} \leq \|f\|^2_{\mathbb{H}^d} \text{ for all } d \in \mathbb{N} \text{ and all } f \in \mathbb{H}^{\otimes d}. \tag{13}$$

By saturation, it suffices to show that there exists an $\mathbb{F} \in {}^*\mathcal{E}(\mathbb{H})$ such that (13) is true and ${}^*E_1, \ldots, {}^*E_k \subseteq \mathbb{F}$ for finitely many $E_1, \ldots, E_k \in \mathcal{E}(\mathbb{H})$: Let $(\mathfrak{e}_1, \ldots, \mathfrak{e}_m)$ be an ONB of $E_1 + \ldots + E_k$ and set $\mathbb{F} := {}^*E_1 + \ldots + {}^*E_k$. Then for each $d \in \mathbb{N}$ and each $f \in \mathbb{H}^{\otimes d}$

$$\|f\|^2_{\mathbb{F}^d} = \sum_{(i_1,\ldots,i_d)\in m^d} {}^*f^2({}^*\mathfrak{e}_{i_1}, \ldots, {}^*\mathfrak{e}_{i_d}) = \sum_{(i_1,\ldots,i_d)\in m^d} f^2(\mathfrak{e}_{i_1}, \ldots, \mathfrak{e}_{i_d}) \leq \|f\|^2_{\mathbb{H}^d}.$$

Now fix an $\mathbb{F} \in {}^*\mathcal{E}(\mathbb{H})$ such that (a) and (13) are true, and fix an ONB $(\mathfrak{e}_i)_{i\in\mathbb{N}}$ of $\mathbb{H}$. Since, for all $d \in \mathbb{N}$, $(\mathfrak{e}_1, \ldots, \mathfrak{e}_d) := ({}^*\mathfrak{e}_1, \ldots, {}^*\mathfrak{e}_d)$ can be extended to an internal ONB of $\mathbb{F}$, by saturation, $(\mathfrak{e}_i)_{i\in\mathbb{N}}$ can be extended to an internal ONB $(\mathfrak{b}_i)_{i\leq\omega}$ of $\mathbb{F}$. This proves (c).

We now prove that (b) is true for $\mathbb{F}$: By (13), $\|{}^*f\|_{\mathbb{F}^d}$ is nearstandard for all $d \in \mathbb{N}$ and all $f \in \mathbb{H}^{\otimes d}$. Moreover, since for all $m \in \mathbb{N}$

$$\sum_{(i_1,\ldots,i_d)\in m^d} f^2(\mathfrak{e}_{i_1}, \ldots, \mathfrak{e}_{i_d}) \leq {}^\circ \sum_{(i_1,\ldots,i_d)\in\omega^d} {}^*f^2(\mathfrak{b}_{i_1}, \ldots, \mathfrak{b}_{i_d}),$$

$\|f\|^2_{\mathbb{H}^d} \leq {}^\circ \|{}^*f\|^2_{\mathbb{F}^d}$. The assertion (b) follows. □

For $d \in \mathbb{N}$ let $\mathbb{F}^{\otimes d}$ be the space of all internal *real-valued multilinear forms on $\mathbb{F}^d$, endowed with the norm $\|\cdot\|_{\mathbb{F}^d}$. The corresponding scalar product is denoted by $\langle\cdot,\cdot\rangle_{\mathbb{F}^d}$, that is

$$\langle F, G\rangle_{\mathbb{F}^d} := \sum_{\mathfrak{e}\in\mathfrak{E}^d} F(\mathfrak{e}) \cdot G(\mathfrak{e}),$$

where $\mathfrak{E}$ is an internal ONB of $\mathbb{F}$. Note that $\mathbb{F}^{\otimes 1} = \mathbb{F}$ and $\|\cdot\|_{\mathbb{F}^1} = \|\cdot\|$ on $\mathbb{F}$. For $F, G \in {}^*\mathbb{H}^{\otimes d} \cup \mathbb{F}^{\otimes d}$ we define

$$F \approx_{\mathbb{F}^d} G \text{ if and only if } \sum_{\mathfrak{e}\in\mathfrak{E}^d} (F - G)^2(\mathfrak{e}) \approx 0.$$

An element $G \in \mathbb{F}^{\otimes d}$ is called **nearstandard** $\left(\textbf{in } \mathbb{H}^{\otimes d}\right)$ if there exists a $g \in \mathbb{H}^{\otimes d}$ such that ${}^*g \approx_{\mathbb{F}^d} G$. Then g is called the **standard part** of G. Let us write $g \approx_{\mathbb{F}^d} G$ instead of ${}^*g \approx_{\mathbb{F}^d} G$. By Proposition 9.8.1 (b), each G has at most one standard part. Therefore, we may denote the standard part of G by ${}^\circ G$ in case it exists.

Corollary 9.8.2 *For each ONB $\mathfrak{E}$ of $\mathbb{F}$,*

$$\sum_{\mathfrak{e} \in \mathfrak{E}^d} \left({}^*({}^\circ G) - G\right)^2 (\mathfrak{e}) \approx 0 \quad \textit{and} \quad \|{}^\circ G\|_{\mathbb{H}^d} \approx \|G\|_{\mathbb{F}^d}.$$

Since the scalar product can be defined in terms of the norm, we obtain for nearstandard $F, G \in \mathbb{F}^{\otimes d}$

$$\langle {}^\circ G, {}^\circ F \rangle_{\mathbb{H}^d} \approx \langle G, F \rangle_{\mathbb{F}^d}.$$

Finally, note that the nearstandard multilinear forms are closed under addition and scalar multiplication with limited elements in ${}^*\mathbb{R}$ and tensor products. We will only show that the nearstandard elements are closed under tensor products. For each $f \in \mathbb{H}^{\otimes d}$ and $g \in \mathbb{H}^{\otimes m}$ define the **tensor product** $f \otimes g : \mathbb{H}^{d+m} \to \mathbb{R}$ by setting

$$f \otimes g(x,y) := f(x) \cdot g(y) \quad \text{for } x \in \mathbb{H}^d, y \in \mathbb{H}^m.$$

Obviously, $f \otimes g \in \mathbb{H}^{\otimes(d+m)}$. In the same way the tensor product between functions in $\mathbb{F}^{\otimes d}$ and $\mathbb{F}^{\otimes m}$ is defined. We obtain the following simple fact:

Proposition 9.8.3 *If $F \in \mathbb{F}^{\otimes d}$ and $G \in \mathbb{F}^{\otimes m}$ are nearstandard, then $F \otimes G$ is nearstandard and*

$${}^\circ(F \otimes G) = {}^\circ F \otimes {}^\circ G.$$

Proof To save indices, we assume that $d = m = 1$. Let $g = {}^\circ G$ and $f = {}^\circ F$. Let $(\mathfrak{e}_i)_{i \in \mathbb{N}}$ be an ONB of $\mathbb{H}$ and $(\mathfrak{e}_i)_{i \leq \omega}$ be an extension of $(\mathfrak{e}_i)_{i \in \mathbb{N}}$ to an internal ONB of $\mathbb{F}$. Then, by Hölder's inequality,

$$\left(\sum_{(i,j) \in \omega^2} \left({}^*(f \otimes g) - (F \otimes G)\right)^2 (\mathfrak{e}_i, \mathfrak{e}_j) \right)^{\frac{1}{2}}$$

$$= \left(\sum_{(i,j) \in \omega^2} \left({}^*f(\mathfrak{e}_i) \cdot {}^*g(\mathfrak{e}_j) - F(\mathfrak{e}_i) \cdot G(\mathfrak{e}_j)\right)^2 \right)^{\frac{1}{2}}$$

$$\leq \left(\sum_{(i,j)\in\omega^2} {}^*f^2(\mathfrak{e}_i)\cdot\left({}^*g-G\right)^2(\mathfrak{e}_j)\right)^{\frac{1}{2}} + \left(\sum_{(i,j)\in\omega^2} G^2(\mathfrak{e}_j)\cdot\left({}^*f-F\right)^2(\mathfrak{e}_i)\right)^{\frac{1}{2}}$$

$$= \left\|{}^*g-G\right\|_{\mathbb{F}}\cdot\left\|{}^*f\right\|_{\mathbb{F}}+\|G\|_{\mathbb{F}}\cdot\left\|{}^*f-F\right\|_{\mathbb{F}}\approx 0,$$

because $\|{}^*f\|_{\mathbb{F}}\approx\|f\|_{\mathbb{H}}$, $\|G\|_{\mathbb{F}}\approx\|{}^*g\|_{\mathbb{F}}\approx\|g\|_{\mathbb{H}}$ and ${}^*g\approx_{\mathbb{F}}G$ and ${}^*f\approx_{\mathbb{F}}F$. Recall that $\|F\|_{\mathbb{F}}^2=\sum_{i\leq\omega}F^2(\mathfrak{e}_i)$. □

9.9 †The Skorokhod topology

In this section we study the Skorokhod topology on the set J of càdlàg functions; these are right continuous functions $f:[0,\infty[\to\mathbb{R}^d$ with left-hand limits. By using the *finite set $T=\left\{\frac{1}{H},\ldots,\frac{H^2}{H}\right\}$, we characterize the Skorokhod topology. The space J is important, because martingales and Lévy processes, defined on the timeline $[0,\infty[$, do have càdlàg versions. Following the work of Lindstrøm [67], in Chapter 15 we will construct from internal Lévy processes defined on T standard Lévy processes on the timeline $[0,\infty[$, whose sample paths are functions in J. Moreover, each Lévy process existing on a suitable probability space can be essentially obtained in this way. Hoover and Perkins [46] and Lindstrøm [64] have constructed standard martingales, whose sample paths are functions in J, from internal martingales, and vice versa. For simplicity we set $d=1$.

A function $f:[0,\infty[\to\mathbb{R}$ is called **càdlàg** if, for each $r\in[0,\infty[$, $f(r^-):=\lim_{x\uparrow r}f(x)$ exists for $r>0$ and $f(r^+):=\lim_{x\downarrow r}f(x)=f(r)$. Set $f(0^-):=0$. Note that the set J of càdlàg functions is a set in the standard model $\mathfrak{V}$. The proof of the following proposition is similar to the proof of Proposition 9.3.1.

Proposition 9.9.1 (Robinson [100]) *Fix $r\in[0,\infty[$ and $b\in\mathbb{R}$.*

(a) $\lim_{x\downarrow r}f(x)=b\Leftrightarrow\forall\alpha\in{}^*[0,\infty[,\alpha\approx r\,(\alpha>r\Rightarrow({}^*f)(\alpha)\approx b)$.
(b) *For $r>0$,*

$$\lim_{x\uparrow r}f(x)=b\Leftrightarrow\forall\alpha\in{}^*[0,\infty[,\alpha\approx r\left(\alpha<r\Rightarrow\left({}^*f\right)(\alpha)\approx b\right).$$

For $k\in\mathbb{N}$ let us denote by C_k the set of all continuous strictly increasing functions from $[0,k]$ onto $[0,k]$. It follows that $\lambda(0)=0$ and $\lambda(k)=k$ if $\lambda\in C_k$. Note that C_k is a set in $\mathfrak{V}$. For each $f,g\in J$ we define

$$d_k(f,g):=\inf\left\{c>0\mid\exists\lambda\in C_k\left(\sup_{t\in[0,k]}\{|\lambda(t)-t|,|f(\lambda(t))-g(t)|\}\leq c\right)\right\}.$$

Note that d_k is a set in $\mathfrak{V}$. We refer the reader to the proof in Billingsley [12] (in his work $k = 1$), which shows that d_k is a metric on the set

$$J_k := \{f \in J \mid f(x) = f(k) \text{ for all } x \in [k, \infty[\}.$$

Note that d is a set in $\mathfrak{V}$ and assigns to each $k \in \mathbb{N}$ the metric d_k on J_k. Thus $^*d$ is well defined and assigns to each $k \in {}^*\mathbb{N}$ the metric $(^*d)_k$ (with values in $^*\mathbb{R}_0^+$) on $^*J_k$. In Billingsley's work, one can find more information about the metric d_k; in particular, the metric space (J_k, d_k) is separable. The reason is, roughly speaking, that it is possible to shift, by means of $\lambda \in C_k$, arbitrary finite sequences r of real numbers in $[0,k]$ to finite sequences q of rationals in $[0,k]$ close to r, which form a countable set. Now define a metric d_∞ on J, the so-called **Skorokhod metric on** J, by

$$d_\infty(f,g) := \sum_{k \in \mathbb{N}} \frac{1}{2^k} \frac{d_k(f,g)}{1 + d_k(f,g)}.$$

We have the following characterization of 'infinitely close' in the Skorokhod metric.

Proposition 9.9.2 *Fix $F, G \in {}^*J$. Then $F \approx_{d_\infty} G$ if and only if for each $k \in \mathbb{N}$ there exists a $\lambda \in {}^*C_k$ such that $\lambda(t) \approx t$ and $F(\lambda(t)) \approx G(t)$ for all $t \in {}^*[0,k]$.*

Proof Suppose $F \approx_{d_\infty} G$, i.e., $\sum_{k \in {}^*\mathbb{N}} \frac{1}{2^k} \frac{(^*d)_k(F,G)}{1+(^*d)_k(F,G)} \approx 0$. It follows that $^*(d_k)(F,G) = (^*d)_k(F,G) < \delta$ for some $\delta \approx 0$ for all standard $k \in \mathbb{N}$. For all $f, g \in J_k$ and for all $\varepsilon > 0$ with $d_k(f,g) < \varepsilon$ there exists a $\lambda \in C_k$ such that $|\lambda(t) - t| < \varepsilon$ and $|f(\lambda(t)) - g(t)| < \varepsilon$ for all $t \in [0,k]$. Since $*$ is an elementary embedding, we have for F, G and δ: there is a $\lambda \in {}^*C_k$ with $|\lambda(t) - t| < \delta$ and $|F(\lambda(t)) - G(t)| < \delta$ for all $t \in {}^*[0,k]$. Since δ is infinitesimal, $\lambda(t) \approx t$ and $F(\lambda(t)) \approx G(t)$ for all $t \in {}^*[0,k]$.

Now assume that for each $k \in \mathbb{N}$ there exists a $\lambda \in {}^*C_k$ such that $\lambda(t) \approx t$ and $F(\lambda(t)) \approx G(t)$ for all $t \in {}^*[0,k]$. Fix a standard $\varepsilon > 0$. There exists an $m \in \mathbb{N}$ such that $\sum_{k \in \mathbb{N}, m<k} \frac{1}{2^k} < \varepsilon$, thus $\sum_{k \in {}^*\mathbb{N}, m<k} \frac{1}{2^k} < \varepsilon$. We obtain

$$\sum_{k \in {}^*\mathbb{N}} \frac{1}{2^k} \frac{(^*d)_k(F,G)}{1+(^*d)_k(F,G)} \leq \sum_{k \in {}^*\mathbb{N}, m<k} \frac{1}{2^k} + \sum_{k \leq m} \frac{1}{2^k} \frac{^*d_k(F,G)}{1+{}^*d_k(F,G)} < 2\varepsilon,$$

because $^*d_k(F,G) \approx 0$ for all $k \leq m$. Therefore, $\sum_{k \in {}^*\mathbb{N}} \frac{1}{2^k} \frac{(^*d)_k(F,G)}{1+(^*d)_k(F,G)} \approx 0$, thus $^*d_\infty(F,G) \approx 0$. This proves that $F \approx_{d_\infty} G$. □

In order to find conditions on internal functions $G : T \to {}^*\mathbb{R}$ under which they can be converted to functions in J, we fix $f \in J$ and $F \in {}^*J$ with $F \approx_{d_\infty} {}^*f$.

By Proposition 9.9.2, for each $k \in \mathbb{N}$ there exists a $\lambda_k \in {}^*C_k$ such that $\lambda_k(t) \approx t$ and $F(\lambda_k(t)) \approx {}^*f(t)$ for all $t \in {}^*[0,k]$. For each $k \in \mathbb{N}$ and each $r \in [k-1,k[$ define

$$g(r) := \min\{s \in T \mid \lambda_k(r) \le s\}.$$

Then $g : [0,\infty[\to T$ and $g(0) = \frac{1}{H}$, because $\lambda_1(0) = 0$. Now fix $r \in [k-1,k[$ and $t \approx r$ with $t \in T$. Since $\lambda_k(r) \approx r$ and $\lambda_k(r) \approx g(r)$, we have $g(r) \approx r$. Since $\lambda_k : {}^*[0,k] \leftrightarrow {}^*[0,k]$ is bijective, there are $a,b,c \in {}^*[0,k]$ with $\lambda_k(a) = t, \lambda_k(b) = g(r) \ge \lambda_k(r)$ and $\lambda_k(c) = g(r) - \frac{1}{H} < \lambda_k(r)$ if $r > 0$. Since λ_k is strictly monotone increasing, $r \le b$ and $c < r$ for $0 < r$.

Case 1: $g(r) \le t$. Then $\lambda_k(r) \le g(r) \le \lambda_k(a)$, thus $r \le a$. Therefore, by Proposition 9.9.1, $F(t) = F(\lambda_k(a)) \approx ({}^*f)(a) \approx f(r) \approx {}^*f(b) \approx F(\lambda_k(b)) = F(g(r))$; in particular, $F(t) \approx f(r)$ and

(SDJ 1) $F(t) \approx F(g(r))$ if $t \in T, t \approx r$ and $g(r) \le t$.

Case 2: $t < g(r)$. Then $\lambda_k(a) = t \le g(r) - \frac{1}{H} < \lambda_k(r)$, thus $a < r$. Therefore, by Proposition 9.9.1, $F(t) = F(\lambda_k(a)) \approx ({}^*f)(a) \approx f(r^-) \approx {}^*f(c) \approx F(\lambda_k(c)) = F\left(g(r) - \frac{1}{H}\right)$; in particular, $F(t) \approx f(r^-)$ and

(SDJ 2) $F(t) \approx F\left(g(r) - \frac{1}{H}\right)$ if $t \in T, t \approx r$ and $t < g(r)$.

Moreover,

(SDJ 3) ${}^\circ F(t)$ exists for all limited $t \in T$.

These results lead to the following definition: Fix an internal $F : T \to {}^*\mathbb{R}$. A function $g : [0,\infty[\to T$ is called a **witness for the SDJ-property of** F if

(i) $g(r) \approx r$ for all $r \in [0,\infty[$ and $g(0) = \frac{1}{H}$.
(ii) g fulfils (SDJ 1) and (SDJ 2).

Note that $F(s) \approx F(t)$ for all limited $s \approx t$ if and only if every function $g : [0,\infty[\to T$ with the property (i) also fulfils property (ii). It follows that in monads of 'jumps' the witnesses are uniquely determined.

Proposition 9.9.3 *Suppose that* $g, \widetilde{g} : [0,\infty[\to T$ *are witnesses for the SDJ-property of* F. *Fix* $r \in]0,\infty[$.

(a) $F(g(r)) \not\approx F\left(g(r) - \frac{1}{H}\right) \Rightarrow g(r) = \widetilde{g}(r)$.
(b) $F(g(r)) \approx F\left(g(r) - \frac{1}{H}\right) \Leftrightarrow F(s) \approx F(t)$ *for all* $s,t \approx r$.

Hoover and Perkins [46] defined an internal function $F : T \to {}^*\mathbb{R}$ as **SDJ** if (SDJ 3) is true and if there exists a witness for the SDJ-property of F. Lindstrøm [64] called SDJ-functions **well behaved**.

Now assume that F is SDJ with witness g. We obtain a càdlàg function ${}^\circ F$ from F, defined for all $r \in [0,\infty[$, by

$$({}^\circ F)(r) := {}^\circ(F(g(r))).$$

By Proposition 9.9.3, the definition of ${}^\circ F$ does not depend on the special witness and we obtain:

Proposition 9.9.4 *Let F be SDJ with witness g. Then ${}^\circ F \in J$ and for $r \in [0,\infty[, r \in [0,\infty]$, respectively,*

$$\begin{aligned}({}^\circ F)(r) &= \lim_{{}^\circ s \downarrow r, s \in T} {}^\circ(F(s)),\\ ({}^\circ F)(r^-) &= {}^\circ\left(F\left(g(r) - \frac{1}{H}\right)\right)\\ &= \lim_{{}^\circ s \uparrow r, s \in T} {}^\circ(F(s)) \text{ for } r > 0.\end{aligned}$$

Proof Assume that ${}^\circ(F(g(r))) = \lim_{{}^\circ s \downarrow r, s \in T} {}^\circ(F(s))$ fails. Then there are an $\varepsilon > 0$ and a sequence $(s_n)_{n \in \mathbb{N}}$ in T with $0 < {}^\circ s_n - r < \frac{1}{n}$ and $|{}^\circ F(g(r)) - {}^\circ F(s_n)| > \varepsilon$, thus $0 < s_n - g(r) < \frac{1}{n}$ and $|F(g(r)) - F(s_n)| > \varepsilon$ for all $n \in \mathbb{N}$. By Proposition 8.2.2, there exists an internal extension $(s_n)_{n \in {}^*\mathbb{N}}$ of $(s_n)_{n \in \mathbb{N}}$ with $0 < s_n - g(r)$ for all $n \in {}^*\mathbb{N}$. Now the set

$$B := \left\{ n \in {}^*\mathbb{N} \mid 0 < s_n - g(r) < \frac{1}{n} \wedge |F(g(r)) - F(s_n)| > \varepsilon \right\}$$

is internal and $\mathbb{N}$ is a subset of B. By Proposition 8.4.1 (b) there exists an unlimited $N \in B$. It follows that $s_N \approx g(r) < s_N$ and $|F(g(r)) - F(s_N)| > \varepsilon$. However, by the SDJ-property, $F(g(r)) \approx F(s_N)$, which is a contradiction. The proof of the equation ${}^\circ\left(F(g(r) - \frac{1}{H})\right) = \lim_{{}^\circ s \uparrow r, s \in T} {}^\circ(F(s))$ is similar. Moreover, note that ${}^\circ F \in J$. □

Note that the witness $g : [0,\infty[\to T$ is an external function, but for each fixed $r \in [0,\infty[$, $g(r)$ is internal as an element of T of $\mathfrak{W}$.

Altogether, we obtain the desired characterization of functions in J, defined on $[0,\infty[$, by internal SDJ-functions, defined on T, which is essentially a result of Hoover and Perkins [46].

Theorem 9.9.5 *A function $f : [0,\infty[\to \mathbb{R}$ belongs to J if and only if there exists an SDJ-function F with witness g such that $f(r) \approx F(g(r))$ for all $r \in [0,\infty[$, in which case $f(r^-) \approx F\left(g(r) - \frac{1}{H}\right)$ with $F(0) = 0$.*

Finally we mention that functions in J can be constructed from functions within a more extensive class (see also Hoover and Perkins [46]). An internal function $F : T \to {}^*\mathbb{R}$ is called **SD with witnesses** $g_1, g_2 : [0,\infty[\to T$, where $r \approx g_1(r) \leq g_2(r) \approx r$ for all $r \in [0,\infty[$ if

(SD 1) $F(t) \approx F(g_2(r))$ for $t \approx g_2(r) \leq t$,
(SD 2) $F(t) \approx F(g_1(r) - \frac{1}{H})$ for $t < g_1(r) \approx t$,
(SD 3) ${}^\circ F(t)$ exists for all limited $t \in T$.

Now the standard part ${}^\circ F$ can be defined again by:

$$({}^\circ F)(r) := {}^\circ(F(g_2(r)) \text{ for } r \in [0,\infty[.$$

Stroyan [111] has pointed out that internal SDJ-processes can be obtained from SD-processes on Loeb spaces by taking a coarser internal discrete subtimeline S of T in such a way that each $r \in [0,\infty[$ is also infinitely close to some $s \in S$.

Exercises

9.1 Let $\left(\mathbb{B}, \left(|\cdot|_j\right)_{j\in\mathbb{N}}\right)$ be a Fréchet space within the standard model $\mathfrak{V}$ and let $a \in {}^*\mathbb{B}$. Then $a \approx 0$ if and only if $|a|_j \approx 0$ in ${}^*\mathbb{R}$ for all $j \in \mathbb{N}$.

9.2 Prove Proposition 9.5.1 part (c).

9.3 Prove Proposition 9.5.1 part (d).

9.4 Prove Proposition 9.9.1.

9.5 Prove that ${}^\circ T := \{{}^\circ t \mid t \in \widetilde{T}\} = [0,\infty[$.

9.6 Find $D \subseteq \mathbb{R}$ and an internal $F : {}^*D \to {}^*\mathbb{R}$ which is ***continuous**, i.e.,

$$\forall a \in {}^*D \, \forall \varepsilon \, {}^*\!> 0 \, \exists \delta \, {}^*\!> 0 \, \forall \alpha \in {}^*D \left({}^*|\alpha - a| \, {}^*\!< \delta \right.$$
$$\left. \Rightarrow {}^*|F(\alpha) - F(a)| \, {}^*\!< \varepsilon\right),$$

but not S-continuous, i.e.,

$$\forall a, \alpha \in {}^*D \, (\alpha \approx a \Rightarrow F(\alpha) \approx F(a)) \text{ fails.}$$

9.7 Find $D \subseteq \mathbb{R}$ and an internal $F : {}^*D \to {}^*\mathbb{R}$ which is S-continuous, but not *continuous.

9.8 Recall that $f : D \to \mathbb{R}$ is continuous if and only if

$$\forall a \in D \, \forall \alpha \in {}^*D \left(\alpha \approx a \Rightarrow {}^*f(\alpha) \approx f(a)\right).$$

Prove that $f : D \to \mathbb{R}$ is uniformly continuous if and only if

$$\forall a \in {}^*D \, \forall \alpha \in {}^*D \left(\alpha \approx a \Rightarrow {}^*f(\alpha) \approx {}^*f(a) \right).$$

9.9 Use Exercise 9.8 and Proposition 9.4.1 to prove that if $f : D \to \mathbb{R}$ is continuous and D is compact, then f is uniformly continuous.

9.10 Use Exercise 9.8 to prove that $x \mapsto x^2$ on $\mathbb{R}$ is not uniformly continuous.

10

Measure and integration on Loeb spaces

In this chapter we give an introduction to measure and integration theory on Loeb spaces (see Loeb and Wolff [73] and Cutland [26]).

Loeb spaces $(\Lambda, L_\mu(\mathcal{C}), \widehat{\mu})$ are σ-additive complete measure spaces in the usual standard sense. They benefit from the following facts.

The σ-algebra $L_\mu(\mathcal{C})$ is generated by an internal algebra $\mathcal{C}$, which, by saturation, is very rich: Each $B \in L_\mu(\mathcal{C})$ is **equivalent** to an $A \in \mathcal{C}$, i.e., $\widehat{\mu}(A \Delta B) = 0$. Moreover, the σ-additive measure $\widehat{\mu}$ on $L_\mu(\mathcal{C})$ is infinitely close to an internal measure μ, defined on the generating set $\mathcal{C}$. In particular, if $\mathcal{C}$ is *finite, then $\widehat{\mu}$ is infinitely close to a counting measure μ. In Theorem 10.5.2 we shall see that, for instance, Lebesgue measure on $\mathbb{R}^n$ is σ-isomorphic to a Loeb counting measure on a *finite set. Therefore, Lebesgue measure can be handled as though it were a counting measure.

The property that each element of $L_\mu(\mathcal{C})$ is equivalent to an element of $\mathcal{C}$ can be extended to measurable and integrable functions.

10.1 The construction of Loeb measures

Let Λ be an internal non-empty set in $\mathfrak{W}$ and let $\mathcal{C}$ be an internal algebra on Λ. By Proposition 8.2.1 (e), $A_1 \cup \ldots \cup A_k \in \mathcal{C}$ and $A_1 \cap \ldots \cap A_k \in \mathcal{C}$ for each $k \in {}^*\mathbb{N}$ and each internal k-tuple $(A_1, \ldots, A_k)$ in $\mathcal{C}$. Assume that μ is an internal finitely additive measure defined on $\mathcal{C}$ (then μ is a set in $\mathfrak{W}$) with values in the limited part of ${}^*[0, \infty[$. By Proposition 8.2.1 (e) again, for each internal k-tuple $(A_1, \ldots, A_k)$ in $\mathcal{C}$, $k \in {}^*\mathbb{N}$, with $A_i \cap A_j = \emptyset$ for $i \neq j$,

$$\mu(A_1 \cup \ldots \cup A_k) = \mu(A_1) + \ldots + \mu(A_k).$$

By Proposition 8.7.1, $^\circ(\mu(A))$ exists. By Proposition 8.7.2,

$$^\circ\mu : \mathcal{C} \ni A \to {}^\circ(\mu(A))$$

is a finitely additive measure on the algebra $\mathcal{C}$. By saturation, $^\circ\mu$ is even σ-additive on the algebra $\mathcal{C}$ (see Remark 7.1.2). By Caratheodory's extension theorem, $^\circ\mu$ can be extended to a measure on the σ-algebra $\sigma(\mathcal{C})$, generated by $\mathcal{C}$. The completion of this measure is called the Loeb measure induced by μ. We want to present a more informative construction of Loeb measures, combining both methods in Loeb [69] and [70]: the **outer measure** μ^{outer} of μ is defined for all subsets D of Λ, setting

$$\mu^{\text{outer}}(D) := \inf\{{}^\circ\mu(A) \mid A \in \mathcal{C} \text{ and } D \subseteq A\}.$$

A subset $N \subseteq \Lambda$ is called a $\widehat{\mu}$-**nullset** if $\mu^{\text{outer}}(N) = 0$. Note the difference between a μ-nullset and a $\widehat{\mu}$-nullset. Set

$$\mathcal{N}_{\widehat{\mu}} := \{N \subseteq \Lambda \mid N \text{ is a } \widehat{\mu}\text{-nullset}\}.$$

Lemma 10.1.1 *The set of $\widehat{\mu}$-nullsets is closed under countable unions.*

Proof Suppose that $N_1, \ldots, N_k, \ldots \in \mathcal{N}_{\widehat{\mu}}$. In order to show that $\bigcup N_k \in \mathcal{N}_{\widehat{\mu}}$, fix an $\varepsilon \in \mathbb{R}^+$. For each $k \in \mathbb{N}$ there exists an $A_k \in \mathcal{C}$ with $N_k \subseteq A_k$ and $^\circ\mu(A_k) < \frac{\varepsilon}{2^k}$, thus $\mu(A_k) < \frac{\varepsilon}{2^k}$. By saturation, there exists an $A \in \mathcal{C}$ with $\bigcup A_k \subseteq A$ and $\mu(A) < \varepsilon$. Since $\bigcup N_k \subseteq \bigcup A_k \subseteq A$, we see that $\bigcup N_k$ is a $\widehat{\mu}$-nullset. □

If $B \subseteq \Lambda$ and $A \in \mathcal{C}$, then A is called a $\widehat{\mu}$-**approximation** of B if $A \Delta B$ is a $\widehat{\mu}$-nullset. We define

$$L_\mu(\mathcal{C}) := \{B \subseteq \Lambda \mid B \text{ has a } \widehat{\mu}\text{-approximation } A \in \mathcal{C}\}, \tag{14}$$

$$\widehat{\mu}(B) := {}^\circ\mu(A) \text{ if } A \text{ is a } \widehat{\mu}\text{-approximation of } B. \tag{15}$$

Theorem 10.1.2 (Loeb [69])

(a) *$\widehat{\mu}$ is well defined, i.e., $\widehat{\mu}$ does not depend on the chosen $\widehat{\mu}$-approximation.*
(b) *$L_\mu(\mathcal{C})$ is a σ-algebra with $\mathcal{C} \subseteq L_\mu(\mathcal{C})$.*
(c) *$\widehat{\mu} : L_\mu(\mathcal{C}) \to \mathbb{R}$ is σ-additive.*
(d) *$L_\mu(\mathcal{C})$ is complete.*
(e) *A subset $B \subseteq \Lambda$ belongs to $L_\mu(\mathcal{C})$ if and only if for each $\varepsilon \in \mathbb{R}^+$ there exist $A, A' \in \mathcal{C}$ such that $A \subseteq B \subseteq A'$ and $\mu(A' \setminus A) < \varepsilon$. Therefore, $\mu^{\text{outer}}(B) = \widehat{\mu}(B)$ for all $B \in L_\mu(\mathcal{C})$.*

Proof (a) Assume that A and A' are $\widehat{\mu}$-approximations of B. Then $A\Delta A' \in \mathcal{N}_{\widehat{\mu}}$, thus, $\mu(A\Delta A') \approx 0$. Since A' is the disjoint union of $(A' \setminus A)$ and $(A \setminus (A \setminus A'))$, we obtain

$$\begin{aligned}\mu(A) &\approx \mu(A) - \mu(A \setminus A') + \mu(A' \setminus A) \\ &= \mu\left(\left(A \setminus (A \setminus A')\right) \cup (A' \setminus A)\right) = \mu(A').\end{aligned}$$

Therefore, $\mu(A) \approx \mu(A')$, thus ${}^\circ\mu(A) = {}^\circ\mu(A')$.

(b) Obviously, $\mathcal{C} \subseteq L_\mu(\mathcal{C})$. Now we will show that $L_\mu(\mathcal{C})$ is a σ-algebra. Since $\Lambda \in \mathcal{C}$, $\Lambda \in L_\mu(\mathcal{C})$. Fix $B, B' \in L_\mu(\mathcal{C})$ with $\widehat{\mu}$-approximations $A, A' \in \mathcal{C}$ of B, B', respectively. Then $A \setminus A'$ is a $\widehat{\mu}$-approximation of $B \setminus B'$ and $A \cap A'$ is a $\widehat{\mu}$-approximation of $B \cap B'$. Fix a sequence $(B_k)_{k\in\mathbb{N}}$ in $L_\mu(\mathcal{C})$ such that $B_i \cap B_j = \emptyset$ for $i \neq j$, and for each B_k fix a $\widehat{\mu}$-approximation A_k. We may assume that $A_i \cap A_j = \emptyset$ for $i \neq j$, because, if C_k is a $\widehat{\mu}$-approximation of B_k, $k \in \mathbb{N}$, then $C_k \setminus (C_1 \cup \ldots \cup C_{k-1})$ is a $\widehat{\mu}$-approximation of B_k. Since $\sum_{i=1}^k {}^\circ\mu(A_i) = {}^\circ\mu(\bigcup_{i=1}^k A_i) \leq {}^\circ\mu(\Lambda) < \infty$ for all $k \in \mathbb{N}$,

$$s := \sum_{k=1}^{\infty} {}^\circ\mu(A_k) < \infty.$$

By saturation, there exists an $A \in \mathcal{C}$ with $\bigcup A_k \subseteq A$ and $\mu(A) < s + \frac{1}{k}$ for each $k \in \mathbb{N}$. It follows that $\mu(A) \approx s$. In order to show that A is a $\widehat{\mu}$-approximation of $\bigcup A_k$, fix $\varepsilon \in \mathbb{R}^+$. We may choose $k \in \mathbb{N}$ such that $s - {}^\circ\mu(A_k') < \varepsilon$, where $A_k' := A_1 \cup \ldots \cup A_k$. Therefore, $s - \mu(A_k') < \varepsilon$. It follows that

$$A\Delta \bigcup A_k = A \setminus \bigcup A_k \subseteq A \setminus A_k' \in \mathcal{C} \quad \text{and} \quad \mu(A \setminus A_k') \approx s - \mu(A_k') < \varepsilon.$$

Moreover, A is also a $\widehat{\mu}$-approximation of $\bigcup B_k$, because, by Lemma 10.1.1,

$$A\Delta \bigcup B_k \subseteq \bigcup (A_k \Delta B_k) \cup \left(A\Delta \bigcup A_k\right) \in \mathcal{N}_{\widehat{\mu}}.$$

This proves that $\bigcup B_k \in L_\mu(\mathcal{C})$.

(c) Now we will show that $\widehat{\mu}$ is σ-additive: Choose B_k and A_k, $k \in \mathbb{N}$, and A as in the proof of (b). Then

$$\widehat{\mu}\left(\bigcup B_k\right) = {}^\circ\mu(A) = s = \sum_{k=1}^{\infty} {}^\circ\mu(A_k) = \sum_{k=1}^{\infty} \widehat{\mu}(B_k).$$

(d) Assume that $\widehat{\mu}(B) = 0$ and $N \subseteq B$. Fix a $\widehat{\mu}$-approximation A of B. Then A is also a $\widehat{\mu}$-approximation of N. It follows that $N \in L_\mu(\mathcal{C})$.

(e) Fix a $B \in L_\mu(\mathcal{C})$, a $\widehat{\mu}$-approximation $C \in \mathcal{C}$ of B and an $\varepsilon \in \mathbb{R}^+$. Then there exists a $D \in \mathcal{C}$ with $C \Delta B \subseteq D$ and $\mu(D) < \varepsilon$. We obtain

$$C \setminus D \subseteq B \subseteq C \cup D \quad \text{and} \quad \mu((C \cup D) \setminus (C \setminus D)) \leq \mu(D) < \varepsilon.$$

Conversely, by the hypothesis, for each $n \in \mathbb{N}$ there exist $A_n, A'_n \in \mathcal{C}$ with $A_n \subseteq B \subseteq A'_n$ and such that $\mu(A'_n \setminus A_n) < \frac{1}{n}$. By saturation, there exists an $A \in \mathcal{C}$ with $A_n \subseteq A \subseteq A'_n$. Fix $\varepsilon \in \mathbb{R}^+$ and $n \in \mathbb{N}$ with $\frac{1}{n} < \varepsilon$. Then

$$A \Delta B \subseteq A'_n \setminus A_n \in \mathcal{C} \quad \text{and} \quad \mu(A'_n \setminus A_n) < \frac{1}{n} < \varepsilon.$$

This proves that A is a $\widehat{\mu}$-approximation of B, i.e., $B \in L_\mu(\mathcal{C})$. □

The measure space $(\Lambda, L_\mu(\mathcal{C}), \widehat{\mu})$ is called the **Loeb space over** $(\Lambda, \mathcal{C}, \mu)$. Although Loeb spaces are measure spaces in the usual sense, they neither belong to the standard model nor to the nonstandard model. Loeb spaces are elements of the surrounding naive model of set theory, in which we have constructed standard and nonstandard models.

Finally, we present a slight modification of Theorem 10.1.2. Let $\mathcal{B}$ be an internal subalgebra of $\mathcal{C}$. We denote by $\mathcal{B} \vee \mathcal{N}_{\widehat{\mu}}$ the smallest σ-algebra containing $\mathcal{B}$ and all $\widehat{\mu}$-nullsets. Note that $L_{\mu \upharpoonright \mathcal{B}}(\mathcal{B}) \subseteq \mathcal{B} \vee \mathcal{N}_{\widehat{\mu}}$ and this inclusion may be strict. The proof of the following result is left to the reader.

Corollary 10.1.3 *A subset C of Λ belongs to $\mathcal{B} \vee \mathcal{N}_{\widehat{\mu}}$ if and only if there is a $B \in \mathcal{B}$ such that $C \Delta B$ is a $\widehat{\mu}$-nullset for $\mathcal{C}$.*

10.2 Loeb measures over Gaussian measures

In this section we present a first example of a Loeb measure and give a useful and intuitive characterization of measurable semi-norms. Fix a separable Hilbert space $(\mathbb{H}, \|\cdot\|)$ inside the standard model $\mathfrak{V}$ and a sequence $\left(|\cdot|_j\right)_{j \in \mathbb{N}}$ of separating semi-norms on $\mathbb{H}$. Let $\mathcal{E}$ be the set of all finite-dimensional subspaces of $\mathbb{H}$. Then γ assigns to each $\sigma \in]0, \infty[$ and $E \in \mathcal{E}$ the Gaussian measure γ_σ^E on E of variance σ. By the transfer principle, ${}^*\gamma$ assigns to each $\sigma \in {}^*]0, \infty[$ and each $E \in {}^*\mathcal{E}$ the internal Gaussian measure $({}^*\gamma)_\sigma^E$ on E of variance σ. In a similar way ${}^*(\gamma_\sigma)$ is defined, provided σ is standard. Moreover, $|\cdot|$ assigns to each $j \in \mathbb{N}$ a semi-norm $|\cdot|_j$, thus ${}^*|\cdot|$ assigns to each $j \in {}^*\mathbb{N}$ a semi-norm $({}^*|\cdot|)_j$.

Proposition 10.2.1 *Let d be a semi-metric on $\mathbb{H}$. The following statements are equivalent:*

(a) *d is measurable.*
(b) *For all $\sigma \in]0,\infty[$ and all $E \in {}^*\mathcal{E}$ with $E \perp *[\mathbb{H}]$,*

$$\widehat{{}^*(\gamma_\sigma)^E}\,\{x \in E \mid d(x,0) \not\approx 0\} = 0.$$

Proof '(a) $\Rightarrow$ (b)' Let d be measurable and let $E \perp *[\mathbb{H}]$. It suffices to show that, for all $\sigma \in]0,\infty[$ and $m \in \mathbb{N}$, ${}^*\gamma_\sigma^E\left\{x \in E \mid {}^*d(x,0) \geq \frac{1}{2^m}\right\} \leq \frac{1}{2^{m+1}}$. There exists an $E_{\sigma,m} \in \mathcal{E}$ such that

$$\forall F \in \mathcal{E}, F \perp E_{\sigma,m}\left(\gamma_\sigma^F\left\{x \in F \mid d(x,0) \geq \frac{1}{2^m}\right\} \leq \frac{1}{2^{m+1}}\right).$$

By transfer, and since $E \perp^* E_{\sigma,m}$ for all $m \in \mathbb{N}$, we obtain the desired result.

'(b) $\Rightarrow$ (a)' Assume (b) is true. Fix $\sigma \in]0,\infty[$ and $m \in \mathbb{N}$. Then, taking $E_{\sigma,m} \in {}^*\mathcal{E}$ with $*[\mathbb{H}] \subseteq E_{\sigma,m}$, according to Theorem 9.8.1, we obtain

$$\exists E_{\sigma,m} \in {}^*\mathcal{E}\forall E \in {}^*\mathcal{E}, E \perp E_{\sigma,m}\left({}^*\gamma_\sigma^E\left\{x \in E \mid {}^*d(x,0) \geq \frac{1}{2^m}\right\} \leq \frac{1}{2^{m+1}}\right).$$

By transfer,

$$\exists E_{\sigma,m} \in \mathcal{E}\forall E \in \mathcal{E}, E \perp E_{\sigma,m}\left(\gamma_\sigma^E\left\{x \in E \mid d(x,0) \leq \frac{1}{2^m}\right\} \leq \frac{1}{2^{m+1}}\right).$$

This proves that d is measurable. □

Now we will prove Proposition 4.3.1.

Corollary 10.2.2 *Let d be a metric on $\mathbb{H}$, generated by a separating sequence $\left(|\cdot|_j\right)_{j\in\mathbb{N}}$ of semi-norms. Then d is measurable for $\sigma > 0$ if and only if $|\cdot|_j$ is measurable for σ for all $j \in \mathbb{N}$.*

Proof Fix $\sigma \in]0,\infty[$ and $E \perp *[\mathbb{H}]$.

Assume that d is measurable for σ. Then we obtain for all $j \in \mathbb{N}$

$$\widehat{{}^*\gamma_\sigma^E}\left\{x \in E \mid |x|_j \not\approx 0\right\}$$

$$\leq \widehat{{}^*\gamma_\sigma^E}\left\{x \in E \mid \sum_{j\in{}^*\mathbb{N}} \frac{1}{2^j}\frac{{}^*|x|_j}{1+{}^*|x|_j} \not\approx 0\right\} = \widehat{{}^*\gamma_\sigma^E}\left\{x \in E \mid {}^*d(x,0) \not\approx 0\right\} = 0.$$

Conversely, assume that $|\cdot|_j$ is measurable for all $j \in \mathbb{N}$. Then we have, for all $j \in \mathbb{N}$, $\widehat{{}^*\gamma_\sigma^E}\left\{x \in E \mid |x|_j \not\approx 0\right\} = 0$. First we will prove the following. If ${}^*d(x,0) \not\approx 0$, then $|x|_j \not\approx 0$ for some $j \in \mathbb{N}$. Assume that $|x|_j \approx 0$ for $j \in \mathbb{N}$. Then there exist $\alpha_j > 0, \alpha_j \approx 0$, with $|x|_j \le \alpha_j$. Then $\frac{1}{\alpha_j}$ is unlimited. By Proposition 8.2.1 (f), there is an unlimited α with $\alpha \le \frac{1}{\alpha_j}$, thus $|x|_j \le \frac{1}{\alpha} \approx 0$ for all $j \in \mathbb{N}$. By Proposition 8.2.2, there exists an internal extension $\left(|x|_j\right)_{j \in {}^*\mathbb{N}}$ of $\left(|x|_j\right)_{j \in \mathbb{N}}$ with $|x|_j \le \frac{1}{\alpha}$ for all $j \in {}^*\mathbb{N}$. Then,

$$ {}^*d(x,0) = \sum_{j \in {}^*\mathbb{N}} \frac{1}{2^j} \frac{{}^*|x|_j}{1 + {}^*|x|_j} \le \frac{1}{\alpha} \approx 0. $$

Now we obtain

$$ \widehat{{}^*\gamma_\sigma^E}\left\{x \in E \mid {}^*d(x,0) \not\approx 0\right\} \le \widehat{{}^*\gamma_\sigma^E} \bigcup_{j \in \mathbb{N}} \left\{x \in E \mid {}^*|x|_j \not\approx 0\right\} = 0. $$

□

10.3 Loeb measurable functions

Fix a Loeb space $(\Lambda, L_\mu(\mathcal{C}), \widehat{\mu})$ over $(\Lambda, \mathcal{C}, \mu)$, where $\mu(\Lambda)$ is limited. We will study Loeb-measurable functions $f : \Lambda \to \mathbb{H}^{\otimes d}$, where $\mathbb{H}$ is a separable Hilbert space and an element of the standard model $\mathfrak{V}$. Then $\mathbb{H}^{\otimes d}$ is also an element of $\mathfrak{V}$. Using a *finite-dimensional representation $\mathbb{F}$ of $\mathbb{H}$, according to Proposition 9.8.1, we will represent f by internal functions F from Λ into the *finite-dimensional space $\mathbb{F}^{\otimes d}$. Such representations are called liftings. Note that we can identify $F : \Lambda \to \mathbb{F}^{\otimes d}$ with $G : \Lambda \times \mathbb{F}^d \to {}^*\mathbb{R}$, where $G(X, a) := F(x)(a)$.

First of all let us mention the Loeb–Anderson lifting result. An internal function $F : \Lambda \to {}^*\mathbb{H}^{\otimes d}$ is called a **lifting** of $f : \Lambda \to \mathbb{H}^{\otimes d}$ if $F(X) \approx_{{}^*\mathbb{H}^d} {}^*(f(X))$ in **the norm of** $\mathbb{H}^{\otimes d}$ for $\widehat{\mu}$-almost all $X \in \Lambda$, i.e., $\sum_{\mathfrak{e} \in \overline{\mathfrak{E}}^d} (F(x)(\mathfrak{e}) - {}^*(f(X))(\mathfrak{e}))^2 \approx 0$, where $\overline{\mathfrak{E}}$ is an internal ONB of ${}^*\mathbb{H}$.

Theorem 10.3.1 (Loeb [69], Anderson [4]) *A function $f : \Lambda \to \mathbb{H}^{\otimes d}$ is $L_\mu(\mathcal{C})$-measurable if and only if f has a $\mathcal{C}$-*simple lifting $F : \Lambda \to {}^*\mathbb{H}^{\otimes d}$.*

Proof '⇒' See the proof of Theorem 10.3.2 '⇒' below. The proof of '⇐' is similar to the proof of Theorem 10.3.2 '⇐' below. □

The following central result is a slight modification of the Loeb–Anderson lifting result. Recall that $\mathbb{H}^{\otimes d} = \mathbb{R}$ and $\mathbb{F}^{\otimes d} = {}^*\mathbb{R}$ in case $d = 0$. Therefore, we also obtain the real-valued Loeb–Anderson lifting result. An internal function $F : \Lambda \to \mathbb{F}^{\otimes d}$ is called a **lifting** of $f : \Lambda \to \mathbb{H}^{\otimes d}$ if $F(X) \approx_{\mathbb{F}^d} f(X)$ for $\widehat{\mu}$-almost all $X \in \Lambda$, i.e.,

$$ \|F(X) - f(X)\|_{\mathbb{F}^d} := \sum_{\mathfrak{e} \in \mathfrak{E}^d} \left(F(x)(\mathfrak{e}) - {}^*(f(X))(\mathfrak{e})\right)^2 \approx 0, $$

where $\mathfrak{E}$ is an internal ONB of $\mathbb{F}$.

Note that ${}^*f$ is not defined for $f : \Lambda \to \mathbb{H}^{\otimes d}$, because f does not belong to the standard model in general. Nevertheless, for simplicity we define ${}^*f : \Lambda \to {}^*\mathbb{H}^{\otimes d}$ by setting

$$\left({}^*f\right)(X) := {}^*(f(X)),$$

where ${}^*(f(X)) : {}^*\mathbb{H}^d \to {}^*\mathbb{R}$ is internal.

Theorem 10.3.2 *A function $f : \Lambda \to \mathbb{H}^{\otimes d}$ is $L_\mu(\mathcal{C})$-measurable if and only if f has a $\mathcal{C}$-*simple lifting $F : \Lambda \to \mathbb{F}^{\otimes d}$, in which case there exists a sequence $(f_n)_{n\in\mathbb{N}}$ of $\mathcal{C}$-simple functions $f_n : \Lambda \to \mathbb{H}^{\otimes d}$, converging to f in measure. By identification of $a \in \mathbb{H}^{\otimes d}$ with ${}^*a \in {}^*\left[\mathbb{H}^{\otimes d}\right]$, f_n is internal.*

Proof '$\Rightarrow$' Assume that f is $L_\mu(\mathcal{C})$-measurable. Then there exists a sequence $(f_k)_{k\in\mathbb{N}}$ of $L_\mu(\mathcal{C})$-simple functions f_k such that for all $\varepsilon > 0$

$$\lim_{k\to\infty} \widehat{\mu}\left\{X \mid \|f_k(X) - f(X)\|_{\mathbb{H}^d} \geq \varepsilon\right\} = 0.$$

Let $f_k = \mathbf{1}_{B_1} \otimes h_1 + \ldots + \mathbf{1}_{B_l} \otimes h_l$ with $B_i \in L_\mu(\mathcal{C})$ and $h_i \in \mathbb{H}^{\otimes d}$. Let $A_i \in \mathcal{C}$ be a $\widehat{\mu}$-approximation of B_i. Set $\overline{F_k} := \mathbf{1}_{A_1} \otimes {}^*h_1 + \ldots + \mathbf{1}_{A_l} \otimes {}^*h_l$. Then $\overline{F}_k$ and $F_k := \overline{F}_k \upharpoonright \Lambda \times \mathbb{F}^d$ are mappings from Λ into ${}^*\mathbb{H}^{\otimes d}$, $\mathbb{F}^{\otimes d}$, respectively, and both functions are $\mathcal{C}$-simple and $\overline{F}_k(X) = {}^*(f_k(X))$, $F_k(X) = {}^*(f_k(X)) \upharpoonright \mathbb{F}^d$ for all $X \notin \bigcup_{i=1}^l (A_i \Delta B_i)$, thus, $\overline{F}_k, F_k$ are liftings of f_k. There exists a function $g : \mathbb{N} \to \mathbb{N}$ such that $\widehat{\mu}\left\{\|f - f_{g(m)}\|_{\mathbb{H}^d} \geq \frac{1}{m}\right\} < \frac{1}{m}$ for all $m \in \mathbb{N}$ and such that for all $k \geq g(m)$

$$\mu\left\{\|\overline{F}_k - \overline{F}_{g(m)}\|_{{}^*\mathbb{H}^d} \geq \frac{1}{m}\right\} < \frac{1}{m}, \tag{16}$$

$$\mu\left\{\|F_k - F_{g(m)}\|_{\mathbb{F}^d} \geq \frac{1}{m}\right\} < \frac{1}{m}. \tag{17}$$

Let $\left(\overline{F}_k\right)_{k\in{}^*\mathbb{N}}$, $(F_k)_{k\in{}^*\mathbb{N}}$ be internal extensions of $\left(\overline{F}_k\right)_{k\in\mathbb{N}}$, $(F_k)_{k\in\mathbb{N}}$, respectively, such that all these functions are $\mathcal{C}$-*simple. By Proposition 9.5.1 (a), there exists an unlimited K such that (16) and (17) are true for all $m \in \mathbb{N}$ if k is replaced by K. Recall that $\|{}^*a\|_{\mathbb{F}^d} \approx \|a\|_{\mathbb{H}^d}$ for $a \in \mathbb{H}^{\otimes d}$. Then we have, for all $m \in \mathbb{N}$ and for $\alpha := \widehat{\mu}\left\{\|F_{g(m)} - {}^*f_{g(m)}\|_{\mathbb{F}^d} \geq \frac{1}{m}\right\} = 0$,

$$\begin{aligned}
&\mu^{\text{outer}}\left\{\|F_K - {}^*f\|_{\mathbb{F}^d} \geq \frac{3}{m}\right\} \\
&\leq {}^\circ\mu\left\{\|F_K - F_{g(m)}\|_{\mathbb{F}^d} \geq \frac{1}{m}\right\} + \alpha + \widehat{\mu}\left\{\|{}^*f_{g(m)} - {}^*f\|_{\mathbb{F}^d} \geq \frac{1}{m}\right\} < \frac{2}{m},
\end{aligned}$$

where $\|{}^*f_{g(m)} - {}^*f\|_{\mathbb{F}^d} \approx \|f_{g(m)} - f\|_{\mathbb{H}^d}$. This proves that F_K is a lifting of f. In the same way one can see that $\overline{F}_K$ is a lifting of f.

'$\Leftarrow$' Now assume that f has a $\mathcal{C}$-*simple lifting $F:\Lambda \to \mathbb{F}^{\otimes d}$. In order to prove that f is $L_\mu(\mathcal{C})$-measurable, it suffices to prove that $\{f(\cdot)(\mathfrak{s}) < c\} \in L_\mu(\mathcal{C})$ for all $c \in \mathbb{R}$ and for all $\mathfrak{s} = (\mathfrak{s}_1,\ldots,\mathfrak{s}_d) \in \mathbb{H}^d$, where the $\mathfrak{s}_i$ are pairwise orthonormal (see Proposition 4.6.3). Since ${}^*\mathfrak{s} = ({}^*\mathfrak{s}_1,\ldots,{}^*\mathfrak{s}_d) \in \mathbb{F}^d$, we obtain $\widehat{\mu}$-a.e.

$$\{f(\cdot)(\mathfrak{s}) < c\} = \bigcup_{k \in \mathbb{N}} \left\{F(\cdot)\left({}^*\mathfrak{s}\right) < c - \frac{1}{k}\right\} \in L_\mu(\mathcal{C}).$$

By the proof of '$\Rightarrow$' f has a $\mathcal{C}$-*simple lifting $G:\Lambda \to {}^*\mathbb{H}^{\otimes d}$. Now we proceed as in the proof of Proposition 7.1 (b) in [71]. There exists a decreasing sequence $(A_n)_{n\in\mathbb{N}}$ in $\mathcal{C}$ such that, for each $n \in \mathbb{N}$, $\mu(A_n) < 1/n$ and $G(X) \approx_{{}^*\mathbb{H}^d} {}^*(f(X))$ for all $X \in \Lambda \setminus A_n$. Fix $n \in \mathbb{N}$. Since $\{G(X) : X \in \Lambda \setminus A_n\}$ is an internal set of nearstandard points of ${}^*\mathbb{H}^{\otimes d}$, by Proposition 9.6.1,

$$K_n = \{f(X) : X \in \Lambda \setminus A_n\} = \{{}^\circ G(X) : X \in \Lambda \setminus A_n\}$$

is compact. We fix a covering of K_n by open neighbourhoods $O_i = U_{\frac{1}{2n}}(y_i)$ of certain $y_i \in \mathbb{H}^{\otimes d}$, $i = 1,\ldots,m \in \mathbb{N}$. By Proposition 9.1.2, $G(X) \in \bigcup_{i=1}^m {}^*O_i$ for all $X \in \Lambda \setminus A_n$. We may construct suitable $\mathcal{C}$-simple functions f_n, as follows. For $X \in A_n$ set $f_n(X) = 0$. For each $X \notin A_n$, set $f_n(X) = y_i$ if $G(X) \in {}^*O_i$ and $G(X) \notin {}^*O_1 \cup \ldots \cup {}^*O_{i-1}$, $1 \le i \le m$. We thus obtain, for all $X \in \Lambda \setminus A_n$,

$$\begin{aligned} &\|f(X) - f_n(X)\|_{\mathbb{H}^d} \\ &\le \left\|{}^*f(X) - G(X)\right\|_{{}^*\mathbb{H}^d} + \left\|G(X) - {}^*f_n(X)\right\|_{{}^*\mathbb{H}^d} < \frac{1}{n}. \end{aligned}$$

Therefore,

$$\widehat{\mu}\left\{\|f_n - f\|_{\mathbb{H}^d} \ge \frac{1}{n}\right\} \le \widehat{\mu}(A_n) = {}^\circ\mu(A_n) \le \frac{1}{n}.$$

It follows that $(f_n)_{n\in\mathbb{N}}$ converges to f in measure. □

10.4 On Loeb product spaces

Fix two internal measure spaces $(\Lambda, \mathcal{C}, \mu)$ and $\left(\Lambda', \mathcal{C}', \mu'\right)$ with limited $\mu(\Lambda)$ and $\mu'\left(\Lambda'\right)$. Let $\mathcal{C} \otimes \mathcal{C}'$ denote the ${}^*\sigma$-algebra generated by the internal algebra of all *finite disjoint unions of sets of the form $A \times A'$ with $A \in \mathcal{C}$ and $A' \in \mathcal{C}'$. The internal product measure of μ and μ' on $\mathcal{C} \otimes \mathcal{C}'$ is denoted by $\mu \otimes \mu'$. We always assume that $\mathcal{C} \otimes \mathcal{C}'$ is augmented by the $\mu \otimes \mu'$-nullsets. For standard product spaces we have analogous notation.

The next result shows that the usual product $L_\mu(\mathcal{C}) \otimes L_{\mu'}(\mathcal{C}')$ of the Loeb σ-algebras $L_\mu(\mathcal{C})$ and $L_{\mu'}(\mathcal{C}')$ is contained in the Loeb product $L_{\mu\otimes\mu'}(\mathcal{C}\otimes\mathcal{C}')$. It often happens that this inclusion is strict. First examples have been given by D. Hoover and D. Norman. Sun [113] proved that the inclusion is always strict when the Loeb spaces are atomless. Results in [9] show how extremely rich the Loeb product $L_{\mu\otimes\mu'}(\mathcal{C}\otimes\mathcal{C}')$ is compared with the usual product $L_\mu(\mathcal{C})\otimes L_{\mu'}(\mathcal{C}')$. We will often apply the following result:

Proposition 10.4.1 (Anderson [4]) *If $B \in L_\mu(\mathcal{C}) \otimes L_{\mu'}(\mathcal{C}')$ then $B \in L_{\mu\otimes\mu'}(\mathcal{C}\otimes\mathcal{C}')$ and $\widehat{\mu\otimes\mu'}(B) = \widehat{\mu}\otimes\widehat{\mu'}(B)$.*

Proof It is easy to see that $\Lambda \times N \in \mathcal{N}_{\widehat{\mu\otimes\mu'}}$ and $N \times \Lambda' \in \mathcal{N}_{\widehat{\mu\otimes\mu'}}$ whenever $N \in \mathcal{N}_{\widehat{\mu'}}$, $N \in \mathcal{N}_{\widehat{\mu}}$, respectively. Suppose that $X \in L_\mu(\mathcal{C})$ and $Y \in L_{\mu'}(\mathcal{C}')$. By (14) in Section 10.1, there exist $U \in \mathcal{C}$ and $V \in \mathcal{C}'$ such that $U \Delta X \in \mathcal{N}_{\widehat{\mu}}$ and $V \Delta Y \in \mathcal{N}_{\widehat{\mu'}}$. Since

$$(U \times V)\Delta(X \times Y) \subseteq \big((U\Delta X)\times\Lambda'\big)\cup(\Lambda\times(V\Delta Y)) \in \mathcal{N}_{\widehat{\mu\otimes\mu'}},$$

we have $X \times Y \in L_{\mu\otimes\mu'}(\mathcal{C}\otimes\mathcal{C}')$ and

$$\widehat{\mu\otimes\mu'}(X\times Y) \approx \mu\otimes\mu'(U\times V) = \mu(U)\cdot\mu'(V) \approx \widehat{\mu}(X)\cdot\widehat{\mu'}(Y) = \widehat{\mu}\otimes\widehat{\mu'}(X\times Y).$$

It follows that $L_\mu(\mathcal{C}) \otimes L_{\mu'}(\mathcal{C}') \subseteq L_{\mu\otimes\mu'}(\mathcal{C}\otimes\mathcal{C}')$ and the measure $\widehat{\mu\otimes\mu'}$ is identical to $\widehat{\mu}\otimes\widehat{\mu'}$ on $L_\mu(\mathcal{C})\otimes L_{\mu'}(\mathcal{C}')$. □

10.5 Lebesgue measure as a counting measure

In a further application, we construct a Loeb counting measure $\widehat{\nu^n}$ on the hyperfinite set T^n, $n \in \mathbb{N}$, which strictly extends the Lebesgue measure λ^n on $[0,\infty[^n$. If we take on T^n the σ-algebra $\mathcal{L}^n$ generated by the standard part map, then $\widehat{\nu^n}$ restricted to $\mathcal{L}^n$ is equivalent to λ^n. Recall that $T := \left\{\frac{1}{H}, \frac{2}{H}, \ldots, H\right\}$, where $H \in {}^*\mathbb{N}$ such that each $k \in \mathbb{N}$ divides H. Let ${}^*\mathcal{P}$ be a function which assigns to each internal set A the internal set of all internal subsets of A. Define for all $A \in {}^*\mathcal{P}(T^n)$:

$$\nu^n(A) := \frac{|A|}{H^n} < \infty \text{ in } {}^*\mathbb{R}.$$

Fix $t \in T$ and set $T_t := \left\{\frac{1}{H}, \frac{2}{H}, \ldots, t\right\}$. Define $\nu_t^n := \nu^n \restriction {}^*\mathcal{P}\left(T_t^n\right)$. The Loeb space over $\left(T_t^n, {}^*\mathcal{P}\left(T_t^n\right), \nu_t^n\right)$ is denoted by $\left(T_t^n, L_{\nu_t^n}, \widehat{\nu_t^n}\right)$, provided t is limited.

Now we use the notion 'lifting' in a slightly modified way again. Fix $k \in \mathbb{N}$, a separable Fréchet space $\mathbb{B}$ and a separable Hilbert space $\mathbb{H}$ with internal representation $\mathbb{F}$ of $\mathbb{H}$, according to Theorem 9.8.1. Moreover, let $F : T_k^n \to \mathbb{F}^{\otimes d}$ or $F : T_k^n \to {}^*\mathbb{B}$ be $\mathcal{C}$-measurable. Then F is called a **lifting** of $f : [0,k]^n \to \mathbb{H}^{\otimes d}$, $f : [0,k]^n \to \mathbb{B}$ respectively if F is a lifting of $f \circ st_k$, where

$$st_k : T_k^n \to [0,k]^n, \quad t \mapsto {}^\circ t,$$

and $t = (t_1, \ldots, t_n)$ and ${}^\circ t = ({}^\circ t_1, \ldots, {}^\circ t_n)$.

The following result shows the well-known close relationship between $\widehat{\nu_k^n}$ and Lebesgue measure λ^n on $[0,k]^n$ (see Albeverio *et al.* [2] Theorem 2.3.4 and Proposition 2.3.5): The image measure of $\widehat{\nu_k^n}$ by the standard part map st_k is the Lebesgue measure on $[0,k]^n$. Denote by Leb^n the set of Lebesgue-measurable sets in $\mathbb{R}^n$.

Lemma 10.5.1 *Fix standard $n, k \in \mathbb{N}$, $d \in \mathbb{N}_0$.*

(a) *A subset $B \subset [0,k]^n$ is Lebesgue measurable if and only if $\mathrm{st}_k^{-1}[B] \in L_{\nu_k^n}(T_k^n)$, in which case*

$$\lambda^n(B) = \widehat{\nu_k^n}(\mathrm{st}^{-1}[B]).$$

(b) *A function $f : [0,k]^n \to \mathbb{H}^{\otimes d}$ is Lebesgue measurable if and only if f has a lifting $F : T_k^n \to \mathbb{F}^{\otimes d}$.*

Proof (a) '$\Rightarrow$' Since, by Proposition 10.4.1, $\widehat{\nu_k^n} = \widehat{\nu_k}^n$ on $\left(L_{\nu_k}(T_k)\right)^n$, it suffices to show that, for each interval $I \subseteq [0,k]$, $\mathrm{st}_k^{-1}[I] \in L_{\nu_k}(T_k)$ and $\widehat{\nu}(\mathrm{st}_k^{-1}[I]) = \lambda(I)$. Let a, b, $a \le b$, be the endpoints of I. Set $]a,b[_{T_k} := \{t \in T_k \mid a < t < b\}$. We will show that $\nu(]a,b[_{T_k}) \approx b - a$. For $a = b$ this is obvious. Let $a < b$. Choose $l, m \in {}^*\mathbb{N}$ such that $\frac{(l-1)}{H} \le a < \frac{l}{H}$ and $\frac{(l+m-1)}{H} < b \le \frac{(l+m)}{H}$. Then

$$\nu(]a,b[_{T_k}) = \frac{m}{H} = \frac{l+m}{H} - \frac{l}{H} \approx b - a.$$

In particular, $\widetilde{c} := \{t \in T_k \mid t \approx c\}$ is a $\widehat{\nu_k}$-nullset for all $c \in [0,k]$. It follows that

$$\mathrm{st}_k^{-1}[I] \,\Delta\,]a,b[_{T_k} \subseteq \widetilde{a} \cup \widetilde{b} \quad \text{with } \widehat{\nu_k}\left(\widetilde{a} \cup \widetilde{b}\right) = 0.$$

Therefore, $\mathrm{st}_k^{-1}[I] \in L_{\nu_k}(T_k)$ and $\widehat{\nu}(\mathrm{st}_k^{-1}[I]) = {}^\circ\nu_k(]a,b[_{T_k}) = b - a = \lambda(I)$. Since $L_{\nu_k}(T_k)$ is complete, $\mathrm{st}_k^{-1}[I] \in L_{\nu_k}(T_k)$ and $\widehat{\nu}(\mathrm{st}_k^{-1}[I]) = \lambda(I)$.

'$\Leftarrow$' Fix $B \subseteq [0,k]^n$ with $\mathrm{st}_k^{-1}[B] \in L_{\nu_k^n}(T_k^n)$. Then, setting $I := [0,k]$,

$$\mathrm{st}_k^{-1}[I^n \setminus B] = T_k^n \setminus \mathrm{st}_k^{-1}[B] \in L_{\nu_k^n}(T_k^n).$$

In order to show that $B \in \mathrm{Leb}(I^n)$, fix $\varepsilon > 0$. By Theorem 10.1.2 (e), there exist internal $A', C' \subseteq T_k^n$ such that

$$A' \subseteq \mathrm{st}_k^{-1}[B] \quad \text{and} \quad \widehat{\nu_k^n}(\mathrm{st}_k^{-1}[B] \setminus A') < \varepsilon,$$
$$C' \subseteq \mathrm{st}_k^{-1}[I^n \setminus B] \quad \text{and} \quad \widehat{\nu_k^n}(\mathrm{st}_k^{-1}[I^n \setminus B] \setminus C') < \varepsilon.$$

By Proposition 9.6.1, $A := \mathrm{st}_k[A']$ and $C := \mathrm{st}_k[C']$ are compact subsets of I^n, thus Borel sets in I^n. Moreover, $A \subseteq B \subseteq I^n \setminus C$. By '$\Rightarrow$', we obtain

$$\begin{aligned}\lambda^n((I^n \setminus C) \setminus A) &= \widehat{\nu_k^n}(\mathrm{st}_k^{-1}[(I^n \setminus C) \setminus A]) \\ &\leq \widehat{\nu_k^n}\left(\mathrm{st}_k^{-1}[I^n \setminus B] \setminus C'\right) + \widehat{\nu_k^n}\left(\mathrm{st}_k^{-1}[B] \setminus A'\right) < 2\varepsilon.\end{aligned}$$

Therefore, for each $\delta > 0$ there exists an open set $G := I^n \setminus C$ in I^n and a closed set A in I^n such that $A \subseteq B \subseteq G$ and $\lambda^n(G \setminus A) < \delta$, which proves that $B \in \mathrm{Leb}[0,k]^n$.

(b) Fix $f : [0,k]^n \to \mathbb{H}^{\otimes d}$. By Part (a), the Lebesgue measurability of f is equivalent to the $L_{\nu_k^n}(T_k^n)$-measurability of $f \circ st_k$. By Theorem 10.3.2, $f \circ st_k$ is $L_{\nu_k^n}(T_k^n)$-measurable if and only if $f \circ st_k$ has a lifting, which means that f has a lifting. □

Now let us extend the preceding result to the whole timelines T^n and $[0,\infty[^n$ by converting ν^n to a standard measure $\widehat{\nu^n}$ on T^n as follows: Denote by L_{ν^n} the set of all **Loeb-measurable** $B \subseteq T^n$, i.e., $B \cap T_k^n \in L_{\nu_k^n}(T_k^n)$ for all $k \in \mathbb{N}$. Define for $B \in L_{\nu^n}$

$$\widehat{\nu^n}(B) := \lim_{k \to \infty} \widehat{\nu_k^n}(B \cap T_k^n) \in [0,\infty].$$

Then $\left(T^n, L_{\nu^n}, \widehat{\nu^n}\right)$ is an infinite measure space and $\widehat{\nu^n}(T^n \setminus \bigcup_{k \in \mathbb{N}} T_k^n) = 0$. It is a strict extension of the Lebesgue space $\left([0,\infty[^n, \mathrm{Leb}^n, \lambda^n\right)$ in the following sense.

Let $\mathcal{L}^n$ denote the set of all $B \subseteq T^n$ such that there exists a $C_B \in \mathrm{Leb}^n$ with $B \cap T_k^n = \left(\mathrm{st}_k^n\right)^{-1}[C_B \cap [0,k]^n]$ for all $k \in \mathbb{N}$, augmented by all $\widehat{\nu^n}$-nullsets. Obviously, C_B is uniquely determined by B. Vice versa, if $C_B = C_{B'}$, then $\widehat{\nu^n}(B \Delta B') = 0$, because $B \cap T_k^n = B' \cap T_k^n$ for all $k \in \mathbb{N}$. In summary, we have the following result.

Theorem 10.5.2 *The mapping from $\mathcal{L}^n$ onto* Leb^n *with $B \mapsto C_B$ is a measure-preserving bijection from $\mathcal{L}^n$ onto* Leb^n*, provided that we identify B and B' in case $B \cap T_k^n = B' \cap T_k^n$ for all $k \in \mathbb{N}$.*

Because of Theorem 10.5.2, and since the standard part ${}^\circ t$ of $t \in T^n$ is $\widehat{\nu^n}$-a.e. well defined, we obtain, using Corollary 5.6.4, the following.

Corollary 10.5.3 *Fix* $p \in [1,\infty[$. *The spaces* $L^p\left([0,\infty[^n, \mathrm{Leb}^n, \lambda^n\right)$ *and* $L^p\left(T^n, \mathcal{L}^n, \widehat{\nu^n}\right)$ *can be identified, because the mapping*

$$\iota : L^p\left([0,\infty[^n, \mathrm{Leb}^n, \lambda^n\right) \to L^p\left(T^n, \mathcal{L}^n, \widehat{\nu^n}\right), \quad \iota(\varphi)(t) := \varphi(^\circ t),$$

is a canonical (basis-independent) isometric isomorphism between these spaces. We call φ *and* $\iota(\varphi)$ ***equivalent*** *and identify both functions.*

Let $\left(\Lambda, L_\mu(\mathcal{C}), \widehat{\mu}\right)$ again be a finite Loeb space. We define here $\mathrm{st} : \Lambda \times T^n \to \Lambda \times [0,\infty[^n$, $(X,t) \mapsto (X, ^\circ t)$. Note that st is $\widehat{\mu \otimes \nu}$-a.e. well defined. The following extension of Theorem 10.3.2 to product spaces is straightforward. See Theorem 10.9.2 below.

Proposition 10.5.4 *Fix standard* $n \in \mathbb{N}$.

(a) *A subset* $B \subset \Lambda \times [0,\infty[^n$ *is* $L_\mu(\mathcal{C}) \otimes \mathrm{Leb}^n$*-measurable if and only if* $\mathrm{st}^{-1}[B] \in L_\mu(\mathcal{C}) \otimes L_{\nu^n}(T^n)$*, in which case*

$$\widehat{\mu} \otimes \lambda^n(B) = \widehat{\mu \otimes \nu^n}(\mathrm{st}^{-1}[B]). \tag{18}$$

(b) *A function* $f : \Lambda \times [0,\infty[^n \to \mathbb{H}^{\otimes d}$ *is* $L_\mu(\mathcal{C}) \otimes \mathrm{Leb}^n$*-measurable if and only if* f *has a* $\mathcal{C} \otimes {}^*\mathcal{P}(T)$*-measurable* ***lifting*** $F : \Lambda \times T^n \to \mathbb{F}^{\otimes d}$*, i.e.,* $F(X,t) \approx_{\mathbb{F}^n} {}^*f(X, ^\circ t)$ *for* $\widehat{\mu \otimes \nu^n}$*-almost all* (X,t).

Moreover, according to Theorem 10.3.2, f *can be approximated by internal functions with finite range iff* f *is* $L_\mu(\mathcal{C}) \otimes \mathrm{Leb}^n$*-measurable*

Often it is more convenient to work with the domain

$$T^n_{\neq} := \left\{(t_1,\ldots,t_n) \in T^n \mid t_i \neq t_j \text{ if } i \neq j\right\}$$

instead of T^n. The proof of the following corollary is left to the reader.

Corollary 10.5.5 $\widehat{\nu^n}(T^n \setminus T^n_{\neq}) = 0$.

Therefore, we can assume and we will assume in general that the liftings $F : T^n \to {}^*\mathbb{R}$ of a Lebesgue-measurable function have the property

$$F(t_1,\ldots,t_n) = 0 \quad \text{if } (t_1,\ldots,t_n) \notin T^n_{\neq}.$$

We end this section with a second simple example of an infinite Loeb counting measure. For each $k \in \mathbb{N}$ we define a finitely additive measure c_k setting

$$c_k : {}^*\mathcal{P}\left({}^*\mathbb{N}\right) \to \{1,\ldots,k\}, \quad A \mapsto |A \cap \{1,\ldots,k\}|.$$

Note that $A \in {}^*\mathcal{P}({}^*\mathbb{N})$ is a c_k-approximation of $B \subseteq {}^*\mathbb{N}$ if $A = B \cap \{1,\ldots,k\}$, thus each subset $B \subseteq {}^*\mathbb{N}$ is in $L_{c_k}({}^*\mathcal{P}({}^*\mathbb{N}))$ for all $k \in \mathbb{N}$. Define

$$\widehat{c}(B) := \lim_{k\to\infty} \widehat{c_k}(B) \in \mathbb{N} \cup \{\infty\}.$$

Since ${}^*\mathbb{N} \setminus \mathbb{N}$ is a $\widehat{c}$-nullset, $\widehat{c}$ is the usual **counting measure** on $\mathbb{N}$, denoted by c.

10.6 Adapted Loeb spaces

Let $(\Lambda, \mathcal{C}, \mu)$ be an internal probability space. According to the standard definition in Section 2.1, an **internal filtration on** $\mathcal{C}$ is an internal H^2-tuple $(\mathcal{C}_t)_{t\in T}$ of internal algebras $\mathcal{C}_t \subseteq \mathcal{C}$ such that $\mathcal{C}_s \subseteq \mathcal{C}_t$ if $s \le t$. Following Keisler's idea, we construct from an internal filtration $(\mathcal{C}_s)_{s\in T}$ on $\mathcal{C}$ a standard filtration $(\mathfrak{c}_t)_{t\in[0,\infty[}$ on $L_\mu(\mathcal{C})$ that fulfils the Doob–Meyer conditions (see Section 2.1).

Fix an internal filtration $(\mathcal{C}_t)_{t\in T}$ on $\mathcal{C}$. For each $r \in [0,\infty[$ set

$$\mathfrak{c}_r := \bigcup_{t\in T, t\approx r} \mathcal{C}_t \vee \mathcal{N}_{\widehat{\mu}}.$$

Note that $L_{\mu\restriction \mathcal{C}_t}(\mathcal{C}_t) \subseteq \mathcal{C}_t \vee \mathcal{N}_{\widehat{\mu}}$.

Theorem 10.6.1 (Keisler [53])

(a) *Each* $\mathfrak{c}_r$ *is a* σ*-subalgebra of* $L_\mu(\mathcal{C})$.
(b) $\mathfrak{c}_r \subseteq \mathfrak{c}_s$ *for all* $r < s$ *in* $[0,\infty[$.
(c) $\mathfrak{c}_r = \bigcap_{s>r} \mathfrak{c}_s$, *i.e., the filtration* $(\mathfrak{c}_r)_{r\in[0,\infty[}$ ***is right continuous***.
(d) *If* $N \in \mathcal{N}_{\widehat{\mu}}$, *then* $N \in \mathfrak{c}_0$, *thus* $N \in \mathfrak{c}_r$ *for each* $r \in [0,\infty[$.

Proof (a) Since $L_\mu(\mathcal{C})$ is complete, $\mathfrak{c}_r \subseteq L_\mu(\mathcal{C})$. Using Corollary 10.1.3, it is easy to see that $\mathfrak{c}_r$ is an algebra. To prove that it is a σ-algebra, fix a non-decreasing sequence $(B_n)_{n\in\mathbb{N}}$ in $\mathfrak{c}_r$. For each $n \in \mathbb{N}$ there exists a $t_n \in T$ with $t_n \approx r$ and an $A'_n \in \mathcal{C}_{t_n}$ such that $\widehat{\mu}(A'_n \,\triangle\, B_n) = 0$. For $A_n := A'_1 \cup \ldots \cup A'_n$ we have also $\widehat{\mu}(A_n \,\triangle\, B_n) = 0$, because $B_1 \subseteq \ldots \subseteq B_n$. Set $s := \lim_{n\to\infty} {}^\circ\mu(A_n)$. By saturation, there exist a $t \in T$ with $r \approx t \ge t_n$ and an $A \in \mathcal{C}_t$ with $\bigcup_{n\in\mathbb{N}_0} A_n \subseteq A$ and $\mu(A) < s + \frac{1}{n}$ for each $n \in \mathbb{N}$. Therefore, $\mu(A) \approx s$. It follows that $\widehat{\mu}(A \,\triangle\, \bigcup_{n\in\mathbb{N}} A_n) = 0$. Since

$$A \,\triangle \bigcup_{n\in\mathbb{N}} B_n \subseteq (A \,\triangle \bigcup_{n\in\mathbb{N}} A_n) \cup \bigcup_{n\in\mathbb{N}} (A_n \,\triangle\, B_n),$$

$\widehat{\mu}(A \,\triangle\, \bigcup_{n\in\mathbb{N}} B_n) = 0$, thus, $\bigcup_{n\in\mathbb{N}} B_n \in \mathfrak{c}_r$.

(b) is true, because $\mathcal{C}_s \subseteq \mathcal{C}_t$ for $s < t$.

(c) By (b), $\mathfrak{c}_r \subseteq \bigcap_{s>r} \mathfrak{c}_s$. Now fix $C \in \bigcap_{s>r} \mathfrak{c}_s$. There exists a decreasing sequence $(t_n)_{n\in\mathbb{N}}$ in T with $r < {}^\circ t_n$ and $\lim_{n\to\infty} {}^\circ t_n = r$. Moreover, for each $n \in \mathbb{N}$ there exists an $A_n \in \mathcal{C}_{t_n}$ such that $\widehat{\mu}(A_n \triangle C) = 0$. By saturation, there exist an $s \in T$ with $r \leq s \leq t_n$ and an $A \in \mathcal{C}_s$ such that $\mu(A \Delta A_1) < \frac{1}{n}$ for each $n \in \mathbb{N}$, thus $\mu(A \Delta A_1) \approx 0$. Since $\widehat{\mu}(A \Delta C) \leq \widehat{\mu}(A \Delta A_1) + \widehat{\mu}(A_1 \Delta C) = 0$ and $r \approx s$, we have $C \in \mathfrak{c}_r$.

(d) Each $N \in \mathcal{N}_{\widehat{\mu}}$ belongs to $\mathfrak{c}_0$, because $\emptyset \in \mathcal{C}_{\frac{1}{H}}$ and $\widehat{\mu}(N \Delta \emptyset) = 0$. □

The filtration $(\mathfrak{c}_r)_{r\in[0,\infty[}$ on $L_\mu(\mathcal{C})$ is called the **standard part** of the internal filtration $(\mathcal{C}_t)_{t\in T}$ on $\mathcal{C}$, and the quadruple $(\Lambda, L_\mu(\mathcal{C}), \widehat{\mu}, (\mathfrak{c}_r)_{r\in[0,\infty[})$ is called the **adapted Loeb space** over $(\Lambda, \mathcal{C}, \mu, (\mathcal{C}_t)_{t\in T})$.

10.7 *S*-integrability and equivalent conditions

Fix a Loeb space $(\Lambda, L_\mu(\mathcal{C}), \widehat{\mu})$ with limited $\mu(\Lambda)$. Assume that $\mathcal{C}$ is a ${}^*\sigma$-algebra. For $\mathcal{C}$-measurable functions $F : \Lambda \to \mathbb{F}^{\otimes n}$ we have two notions of integrability, namely μ-integrability and S_μ-integrability, where **μ-integrability** is nothing but the usual Bochner integrability 'copied' from the model $\mathfrak{V}$ to the model $\mathfrak{W}$ by transfer. More important than μ-integrability is the stronger notion of S_μ-integrability, which is close to standard integrability. An internal $\mathcal{C}$-measurable function $F : \Lambda \to {}^*\mathbb{R}_0^+$ is called **S_μ-integrable** if for all unlimited $K \in {}^*\mathbb{N}$

$$\int_{\{F \geq K\}} F d\mu \approx 0.$$

A $\mathcal{C}$-measurable function $F : \Lambda \to \mathbb{F}^{\otimes n}$ is called **S_μ-integrable** if $\|F\|_{\mathbb{F}^n}$ is S_μ-integrable. Fix a standard $p \in [1, \infty[$ and define

$$SL^p(\mu, \mathbb{F}^{\otimes n}) := \left\{ F : \Lambda \to \mathbb{F}^{\otimes n} \mid \|F\|_{\mathbb{F}^n}^p \text{ is } S_\mu\text{-integrable} \right\}.$$

If there is no risk of confusion, we shall write $SL^p(\mu)$ instead of $SL^p(\mu, \mathbb{F}^{\otimes n})$, in particular, if $\mathbb{F}^{\otimes n} = {}^*\mathbb{R}$. In the case $p = 2$, we call the elements of $SL^2(\mu)$ **S_μ-square-integrable**.

Lemma 10.7.1 *Assume that $F : \Lambda \to {}^*\mathbb{R}_0^+$ is $\mathcal{C}$-measurable.*

(a) *If $\int_\Lambda F^p \, d\mu$ is limited, then F is limited $\widehat{\mu}$-a.e.*
(b) *If $\int_\Lambda F^p \, d\mu \approx 0$, then $F \approx 0$ $\widehat{\mu}$-a.e.*

Proof (a) Since $U := \{F \text{ is unlimited}\} = \bigcap_{n\in\mathbb{N}} \{F \geq n\}$, we have $U \in L_\mu(\mathcal{C})$. Assume that $\varepsilon := \widehat{\mu}(U) \in \mathbb{R}^+$. Then $\mu(\{F \geq n\}) > \frac{\varepsilon}{2}$ for each $n \in \mathbb{N}$. By

Proposition 8.4.1 (b), there is an unlimited $K \in {}^*\mathbb{N}$ with $\mu(\{F \geq K\}) > \frac{\varepsilon}{2}$. We obtain

$$\int_\Lambda F^p d\mu \geq \int_{\{F \geq K\}} F^p d\mu > K^p \cdot \frac{\varepsilon}{2} \text{ is unlimited,}$$

which proves (a).

(b) Assume that $\int_\Lambda F^p d\mu \approx 0$. Then $\int_\Lambda (nF)^p \, d\mu = n^p \int_\Lambda F^p d\mu \leq 1$ for each $n \in \mathbb{N}$. By Proposition 8.4.1 (b), there exists an unlimited $K \in {}^*\mathbb{N}$ such that $\int_\Lambda (KF)^p \, d\mu \leq 1$. By (a), $K \cdot F$ is limited $\widehat{\mu}$-a.e. Since K is unlimited, $F \approx 0$ $\widehat{\mu}$-a.e. □

Proposition 10.7.2 (Anderson [4], Loeb [69]) *Let $F : \Lambda \to {}^*\mathbb{R}_0^+$ be $\mathcal{C}$-measurable. The following statements (a)–(e) are equivalent:*

(a) *F is S_μ-integrable.*
(b) $\lim_{n \to \infty} {}^\circ\!\int_{\{F \geq n\}} F d\mu = 0$.
(c) *For each $A \in \mathcal{C}$*

$$\int_A F d\mu \begin{cases} \text{is limited} & \\ \approx 0 & \text{if } \mu(A) \approx 0. \end{cases}$$

(d) *$\int_\Lambda F d\mu$ is limited and for each $\varepsilon \in \mathbb{R}^+$ there exists a $\delta \in \mathbb{R}^+$ such that $\int_A F d\mu < \varepsilon$ for all $A \in \mathcal{C}$ with $\mu(A) < \delta$.*
(e) *There exists a function $g : \mathbb{N} \to \mathbb{N}$ such that for all $n \in \mathbb{N}$*

$$\int_{\{F \geq g(n)\}} F d\mu < \frac{1}{n}.$$

Proof '$(a) \Rightarrow (b)$' Assume that (b) is not true. Then there is an $\varepsilon \in \mathbb{R}^+$ such that $\left\{ n \in \mathbb{N} \mid \int_{\{F \geq n\}} F d\mu \geq \varepsilon \right\}$ is unbounded. By Proposition 8.4.1 (b), there exists an unlimited $K \in {}^*\mathbb{N}$ such that $\int_{\{F \geq K\}} F d\mu \geq \varepsilon$, contradicting (a).

'$(b) \Rightarrow (c)$' Fix $\varepsilon \in \mathbb{R}^+$. By (b), we have $\int_{\{F \geq n\}} F d\mu < \varepsilon$ for some $n \in \mathbb{N}$. We obtain for each $A \in \mathcal{C}$

$$\int_A F d\mu \leq \int_{\{F \geq n\}} F d\mu + \int_{A \cap \{F < n\}} F d\mu$$

$$< \varepsilon + n\mu(A) \begin{cases} \text{is limited} & \\ < 2\varepsilon & \text{if } \mu(A) \approx 0. \end{cases}$$

Since ε is arbitrary, (c) follows.

'$(c) \Rightarrow (d)$' By (c), $\int_\Lambda F d\mu$ is limited. Assume that (d) is not true. Then there exists an $\varepsilon \in \mathbb{R}^+$ such that for each $n \in \mathbb{N}$ there is an $A_n \in \mathcal{C}$ with $\mu(A_n) < \frac{1}{n}$

and $\int_{A_n} F d\mu \geq \varepsilon$. By saturation, there exists an $A \in \mathcal{C}$ such that $\mu(A) \approx 0$, and $\int_A F d\mu \geq \varepsilon$. It follows that (c) fails.

'(d) $\Rightarrow$ (e)' Suppose that (d) is true. Since $\int_\Lambda F d\mu$ is limited, by Lemma 10.7.1 (a), $\mu\{F \geq K\} \approx 0$ for each unlimited $K \in {}^*\mathbb{N}$. Fix $n \in \mathbb{N}$. Then $\int_{\{F \geq K\}} F d\mu < \frac{1}{n}$ for all unlimited $K \in {}^*\mathbb{N}$. Therefore, by Proposition 8.4.1 (a), there exists an $m \in \mathbb{N}$ such that $\int_{\{F \geq m\}} F d\mu < \frac{1}{n}$. Set $g(n) := \min\left\{m \in \mathbb{N} \mid \int_{\{F \geq m\}} F d\mu < \frac{1}{n}\right\}$.

'(e) $\Rightarrow$ (a)' By (e), for each $n \in \mathbb{N}$ and each unlimited $K \in {}^*\mathbb{N}$,

$$\int_{\{|F| \geq K\}} F d\mu \leq \int_{\{|F| \geq g(n)\}} F d\mu < \frac{1}{n}.$$

Since n is arbitrary, $\int_{\{F \geq K\}} F d\mu \approx 0$. □

Note that, since the integral of an S_μ-integrable function is limited, S_μ-integrability implies μ-integrability. But, for example, a constant function on Λ having an unlimited value is μ-integrable, but not S_μ-integrable (as long as $\mu(\Lambda)$ is not infinitesimal).

The following results are simple application of the previous one and are left to the reader.

Corollary 10.7.3 *Fix* $p \in [1, \infty[$.

(α) *Fix* $G, F \in SL^p\left(\mu, \mathbb{F}^{\otimes d}\right)$ *and a limited* $a \in {}^*\mathbb{R}$. *Then* $F + G \in SL^p\left(\mu, \mathbb{F}^{\otimes d}\right)$ *and* $a \cdot F \in SL^p\left(\mu, \mathbb{F}^{\otimes d}\right)$.

(β) $SL^p(\mu, \mathbb{F}^{\otimes d}) \subseteq SL^q(\mu, \mathbb{F}^{\otimes d})$ *if* $q \in [1, p]$.

(γ) *A* $\mathcal{C}$*-measurable function* F *belongs to* $SL^p(\mu, \mathbb{F}^{\otimes d})$ *if and only if there exists a sequence* $(G_n)_{n \in \mathbb{N}}$ *in* $SL^p\left(\mu, \mathbb{F}^{\otimes d}\right)$ *with* $\int_\Lambda \|F - G_n\|^p_{\mathbb{F}^d} d\mu < \frac{1}{n}$ *for each* $n \in \mathbb{N}$.

10.8 Bochner integrability and S-integrability

In this section we will prove slight modifications of the Loeb–Anderson lifting theorem for Bochner integrable functions (see Loeb [69], Anderson [4] and also [71]). It shows that Bochner integrable functions can be characterized by their S-integrable liftings. Since $\mathbb{H}^{\otimes 0} = \mathbb{R}$ and $\mathbb{F}^{\otimes 0} = {}^*\mathbb{R}$, we obtain the real-valued Loeb–Anderson result on S-integrability in this case. We use the notation in Section 10.7.

Theorem 10.8.1 *Fix an* $L_\mu(\mathcal{C})$*-measurable function* $f : \Lambda \to \mathbb{H}^{\otimes d}$ *and* $p \in [1, \infty[$.

(a) *Then f is Bochner integrable if and only if f has an S_μ-integrable $\mathcal{C}$-*simple lifting $F:\Lambda\to\mathbb{F}^{\otimes d}$, in which case*

$$\int_\Lambda f d\widehat{\mu} = {}^\circ\!\int_\Lambda F d\mu \quad \textit{(see Section 9.8).}$$

(b) *Let $F:\Lambda\to\mathbb{F}^{\otimes d}$ be a $\mathcal{C}$-measurable lifting of $f:\Lambda\to\mathbb{H}^{\otimes d}$. Suppose that $\|f\|^p_{\mathbb{H}^d}$ is $\widehat{\mu}$-integrable. Then there exists an unlimited $N\in{}^*\mathbb{N}$ such that $\mathbf{1}_{\{\|F\|_{\mathbb{F}^d}\le N\}}\cdot F\in SL^p(\mu,\mathbb{F}^{\otimes d})$. This function remains a lifting of f.*

(c) *f is Bochner $\widehat{\mu}$-integrable if and only if $\|f\|_{\mathbb{H}^d}$ is $\widehat{\mu}$-integrable.*

(d) *$f\in L^p\left(\widehat{\mu},\mathbb{H}^{\otimes d}\right)$ if and only if f has a lifting $F\in SL^p\left(\mu,\mathbb{F}^{\otimes d}\right)$, in which case*

$$\int_\Lambda \|f\|^p_{\mathbb{H}^d}\, d\widehat{\mu} = {}^\circ\!\int_\Lambda \|F\|^p_{\mathbb{F}^d}\, d\mu.$$

Proof (a) '$\Rightarrow$' Choose a witness $(f_k)_{k\in\mathbb{N}}$ for the integrability of f and let F_k be a $\mathcal{C}$-simple lifting of f_k, according to the proof of Theorem 10.3.2. Then $\int_\Lambda F_k d\mu \approx_{\mathbb{F}^d} \int_\Lambda f_k d\widehat{\mu}$. Since $\|F_k\|_{\mathbb{F}^d}$ is limited, $F_k\in SL^1\left(\mu,\mathbb{F}^{\otimes d}\right)$. Let $G:\Lambda\to\mathbb{F}^{\otimes d}$ be a $\mathcal{C}$-*simple lifting of f according to Theorem 10.3.2. Then there exists a subsequence $\left(F_{g(m)}\right)_{m\in\mathbb{N}}$ of $(F_m)_{m\in\mathbb{N}}$ such that for all $m\in\mathbb{N}$ and $k\ge g(m)$

$$^\circ\!\int_\Lambda \left\|F_k - F_{g(m)}\right\|_{\mathbb{F}^d} d\mu < \frac{1}{m}, \tag{19}$$

and

$$^\circ\mu\left\{\|F_k - G\|_{\mathbb{F}^d} \ge \frac{1}{m}\right\} < \frac{1}{m}. \tag{20}$$

Let $(F_k)_{k\in{}^*\mathbb{N}}$ be an internal extension of $(F_k)_{k\in\mathbb{N}}$ such that all $F_k:\Lambda\to\mathbb{F}^{\otimes d}$ are $\mathcal{C}$-*simple. By Proposition 9.5.1 (a), there is an unlimited $K\in{}^*\mathbb{N}$ such that (19) and (20) become true for all $m\in\mathbb{N}$ if we drop $^\circ$ and replace k with K. By (20), $F:=F_K$ is a lifting of f. By (19) and Corollary 10.7.3 (γ), $F\in SL^1(\mu)$. Since

$$\int_\Lambda F_k d\mu \approx_{\mathbb{F}^d} \int_\Lambda f_k d\widehat{\mu} \to_{k\to\infty} \int_\Lambda f d\widehat{\mu},$$

we obtain, omitting $*$ at the integrals,

$$\left\|\int_\Lambda f d\widehat{\mu} - \int_\Lambda F d\mu\right\|_{\mathbb{F}^d} \le \left\|\int_\Lambda f - f_{g(k)} d\widehat{\mu}\right\|_{\mathbb{F}^d}$$
$$+\left\|\int_\Lambda f_{g(k)} d\widehat{\mu} - \int_\Lambda F_{g(k)} d\mu\right\|_{\mathbb{F}^d} + \int_\Lambda \left\|F_{g(k)} - F\right\|_{\mathbb{F}^d} d\mu,$$

which can be made arbitrarily small from the standard point of view. This proves that $\int_\Lambda f d\widehat{\mu} = {}^\circ\int_\Lambda F d\mu$.

'$\Leftarrow$' Assume that f has a lifting $F : \Lambda \to \mathbb{F}^{\otimes d} \in SL^1(\mu, \mathbb{F}^{\otimes d})$. By Theorem 10.3.2, there is a sequence $(f_n)_{n\in\mathbb{N}}$ of $\mathcal{C}$-simple functions $f_n : \Lambda \to \mathbb{H}^{\otimes d}$, converging to f in measure. In order to show that $\left(\int_\Lambda \|f_n - f_k\|_{\mathbb{H}^d}\, d\widehat{\mu}\right)_{n,k\in\mathbb{N}}$ converges to 0, fix an $\varepsilon \in \mathbb{R}$. Since $F \in SL^1(\mu, \mathbb{F}^{\otimes d})$, by Proposition 10.7.2 (d), there exists a $\delta \in \mathbb{R}^+$ such that $\int_A \|F\|_{\mathbb{F}^d}\, d\mu < \varepsilon$ for each $A \in \mathcal{C}$ with $\mu(A) < \delta$. Choose $n_0 \in \mathbb{N}$ with $\frac{1}{n_0} < \delta$. Let A_n be defined as in the proof of Theorem 10.3.2 '$\Leftarrow$'. Then we obtain, for each $n \in \mathbb{N}$, $n \geq n_0$, because $f_n \restriction A_n = 0$ and $\mu(A_n) < \delta$ and $\|F(X) - {}^*f_n(X)\|_{\mathbb{F}^d} < \frac{1}{n}$ for $X \notin A_n$,

$$\begin{aligned}\int_\Lambda \|{}^*f_n - F\|_{\mathbb{F}^d}\, d\mu &= \int_{A_n} \|{}^*f_n - F\|_{\mathbb{F}^d}\, d\mu + \int_{\Lambda\setminus A_n} \|{}^*f_n - F\|_{\mathbb{F}^d}\, d\mu \\ &= \int_{A_n} \|F\|_{\mathbb{F}^d}\, d\mu + \int_{\Lambda\setminus A_n} \|{}^*f_n - F\|_{\mathbb{F}^d}\, d\mu \leq \varepsilon + \frac{1}{n}\mu(\Lambda).\end{aligned}$$

Therefore, for $n, k \in \mathbb{N}$ with $n, k \geq n_0$ we obtain

$$\int_\Lambda \|f_n - f_k\|_{\mathbb{H}^d}\, d\widehat{\mu} = \int_\Lambda {}^\circ \|{}^*f_n - {}^*f_k\|_{\mathbb{F}^d}\, d\widehat{\mu} \leq 2\varepsilon + \left(\frac{1}{n} + \frac{1}{k}\right){}^\circ\mu(\Lambda),$$

whence, $\int_\Lambda \|f_n - f_k\|_{\mathbb{H}^d}\, d\widehat{\mu} \to_{n,k\to\infty} 0$.

(b) Set $a := \|f\|^p_{\mathbb{H}^d}$ and $A := \|F\|^p_{\mathbb{F}^d}$. Then A is a lifting of a. Since A is therefore limited $\widehat{\mu}$-a.e., $\lim_{n\to\infty} \mathbf{1}_{\{A\leq n\}} a = a$ $\widehat{\mu}$-a.e. It follows that

$$\lim_{n\to\infty} \int_\Lambda a - \mathbf{1}_{\{A\leq n\}} a d\widehat{\mu} = 0 \quad \text{and} \quad \lim_{n,m\to\infty} \left| \int_\Lambda \mathbf{1}_{\{A\leq n\}} a - \mathbf{1}_{\{A\leq m\}} a d\widehat{\mu} \right| = 0.$$

Since $\mathbf{1}_{\{A\leq n\}} A$ is an S_μ-integrable lifting of $\mathbf{1}_{\{A\leq n\}} a$ for all $n \in \mathbb{N}$, by (a), $\lim_{n,m\to\infty} {}^\circ\left|\int_\Lambda \mathbf{1}_{\{A\leq n\}} A - \mathbf{1}_{\{A\leq m\}} A d\mu\right| = 0$. By Proposition 9.5.1 (a), there exists a function $g : \mathbb{N} \to \mathbb{N}$ and an unlimited $M \in {}^*\mathbb{N}$ such that $\left|\int_\Lambda \mathbf{1}_{\{A\leq M\}} A - \mathbf{1}_{\{A\leq g(m)\}} A d\mu\right| < \frac{1}{m}$ for all $m \in \mathbb{N}$. By Corollary 10.7.3 (γ), $\mathbf{1}_{\{A\leq M\}} A$ is S_μ-integrable, thus $\mathbf{1}_{\left\{\|F\|_{\mathbb{F}^d} \leq N\right\}} F \in SL^p\left(\mu, \mathbb{F}^{\otimes d}\right)$, where $N \in {}^*\mathbb{N}$ is unlimited with $N \leq \sqrt[p]{M}$. This function remains a lifting of f, because N is unlimited and f is standard.

(c) Obviously, $\|f\|_{\mathbb{H}^d}$ is $\widehat{\mu}$-integrable if f is Bochner $\widehat{\mu}$-integrable. Conversely, assume that $\|f\|_{\mathbb{H}^d}$ is $\widehat{\mu}$-integrable. Since f is $L_\mu(\mathcal{C})$-measurable, f has a $\mathcal{C}$-measurable lifting F. By (b), there exists an unlimited $N \in {}^*\mathbb{N}$ such that $\mathbf{1}_{\left\{\|F\|_{\mathbb{F}^d} \leq N\right\}} F$ is an S_μ-integrable lifting of f. By (a), f is Bochner $\widehat{\mu}$-integrable.

(d) Let f be p-times Bochner integrable and let G be a $\mathcal{C}$-*simple lifting of f. Then, by (b), there exists an unlimited $N \in {}^*\mathbb{N}$ such that $F := \mathbf{1}_{\left\{\|F\|_{\mathbb{F}^d} \leq N\right\}} \cdot G$ is a lifting of f in $SL^p\left(\mu, \mathbb{F}^{\otimes d}\right)$. Therefore,

$$\int_\Lambda \|f\|^p_{\mathbb{H}^d}\, d\widehat{\mu} = \int_\Lambda {}^\circ\|{}^*f\|^p_{\mathbb{F}^d}\, d\widehat{\mu} = \int_\Lambda {}^\circ\|F\|^p_{\mathbb{F}^d}\, d\mu = {}^\circ\int_\Lambda \|F\|^p_{\mathbb{F}^d}\, d\mu.$$

□

The following corollary tells us that an internal ${}^*\mathbb{R}_0^+$-valued function F is S-integrable if and only if the standard part of the integral of F equals the integral of the standard part of F.

Corollary 10.8.2 (Anderson [4], Loeb [69]) *Let $F : \Lambda \to {}^*\mathbb{R}_0^+$ be $\mathcal{C}$-measurable. Set ${}^\circ F(x) := a \in \mathbb{R}_0^+$ if $F(x)$ is limited and set ${}^\circ F(x) := \infty$ if $F(x)$ is unlimited. Then*

(a) $\int_\Lambda {}^\circ F d\widehat{\mu} \leq {}^\circ\int_\Lambda F d\mu$,
(b) *F is S_μ-integrable if and only if* ${}^\circ\int_\Lambda F d\mu = \int_\Lambda {}^\circ F d\widehat{\mu} < \infty$.

Proof (a) First assume that $\widehat{\mu}\{F \text{ is unlimited}\} = \varepsilon \neq 0$. Then $\int_\Lambda {}^\circ F d\widehat{\mu} = \infty$ and we have, for all $n \in \mathbb{N}$,

$${}^\circ\int_\Lambda F d\mu \geq {}^\circ\int_{\{F \geq n\}} F d\mu \geq n \cdot {}^\circ\mu\{F \geq n\} \geq n \cdot \varepsilon.$$

Thus, ${}^\circ\int_\Lambda F d\mu = \infty$. Now assume that F is limited $\widehat{\mu}$-a.e. Then

$$\int_\Lambda {}^\circ F d\widehat{\mu} = \lim_{n\to\infty} \int_\Lambda \mathbf{1}_{\{F<n\}} {}^\circ F d\widehat{\mu} = \lim_{n\to\infty} {}^\circ\int_\Lambda \mathbf{1}_{\{F<n\}} F d\mu \leq {}^\circ\int_\Lambda F d\mu.$$

(b) '$\Rightarrow$' follows from Theorem 10.8.1 with $d = 0$. To prove '$\Leftarrow$', suppose that $\int_\Lambda F d\mu \approx \int_\Lambda {}^\circ F d\widehat{\mu} < \infty$. Then, applying '$\Rightarrow$' to the S_μ-integrable function $\mathbf{1}_{\{F<n\}} F$ for $n \in \mathbb{N}$, we obtain

$$\begin{aligned}
\lim_{n\to\infty} {}^\circ\int_\Lambda \left|F - \mathbf{1}_{\{F<n\}} F\right| d\mu &= {}^\circ\int_\Lambda F d\mu - \lim_{n\to\infty} {}^\circ\int_\Lambda \mathbf{1}_{\{F<n\}} F d\mu \\
&= \int_\Lambda {}^\circ F d\widehat{\mu} - \lim_{n\to\infty} \int_\Lambda \mathbf{1}_{\{F<n\}} {}^\circ F d\widehat{\mu} \\
&= \int_\Lambda {}^\circ F d\widehat{\mu} - \int_\Lambda {}^\circ F d\widehat{\mu} = 0.
\end{aligned}$$

Therefore, $F \in SL^1(\mu)$, by Corollary 10.7.3 (γ). □

Let us end this section with the following remark concerning the connection between S_ν-integrability and S-continuity, which plays an important role later.

Remark 10.8.3 Fix an internal $F: T \to {}^*\mathbb{R}$ such that $F \upharpoonright T_k$ is S_{ν_k}-integrable for all $k \in \mathbb{N}$.

(a) Then $F_I : t \mapsto \int_{\{s \in T \mid s \le t\}} F d\nu$ is S-continuous. Therefore, we can define a continuous function $\int F d\nu : [0,\infty[\to \mathbb{R}$ by setting

$$\int F d\nu : {}^\circ t \mapsto {}^\circ F_I(t)$$

for all limited $t \in T$.

(b) If F is a lifting of a Lebesgue-measurable function $f : [0,\infty[\to \mathbb{R}$, then $\int_0^r f d\lambda = \int F d\nu(r)$ for all $r \in [0,\infty[$.

10.9 Integrable functions defined on $\mathbb{N}^n \times \Lambda \times [0,\infty[^m$

We extend Theorem 10.8.1 and Lemma 10.5.1 to functions $f : \mathbb{N}^n \times \Lambda \times [0,\infty[^m \to \mathbb{H}^{\otimes d}$, which play an important role in the Malliavin calculus for Lévy processes. They are called ***p*-summable** if $\sum_{k \in \mathbb{N}^n} \int_{\Lambda \times [0,\infty[^m} \|f_k\|^p_{\mathbb{H}^d} \, d\widehat{\mu} \otimes \lambda^m < \infty$. Then we write $f \in L^p\left(c^n \otimes \widehat{\mu} \otimes \lambda^m, \mathbb{H}^{\otimes d}\right)$. If $\mathcal{D}$ is a σ-subalgebra of $L_\mu(\mathcal{C})$ and $\mathcal{E}$ a σ-subalgebra of Leb^m and if all the f_k are, in addition, $\mathcal{D}\otimes\mathcal{E}$-measurable, then we write $f \in L^p_{\mathcal{D}\otimes\mathcal{E}}\left(c^n \otimes \widehat{\mu} \otimes \lambda^m, \mathbb{H}^{\otimes d}\right)$. According to Corollary 10.5.3, we can identify Lebesgue-measurable functions f with $\mathcal{L}^m$-measurable functions g in the following sense:

Corollary 10.9.1 *Fix $p \in [1,\infty[$. Then*

$$\iota : L^p_{L_\mu(\mathcal{C})\otimes \mathrm{Leb}^m}(c^n \otimes \widehat{\mu} \otimes \lambda^m, \mathbb{H}^{\otimes d}) \to L^p_{L_\mu(\mathcal{C})\otimes \mathcal{L}^m}(c^n \otimes \widehat{\mu \otimes \nu^m}, \mathbb{H}^{\otimes d}),$$

defined by $\iota(f)(k,X,t) = f(k,X,{}^\circ t)$ for all $k \in \mathbb{N}^n$ and $\widehat{\mu \otimes \nu^m}$-almost all $(X,t) \in \Lambda \times T^m$, is a canonical isomorphic isometry between both spaces.

Since non-finite measures on $\mathbb{N}^n$ and $[0,\infty[^m$ are involved, we need an extension of the notion of S-integrability. Fix a $\mathcal{C}$**-measurable** internal $F: {}^*\mathbb{N}^n \times \Lambda \times T^m \to \mathbb{F}^{\otimes d}$, i.e., $F(k,\cdot,t)$ is $\mathcal{C}$-measurable for all $k \in {}^*\mathbb{N}^n, t \in T^m$. Fix $p \in [1,\infty[$. We call F **locally $S^p_{c^n\otimes\mu\otimes\nu^m}$-integrable** if $F_k \upharpoonright \Lambda \times T^m_\sigma \in SL^p\left(\mu \otimes \nu^m, \mathbb{F}^{\otimes d}\right)$ for all $k \in \mathbb{N}^n$ and all $\sigma \in \mathbb{N}$. The whole F_k is called **$S^p_{\mu\otimes\nu^m}$-integrable**, in which case we write $F_k \in SL^p\left(\mu \otimes \nu^m, \mathbb{F}^{\otimes d}\right)$, if

(i) F_k is locally $S^p_{\mu\otimes\nu^m}$-integrable,

(ii) $\int_{\Lambda \times (T^m \setminus T^m_S)} \|F_k\|^p_{\mathbb{F}^d} \, d\mu \otimes \nu^m \approx 0$ for all unlimited $S \le H$.

The function F is called $S^p_{c^n\otimes\mu\otimes\nu^m}$**-integrable**, in which case we write $F \in SL^p\left(c^n \times \mu \otimes \nu^m, \mathbb{F}^{\otimes d}\right)$, if

(iii) $F_k \in SL^p\left(\mu \otimes \nu^m, \mathbb{F}^{\otimes d}\right)$ for all $k \in \mathbb{N}^n$,
(iv) $\sum_{k\in {}^*\mathbb{N}^n\setminus S^n} \int_{\Lambda\times T^m} \|F_k\|^p_{\mathbb{F}^d}\, d\mu\otimes\nu^m \approx 0$ for all unlimited $S \in {}^*\mathbb{N}$.

Note that $\sum_{k\in {}^*\mathbb{N}^n} \int_{\Lambda\times T^m} \|F_k\|^p_{\mathbb{F}^d}\, d\mu\otimes\nu^m$ is limited if the integrand F belongs to $SL^p\left(c^n \otimes \mu \otimes \nu^m, \mathbb{F}^{\otimes d}\right)$.

The following result is a slight extension of the Loeb–Anderson lifting theorem and shows that functions $f : \mathbb{N}^n \times \Lambda \times [0,\infty[^m \to \mathbb{H}^{\otimes d}$ can represented by *finite functions with values in *finite-dimensional spaces.

Theorem 10.9.2 *Fix* $f : \mathbb{N}^n \times \Lambda \times [0,\infty[^m \to \mathbb{H}^{\otimes d}$.

(a) f_k *is* $L_\mu(\mathcal{C})\otimes\mathrm{Leb}^m$*-measurable for all* $k \in \mathbb{N}^n$ *if and only if* f *has a* $\mathcal{C}$*-*simple* ***lifting*** $F : {}^*\mathbb{N}^n \times \Lambda \times T^m \to \mathbb{F}^{\otimes d}$, *i.e.,* F_k *is a lifting of* f_k *for all* $k \in \mathbb{N}^n$.
(b) f *belongs to* $L^p_{L_\mu(\mathcal{C})\otimes\mathrm{Leb}^m}\left(c^n \otimes \widehat{\mu} \otimes \lambda^m, \mathbb{H}^{\otimes d}\right)$ *if and only if* f *has a* $\mathcal{C}$*-*simple lifting* F *in* $SL^p\left(c^n \otimes \mu \otimes \nu^m, \mathbb{F}^{\otimes d}\right)$, *in which case*

$$\sum_{k\in {}^*\mathbb{N}^n} \int_{\Lambda\times T^m} \|F_k\|^p_{\mathbb{F}^n}\, d\mu\otimes\nu^m \approx \sum_{k\in\mathbb{N}^n} \int_{\Lambda\times \mathrm{Leb}^m} \|f_k\|^p_{\mathbb{H}^n}\, d\widehat{\mu}\otimes\lambda^m.$$

If $p = 1$, *then*

$$\sum_{k\in {}^*\mathbb{N}^n} \int_{\Lambda\times T^m} F_k d\mu\otimes\nu^m \approx_{\mathbb{F}^d} {}^*\left(\sum_{k\in\mathbb{N}^n} \int_{\Lambda\times T^m} f_k d\widehat{\mu}\otimes\lambda^m\right).$$

In case $d = 0$ *we have* $\mathbb{H}^{\otimes d} = \mathbb{R}$ *and* $\mathbb{F}^{\otimes d} = {}^*\mathbb{R}$. *Moreover, (a) and (b) are equivalent to statements which result by replacing* Leb^m *with* $\mathcal{L}^m$ *and* $\widehat{\mu}\otimes\lambda^m$, *with* $\widehat{\mu\otimes\nu^m}$. *We shall use this identification again and again.*

Proof In order to save indices, set $m = n = d = 1$. For functions f defined on $\Lambda \times T$, and for $\sigma \in {}^*\mathbb{N}$, set $f_\sigma := \mathbf{1}_{\Lambda\times T_\sigma} \cdot f$.

(a) Suppose that $f_k := f(k,\cdot,\cdot)$ is $L_\mu(\mathcal{C})\otimes\mathcal{L}$-measurable for all $k \in \mathbb{N}$. Then f_k has a $\mathcal{C}$-*simple lifting F_k. Let $(F_k)_{k\in {}^*\mathbb{N}}$ be an internal extension of $(F_k)_{k\in \mathbb{N}}$ such that all F_k are $\mathcal{C}$-*simple. Then $(F_k)_{k\in {}^*\mathbb{N}}$ is a $\mathcal{C}$-*simple lifting of f. The converse follows immediately from Proposition 10.5.4.

(b) Fix $f \in L^p_{L_\mu(\mathcal{C})\otimes\mathcal{L}}\left(c \otimes \widehat{\mu\otimes\nu}, \mathbb{H}\right)$ and $\sigma \in \mathbb{N}$. According to Theorem 10.8.1 (d), there exists a $\mathcal{C}$-*simple lifting $F_{k,\sigma} : \Lambda \times T_\sigma \to \mathbb{F} \in SL^p(\mu\otimes\nu, \mathbb{F})$ of $f_{k,\sigma}$. We may assume that $F_{k,\sigma}(x) = F_{k,\sigma+1}(x)$ for all $x \in \Lambda \times T_\sigma$. Let $\left(F_{k,\sigma}\right)_{\sigma\le H}$ be an internal extension of $\left(F_{k,\sigma}\right)_{\sigma\in\mathbb{N}}$ with $F_{k,\sigma}(x) = F_{k,\sigma+1}(x)$ for all $x \in \Lambda \times T_\sigma$

and all $\sigma \leq H$ and such that all $F_{k,\sigma}$ are $\mathcal{C}$-*simple. Since

$$\begin{aligned}
0 &= \lim_{\sigma,\sigma'\to\infty} \int_{\Lambda\times T} \|f_{k,\sigma} - f_{k,\sigma'}\|_{\mathbb{H}}^p \, d\widehat{\mu\otimes\nu} \\
&= \lim_{\sigma,\sigma'\to\infty} \int_{\Lambda\times T} {}^\circ \left\|{}^*\left(f_{k,\sigma}(X,r)\right) - {}^*\left(f_{k,\sigma'}(X,r)\right)\right\|_{\mathbb{F}}^p \, d\widehat{\mu\otimes\nu}(X,r) \\
&= \lim_{\sigma,\sigma'\to\infty} \int_{\Lambda\times T} {}^\circ \|F_{k,\sigma} - F_{k,\sigma'}\|_{\mathbb{F}}^p \, d\widehat{\mu\otimes\nu} \\
&= \lim_{\sigma,\sigma'\to\infty} {}^\circ\!\int_{\Lambda\times T} \|F_{k,\sigma} - F_{k,\sigma'}\|_{\mathbb{F}}^p \, d\mu\otimes\nu,
\end{aligned}$$

by Propositions 9.5.1, there exists a function $g:\mathbb{N}\to\mathbb{N}$ and an unlimited $S_\infty \leq H$ such that for all unlimited $S \leq S_\infty$ and for all $j \in \mathbb{N}$

$$\int_{\Lambda\times T} \|F_{k,g(j)} - F_{k,S}\|_{\mathbb{F}}^p \, d\mu\otimes\nu < \frac{1}{j}.$$

Set $F_k := F_{k,S_\infty}$. We obtain $\int_{\Lambda\times(T\setminus T_S)} \|F_k\|_{\mathbb{F}}^p \, d\mu\otimes\nu \approx 0$ for all unlimited $S \leq H$. This proves that $F_k \in SL^p(\mu\otimes\nu,\mathbb{F})$. In the same way we see that

$$\begin{aligned}
\int_{\Lambda\times T} \|f_k\|_{\mathbb{H}}^p \, d\widehat{\mu\otimes\nu} &= \lim_{\sigma\to\infty} \int_{\Lambda\times T} \|f_{k,\sigma}\|_{\mathbb{H}}^p \, d\widehat{\mu\otimes\nu} \\
&= \lim_{\sigma\to\infty} {}^\circ\!\int_{\Lambda\times T} \|F_{k,\sigma}\|_{\mathbb{F}}^p \, d\mu\otimes\nu \\
&= {}^\circ\!\int_{\Lambda\times T} \|F_k\|_{\mathbb{F}}^p \, d\mu\otimes\nu.
\end{aligned}$$

In case $p = 1$, we obtain, by Theorem 10.8.1,

$$\begin{aligned}
\int_{\Omega\times T} f_k \, d\widehat{\mu\otimes\nu} &= \lim_{\sigma\to\infty} \int_{\Omega\times T} f_{k,\sigma} \, d\widehat{\mu\otimes\nu} \\
&= \lim_{\sigma\to\infty} {}^\circ\!\int_{\Omega\times T} F_{k,\sigma} \, d\mu\otimes\nu = {}^\circ\!\int_{\Omega\times T} F_k \, d\mu\otimes\nu,
\end{aligned}$$

which can be seen as follows. Assume that the third equality is not true. Then there exists a standard $\varepsilon > 0$ such that for infinitely many σ

$$\varepsilon \leq \left\|\int_{\Omega\times T} (F_k - F_{k,\sigma}) \, d\mu\otimes\nu\right\|_{\mathbb{F}} \leq \int_{\Omega\times T} \|F_k - F_{k,\sigma}\|_{\mathbb{F}} \, d\mu\otimes\nu.$$

By Proposition 8.4.1 (b), there exists an unlimited $S \leq S_\infty$ in ${}^*\mathbb{N}$ with

$$\varepsilon \leq \int_{\Omega\times T} \left\| F_k - F_{k,S} \right\|_{\mathbb{F}} d\mu \otimes \nu = \int_{\Omega\times(T\setminus T_S)} \|F_k\|_{\mathbb{F}}\, d\mu \otimes \nu,$$

which contradicts the S-integrability of F_k.

Let $(F_k)_{k\in {}^*\mathbb{N}}$ be an internal extension of $(F_k)_{k\in \mathbb{N}}$. We proceed in the same manner as before and obtain an $F := (F_k)_{k\leq K_\infty}$ for some unlimited K_∞ with $\sum_{k\in {}^*\mathbb{N}\setminus K} \int_{\Lambda\times T} \|F_k\|_{\mathbb{F}}^p\, d\mu\otimes\nu \approx 0$ for all unlimited $K \leq K_\infty$. We set $F_k = 0$ for $k > K_\infty$. It follows that $F \in SL^p\,(c\otimes\mu\otimes\nu)$ and

$$\sum_{k\in {}^*\mathbb{N}} \int_{\Lambda\times T} \|F_k\|_{\mathbb{F}}^p\, d\mu\otimes\nu \approx \sum_{k\in \mathbb{N}} \int_{\Lambda\times T} \|f_k\|_{\mathbb{H}}^p\, d\widehat{\mu\otimes\nu},$$

$$\sum_{k\in {}^*\mathbb{N}} \int_{\Lambda\times T} F_k d\mu\otimes\nu \approx_{\mathbb{F}} \sum_{k\in \mathbb{N}} \int_{\Lambda\times T} f_k d\widehat{\mu\otimes\nu} \text{ for } p=1,$$

where we identify $a\in\mathbb{H}$ with ${}^*a\in\mathbb{F}$.

Vice versa, fix a $\mathcal{C}$-*simple lifting $F\in SL^p\,(c\otimes\mu\otimes\nu,\mathbb{F})$ of f and $k,\sigma\in\mathbb{N}$. Since $F_{k,\sigma}\in SL^p\,(\mu\otimes\nu,\mathbb{F})$ is a lifting of $f_{k,\sigma}$, we have $f_{k,\sigma}\in L^p\left(\widehat{\mu\otimes\nu},\mathbb{H}\right)$ and is $L_\mu(\mathcal{C})\otimes\mathcal{L}$-measurable. In order to prove that $f_k\in L^p\left(\widehat{\mu\otimes\nu},\mathbb{H}\right)$, we have to prove that $\lim_{\sigma,\sigma'\to\infty}\int_{\Lambda\times(T_\sigma\setminus T_{\sigma'})} \|f_k\|_{\mathbb{H}}^p\, d\widehat{\mu\otimes\nu} = 0$. Assume that this is not true. Then there exists an $\varepsilon > 0$ such that for all $n\in\mathbb{N}$ there are $\sigma,\sigma'\geq n$ such that

$$\varepsilon \leq \int_{\Lambda\times(T_{\sigma'}\setminus T_\sigma)} \|f_k\|_{\mathbb{H}}^p\, d\widehat{\mu\otimes\nu} \approx \int_{\Lambda\times(T_{\sigma'}\setminus T_\sigma)} \|F_k\|_{\mathbb{F}}^p\, d\mu\otimes\nu =: A_{\sigma,\sigma'}.$$

By Proposition 8.4.1, there exist unlimited S,S' with $A_{S,S'}\geq\varepsilon$, contradicting condition (ii) of the S-integrability of $\|F_k\|_{\mathbb{F}}$. It follows that $f_k\in L^p\left(\widehat{\mu\otimes\nu},\mathbb{H}\right)$. The proof that f is p-summable is similar. Obviously, f is $L_\mu(\mathcal{C})\otimes\mathcal{L}$-measurable. □

The preceding theorem implies that any Cauchy sequence $(F_n)_{n\in\mathbb{N}}$ of near-standard elements in a complete space produces a limit ${}^\circ F$, where F is an F_N of an internal extension $(F_n)_{n\in {}^*\mathbb{N}}$ of $(F_n)_{n\in\mathbb{N}}$. Important is that F inherits the common internal properties of the F_n. Here are the details.

Corollary 10.9.3 *Fix $d\in\mathbb{N}_0$ and a σ-algebra $\mathcal{F}\subseteq L_\mu(\mathcal{C})\otimes\mathcal{L}^m$ containing all $\widehat{\mu\otimes\nu^m}$-nullsets. Let $(F_i)_{i\in\mathbb{N}}$ be a sequence of $\mathcal{C}\otimes{}^*\mathcal{P}(T)$-measurable functions $F_i\colon {}^*\mathbb{N}^n\times\Lambda\times T^m\to\mathbb{F}^{\otimes d}\in SL^p(c^n\otimes\mu\otimes\nu^m,\mathbb{F}^{\otimes d})$ such that ${}^\circ F_i$ exists in*

$L^p_{\mathcal{F}}\left(c^n \otimes \widehat{\mu \otimes \nu^m}, \mathbb{H}^{\otimes d}\right)$. *Moreover, suppose that*

$$\lim_{i,j\to\infty} {}^\circ \sum_{k\in {}^*\mathbb{N}^n} \int_{\Lambda\times T^m} \|F_i(k,\cdot) - F_j(k,\cdot)\|^p_{\mathbb{F}^d}\, d\mu\otimes\nu^m = 0.$$

Let $(F_i)_{i\in {}^*\mathbb{N}}$ *be an internal extension of* $(F_i)_{i\in\mathbb{N}}$. *Then there exists an unlimited* $I_\infty \in {}^*\mathbb{N}$ *such that for all unlimited* $I \in {}^*\mathbb{N}$ *with* $I \leq I_\infty$:

(a) $F := F_I \in SL^p(c^n \otimes \mu \otimes \nu^m, \mathbb{F}^{\otimes d})$,
(b) F *is nearstandard* $c^n \otimes \widehat{\mu \otimes \nu^m}$*-a.e. and* ${}^\circ F$ *is* $\mathcal{F}$*-measurable,*
(c) $({}^\circ F_i)_{i\in\mathbb{N}}$ *converges to* ${}^\circ F$ *in* $L^p_{\mathcal{F}}\left(c^n \otimes \widehat{\mu \otimes \nu^m}, \mathbb{H}^{\otimes d}\right)$.

Moreover, F *inherits the common internal properties of the* F_i.

Proof In order to save indices, set $n = m = d = 1$. By the assumption and Proposition 9.5.1 (a), there exists a subsequence $\left(F_{g(i)}\right)_{i\in\mathbb{N}}$ of (F_i) and an unlimited $I \in {}^*\mathbb{N}$ such that for $F := F_I$

$$\sum_{k\in {}^*\mathbb{N}} \int_{\Lambda\otimes T} \|F(k,\cdot) - F_{g(i)}(k,\cdot)\|^p_{\mathbb{F}}\, d\mu\otimes\nu < \frac{1}{i^p}. \tag{21}$$

Using the triangle equality and the fact that $F_{g(i)}$ belong to $SL^p(c\otimes\mu\otimes\nu, \mathbb{F})$, it is easy to see that F is also in $SL^p(c\otimes\mu\otimes\nu, \mathbb{F})$. Since

$$\sum_{k\in\mathbb{N}} \int_{\Lambda\times T} \|{}^\circ F_i(k,\cdot) - {}^\circ F_j(k,\cdot)\|^p_{\mathbb{H}}\, d\widehat{\mu\otimes\nu}$$
$$= {}^\circ \sum_{k\in {}^*\mathbb{N}} \int_{\Lambda\times T} \|F_i(k,\cdot) - F_j(k,\cdot)\|^p_{\mathbb{F}}\, d\mu\otimes\nu \to_{i,j\to\infty} 0,$$

there is a limit f of $({}^\circ F_i)_{i\in\mathbb{N}}$ in $L^p_{\mathcal{F}}(c\otimes\widehat{\mu\otimes\nu}, \mathbb{H})$. Let $G \in SL^p(c\otimes\mu\otimes\nu, \mathbb{F})$ be a lifting of f, according to Theorem 10.9.2. Using (21),

$$\sum_{k\in {}^*\mathbb{N}} \int_{\Lambda\times T} \|F(k,\cdot) - G(k,\cdot)\|^p_{\mathbb{F}}\, d\mu\otimes\nu \approx 0,$$

thus F is a lifting of f, □

10.10 Standard part of the conditional expectation

In an application of the preceding sections we give an answer to the following question: What is the standard part of the internal conditional expectation of $F : \Lambda \to \mathbb{F}^{\otimes n}$ if $f : \Lambda \to \mathbb{H}^{\otimes n}$ is the standard part of F?

We need **conditional expectation for Hilbert space valued random variables**. Fix a separable Hilbert space $\mathbb{G}$ and an ONB $(\mathfrak{e}_i)_{i\in\mathbb{N}}$ of $\mathbb{G}$. Let $(\Lambda, L_\mu(\mathcal{C}), \widehat{\mu})$ be a Loeb probability space and let $\mathcal{F}$ be a sub σ-algebra of $L_\mu(\mathcal{C})$. Reducing the conditional expectation for $\mathbb{G}$-valued random variables to scalar-valued ones, we define for any $\varphi \in L^2(\widehat{\mu}, \mathbb{G})$

$$\mathbb{E}^{\mathcal{F}}(\varphi) := \sum_{i=1}^{\infty} \left(\mathbb{E}^{\mathcal{F}} \langle \varphi, \mathfrak{e}_i \rangle\right) \mathfrak{e}_i.$$

Note that $\mathbb{E}^{\mathcal{F}}(\varphi) \in L^2_{\mathcal{F}}(\widehat{\mu}, \mathbb{G})$ is well defined in particular, it does not depend on the ONB.

Theorem 10.10.1 *Let $p \geq 1$, let $\mathcal{F}$ be an internal subalgebra of $\mathcal{C}$ and assume that $F \in SL^p(\mu, \mathbb{F}^{\otimes n})$.*

(a) *Then the conditional expectation $\mathbb{E}^{\mathcal{F}} F$ of F under $\mathcal{F}$ belongs to $SL^p(\mu \upharpoonright \mathcal{F}, \mathbb{F}^{\otimes n})$.*
(b) *Let $F \in SL^1(\mu, \mathbb{F}^{\otimes n})$ be a lifting of $f \in L^1(\widehat{\mu}, \mathbb{H}^{\otimes n})$. Then $\mathbb{E}^{\mathcal{F}} F$ is a lifting of $\mathbb{E}^{\mathcal{F}\vee\mathcal{N}_{\widehat{\mu}}} f$.*

Proof (a) By Jensen's inequality, $\left\|\mathbb{E}^{\mathcal{F}} F\right\|^p_{\mathbb{F}^n} \leq \mathbb{E}^{\mathcal{F}} \|F\|^p_{\mathbb{F}^n}$. Part (a) follows.

(b) Let M be the set of all $g \in L^1(\widehat{\mu}, \mathbb{H}^{\otimes n})$ such that g has a lifting $G \in SL^1(\mu, \mathbb{F}^{\otimes n})$ with $\mathbb{E}^{\mathcal{F}} G$ is a lifting of $\mathbb{E}^{\mathcal{F}\vee\mathcal{N}_{\widehat{\mu}}} g$. Obviously, M is a linear space. Fix $\widetilde{D} \in L_\mu(\mathcal{C})$ and $a \in \mathbb{H}^{\otimes n}$. Let $D \in \mathcal{C}$ be a μ-approximation of $\widetilde{D}$. Then $L := \mathbf{1}_D \otimes {}^*a \upharpoonright \mathbb{F}^n : \Lambda \to \mathbb{F}^{\otimes n} \in SL^1(\mu, \mathbb{F}^{\otimes n})$ is a lifting of $l := \mathbf{1}_{\widetilde{D}} \otimes a : \Lambda \to \mathbb{H}^{\otimes n} \in L^1(\widehat{\mu}, \mathbb{H}^{\otimes n})$. To prove that $\mathbb{E}^{\mathcal{F}} \mathbf{1}_D$ is a lifting of $\mathbb{E}^{\mathcal{F}\vee\mathcal{N}_{\widehat{\mu}}} \mathbf{1}_{\widetilde{D}}$, first note that ${}^\circ\mathbb{E}^{\mathcal{F}} \mathbf{1}_D$ is $\mathcal{F} \vee \mathcal{N}_{\widehat{\mu}}$-measurable. Fix an $\widetilde{E} \in \mathcal{F} \vee \mathcal{N}_{\widehat{\mu}}$ and a μ-approximation $E \in \mathcal{F}$ of $\widetilde{E}$. Then we obtain

$$\begin{aligned}
\int_{\widetilde{E}} \mathbb{E}^{\mathcal{F}\vee\mathcal{N}_{\widehat{\mu}}}(\mathbf{1}_{\widetilde{D}}) d\widehat{\mu} &= \widehat{\mu}(\widetilde{D} \cap \widetilde{E}) = {}^\circ\mu(D \cap E) \\
&= {}^\circ\int_E \mathbb{E}^{\mathcal{F}}(\mathbf{1}_D) d\mu = \int_E {}^\circ\mathbb{E}^{\mathcal{F}}(\mathbf{1}_D) d\widehat{\mu} \\
&= \int_{\widetilde{E}} {}^\circ\mathbb{E}^{\mathcal{F}}(\mathbf{1}_D) d\widehat{\mu}.
\end{aligned}$$

It follows that $\mathbb{E}^{\mathcal{F}\vee\mathcal{N}_{\widehat{\mu}}} \mathbf{1}_{\widetilde{D}} = {}^\circ\mathbb{E}^{\mathcal{F}}(\mathbf{1}_D)$ $\widehat{\mu}$-a.e., thus, $\mathbb{E}^{\mathcal{F}}(L)$ is a lifting of $\mathbb{E}^{\mathcal{F}\vee\mathcal{N}_{\widehat{\mu}}} l$. Now we will prove that M is complete. Let (g_n) be a Cauchy sequence in M with $\lim_{n\to\infty} g_n = g \in L^1(\widehat{\mu}, \mathbb{H}^{\otimes n})$. Let $G_n \in SL^1(\mu, \mathbb{F}^{\otimes n})$ be a lifting of g_n with $\mathbb{E}^{\mathcal{F}} G_n$ being a lifting of $\mathbb{E}^{\mathcal{F}\vee\mathcal{N}_{\widehat{\mu}}} g_n$. Let $G \in SL^1(\mu, \mathbb{F}^{\otimes n})$ be a lifting of g. By

Jensen's inequality, we obtain:

$$\begin{aligned} {}^{\circ}\mathbb{E}_{\mu}\left\|\mathbb{E}^{\mathcal{F}}G_n-\mathbb{E}^{\mathcal{F}}G\right\|_{\mathbb{F}^n} &\leq {}^{\circ}\mathbb{E}_{\mu}\left\|G_n-G\right\|_{\mathbb{F}^n} \\ &= \mathbb{E}_{\widehat{\mu}}\,{}^{\circ}\left\|G_n-G\right\|_{\mathbb{F}^n} = \mathbb{E}_{\widehat{\mu}}\left\|g_n-g\right\|_{\mathbb{H}^n} \to_{n\to\infty} 0 \end{aligned}$$

and

$$\lim_{n\to\infty}\mathbb{E}_{\widehat{\mu}}\left\|\mathbb{E}^{\mathcal{F}\vee\mathcal{N}_{\widehat{\mu}}}g_n-\mathbb{E}^{\mathcal{F}\vee\mathcal{N}_{\widehat{\mu}}}g\right\|_{\mathbb{H}^n}=0.$$

It follows that

$$\mu^{\text{outer}}\left\{\left\|\mathbb{E}^{\mathcal{F}}G-{}^{*}\mathbb{E}^{\mathcal{F}\vee\mathcal{N}_{\widehat{\mu}}}g\right\|_{\mathbb{F}^n}\geq\frac{1}{m}\right\}\leq\alpha+\beta+\gamma$$

can be made arbitrarily small, where

$$\begin{aligned} \alpha &= {}^{\circ}\mu\left\{\left\|\mathbb{E}^{\mathcal{F}}G-\mathbb{E}^{\mathcal{F}}G_n\right\|_{\mathbb{F}^n}\geq\frac{1}{3m}\right\}\leq 3m\,{}^{\circ}\mathbb{E}_{\mu}\left\|\mathbb{E}^{\mathcal{F}}G_n-\mathbb{E}^{\mathcal{F}}G\right\|_{\mathbb{F}^n}, \\ \beta &= \widehat{\mu}\left\{\left\|\mathbb{E}^{\mathcal{F}}(G_n)-{}^{*}\mathbb{E}^{\mathcal{F}\vee\mathcal{N}_{\widehat{\mu}}}g_n\right\|_{\mathbb{F}^n}\geq\frac{1}{3m}\right\}=0, \\ \gamma &= \widehat{\mu}\left\{\left\|{}^{*}\mathbb{E}^{\mathcal{F}\vee\mathcal{N}_{\widehat{\mu}}}g_n-{}^{*}\mathbb{E}^{\mathcal{F}\vee\mathcal{N}_{\widehat{\mu}}}g\right\|_{\mathbb{F}^n}\geq\frac{1}{3m}\right\} \\ &\leq \widehat{\mu}\left\{\left\|\mathbb{E}^{\mathcal{F}\vee\mathcal{N}_{\widehat{\mu}}}g_n-\mathbb{E}^{\mathcal{F}\vee\mathcal{N}_{\widehat{\mu}}}g\right\|_{\mathbb{H}^n}\geq\frac{1}{3m}\right\} \\ &\leq 3m\cdot\mathbb{E}_{\widehat{\mu}}\left\|\mathbb{E}^{\mathcal{F}\vee\mathcal{N}_{\widehat{\mu}}}g_n-\mathbb{E}^{\mathcal{F}\vee\mathcal{N}_{\widehat{\mu}}}g\right\|_{\mathbb{H}^n}. \end{aligned}$$

It follows that $\mathbb{E}^{\mathcal{F}}G$ is a lifting of $\mathbb{E}^{\mathcal{F}\vee\mathcal{N}_{\widehat{\mu}}}g$, thus M is complete. By Proposition 5.6.1, $M=L^1\left(\widehat{\mu},\mathbb{H}^{\otimes n}\right)$. □

10.11 Witnesses of S-integrability

The following characterization of S-integrability, using convex functions, due to Hoover and Perkins [46], is very useful. In particular, it provides a powerful tool in martingale theory, because a convex function applied to a martingale still results in a submartingale. We assume that $\mu(\Lambda)$ is limited and fix a $\mathcal{C}$-measurable mapping F from Λ into ${}^{*}[0,\infty[$.

We will call an internal function $\Phi:\ {}^{*}[0,\infty[\to{}^{*}[0,\infty[$ a **witness for the S_{μ}-integrability of** F if the following conditions are true.

(a) $\Phi(2x) \leq 4\Phi(x)$.
(b) Φ is monotone increasing, $\Phi(0) = 0$ and Φ is ***convex**. This means $\Phi(x + \alpha(y-x)) \leq \Phi(x) + \alpha(\Phi(y) - \Phi(x))$ for all $x \leq y$ in ${}^*[0,\infty[$ and all $\alpha \in {}^*[0,1]$.
(c) $\sup_{M \leq x} \frac{x}{\Phi(x)} \approx 0$ for each unlimited $M \in {}^*\mathbb{N}_0$.
(d) $\int_\Lambda \Phi \circ F d\mu$ is limited.

Theorem 10.11.1 (Hoover and Perkins [46]) *F is S_μ-integrable if and only if there exists a witness Φ for the S_μ-integrability of F. The proof tells us that for S_μ-integrable F a witness Φ may have the following form:*

(WP) *There exists an internal sequence $(a_n)_{n \in {}^*\mathbb{N}}$ in ${}^*[0,\infty[$ such that*
(i) *$a_n \in \mathbb{R}$ if and only if $n \in \mathbb{N}$,*
(ii) *$a_0 = 0$, $a_1 > 1$ and $4a_{n-1} < a_n$ for all $n \in {}^*\mathbb{N}$,*
(iii) *$\Phi(x) = (n+1)x - (a_1 + \ldots + a_n)$ for all $x \in [a_n, a_{n+1}[$.*

We need this property (WP) in order to apply Corollary 2.14.2 in the proof of Theorem 10.13.1.

Proof First we prove that the conditions on Φ are sufficient for the S_μ-integrability of F. Fix an unlimited $M \in {}^*\mathbb{N}$. Then, because of (d),

$$\int_{\{F \geq M\}} F d\mu = \int_{\{F \geq M\}} \frac{F}{\Phi \circ F} \cdot \Phi \circ F d\mu \leq \sup_{x \geq M} \frac{x}{\Phi(x)} \int_\Lambda \Phi \circ F d\mu \approx 0.$$

Now we will show that the conditions on Φ are also necessary. Let F be S_μ-integrable. If $\int_\Lambda F d\mu \approx 0$, then there exists an unlimited $K \in {}^*\mathbb{N}$ such that $\int_\Lambda F d\mu < \frac{1}{K}$. Define $\Phi(x) := K \cdot x$. Then Φ fulfils the conditions of the theorem. Now assume that $\int_\Lambda F d\mu \not\approx 0$. Since $\lim_{k \to \infty} {}^\circ\int_{\{k \leq F\}} F d\mu = 0$, there exists a standard sequence $(a_n)_{n \in \mathbb{N}}$ with $a_1 = 0$, $1 < a_2$ and such that for all $n > 1$

$$4a_{n-1} < a_n \quad \text{and} \quad \int_{\{a_n \leq F\}} F d\mu \leq \frac{1}{2^n} \int_\Lambda F d\mu. \tag{22}$$

By Proposition 8.2.2, $(a_n)_{n \in \mathbb{N}}$ can be extended to an internal sequence $(a_n)_{n \in {}^*\mathbb{N}}$. By Proposition 8.4.1 (b), there exists an unlimited $K \in {}^*\mathbb{N}$ such that (22) is true for all $n \leq K$. Since F is μ-integrable, we may assume that (22) is even true for all $n \in {}^*\mathbb{N}$. Now set

$$\Phi(x) = (n+1) \cdot x - (a_1 + \ldots + a_n) \quad \text{if } x \in [a_n, a_{n+1}[$$

It is easy to see that conditions (a) and (b) are true. Moreover, Φ is *continuous. It remains to prove that conditions (c) and (d) are true. Note that for each unlimited $M \in {}^*\mathbb{N}$ and all $x \geq M$ with $x \in [a_n, a_{n+1}[$ (thus n is unlimited):

$$\frac{\Phi(x)}{x} = \frac{(n+1)\cdot x - (a_1 + \ldots + a_n)}{x}$$
$$\geq n+1 - \frac{a_1 + \ldots + a_n}{a_n} \geq n+1 - \sum_{n=0}^{\infty} (\frac{1}{4})^n$$

is unlimited, thus

$$\frac{x}{\Phi(x)} \approx 0.$$

Moreover

$$\int_\Lambda \Phi \circ F d\mu = \sum_{n \in {}^*\mathbb{N}} \int_{\{a_n \leq F < a_{n+1}\}} (n+1)\cdot F - (a_1 + \ldots + a_n) d\mu$$
$$\leq \sum_{n \in {}^*\mathbb{N}} \int_{\{a_n \leq F\}} (n+1)\cdot F d\mu \leq \sum_{n \in {}^*\mathbb{N}} (n+1) \frac{1}{2^n} \int_\Lambda F d\mu.$$

Since the series $\sum_{n \in {}^*\mathbb{N}} (n+1) \frac{1}{2^n}$ converges to an element in $\mathbb{R}$ and $\int_\Lambda F d\mu$ is limited, $\int_\Lambda \Phi \circ F d\mu$ is limited. □

For example, if $\Phi(x) = x^p$ and $\int_\Lambda \Phi \circ F d\mu = \int_\Lambda F^p d\mu$ is limited for some standard $p > 1$, then F is S_μ-integrable. (This result could also be obtained by using Hölder's inequality, but Theorem 10.11.1 is substantially stronger.)

10.12 Keisler's Fubini theorem

A nice application of the preceding result is a simple proof of the internal version of Keisler's Fubini theorem, which will be used several times. Since the Loeb space over the internal product of two internal measure spaces may be, by Sun's result, a strict extension of the usual product of the associated Loeb spaces, Keisler proved a new Fubini theorem for the 'Loeb product'. Let $(\Lambda, \mathcal{C}, \mu)$ and $(\Lambda', \mathcal{C}', \mu')$ be internal measure spaces such that $\mu(\Lambda)$ and $\mu'(\Lambda')$ are limited. Recall that the Loeb space over $(\Lambda \times \Lambda', \mathcal{C} \otimes \mathcal{C}', \mu \otimes \mu')$ is denoted by $\left(\Lambda \times \Lambda', L_{\mu \otimes \mu'}(\mathcal{C} \otimes \mathcal{C}'), \widehat{\mu \otimes \mu'}\right)$.

Theorem 10.12.1 (Keisler [53]) *Suppose $F : \Lambda \times \Lambda' \to {}^*\mathbb{R}_0^+$ is $\mathcal{C} \otimes \mathcal{C}'$-measurable and $S_{\mu \otimes \mu'}$-integrable. Then the following hold.*

(a) *For $\widehat{\mu}$-almost all $\omega \in \Lambda$, $F(\omega, \cdot)$ is $S_{\mu'}$-integrable.*
(b) *The function $\omega \mapsto \int_{\Lambda'} F(\omega, \cdot) d\mu'$ is S_μ-integrable.*

(c) $\int_{\Lambda\times\Lambda'} F d\mu\otimes\mu' = \int_\Lambda \int_{\Lambda'} F(\omega,\cdot)d\mu' d\mu(\omega) = \int_{\Lambda'}\int_\Lambda F(\omega,\cdot)d\mu(\omega)d\mu'$.

Proof We may assume that $\mu(\Lambda), \mu'(\Lambda') \not\approx 0$. By normalizing, we may also assume that μ and μ' are probability measures.

(c) follows from Fubini's theorem and the transfer principle. Now let Φ be a witness for the $S_{\mu\otimes\mu'}$-integrability of F. Since Φ is *convex, Φ is *continuous, thus $\Phi\circ F$ is $\mathcal{C}\otimes\mathcal{C}'$-measurable. The equation

$$\int_\Lambda\int_{\Lambda'} \Phi\circ F(\omega,\omega')d\mu'(\omega')d\mu(\omega) = \int_{\Lambda\times\Lambda'} \Phi\circ F d\mu\otimes\mu'$$

is limited and Lemma 10.7.1 (a) implies that $\int_{\Lambda'}\Phi\circ F(\omega,\cdot)d\mu'$ is limited for $\widehat{\mu}$-almost all $\omega\in\Lambda$, thus Φ is a witness for the $S_{\mu'}$-integrability of $F(\omega,\cdot)$. Part (a) follows. Since Φ is *convex, and μ' is a probability measure, by Jensen's inequality (see Ash [5]),

$$\int_\Lambda \Phi\left(\int_{\Lambda'} F(\omega,\omega')d\mu'(\omega')\right)d\mu(\omega) \leq \int_\Lambda\int_{\Lambda'} \Phi\circ F(\omega,\omega')d\mu'(\omega')d\mu(\omega)$$

is limited. Part (b) follows. □

The following kinds of function are important in the analysis on the Wiener space. Fix $k\in\mathbb{N}$, and for $i\in\{1,\ldots,k\}$ let f_i be an $\mathbb{R}$- or ${}^*\mathbb{R}$-valued function on a set Λ_i. Then the **tensor product** $\otimes_{i=1}^k f_i$ of the f_i is defined on $\Lambda_1\times\ldots\times\Lambda_k$ by setting

$$\otimes_{i=1}^k f_i := f_1\otimes\ldots\otimes f_k : (x_1,\ldots,x_k)\mapsto f_1(x_1)\cdot\ldots\cdot f_k(x_k).$$

If $f := f_1 = \ldots = f_k$, then we will write $f^{\otimes k}$ instead of $f_1\otimes\ldots\otimes f_k$.

Lemma 10.12.2

(a) *Suppose that $N\subseteq\Lambda\times\Lambda'$ is a $\widehat{\mu\otimes\mu'}$-nullset. Then for $\widehat{\mu}$-almost all $\omega\in\Lambda$, the cut*

$$N(\omega,\cdot) := \left\{\omega' \mid (\omega,\omega')\in N\right\} \text{ is a } \widehat{\mu'}\text{-nullset.}$$

(b) *Let $F:\Lambda\to\mathbb{F}^{\otimes n}$ and $G:\Lambda'\to\mathbb{F}^{\otimes m}$ be S_μ-integrable, $S_{\mu'}$-integrable, respectively. Then $F\otimes G$ is $S_{\mu\otimes\mu'}$-integrable.*

Proof (a) We have to prove that

$$\mu^{\text{outer}}\left\{\omega \mid (\mu')^{\text{outer}}(N(\omega,\cdot))\neq 0\right\} = 0,$$

thus it suffices to prove that $\mu^{\text{outer}}\left\{\omega \mid (\mu')^{\text{outer}}(N(\omega,\cdot))\geq\frac{1}{k}\right\} < \frac{1}{k}$ for all $k\in\mathbb{N}$. (See the proof of Lemma 10.1.1.) Fix $k\in\mathbb{N}$. Since, N is a $\widehat{\mu\otimes\mu'}$-nullset, there exists an $A \in \mathcal{C} \otimes \mathcal{C}'$ with $N \subseteq A$ and $\mu \otimes \mu'(A) < \frac{1}{k^2}$. Then, by Theorem 10.12.1 (c),

$$\mu^{\text{outer}} \left\{ \omega \mid (\mu')^{\text{outer}} (N(\omega,\cdot)) \geq \frac{1}{k} \right\}$$
$$\leq \mu \left\{ \omega \mid \mu'(A(\omega,\cdot)) \geq \frac{1}{k} \right\} \leq \int_{\{\omega \mid k\mu'(A(\omega,\cdot)) \geq 1\}} k\mu'(A(\omega,\cdot))\, d\mu$$
$$\leq \int_{\Omega} \int_{A(\omega,\cdot)} k\, d\mu' d\mu(\omega) = k \int_A 1\, d\mu \otimes \mu' = k\mu \otimes \mu'(A) < \frac{1}{k}.$$

(b) We may understand F and G as functions on $\Lambda \times \Lambda'$ by setting $F(\cdot,y) = F$ and $G(x,\cdot) = G$. Since the product of two $\mathcal{C} \otimes \mathcal{C}'$-measurable functions is $\mathcal{C} \otimes \mathcal{C}'$-measurable, $F \otimes G$ is $\mathcal{C} \otimes \mathcal{C}'$-measurable. Let $N \in \mathcal{C} \otimes \mathcal{C}'$ be a $\widehat{\mu \otimes \mu'}$-nullset. Then, by (a), N_ω is a $\widehat{\mu}'$-nullset for $\widehat{\mu}$-almost all $\omega \in \Lambda$. Since $\|F\|_{\mathbb{F}^n}$ is limited $\widehat{\mu}$-a.e. and G is $S_{\mu'}$-integrable,

$$\alpha : \Lambda \to {}^*\mathbb{R}, \omega \mapsto \int_{N_\omega} \|G\|_{\mathbb{F}^m}\, d\mu' \cdot \|F\|_{\mathbb{F}^n}(\omega) \approx 0 \; \widehat{\mu}\text{-a.e.}$$

In order to show that α is S_μ-integrable, set $s := \max\{1, \int_{\Lambda'} \|G\|_{\mathbb{F}^m}\, d\mu'\}$. Then s is limited and thus for all unlimited $K \in {}^*\mathbb{N}$

$$\int_{\{\omega \in \Lambda \mid \alpha(\omega) \geq K\}} \alpha\, d\mu \leq \int_{\left\{\omega \in \Lambda \mid \|F\|_{\mathbb{F}^n}(\omega) \geq \frac{K}{s}\right\}} s \cdot \|F\|_{\mathbb{F}^n}\, d\mu \approx 0,$$

because F is S_μ-integrable. This proves the S_μ-integrability of α. We obtain, by Theorem 10.12.1 (c) and Corollary 10.8.2 (b),

$$\int_N \|F \otimes G\|_{\mathbb{F}^{n+m}}\, d\mu \otimes \mu' = \int_\Lambda \left(\int_{N_\omega} \|G\|_{\mathbb{F}^m}\, d\mu' \right) \|F\|_{\mathbb{F}^n}(\omega) d\mu = \int_\Lambda \alpha\, d\mu \approx 0.$$

□

Using Lemma 10.12.2 and Theorems 10.8.1, 10.12.1, we obtain

Corollary 10.12.3 (Keisler [53]) *Suppose that $f : \Lambda \times \Lambda' \to \mathbb{R}$ is $\widehat{\mu \otimes \mu'}$-integrable. Then the following hold.*

(a) *For $\widehat{\mu}$-almost all $\omega \in \Lambda$, $f(\omega,\cdot)$ is $\widehat{\mu}'$-integrable.*
(b) *$\omega \mapsto \int_{\Lambda'} f(\omega,\cdot) d\widehat{\mu}'$ is $\widehat{\mu}$-integrable.*
(c) *$\int_\Lambda \int_{\Lambda'} f(\omega,\cdot) d\widehat{\mu}' d\widehat{\mu}(\omega) = \int_{\Lambda \times \Lambda'} f d\widehat{\mu \otimes \mu'}$.*

The final result in this section is a Keisler–Fubini theorem for Bochner integrable functions with values in $\mathbb{H}^{\otimes n}$. By Theorem 10.8.1 and Keisler's Fubini theorem (Theorem 10.12.3), we obtain

Theorem 10.12.4 *Fix a Bochner integrable* $f : \Lambda \times \Lambda' \to \mathbb{H}^{\otimes n}$. *By Theorem 10.8.1, there exists a lifting* $F : \Lambda \times \Lambda' \to \mathbb{F}^{\otimes n} \in SL^1(\mu \otimes \mu', \mathbb{F}^{\otimes n})$ *of* f.

(a) *For* $\widehat{\mu}$*-almost all* $X \in \Lambda$, $F(X, \cdot) \in SL^1(\mu', \mathbb{F}^{\otimes n})$ *is a lifting of* $f(X, \cdot)$, *thus* $f(X, \cdot)$ *is Bochner integrable.*
(b) *The function* $X \mapsto \int_{\Lambda'} f(X, \cdot) d\widehat{\mu'}$ *is Bochner integrable and the internal function* $\int_{\Lambda'} F(\cdot, X') d\mu'(X')$ *belongs to* $SL^1(\mu, \mathbb{F}^{\otimes n})$ *and is a lifting of* $X \mapsto \int_{\Lambda'} f(X, \cdot) d\widehat{\mu'}$.
(c) $\int_{\Lambda \times \Lambda'} f d\widehat{\mu \otimes \mu'} = \int_\Lambda \int_{\Lambda'} f(X, \cdot) d\widehat{\mu'} d\widehat{\mu}(X)$.

10.13 S-integrability of internal martingales

Let us fix an internal adapted probability space $(\Lambda, \mathcal{C}, (\mathcal{C}_t)_{t \in T}, \mu)$. By the transfer principle all *transforms of properties of martingales, developed in Chapter 2, remain valid for internal $(\mathcal{C}_t)_{t \in T}$-martingales. However, we also obtain some important results on internal martingales concerning external properties, like S-integrability and S-continuity.

Theorem 10.13.1 *For each internal* $(\mathcal{C}_t)_{t \in T}$ *martingale* M *and each* $p \in [1, \infty[$ *we obtain for all* $\sigma \in \mathbb{N}$

$$[M]_\sigma^{\frac{p}{2}} \in SL^1(\mu) \text{ if and only if } (M_\sigma^\sim)^p := \max_{s \in T_\sigma} |M_s|^p \in SL^1(\mu).$$

Proof Suppose that $[M]_\sigma^{\frac{p}{2}} \in SL^1(\mu)$. By Theorem 10.11.1, there exists a witness Φ for the S_μ-integrability of $(24p)^p [M]_\sigma^{\frac{p}{2}}$ with property (WP). By Corollary 2.14.2,

$$\mathbb{E}\Phi \circ (M_\sigma^\sim)^p \leq \mathbb{E}\Phi \circ (24p)^p \cdot [M]_\sigma^{\frac{p}{2}}.$$

Therefore, Φ is also a witness for the S_μ-integrability of $(M_\sigma^\sim)^p$. It follows that $(M_\sigma^\sim)^p \in SL^1(\mu)$. The proof of the reverse implication is similar. □

10.14 S-continuity of internal martingales

In this section we will prove an important result, due to Hoover and Perkins [46], namely that, under a mild condition, a martingale is S-continuous a.e. if its quadratic is S-continuous a.e. Here we denote by $\widetilde{\mathbb{N}}$ the set of all $n \in {}^*\mathbb{N}$ such that $\frac{n}{H}$ is limited. An internal function $F : {}^*\mathbb{N} \to {}^*\mathbb{R}$ is called **pre-S-continuous** if $F(n) \approx F(m)$ for all $n, m \in \widetilde{\mathbb{N}}$ with $\frac{n}{H} \approx \frac{m}{H}$.

Lemma 10.14.1 (Hoover and Perkins [46]) *Fix an internal increasing sequence $(\tau_n)_{n\in {}^*\mathbb{N}}$ of stopping times and an internal $(\mathcal{C}_{\tau_n})_{n\in {}^*\mathbb{N}}$-martingale $N : \Lambda \times {}^*\mathbb{N} \to {}^*\mathbb{R}$. Suppose that there exists an $\varepsilon \in {}^*\mathbb{R}$ with $0 < \varepsilon \approx 0$ such that*

$$\mathbb{E}\left([N]_n - [N]_m\right)^2 \cdot \frac{H}{n-m} \approx 0 \quad \textit{for all } n, m \in \widetilde{\mathbb{N}} \textit{ with } \varepsilon < \frac{n-m}{H} \approx 0. \tag{23}$$

Then N is pre-S-continuous $\widehat{\mu}$-a.e.

Proof By (23) and Lemma 2.6.1 (b), $\mathbb{E}(N_t - N_s)^4 \in {}^*\mathbb{R}$ for all $t, s \in \widetilde{\mathbb{N}}$. For all $i \in \mathbb{N}$ and for all $d \in {}^*\mathbb{N}$ such that d divides H, set

$$G_{i,d} := \left\{ \max_{|n-m|\le d, n,m\le H\cdot i} |N_n - N_m| > \frac{2}{i} \right\}.$$

Then

$$\begin{aligned}
\mu(G_{i,d}) &\le \mu\left\{ \max_{1\le k\le \frac{H}{d}i} \max_{(k-1)d<n\le kd} \left|N_n - N_{(k-1)d}\right| > \frac{1}{i} \right\} \\
&\le \sum_{1\le k\le \frac{H}{d}i} \mu\left\{ i^4 \max_{(k-1)d<n\le kd} \left|N_n - N_{(k-1)d}\right|^4 > 1 \right\} \\
&\le \sum_{1\le k\le \frac{H}{d}i} i^4 \mathbb{E} \max_{(k-1)d<n\le kd} \left|N_n - N_{(k-1)d}\right|^4 \\
&\le c\frac{H}{d} i^5 \cdot \max_{1\le k\le \frac{H}{d}i} \mathbb{E}\left([N]_{kd} - [N]_{(k-1)d}\right)^2
\end{aligned}$$

for some constant $c \in \mathbb{R}$ (by 2.13.2), because $t \mapsto N_t - N_{(k-1)d}$ with $(k-1)d < t \le kd$ is a $(\mathcal{C}_{\tau_t})_{(k-1)d<t\le kd}$-martingale. This constant c does not depend on i and d. By the assumption, $\mu(G_{i,d}) \approx 0$ for all $(i,d) \in \mathbb{N} \times {}^*\mathbb{N}$ with $\varepsilon < \frac{d}{H} \approx 0$ and $d \mid H$. Therefore, by Corollary 8.5.2, for each $i \in \mathbb{N}$ there exists a $d_i \in {}^*\mathbb{N}$ such that $\frac{d_i}{H} \not\approx 0$ and $\mu(G_{i,d_i}) < \frac{1}{2^i}$, thus $\widehat{\mu}(G_{i,d_i}) \le \frac{1}{2^i}$. Since $\sum_{i=0}^{\infty} \frac{1}{2^i} < \infty$, by the Borel–Cantelli lemma, $\widehat{\mu}(U) = 1$, where $U := \bigcup_{k\in\mathbb{N}} \bigcap_{i\ge k} \Lambda \setminus G_{i,d_i}$. It is easy to see that $N(X)$ is pre-S-continuous for each $X \in U$. □

Our aim now is to prove the following important theorem due to Hoover and Perkins [46], where $\left(\Lambda, \mathcal{C}, (\mathcal{C}_t)_{t\in T}, \mu\right)$ is an adapted probability space.

Theorem 10.14.2 (Hoover and Perkins [46]) *Fix a $(\mathcal{C}_t)_{t\in T}$-martingale M such that $\mathbb{E}M_S^2$ is limited for all $S \in \mathbb{N}$. If $[M]$ is S-continuous $\widehat{\mu}$-a.e., then M is S-continuous $\widehat{\mu}$-a.e.*

Hoover and Perkins also prove the reverse implication, but since we do not need this result, we omit the proof.

Proof Assume that $[M]$ is S-continuous $\widehat{\mu}$-a.s. It suffices to prove that $M \upharpoonright \Lambda \times T_S$ is S-continuous $\widehat{\mu}$-a.s. for each $S \in \mathbb{N}$. Since, by Theorem 2.4.2, $\mathbb{E}\max_{t\le S} M_t^2 \le 4\mathbb{E}M_S^2$ is limited, $\max_{t\le S} M_s^2$ is limited $\widehat{\mu}$-a.s. Thus, M_t is nearstandard for all $t \le S$ $\widehat{\mu}$-a.s. Moreover,

$$\max_{t\le S} |M_t| \in SL^1(\mu). \tag{24}$$

It remains for us to prove that $\widehat{\mu}$-a.s.: $M_s \approx M_t$ for all $s,t \in T_S$ with $s \approx t$.

Case 1: Suppose that $u := \max_{\frac{1}{H}<t\le S} |\Delta M_t|$ is limited μ-a.s. We define an internal sequence $(\tau_n)_{n\in {}^*\mathbb{N}}$ of stopping times by

$$\tau_n := \inf\left\{t \in T_S \mid [M]_t + t \ge \frac{n}{H}\right\}, \text{ setting } \inf\emptyset := S^+ := S + \frac{1}{H}.$$

Note that $\tau_n \le \tau_{n+1} \le \tau_n + \frac{1}{H}$. Set $N_n(X) := M(X, \tau_n(X) \wedge S)$. By Proposition 2.5.2, $N := (N_n)_{n\in {}^*\mathbb{N}}$ is an $\left(\mathcal{C}_{\tau_n}\right)_{n\in {}^*\mathbb{N}}$-martingale. We obtain for all $n, m \in {}^*\mathbb{N}$ with $\tau_m < \tau_n$

$$\begin{aligned}
[N]_n - [N]_m &\le [M]_{\tau_n \wedge S} - [M]_{\tau_m \wedge S} \\
&\le [M]_{\tau_n - \frac{1}{H}} + \left(M_{\tau_n\wedge S} - M_{\tau_n - \frac{1}{H}}\right)^2 - [M]_{\tau_m \wedge S} \\
&\le \frac{n}{H} - \left(\tau_n - \frac{1}{H}\right) + \max_{\frac{1}{H}<s\le S} (M_s - M_{s^-})^2 + \tau_m - \frac{m}{H} \\
&\le \frac{n-m}{H} + \max_{\frac{1}{H}<s\le S} (\Delta M_s)^2 + \frac{1}{H} \le \frac{2(n-m)}{H} + \max_{\frac{1}{H}<s\le S} (\Delta M_s)^2 .
\end{aligned}$$

This proves that

$$[N]_n - [N]_m \le \frac{2(n-m)}{H} + \max_{\frac{1}{H}<s\le S} (\Delta M_s)^2 .$$

Since $[M]$ is S-continuous $\widehat{\mu}$-a.s. and u is limited μ-a.s., $\max_{\frac{1}{H}<s\le S} |\Delta M_s|^p$ is an S_μ-integrable lifting of 0 for all $p \ge 1$ in $\mathbb{R}$. Thus, $\mathbb{E}\max_{\frac{1}{H}<s\le S} |\Delta M_s|^p \approx 0$.

Set $\varepsilon := \sqrt{i}$ with $i := \mathbb{E}\max_{\frac{1}{H}<s\leq S}|\Delta M_s|^4$. By the previous computation we obtain for all $n,m \in {}^*\mathbb{N}$ with $m<n$ such that $\varepsilon < d := \frac{n-m}{H} \approx 0$:

$$\begin{aligned}\mathbb{E}([N]_n-[N]_m)^2\frac{1}{d} &\leq \frac{1}{d}\mathbb{E}\left(2d+\max_{\frac{1}{H}<s\leq S}(\Delta M_s)^2\right)^2\\ &= \frac{1}{d}\left(4d^2+4d\mathbb{E}\max_{\frac{1}{H}<s\leq S}(\Delta M_s)^2+i\right)\\ &= 4d+4\mathbb{E}\max_{\frac{1}{H}<s\leq S}(\Delta M_s)^2+\frac{i}{d}.\end{aligned}$$

Since $d \approx 0 \approx \mathbb{E}\max_{\frac{1}{H}<s\leq S}(\Delta M_s)^2$ and $\frac{i}{d}=\frac{\varepsilon^2}{d}<d\approx 0$, by Lemma 10.14.1, N is pre-S-continuous $\widehat{\mu}$-a.s. Fix $X \in \Lambda$ such that $N(X)$ and $[M](X)$ are pre-S-continuous. Since this happens for $\widehat{\mu}$-almost all $X \in \Lambda$, it remains for us to prove that $M(X)$ is pre-S-continuous: Fix $s<t$ in T_S with $s \approx t$. There exist $m,n \in {}^*\mathbb{N}$ with

$$[M]_{s-\frac{1}{H}}(X)+s-\frac{1}{H}<\frac{m}{H}\leq [M]_s(X)+s,$$

$$[M]_{t-\frac{1}{H}}(X)+t-\frac{1}{H}<\frac{n}{H}\leq [M]_t(X)+t,$$

thus $\tau_m(X)=s$ and $\tau_n(X)=t$. Now $\frac{n}{H}\approx\frac{m}{H}$, because $[M]_s(X)\approx[M]_t(X)$ and $s\approx t$. It follows that

$$M_s(X)=M_{\tau_m(X)}(X)=N_m(X)\approx N_n(X)=M_t(X).$$

This proves the theorem for Case 1.

Case 2: Now we turn to the general case. Set

$$V_0 := \left\{[M] \text{ is } S\text{-continuous and } \max_{t\leq S}|M_t| \text{ is limited}\right\}.$$

Then, by (24), $\widehat{\mu}(V_0)=1$. For all $t \in T_S \cup \{S+\frac{1}{H}\}$ set

$$Z_t := \sum_{\frac{1}{H}<s\leq t}\mathbf{1}_{\{|\Delta M_s|>1\}}|\Delta M_s| \text{ and } \sigma_1 := \inf\{t\in T_S \mid Z_t \geq 1\},$$

where $Z_{S+\frac{1}{H}} := Z_S$. Note that $\max_{t\leq S}Z_t(X)=0$ and $\sigma_1(X)=S+\frac{1}{H}$ for all $X\in V_0$. Moreover,

$$Z_{\sigma_1}\leq Z_{\sigma_1-\frac{1}{H}}+\left|\Delta Z_{\sigma_1}\right|\leq 1+2\max_{t\leq S}|M_t|\in SL^1(\mu) \quad \text{(by (24))}. \tag{25}$$

Set $G := \sum_{\frac{1}{H} < s \leq \sigma_1} \left| \mathbb{E}^{\mathcal{C}_{s^-}} \left(\mathbf{1}_{\{|\Delta M_s| \leq 1\}} \Delta M_s \right) \right|$. Since M is a martingale, we obtain

$$
\begin{aligned}
{}^\circ\mathbb{E}G &= {}^\circ\mathbb{E} \sum_{\frac{1}{H} < s \leq \sigma_1} \left| \mathbb{E}^{\mathcal{C}_{s^-}} (\mathbf{1}_\Lambda \cdot \Delta M_s) - \mathbb{E}^{\mathcal{C}_{s^-}} \left(\mathbf{1}_{\{|\Delta M_s| > 1\}} \cdot \Delta M_s \right) \right| \\
&= {}^\circ\mathbb{E} \sum_{\frac{1}{H} < s \leq \sigma_1} \left| \mathbb{E}^{\mathcal{C}_{s^-}} \left(\mathbf{1}_{\{|\Delta M_s| > 1\}} \Delta M_s \right) \right| \leq {}^\circ\mathbb{E} Z_{\sigma_1} = \mathbb{E}\,{}^\circ Z_{\sigma_1} \text{ (by (25))} = 0.
\end{aligned}
$$

It follows that $\widehat{\mu}$-a.s.

$$
\max_{\frac{1}{H} < s \leq S} \left| \mathbb{E}^{\mathcal{C}_{s^-}} \left(\mathbf{1}_{\{|\Delta M_s| \leq 1\}} \Delta M_s \right) \right| \leq \sum_{\frac{1}{H} < s \leq S} \left| \mathbb{E}^{\mathcal{C}_{s^-}} \left(\mathbf{1}_{\{|\Delta M_s| \leq 1\}} \Delta M_s \right) \right| \approx 0.
$$

Set

$$
L := (L_t)_{t \in T_S} := \left(\sum_{\frac{1}{H} < s \leq t} \mathbf{1}_{\{|\Delta M_s| \leq 1\}} \Delta M_s - \mathbb{E}^{\mathcal{C}_{s^-}} \left(\mathbf{1}_{\{|\Delta M_s| \leq 1\}} \Delta M_s \right) \right)_{t \in T_S}.
$$

By Example 2.1.1 (v), L is an $(\mathcal{C}_t)_{t \in T_S}$-martingale. Set

$$
V_1 := \left\{ \sum_{\frac{1}{H} < s \leq S} \left| \mathbb{E}^{\mathcal{C}_{s^-}} \left(\mathbf{1}_{\{|\Delta M_s| \leq 1\}} \Delta M_s \right) \right| \approx 0 \right\} \cap V_0.
$$

Then $\widehat{\mu}(V_1) = 1$ and we obtain for all $X \in V_1$ and all $s < t$ in T_S

$$
\begin{aligned}
&\sum_{s < \rho \leq t} (\Delta L_\rho(X))^2 \\
&= \sum_{s < \rho \leq t} \left(\mathbf{1}_{\{|\Delta M_\rho| \leq 1\}} \Delta M_\rho - \mathbb{E}^{\mathcal{C}_{\rho^-}} \left(\mathbf{1}_{\{|\Delta M_\rho| \leq 1\}} \Delta M_\rho \right) \right)^2 (X) \\
&\leq \alpha + 2\beta + \gamma \quad \text{with } \alpha = [M]_t(X) - [M]_s(X)
\end{aligned}
$$

and

$$
\beta = \left| \sum_{s < \rho \leq t} \left(\mathbf{1}_{\{|\Delta M_\rho| \leq 1\}} \Delta M_\rho(X) \right) \left(\mathbb{E}^{\mathcal{C}_{\rho^-}} \left(\mathbf{1}_{\{|\Delta M_\rho| \leq 1\}} \Delta M_\rho \right) (X) \right) \right| \leq \sqrt{A} \cdot \sqrt{B}
$$

with

$$
A = \sum_{s < \rho \leq t} \left(\mathbb{E}^{\mathcal{C}_{\rho^-}} \left(\mathbf{1}_{\{|\Delta M_\rho| \leq 1\}} \Delta M_\rho \right) (X) \right)^2 = \gamma
$$

$$\leq \max_{\frac{1}{H}<s\leq S} \left|\mathbb{E}^{\mathcal{C}_{s^-}}\left(\mathbf{1}_{\{|\Delta M_s|\leq 1\}}\Delta M_s\right)\right|(X) \sum_{s<\rho\leq t} \left|\mathbb{E}^{\mathcal{C}_{s^-}}\left(\mathbf{1}_{\{|\Delta M_\rho|\leq 1\}}\Delta M_\rho\right)\right|(X) \approx 0.$$

$$B = \sum_{s<\rho\leq t} \left(\mathbf{1}_{\{|\Delta M_\rho|\leq 1\}}\Delta M_\rho(X)\right)^2 \leq [M]_t(X) - [M]_s(X) \approx 0$$

for $s \approx t$. It follows that $[L](X)$ is pre-S-continuous. Since $\Delta L_t(Y) \leq 2$ for all $t > \frac{1}{H}$ μ-a.s., by Case 1, L is pre-S-continuous $\widehat{\mu}$-a.s. Since we have $\widehat{\mu}$-a.s.

$$L_t \approx \sum_{\frac{1}{H}<s\leq t} \Delta M_s(X) = M_t(X) - M_{\frac{1}{H}}(X) \quad \text{for all } t \in T_S,$$

we obtain for $\widehat{\mu}$-almost all $X \in \Lambda$ and all $s < t$ in T_S with $t \approx s$

$$\begin{aligned} |M_t(X) - M_s(X)| &\leq \left|M_t(X) - M_{\frac{1}{H}}(X) - \left(M_s(X) - M_{\frac{1}{H}}(X)\right)\right| \\ &\approx |L_t(X) - L_s(X)| \approx 0. \end{aligned}$$

The proof is finished. □

10.15 On symmetric functions

Since the difference of T^n and $T^n_{\neq}$ is a $\widehat{\nu^n}$-nullset (see Corollary 10.5.5), we may assume tacitly that F is identical to 0 outside of $T^n_{\neq}$.

A multilinear form $G \in \mathbb{F}^{\otimes n}$ or $G \in \mathbb{H}^{\otimes n}$ is called **symmetric** if $G(a_1, \ldots, a_n) = G(a_{\sigma_1}, \ldots, a_{\sigma_n})$ for all permutations σ of $\{1, \ldots, n\}$. We have a subtle notion of symmetry for functions $F : T^n \to \mathbb{F}^{\otimes n}$. For each permutation σ on $\{1, \ldots, n\}$ set

$$F_\sigma(t_1, \ldots, t_n)(a_1, \ldots, a_n) = F(t_{\sigma_1}, \ldots, t_{\sigma_n})(a_{\sigma_1}, \ldots, a_{\sigma_n}).$$

F is called **symmetric** if $F_\sigma = F$ for all permutations σ of $\{1, \ldots, n\}$. Analogous notation is used for functions with values in $\mathbb{H}^{\otimes n}$ or ${}^*\mathbb{H}^{\otimes n}$. We have the following examples of symmetric functions.

Examples

(i) Let $F : T \to \mathbb{F}$ be internal. Then $F^{\otimes n} : T^n_{\neq} \to \mathbb{F}^{\otimes n}$ is symmetric. Recall that $F^{\otimes n}_{(t_1,\ldots,t_n)}(a_1, \ldots, a_n) := F_{t_1}(a_1) \cdot \ldots \cdot F_{t_n}(a_n)$.

(ii) Suppose that $F : T^n \to \mathbb{F}^{\otimes n}$ is a constant G. Then F is symmetric if and only if $G \in \mathbb{F}^{\otimes n}$ is symmetric.

(iii) Let $F : T^n \to \mathbb{F}^{\otimes n}$ be internal. If F is symmetric, then the function $t \mapsto \|F(t)\|_{\mathbb{F}^n} \in {}^*\mathbb{R}$ is symmetric.

Proposition 10.15.1 *Fix a standard $n \in \mathbb{N}$. Let $f : T^n \to \mathbb{H}^{\otimes n}$ be $L_{\nu^n}(T^n)$-measurable. Then f is symmetric $\widehat{\nu^n}$-a.e. if and only if f has a symmetric lifting $F : T^n \to \mathbb{F}^{\otimes n}$.*

Indeed, fix a lifting F of f. If f is symmetric $\widehat{\nu^n}$-a.e., then $\frac{1}{n!}\sum_\sigma F_\sigma$ is a symmetric lifting of f, where σ runs through the permutations of $\{1,\ldots,n\}$. Moreover, if F is symmetric, then $f = \frac{1}{n!}\sum_\sigma f_\sigma$ $\widehat{\nu^n}$-a.e., and of course $\frac{1}{n!}\sum_\sigma f_\sigma$ is symmetric.

Proof It suffices to prove the results for functions defined on T_k^n with $k \in \mathbb{N}$. Let σ be a permutation of $\{1,\ldots,n\}$. For each internal $A \subseteq T_k^n$ we set

$$A_\sigma := \{(t_1,\ldots,t_n) \mid (t_{\sigma(1)},\ldots,t_{\sigma(n)}) \in A\}.$$

Since $|A_\sigma| = |A|$, $\nu_k^n(A) = \nu_k^n(A_\sigma)$. Now fix a lifting $F : T_k^n \to \mathbb{F}^{\otimes n}$ of f. Then, by Theorem 10.1.2 (e), for each $\varepsilon \in \mathbb{R}^+$ there exists an internal $A \subset T_k^n$ such that

$$\left\{(t_1,\ldots,t_n) \in T_k^n \mid F(t_1,\ldots,t_n) \not\approx_{\mathbb{F}^n} f(t_1,\ldots,t_n)\right\} \subset A \quad \text{and} \quad \nu_k^n(A) < \varepsilon.$$

Since

$$\left\{(t_1,\ldots,t_n) \in T_k^n \mid F_\sigma(t_1,\ldots,t_n) \not\approx_{\mathbb{F}^n} f_\sigma(t_1,\ldots,t_n)\right\} \subset A_\sigma$$

and $\nu_k^n(A) = \nu_k^n(A_\sigma)$, F_σ is a lifting of f_σ.

Now let f be symmetric $\widehat{\nu_k^n}$-a.e. Then for $\widehat{\nu_k^n}$-almost all $(t_1,\ldots,t_n) \in T_k^n$

$$\frac{1}{n!}\sum_\sigma F_\sigma(t_1,\ldots,t_n) \approx_{\mathbb{F}^n} \frac{1}{n!}\sum_\sigma f_\sigma(t_1,\ldots,t_n) = f(t_1,\ldots,t_n).$$

This proves that $\frac{1}{n!}\sum_\sigma F_\sigma$ is a symmetric lifting of f.

For the converse, let F be symmetric. Then we obtain for $\widehat{\nu_k^n}$-almost all $(t_1,\ldots,t_n) \in T_k^n$

$$f_\sigma(t_1,\ldots,t_n) \approx_{\mathbb{F}^n} F_\sigma(t_1,\ldots,t_n) = F(t_1,\ldots,t_n) \approx_{\mathbb{F}^n} f(t_1,\ldots,t_n).$$

□

10.16 The standard part of internal martingales

Our aim now is to convert, under mild conditions, internal martingales defined on the *finite set T to càdlàg standard martingales defined on the continuous timeline $[0,\infty[$. Conversely, we lift standard martingales to internal ones.

We start with a lemma which constructs from a certain internal function $F: T \to {}^*\mathbb{R}$ a càdlàg function ${}^\circ F: [0,\infty[\to \mathbb{R}$, which is called the **standard part** of F.

Lemma 10.16.1 *Fix an internal $F: T \to {}^*\mathbb{R}$ such that F_t is limited for all limited $t \in T$. Set $F_0 := 0$. Fix $r \in [0,\infty[$.*

(a) *Then $\lim_{{}^\circ s \downarrow r} {}^\circ(F_s)$ exists if and only if*

$$\lim_{k\to\infty} \sup_{r < {}^\circ s \leq r+\frac{1}{k}} {}^\circ(F_s) \leq \lim_{k\to\infty} \inf_{r < {}^\circ s \leq r+\frac{1}{k}} {}^\circ(F_s).$$

In the following assertions assume now that $\lim_{{}^\circ s \downarrow r} {}^\circ(F_s)$ exists. Define ${}^\circ F: [0,\infty[\to \mathbb{R}$ by

$$({}^\circ F)_r := \lim_{{}^\circ s \downarrow r} {}^\circ(F_s).$$

Similar results are obtained for the left-hand limits, where we define for $r > 0$ in the case $\lim_{{}^\circ s \uparrow r} {}^\circ(F_s)$ exists

$$\left({}^{ol}F\right)_{r^-} := \lim_{{}^\circ s \uparrow r} {}^\circ(F_s) \quad \text{and} \quad \left({}^{ol}F\right)_{0^-} := 0.$$

Notice the difference between $\left({}^{ol}F\right)_{r^-}$ and

$$({}^\circ F)_{r^-} := \lim_{s \uparrow r} ({}^\circ F)_s.$$

(b) *There exists an $\widetilde{r} \in {}^*]r,\infty[$, $\widetilde{r} \approx r$, such that $({}^\circ F)_r \approx F_s$ for all $s \in T$ with $s \geq \widetilde{r}$ and $s \approx r$.*
If $\left({}^{ol}F\right)_{r^-}$ exists with $r > 0$, then there exists an $\widetilde{r} \in {}^*]0,r[$, $\widetilde{r} \approx r$, such that $\left({}^{ol}F\right)_{r^-} \approx F_s$ for all $s \in T$ with $s \leq \widetilde{r}$ and $s \approx r$.

(c) *${}^\circ F$ is càdlàg, provided ${}^{ol}F$ also exists for all $r \in]0,\infty[$. Then*

$$\left({}^{ol}F\right)_{r^-} = ({}^\circ F)_{r^-}.$$

Proof The proof of part (a) is straightforward and is left to the reader.

(b) By the assumption and (a), there exist strictly monotone increasing functions $g,h: \mathbb{N} \to \mathbb{N}$ with $g < h$ such that for all $n \in \mathbb{N}$ and for all $k \in \mathbb{N}$, with $k > n$,

$$\forall s \in T \left(r + \frac{1}{{}^*h(k)} \leq s \leq r + \frac{1}{{}^*g(n)} \Rightarrow |F_s - ({}^\circ F)(r)| < \frac{1}{n} \right). \tag{26}$$

By Propositions 8.4.1 (b) and 8.2.1 (f), there exists an unlimited $N_\infty \in {}^*\mathbb{N}$ such that (26) is true for all $n \in \mathbb{N}$, when we replace k by any unlimited $N \leq N_\infty$. Set

$\widetilde{r} := r + \frac{1}{{}^*h(N_\infty)}$. Note that (b) is true for this $\widetilde{r}$. The proof of the second statement of part (b) is similar. In the case $r = 0$, set $\widetilde{r} := 0$.

(c) To prove that ${}^\circ F$ is right continuous in r, fix a sequence $(r_n)_{n\in\mathbb{N}}$ in $]r,\infty[$, converging to r. Fix $\widetilde{r} \approx r$, as in (b). By (b), there are $t_n \in T$ with $t_n \approx r_n$ and $F_{t_n} \approx ({}^\circ F)_{r_n}$. Let $(t_n)_{n\in {}^*\mathbb{N}}$ be an internal extension of $(t_n)_{n\in\mathbb{N}}$ such that $\widetilde{r} \le t_n$ for all $n \in {}^*\mathbb{N}$. There exists an unlimited $N_\infty \in {}^*\mathbb{N}$ such that $t_n \approx r$ for all unlimited $n \le N_\infty$. Assume that $\left({}^\circ F_{r_n}\right)_{n\in\mathbb{N}}$ does not converge to ${}^\circ F_r$. Then there exists a $k \in \mathbb{N}$ such that $\left|F_{t_n} - {}^\circ F_r\right| \ge \frac{1}{k}$ for infinitely many $n \in \mathbb{N}$. By Proposition 8.4.1 (b), there exists an unlimited $N \le N_\infty$ with $\left|F_{t_N} - {}^\circ F_r\right| \ge \frac{1}{k}$ and $t_N \approx r$. However, $\widetilde{r} \le t_N$, thus $F_{t_N} \approx {}^\circ F_r$, which is a contradiction. The proof that ${}^\circ F$ has left-hand limits in $]0,\infty[$ with $\left({}^{\circ l}F\right)_{r^-} = ({}^\circ F)_{r^-}$ is similar. □

Theorem 10.16.2 (Hoover and Perkins [46]). *Fix a Loeb probability space $\left(\Lambda, L_\mu(\mathcal{C}), \widehat{\mu}\right)$ and an internal filtration $(\mathcal{C}_t)_{t\in T}$ on Λ. Let $M : \Lambda \times T \to {}^*\mathbb{R}$ be an internal $(\mathcal{C}_t)_{t\in T}$-martingale such that $\mathbb{E}_\mu |M_t|$ is limited for all limited $t \in T$.*

(a) *Then there exists a set U of $\widehat{\mu}$-measure 1 such that $({}^\circ M)_r(X)$, $\left({}^{\circ l}M\right)_{r^-}(X)$ exist for all $X \in U$ and all $r \in [0,\infty[$. Moreover, ${}^\circ M$ is a càdlàg process.*
(b) *Fix $r \in [0,\infty[$. Then there exist $r_l, r_r \in {}^*[0,\infty[$, $r_l, r_r \approx r$, such that M_t is a lifting of $({}^\circ M)_r$ for all $t \in T$ with $t \ge r_r, t \approx r$ and M_t is a lifting of $({}^\circ M)_{r^-}$ for all $t \in T$ with $t \le r_l, t \approx r$.*

Proof (a) Fix $\sigma \in \mathbb{N}$. By Proposition 2.3.1, we have for all $n \in \mathbb{N}$

$$\mu\left\{\max_{s\in T_\sigma} |M_s| \ge n\right\} \le \frac{1}{n}\mathbb{E}_\mu |M_\sigma|.$$

It follows that $\widehat{\mu}\left\{\max_{s\in T_\sigma} |M_s| \text{ is unlimited}\right\} = 0$. Let U_σ be the set of all X such that $M_s(X)$ is limited for all $s \le \sigma$. Fix $p < q$ in $\mathbb{Q}$ and let $U_{p,q}$ be the number of upcrossings of $]p,q[$ by $(M_s)_{s\le\sigma}$. See Ash [5], Section 7.4. By Theorem 7.4.2 there, $\mathbb{E}_\mu U_{p,q} \le \frac{1}{q-p}\mathbb{E}_\mu\left((M_\sigma - p)^+\right)$ with $(M_\sigma - p)^+ = (M_\sigma - p) \vee 0$. Since $\mathbb{E}_\mu\left((M_\sigma - p)^+\right)$ is limited, $U_{p,q}$ is standard finite $\widehat{\mu}$-a.s. Therefore, we may choose U_σ such that $U_{p,q}(X)$ is finite for all $X \in U_\sigma$ and for all $p < q$ in $\mathbb{Q}$. Set $U = \bigcap_{\sigma\in\mathbb{N}} U_\sigma$. Now fix $r \in [0,\infty[$ and $X \in U$. Assume that $\lim_{{}^\circ s\downarrow r} {}^\circ(M_s(X))$ does not exist. By Lemma 10.16.1 (b) there exist $p < q$ in $\mathbb{Q}$ such that

$$\lim_{k\to\infty} \inf_{r < {}^\circ s \le r+\frac{1}{k}} {}^\circ(M_s(X)) < p < q < \lim_{k\to\infty} \sup_{r < {}^\circ s \le r+\frac{1}{k}} {}^\circ(M_s(X)).$$

It follows that $U_{p,q}(X)$ is infinite, which is a contradiction.

In the same way we see that $\lim_{°s\uparrow r}{}^{\circ}(M_s)$ exists $\widehat{\mu}$-a.s. for $r>0$. Lemma 10.16.1 (c) implies that ${}^{\circ}M$ is a càdlàg process.

(b) Let Y be a $\mathcal{C}$-measurable lifting of $({}^{\circ}M)_r$. Since $\lim_{°s\downarrow r}{}^{\circ}(M_s) = {}^{\circ}Y$ $\widehat{\mu}$-a.s., we have, for all $m\in\mathbb{N}$, $\lim_{°s\downarrow r}\widehat{\mu}\left\{|{}^{\circ}Y - {}^{\circ}(M_s)| \geq \frac{1}{m}\right\}=0$. It follows that

$$\lim_{°s\uparrow r}{}^{\circ}\mu\left\{|Y-M_s|\geq\frac{1}{m}\right\}=0.$$

Now we proceed in the same way as in the proof of Lemma 10.16.1 and find strictly monotone increasing functions $g,h:\mathbb{N}\to\mathbb{N}$ with $g<h$ and an unlimited N_∞ such that for all unlimited $N\leq N_\infty, \varepsilon>0$ and $n\in\mathbb{N}$

$$r+\frac{1}{{}^*h(N)}<s\leq r+\frac{1}{g(n)}\Rightarrow\mu\{|Y-M_s|\geq\varepsilon\}<\frac{1}{n}.$$

It follows that $\widehat{\mu}\{Y\not\approx M_s\}=0$ for all $s>r+\frac{1}{{}^*h(N)}$ with $s\approx r$. All these M_s are liftings of $({}^{\circ}M)_r$, thus $r_r:=r+\frac{1}{{}^*h(N)}$ fulfils part (b). The proof of the existence of r_l is similar. □

Under the assumptions of Theorem 10.16.2

$${}^{\circ}M:\Lambda\times[0,\infty[\to\mathbb{R},\ (X,r)\longmapsto\lim_{°s\downarrow r}{}^{\circ}(M(X,s))$$

is called the **standard part** of M. ${}^{\circ}M(X,\cdot)$ exists for $\widehat{\mu}$-almost all X.

Corollary 10.16.3 *Let $M:\Lambda\times T\to{}^*\mathbb{R}$ be an internal $(\mathcal{C}_t)_{t\in T}$-martingale such that $M_t\in SL^1(\mu)$ for all limited t. Let $(\mathfrak{c}_r)_{r\in[0,\infty[}$ be the standard part of $(\mathcal{C}_t)_{t\in T}$ (see Theorem 10.6.1). Then ${}^{\circ}M$ is a càdlàg $(\mathfrak{c}_r)_{r\in[0,\infty[}$-martingale.*

Proof Fix $r\in[0,\infty[$ and an $\widetilde{r}\approx r$ such that for all $s\approx r,\widetilde{r}\leq s$, M_s is a lifting of ${}^{\circ}M_r$. Now ${}^{\circ}M_r$ is $\widehat{\mu}$-integrable, because all these M_s are S-integrable. It also follows that ${}^{\circ}M_r$ is $\mathfrak{c}_r$-measurable. In order to prove the martingale property, fix $u\in[0,\infty[$ with $u>r$, and $B\in\mathfrak{c}_r$. We have to prove that $\int_B{}^{\circ}M_u d\widehat{\mu}=\int_B{}^{\circ}M_r d\widehat{\mu}$. We may choose s with $s\approx r,\widetilde{r}\leq s$, such that there is an $A\in\mathcal{C}_s$ with $\widehat{\mu}(A\Delta B)=0$ (see Corollary 10.1.3). Fix $t\approx u$ such that M_t is a lifting of ${}^{\circ}M_u$. Then we obtain

$$\int_B{}^{\circ}M_u d\widehat{\mu}=\int_A{}^{\circ}M_t d\widehat{\mu}\approx\int_A M_t d\mu=\int_A M_s d\mu\approx\int_B{}^{\circ}M_r d\widehat{\mu}.$$

□

Now we construct from certain $(\mathfrak{c}_r)_{r\in[0,\infty[}$-martingales m internal martingales M whose standard part ${}^{\circ}M$ is m. In particular, ${}^{\circ}M$ is a càdlàg **version** of m, i.e., for each $r \in [0,\infty[$ there exists a set U_r of $\widehat{\mu}$-measure 1 such that $m_r(X) = ({}^\circ M)_r(X)$ for all $X \in U_r$.

Theorem 10.16.4 *Suppose that $m: \Lambda \to \mathbb{R}$ is integrable. Let $M \in SL^1(\mu)$ be a lifting of m. Define $m_r := \mathbb{E}^{\mathfrak{c}_r} m$ for $r \in [0,\infty[$ and $M_t := \mathbb{E}^{\mathcal{C}_t} M$ for $t \in T$. Then for all $r \in [0,\infty[$*

$$({}^\circ M)_r := \lim_{{}^\circ s \downarrow r} {}^\circ M_s = m_r \ \widehat{\mu}\text{-a.s.}$$

Proof Note that $(m_r)_{r\in[0,\infty[}$ and $(M_t)_{t\in T}$ are $(\mathfrak{c}_r)_{r\in[0,\infty[}$-martingales, internal $(\mathcal{C}_t)_{t\in T}$-martingales, respectively, and $M_t \in SL^1(\mu)$ for all $t \in T$. By Theorem 10.16.2, $\lim_{{}^\circ s\downarrow r} {}^\circ(M_s) = ({}^\circ M)_r$ exists for all $r \in [0,\infty[$ $\widehat{\mu}$-a.s. and defines a càdlàg martingale. It remains for us to prove that $({}^\circ M)_r = m_r$ $\widehat{\mu}$-a.s. Let $B \in \mathfrak{c}_r$. In the previous theorem we have seen that there is an $s \approx r$ such that M_s is a lifting of $({}^\circ M)_r$ and such that there exists an $A \in \mathcal{C}_s$ with $\widehat{\mu}(A \Delta B) = 0$. We obtain

$$\int_B ({}^\circ M)_r d\widehat{\mu} \approx \int_A M_s d\mu = \int_A M d\mu \approx \int_A m d\widehat{\mu} = \int_B m_r d\widehat{\mu}.$$

□

Exercises

10.1 Prove that $\left(T^n, L_{\nu^n}, \widehat{\nu^n}\right)$ is a measure space.
Fix a Loeb space $\left(\Lambda, L_\mu(\mathcal{C}), \widehat{\mu}\right)$ with $\widehat{\mu}(\Lambda) < \infty$.

10.2 A sequence $(F_n)_{n\in\mathbb{N}}$ of S_μ-integrable functions $F_n : \Lambda \to {}^*\mathbb{R}$ can be extended to an internal sequence $(F_n)_{n\in{}^*\mathbb{N}}$ of S_μ-integrable functions if and only if

$$\lim_{k\to\infty} {}^\circ\int_{\{|F_n|\geq k\}} |F_n|\, d\mu = 0 \text{ uniformly in } n \in \mathbb{N}.$$

10.3 Give a nonstandard proof of Vitali's extension of the Lebesgue convergence theorem: Let $(f_n)_{n\in\mathbb{N}}$ be a sequence of $\widehat{\mu}$-integrable functions $f_n : \Lambda \to \mathbb{R}$, converging to f in measure. Moreover, assume that $\lim_{k\to\infty} \int_{\{|f_n|\geq k\}} |f_n|\, d\widehat{\mu} = 0$ uniformly in $n \in \mathbb{N}$. Then f is $\widehat{\mu}$-integrable and

$$\lim_{n\to\infty} \int_\Lambda f_n d\widehat{\mu} = \int_\Lambda f d\widehat{\mu}.$$

10.4 Prove Corollary 10.7.3.

10.5 Prove the second assertion of Theorem 10.9.2 part (b).

10.6 Prove that the function Φ defined in Section 10.12 is *convex.

10.7 Prove '$\Leftarrow$' of Theorem 10.13.1.

10.8 Prove Lemma 10.16.1 part (a).

10.9 Prove the second part of Lemma 10.16.1 part (b).

10.10 Prove that $^\circ F$ has left-hand limits if $\lim_{^\circ s \uparrow r} {^\circ}(F_s)$ exists for all $r > 0$.

10.11 Prove that $\widehat{\nu^n}\left(\left\{T^n_<\right\} \Delta \left\{t \in T^n_< \mid {^\circ}t_1 < {^\circ}t_2 < \ldots < {^\circ}t_n\right\}\right) = 0$.

10.12 Prove that $\mathbf{1}_{T^n_<}$ is $\mathcal{L}^n$-measurable.

PART II

An introduction to finite- and infinite-dimensional stochastic analysis

Introduction

In the preceding chapters we have focused on theories that are crucial in dealing with Malliavin calculus on abstract Wiener spaces and for Lévy processes, where we use a unified framework. Since, as we have seen, Hilbert spaces play an important role, we fix a separable real finite- or infinite-dimensional Hilbert space $\mathbb{H}$ and construct a sample space Ω, depending only on $\mathbb{H}$. We also take the timeline $T := \left\{\frac{1}{H}, \ldots, H\right\}$, extensively used in Sections 9.9 and 10.5.

Let $\mathbb{F}$ be a *finite representation of $\mathbb{H}$, according to Theorem 9.8.1. If $\mathbb{H}$ is infinite-dimensional, then $\mathbb{F}$ is the ω-dimensional space, constructed there. If $\mathbb{H}$ is finite-dimensional, i.e., $\mathbb{H} = \mathbb{R}^d$ for some $d \in \mathbb{N}$, then $\mathbb{F}$ is simply ${}^*\mathbb{R}^d$. Recall that for all $t \in T$, $T_t = \left\{\frac{1}{H}, \ldots, t\right\}$ and $\mathbb{F}^{T_t}$ is the space of all internal functions $X : T_t \to \mathbb{F}$. In the remainder of this book we fix

$$\Omega := \mathbb{F}^T, \quad B : \Omega \times T \to \mathbb{F}, (X,t) \mapsto \sum_{s \le t} X_s = \sum_{s \le t} \sum_{\mathfrak{e} \in \mathfrak{E}} \langle X_s, \mathfrak{e} \rangle\, \mathfrak{e},$$

where $\mathfrak{E}$ is an ONB of $\mathbb{F}$. The components of the vector $X = (X_{\frac{1}{H}}, \ldots, X_H) \in \Omega$ are the increments of B. On Ω we use the filtration $(\mathcal{B}_t)_{t \in T} := \left(\mathcal{B}_t^B\right)_{t \in T}$, generated by B. Recall that

$$\mathcal{B}_t := \left\{A \times \mathbb{F}^{T \setminus T_t} \mid A \text{ is an internal Borel set in } \mathbb{F}^{T_t}\right\}, \text{ setting } \mathcal{B}_0 := \{\emptyset, \Omega\}.$$

We will often use the filtration $(\mathcal{B}_{t^-})_{t \in T}$, where $t^- := t - \frac{1}{H}$. Let $(\mathfrak{b}_r)_{r \in [0,\infty[}$ be the standard part of $(\mathcal{B}_t)_{t \in T}$ (see Theorem 10.6.1). Then $(\mathfrak{b}_r)_{r \in [0,\infty[}$ is also the standard part of $(\mathcal{B}_{t^-})_{t \in T}$.

Moreover, fix an internal probability measure μ^1 on ${}^*\mathbb{R}$ and let μ be the $d \cdot H^2$-fold product of μ^1 on the internal Borel sets of Ω, where $d \in \mathbb{N}$ or $d = \omega$.

If $\mathbb{F}$ is d-dimensional with $d \in \mathbb{N}$, then B is a mapping from $\Omega \times T$ into ${}^*\mathbb{R}^d$. In Chapter 15 we shall see, using Lindstrøm's work [67], that each d-dimensional

Lévy process L lives on Ω and is the standard part ${}^{\circ}B$ of B. Recall that

$$L = {}^{\circ}B : \Omega \times [0, \infty[\to \mathbb{R}^d, \quad L(\cdot, r) = \lim_{{}^{\circ}s \downarrow r} {}^{\circ}(B(\cdot, s)) \ \widehat{\mu}\text{-a.s.}$$

In the finite-dimensional situation, we assume, for simplicity, that $\mathbb{F}$ is one-dimensional, i.e., $\mathbb{F} = {}^*\mathbb{R}$ and $\mathbb{H} = \mathbb{R}$.

In summary we have two dimensions d: $d = 1$ in the case of Lévy processes and $d = \omega$ in the case of infinite-dimensional Brownian motion.

11

From finite- to infinite-dimensional Brownian motion

Following [84], we construct from the internal discrete Brownian motion B a standard Brownian motion with values in any abstract Wiener space. Fix an abstract Wiener–Fréchet space $\mathbb{B}$ over an infinite-dimensional separable Hilbert space $\mathbb{H}$. Recall that $\gamma_{\frac{1}{H}}$ denotes the centred Gaussian measure on ${}^*\mathbb{R}$ of variance $\frac{1}{H}$. Let Γ be the $\omega \cdot H^2$-fold product of $\gamma_{\frac{1}{H}}$ on the internal Borel algebra $\mathcal{B}$ of $\Omega = \mathbb{F}^T$. Following [84], we shall see now that the process $b_{\mathbb{B}}$, $\widehat{\Gamma}$-a.s. well defined by

$$b_{\mathbb{B}} : \Omega \times [0,\infty[\to \mathbb{B}, \quad b_{\mathbb{B}}(\cdot, r) = {}^{\circ}(B(\cdot, s)) \quad \text{for all } s \approx r, s \in T,$$

is a continuous Brownian motion such that for each continuous function $f : [0,\infty[\to \mathbb{B}$ there exists an $X \in \Omega$ with $b_{\mathbb{B}}(X) = f$. Here ${}^{\circ}B$ is the standard part of B in the topology of $\mathbb{B}$. We use the notation of Section 4.3 and assume that the metric d on $\mathbb{B}$ is generated by a separating sequence $\left(|\cdot|_j\right)_{j \in \mathbb{N}}$ of measurable seminorms. The definition of a $\mathbb{B}$-valued Brownian motion is given in Section 4.8.

In [65] T. Lindstrøm has defined a Brownian motion with values in abstract Wiener spaces as the standard part of a hyperfinite random walk extending Anderson's approach [4]. To be able to differentiate, we construct Brownian motion in a smooth way, extending some of Cutland's [23] results on finite-dimensional Brownian motion.

11.1 On the underlying probability space

The measure Γ just introduced has the following form. Fix an internal ONB $\mathfrak{E} := (\mathfrak{e}_i)_{i \in \omega}$ of $\mathbb{F}$. Then for each $B \in \mathcal{B}$

$$\Gamma(B) = \int_B e^{-\frac{H}{2} \sum_{i \leq \omega, s \in T} \langle X_s, \mathfrak{e}_i \rangle^2} dX \sqrt{\frac{H}{2\pi}}^{\omega \cdot H^2}.$$

By Lemma 4.2.1, $\Gamma(B)$ does not depend on the ONB of $\mathbb{F}$. Note that for each $\mathcal{B}$-measurable function $F:\Omega\to {}^*\mathbb{R}$

$$\int_{\Omega} F d\Gamma = \int_{\Omega} F(X) e^{-\frac{H}{2}\sum_{i\in\omega,s\in T}\langle X_s,\mathfrak{e}_i\rangle^2} dX \sqrt{\frac{H}{2\pi}}^{\omega\cdot H^2}$$

if one of the integral exists. Define

$$F^{\mathfrak{E}}: {}^*\mathbb{R}^{\omega\cdot H^2} \to {}^*\mathbb{R}, (X_s(i))_{i\in\omega,s\in T} \mapsto F\left(\sum_{i\in\omega} X_s(i)\mathfrak{e}_i\right)_{s\in T}.$$

For example, if $F(X) = \langle X_t, \mathfrak{e}_j\rangle$, then $F^{\mathfrak{E}}((X_s(i))_{i\in\omega,s\in T}) = X_t(j)$. By the substitution rule we obtain the following.

Lemma 11.1.1 *Let $F:\Omega\to {}^*\mathbb{R}$ be $\mathcal{B}$-measurable. Then*

$$\begin{aligned}\int_{\Omega} F d\Gamma &= \int_{{}^*\mathbb{R}^{\omega\cdot H^2}} F^{\mathfrak{E}}(X_s(i))_{i\in\omega,s\in T} d\gamma_{\frac{1}{H}}^{\omega\cdot H^2}\left(X_s(i)_{i\in\omega,s\in T}\right)\\ &= \int_{{}^*\mathbb{R}^{\omega\cdot H^2}} F^{\mathfrak{E}}(X_s(i))_{i\in\omega,s\in T} e^{-\frac{H}{2}\sum_{i\in\omega,s\in T} X_s(i)^2} d\left(X_s(i)_{i\in\omega,s\in T}\right)\sqrt{\frac{H}{2\pi}}^{\omega\cdot H^2},\end{aligned}$$

if one of the three integrals exists.

Note that, since the variance of $\gamma_{\frac{1}{H}}$ is infinitesimal, the 'standard part' of the density of $\gamma_{\frac{1}{H}}$ is the Dirac δ-function. The proof of the following Lemma about the filtration $(\mathcal{B}_t)_{t\in T\cup\{0\}}$ makes use of the fact that a subset B of $\mathbb{F}^{T_t}$ is a Borel set in $\mathbb{F}^{T_t}$ if $B\times\mathbb{F}^{T\setminus T_t}$ is a Borel set in Ω.

Lemma 11.1.2 *Fix $t\in T\cup\{0\}$ and an internal Borel measurable function F: $\Omega\to\mathbb{F}$. Then F is $\mathcal{B}_t$-measurable if and only if $F(X) = F(Y)$ for all $X, Y\in\Omega$ with $(X_s)_{s\in T_t} = (Y_s)_{s\in T_t}$, where $T_0 = \emptyset$. Note that $(X_s)_{s\in T_0} = (Y_s)_{s\in T_0}$ for all $X, Y\in\Omega$.*

Using Lemma 11.1.1, partial integration and Fubini's Theorem, we obtain the following simple but important properties of the measure Γ, where we use the following notation. Let $n\in\mathbb{N}$ be even. Then

$$n!! := (n-1)\cdot(n-3)\cdot\ldots\cdot 3\cdot 1.$$

Let us identify X_t with the projection $\Omega\to\mathbb{F}, X\mapsto X_t$.

Lemma 11.1.3 *Fix an ONB $(\mathfrak{e}_i)_{i\in\omega}$ of $\mathbb{F}$. Then we obtain Γ-a.s. for each $t\in T$:*

(a) $\mathbb{E}^{\mathcal{B}_{t^-}} \langle X_t, \mathfrak{e}_i \rangle^{2k-1} = 0$ *for all* $k \in {}^*\mathbb{N}$.
(b) $\mathbb{E}^{\mathcal{B}_{t^-}} \left(\langle X_t, \mathfrak{e}_i \rangle \cdot \langle X_t, \mathfrak{e}_j \rangle \right) = 0$ *if* $i \neq j$.
(c) $\mathbb{E}^{\mathcal{B}_{t^-}} \left(\langle X_t, \mathfrak{e}_i \rangle \cdot \langle X_s, \mathfrak{e}_j \rangle \right) = 0$ *if* $s < t$. *This is true for all* $(i,j) \in \omega^2$.
(d) $\mathbb{E}^{\mathcal{B}_{t^-}} \left(\langle X_t, \mathfrak{e}_i \rangle^{2n} \right) = \frac{(2n)!!}{H^n}$.
(e) $\mathbb{E}^{\mathcal{B}_{t^-}} \left(\langle X_t, \mathfrak{e}_i \rangle^2 - \frac{1}{H} \right)^2 = \frac{2}{H^2}$.
(f) $\mathbb{E}^{\mathcal{B}_{t^-}} \left(\langle X_t, \mathfrak{e}_i \rangle^2 - \frac{1}{H} \right) \left(\langle X_t, \mathfrak{e}_j \rangle^2 - \frac{1}{H} \right) = 0$ *for* $i \neq j$.
(g) $\mathbb{E}^{\mathcal{B}_{t^-}} \left(\langle X_t, \mathfrak{e}_i \rangle^2 - \frac{1}{H} \right) \left(\langle X_s, \mathfrak{e}_j \rangle^2 - \frac{1}{H} \right) = 0$ *for* $s < t$.

11.2 The internal Brownian motion

Recall the definition of $B : \Omega \times T \to \mathbb{F}$ at the very beginning of Part II. The measure Γ and the process B have been introduced by Cutland [23] in the case $\mathbb{H} = \mathbb{R}$ and $\mathbb{F} = {}^*\mathbb{R}$. Our aim is to show that, for each separating sequence $(|\cdot|_j)$ of measurable semi-norms on $\mathbb{H}$, the standard part ${}^\circ B$ of B exists in the completion $(\mathbb{B}, (|\cdot|_j))$ of $(\mathbb{H}, (|\cdot|_j))$ and is a continuous Brownian motion, and that each $\mathbb{B}$-valued continuous function defined on $[0, \infty[$ appears as a path of ${}^\circ B$.

The following result shows that B is already an internal Brownian motion, in contrast to Anderson's random walk A, defined in the problems below. From Exercise 11.5 it follows that A is almost an internal Brownian motion.

Proposition 11.2.1 *Fix* $a, b \in \mathbb{F}$ *with* $\|a\| = \|b\| = 1$ *and* $\lambda \in {}^*\mathbb{R}$.

(a) *The internal process* $\left(e^{\lambda \langle a, B_t \rangle - \frac{1}{2}\lambda^2 t} \right)_{t \in T}$ *is a* $(\mathcal{B}_t)_{t \in T}$*-martingale. It follows that* $\langle a, B_t \rangle$ *is* $\mathcal{N}(0,t)$*-distributed, and* $\langle a, B_t - B_s \rangle$ *is independent of* $\mathcal{B}_s$ *for* $s < t$ *in* T *(see Theorem 3.5.2).*
(b) *Fix* $s \in T$, $\lambda \in {}^*\mathbb{R}$, *both limited. Then* $e^{\lambda \langle a, B_s \rangle - \frac{1}{2}\lambda^2 s}$ *is* S_Γ*-integrable.*
(c) $\mathbb{E}_\Gamma (\langle a, B_t \rangle \cdot \langle b, B_t \rangle) = \langle a, b \rangle \cdot t$.
(d) *For each* **finite-dimensional space* $E \subseteq \mathbb{F}$ *and each* $\sigma \in T$, γ_σ^E *is the image measure of* Γ *by* $\pi_{\mathbb{F}E}(B_\sigma(\cdot))$.

Recall that γ_σ^E *denotes the internal centred Gaussian distribution on* $\mathcal{B}(E)$ *with variance* σ *and* $\pi_{\mathbb{F}E}$ *denotes the orthogonal projection from* $\mathbb{F}$ *onto* E.

Proof First extend a to an internal ONB $(\mathfrak{e}_i)_{i \in \omega}$ of $\mathbb{F}$, thus $\mathfrak{e}_1 = a$.

(a) By Lemma 11.1.2, $e^{\lambda \langle a, B_t \rangle - \frac{1}{2}\lambda^2 t}$ is $\mathcal{B}_t$-measurable for each $t \in T$. Moreover,

$$\int_\Omega e^{\lambda \langle a, B_t \rangle - \frac{1}{2}\lambda^2 t} d\Gamma = \int_{{}^*\mathbb{R}^{Ht}} e^{\sum_{s \le t} - \frac{H}{2} (X_s - \frac{\lambda}{H})^2} d\,(X_s)_{s \le t} \sqrt{\frac{H}{2\pi}}^{Ht} = 1.$$

Therefore, by Proposition 3.2.3, $\langle a, B_t\rangle$ is $\mathcal{N}(0,t)$-distributed. To prove the martingale property, fix $s \in T$ with $s < H$ and $F \times \mathbb{F}^{T\setminus T_s} \in \mathcal{B}_s$, where $F \in \mathcal{B}\left(\mathbb{F}^{T_s}\right)$. Set

$$F^{\mathfrak{E}} := \left\{ X_j(i)_{i\in\omega, j\leq s} \mid \left(\sum_{i\in\omega} X_j(i)\mathfrak{e}_i\right)_{j\leq s} \in F \right\}.$$

Then

$$\begin{aligned}
&\int_{F\times\mathbb{F}^{T\setminus T_s}} \left(e^{\lambda\langle a, B_{s^+}\rangle - \frac{1}{2}\lambda^2 s^+}\right) d\Gamma \\
&= \int_{F^{\mathfrak{E}}\times {}^*\mathbb{R}^{\omega(T\setminus T_s)}} e^{\sum_{j\leq s^+}\left(\lambda X_j(1) - \frac{1}{2}\lambda^2\frac{1}{H}\right)} d\gamma_{\frac{1}{H}}^{\omega\cdot H^2}\left(X_j(i)\right)_{i\in\omega, j\in T} \\
&= \alpha\cdot\beta\cdot\gamma\cdot\delta
\end{aligned}$$

with

$$\begin{aligned}
\alpha &= \int_{F^{\mathfrak{E}}} e^{\sum_{j\leq s}\left(\lambda X_j(1) - \frac{1}{2}\lambda^2\frac{1}{H}\right) + \sum_{i\in\omega, j\leq s} - \frac{H}{2}X_j(i)^2} d\left(X_j(i)\right)_{i\in\omega, j\leq s}\sqrt{\frac{H}{2\pi}}^{Hs\omega}, \\
\beta &= \int_{{}^*\mathbb{R}} e^{\left(\lambda x - \frac{1}{2}\lambda^2\frac{1}{H}\right) - \frac{H}{2}x^2} dx\sqrt{\frac{H}{2\pi}} = \int_{{}^*\mathbb{R}} e^{-\frac{H}{2}(x-\frac{\lambda}{H})^2} dx\sqrt{\frac{H}{2\pi}} = 1 \\
= \gamma &= \int_{{}^*\mathbb{R}^{\omega-1}} e^{\sum_{1<i\in\omega} - \frac{H}{2}X(i)^2} d\left(X(i)\right)_{1<i\in\omega}\sqrt{\frac{H}{2\pi}}^{\omega-1}, \\
\delta &= \int_{{}^*\mathbb{R}^{\omega H(H-s^+)}} e^{\sum_{i\in\omega, s^+<j\leq H} - \frac{H}{2}X_j(i)^2} d\left(X_j(i)\right)_{i\in\omega, s^+<j\leq H}\sqrt{\frac{H}{2\pi}}^{H(H-s^+)\omega} = 1.
\end{aligned}$$

We obtain

$$\int_{F\times\mathbb{F}^{T\setminus T_s}} \left(e^{\lambda\langle a, B_{s^+}\rangle - \frac{1}{2}\lambda^2 s^+}\right) d\Gamma = \alpha = \int_{F\times\mathbb{F}^{T\setminus T_s}} \left(e^{\lambda\langle a, B_s\rangle - \frac{1}{2}\lambda^2 s}\right) d\Gamma.$$

(b) It suffices to show that $\mathbb{E}_\Gamma\left(e^{\lambda\langle a, B_s\rangle - \frac{\lambda^2}{2}s}\right)^2$ is limited (see Theorem 10.11.1): the previous computation also shows that for $s \in T$ and $\lambda \in {}^*\mathbb{R}$, both limited,

$$\begin{aligned}
\mathbb{E}_\Gamma\left(e^{\lambda\langle a, B_s\rangle - \frac{\lambda^2}{2}s}\right)^2 &= \int_{{}^*\mathbb{R}^{Hs}} \left(e^{\sum_{t\leq s}\left(2\lambda X_t - \frac{\lambda^2}{H} - \frac{H}{2}X_t^2\right)}\right)\sqrt{\frac{H}{2\pi}}^{Hs} d\left(X_t\right)_{t\leq s} \\
&= \prod_{n=1}^{Hs}\int_{{}^*\mathbb{R}} e^{2\lambda x - \frac{\lambda^2}{H} - \frac{H}{2}x^2}\cdot\sqrt{\frac{H}{2\pi}}dx
\end{aligned}$$

$$= \prod_{n=1}^{Hs} \int_{^*\mathbb{R}} e^{-\frac{H}{2}\left(x-\frac{2\lambda}{H}\right)^2+\frac{\lambda^2}{H}} \cdot \sqrt{\frac{H}{2\pi}} dx = e^{\lambda^2 s} \text{ is limited.}$$

(c) Define $b_0 := b - \langle a,b\rangle a$. Then $b = \langle a,b\rangle a + b_0$ and $b_0 \perp a$. By Lemma 11.1.3 (b) and (c), $\mathbb{E}(\langle b_0, B_t\rangle \langle a, B_t\rangle) = 0$. By Lemma 11.1.3 (c) and (d), we obtain

$$\begin{aligned}
\mathbb{E}(\langle b, B_t\rangle \langle a, B_t\rangle) &= \langle a,b\rangle \mathbb{E}\left(\langle a, B_t\rangle^2\right) + \mathbb{E}(\langle b_0, B_t\rangle \cdot \langle a, B_t\rangle) \\
&= \langle a,b\rangle \mathbb{E} \sum_{j \le t} \langle a, \cdot_j\rangle^2 = \langle a,b\rangle \sum_{j \le t} \frac{1}{H} = \langle a,b\rangle\, t.
\end{aligned}$$

(d) follows from (a) and Proposition 3.1.2. □

11.3 *S*-integrability of the internal Brownian motion

Using Theorem 10.13.1, we obtain:

Proposition 11.3.1 *Fix $a \in \mathbb{F}$ with $\|a\| = 1$.*

(a) *The process $\langle B, a\rangle$ is a Γ-square-integrable $(\mathcal{B}_t)_{t\in T}$-martingale.*
(b) *Fix a limited $s \in T$. Then $\max_{t \in T_s} \langle B_t, a\rangle^2$ is S_Γ-integrable.*

Proof (a) By Lemma 11.1.2 (c), $\langle B_t, a\rangle$ is $\mathcal{B}_t$-measurable for each $t \in T$. By Lemma 11.1.3 and the fact that a can be extended to an ONB $(\mathfrak{e}_i)_{i\in\omega}$ of $\mathbb{F}$:

$$\mathbb{E}\langle B_t, a\rangle^2 = \mathbb{E}\left(\sum_{s \le t} \langle X_s, a\rangle\right)^2 = \mathbb{E}\sum_{s \le t} \langle X_s, a\rangle^2 = \sum_{s \le t} \frac{1}{H} = t < \infty \text{ in } ^*\mathbb{R}.$$

In order to prove the martingale property, fix $s, t \in T$ with $s < t$ and $A \times \mathbb{F}^{T \setminus T_s} \in \mathcal{B}_s$ with $A \in \mathcal{B}\left(\mathbb{F}^{T_s}\right)$. Then, by Lemma 11.1.3, we see that $\int_{A \times \mathbb{F}^{T\setminus T_s}} \langle B_t, a\rangle\, d\Gamma = \int_{A \times \mathbb{F}^{T\setminus T_s}} \langle B_s, a\rangle\, d\Gamma$.

(b) By Theorem 10.13.1, it suffices to show that the quadratic variation $[\langle B,a\rangle]_s$ of $\langle B,a\rangle$ at s is S_Γ-integrable. Because $\frac{Hs(Hs-1)}{2}$ is the number of pairs (r,t) in T_s with $r < t$, by Lemma 11.1.3, we obtain

$$\mathbb{E}[\langle B,a\rangle]_s^2 = \mathbb{E}\left(\sum_{r \in T_s} \langle X_r, a\rangle^2\right)^2$$

$$= \mathbb{E}\left(\sum_{r\in T_s}\langle X_r,a\rangle^4 + 2\sum_{1\le r<t\le s}\langle X_r,a\rangle^2\cdot\langle X_t,a\rangle^2\right)$$

$$= \frac{3sH}{H^2} + \frac{sH(sH-1)}{H^2} = s^2 + \frac{2s}{H} \text{ is limited.}$$

By Theorem 10.11.1, $[\langle B,a\rangle]_s$ is S_Γ-integrable. □

11.4 The S-continuity of the internal Brownian motion

Lemma 11.4.1 *Fix again an element $a \in \mathbb{F}$ with $\|a\| = 1$. The quadratic variation $[\langle B,a\rangle]$ of $\langle B,a\rangle$ is S-continuous $\widehat{\Gamma}$-a.s.*

Proof Fix $k \in \mathbb{N}$. In the preceding proof we have seen that $\mathbb{E}[\langle B,a\rangle]_k$ is limited. By Lemma 10.7.1 (a), there exists a set U_k^0 of $\widehat{\Gamma}$-measure 1 such that $[\langle B,a\rangle]_k(X)$ is limited for all $X \in U_k^0$.

Now we apply Josef Berger's trick [7]: Set $F(X,t) := H\langle X_t,a\rangle^2$ for all $X \in \Omega$ and all $t \in T_k$. Since $\int_{\Omega\times T_k} F^2 d\Gamma\otimes\nu$ is limited, by Theorem 10.11.1, $F \in SL^1(\Gamma\otimes\nu_k)$. By Theorem 10.12.1, there exists a set U_k^1 of $\widehat{\Gamma}$-measure 1 such that $F(X,\cdot) \in SL^1(\nu_k)$ for all $X \in U_k^1$. Fix $X \in U_k^1$ and $s,t \in T_k$ with $s<t$ and $s \approx t$ and set $N := \{j \in T_k \mid s < j \le t\}$. Since $\nu(N) \approx 0$, we obtain

$$[\langle B,a\rangle]_t(X) - [\langle B,a\rangle]_s(X) = \int_N F(X,\cdot)d\nu \approx 0.$$

It follows that $[\langle B,a\rangle](X)$ is S-continuous for all $X \in \bigcap_{k\in\mathbb{N}}\left(U_k^0 \cap U_k^1\right)$. □

The following result is due to Cutland [23]. Here it follows immediately from Proposition 11.3.1, Lemma 11.4.1 and Theorem 10.14.2.

Theorem 11.4.2 *For each $a \in \mathbb{F}$ with $\|a\| = 1$, $\langle B,a\rangle$ is S-continuous $\widehat{\Gamma}$-a.s.*

11.5 One-dimensional Brownian motion

In this section we will prove that for each $a \in \mathbb{F}$ with $\|a\| = 1$ the standard part of $\langle B,a\rangle$ is a one-dimensional Brownian motion. This result is due to Cutland [23]. By Theorem 11.4.2 and Proposition 9.7.1, the function $b^a : \Omega \times [0,\infty[\to \mathbb{R}$,

$$b^a_{{}^\circ t}(X) := \begin{cases} {}^\circ\langle B_t,a\rangle(X) & \text{if } t \mapsto \langle B_t,a\rangle(X) \text{ is } S\text{-continuous} \\ 0 & \text{otherwise,} \end{cases}$$

is well defined and continuous for all $X \in \Omega$. Moreover, $\widehat{\Gamma}$-a.s.:

$$b^a_{{}^\circ t} = {}^\circ\langle B_t, a\rangle \text{ for all limited } t \in T.$$

We shall now see that b^a is a Brownian motion on Ω, using the fact that for all $k \in \mathbb{N}$, $(e^{\lambda\langle B_s,a\rangle - \frac{\lambda^2}{2}\cdot s})_{s\in T_k}$ is an S_Γ-integrable $(\mathcal{B}_s)_{s\in T_k}$-martingale (see Proposition 11.2.1).

Theorem 11.5.1 (Cutland [23]) *For each $a \in \mathbb{F}$ with $\|a\| = 1$, the standard part b^a of $\langle B, a\rangle$ is a Brownian motion under $(\mathfrak{b}_t)_{t\in[0,\infty[}$ (recall that $(\mathfrak{b}_t)_{t\in[0,\infty[}$ is the standard part of the internal filtration $(\mathcal{B}_t)_{t\in T}$ (see Theorem 10.6.1)).*

Proof $b^a(X)$ is continuous for all $X \in \Omega$. Since $\mathbb{E}\left\langle X_{\frac{1}{H}}, a\right\rangle^2 \approx 0$, $\left\langle X_{\frac{1}{H}}, a\right\rangle \approx 0$ for $\widehat{\Gamma}$-almost all $X \in \Omega$ by Lemma 10.7.1 (b). Therefore, $b^a_0 = 0$ $\widehat{\Gamma}$-a.s. Fix $r \in [0, \infty[$ and $t \in T$ with $r \approx t$. Since for all $c \in \mathbb{R}$ we have $\widehat{\Gamma}$-a.s.

$$\{b^a_r \le c\} = \bigcap_{k\in\mathbb{N}} \left\{ \langle B_t, a\rangle \le c + \frac{1}{k} \right\} \in \mathfrak{b}_r,$$

b^a_r and therefore $e^{\lambda b^a_r - \frac{\lambda^2}{2} r}$ is $\mathfrak{b}_r$-measurable. Now we apply Theorem 3.5.2. Let $\lambda \in \mathbb{R}$, $r < q$ in $[0, \infty[$ and $B \in \mathfrak{b}_r$. There exists an $A \in \mathcal{B}_t$ such that $\widehat{\Gamma}(A \Delta B) = 0$. Let $s \in T$ with $s \approx q$. Since $e^{\lambda\langle B_t,a\rangle - \frac{\lambda^2}{2} t}$ and $e^{\lambda\langle B_s,a\rangle - \frac{\lambda^2}{2} s}$ are S_Γ-integrable liftings of $e^{\lambda b^a_r - \frac{\lambda^2}{2} r}$, $e^{\lambda b^a_q - \frac{\lambda^2}{2} q}$, respectively, we obtain, using Proposition 11.2.1 (a),

$$\int_B e^{\lambda b^a_q - \frac{\lambda^2}{2} q} d\widehat{\Gamma} \approx \int_A e^{\lambda\langle B_s,a\rangle - \frac{\lambda^2}{2} s} d\Gamma$$
$$= \int_A e^{\lambda\langle B_t,a\rangle - \frac{\lambda^2}{2} t} d\Gamma \approx \int_B e^{\lambda b^a_t - \frac{\lambda^2}{2} t} d\widehat{\Gamma}.$$

This proves that $(e^{\lambda b^a_r - \frac{\lambda^2}{2} r})_{r\in[0,\infty[}$ is a $(\mathfrak{b}_r)_{r\in[0,\infty[}$-martingale. By Theorem 3.5.2, b^a is a Brownian motion. □

11.6 Lévy's inequality

Our aim now is to construct from the internal $\mathbb{F}$-valued Brownian motion B a $\mathbb{B}$-valued Brownian motion $b_{\mathbb{B}}$ by taking the standard part of B. Here we need the S-continuity of B, which is an application of Lévy's inequality (see Ledoux and Talagrand [61]), a vector-valued reflection principle, tailor-made for our purposes. Let E be a *finite-dimensional subspace of $\mathbb{F}$ with ONB $(e_i)_{i\in\delta}$. For each $X = (X_{\frac{1}{H}}, \ldots, X_H)$ and $s \in T$ let x_s be the orthogonal projection of X_s to E, i.e.,

$$x_s = \sum_{i \in \delta} \langle X_s, \mathfrak{e}_i \rangle \, \mathfrak{e}_i.$$

Let $\mathfrak{E} := (\mathfrak{e}_i)_{i \in \omega}$ be an extension of $(\mathfrak{e}_i)_{i \in \delta}$ to an internal ONB of $\mathbb{F}$. For each internal $\varsigma : T \to \{1, -1\}$and each $s \in T$ define $S_{\varsigma,s} : \Omega \to E^{T_{s+\frac{1}{H}}}$ by

$$S_{\varsigma,s}(X) = \left(\left(\sum_{j \leq t} x_j \right)_{t \leq s}, \sum_{j \leq s} x_j + \sum_{j > s} \varsigma_j \cdot x_j \right).$$

Since

$$S_{\varsigma,s}\left(\sum_{i \in \omega} \alpha_j(i) \mathfrak{e}_i \right)_{j \in T} = \sum_{i \in \delta} \left(\left(\sum_{j \leq t} \alpha_j(i) \right)_{t \leq s}, \sum_{j \leq s} \alpha_j(i) + \sum_{j > s} \varsigma_j \alpha_j(i) \right) \mathfrak{e}_i,$$

the function $S_{\varsigma,s}$ corresponds to the function $S^{\mathfrak{E}}_{\varsigma,s} : {}^*\mathbb{R}^{\omega \cdot H^2} \to {}^*\mathbb{R}^{\delta \cdot (Hs+1)}$ with

$$S^{\mathfrak{E}}_{\varsigma,s} \left(y_j(i) \right)_{i \in \omega, j \leq H^2} := \left(\left(\sum_{j \leq k} y_j(i) \right)_{k \leq Hs}, \sum_{j \leq Hs} y_j(i) + \sum_{j > Hs} \varsigma_j y_j(i) \right)_{i \leq \delta}.$$

The following lemma follows easily from the fact that the density of Γ is an even function.

Lemma 11.6.1 *Fix two internal mappings $\varsigma, \rho : T \to \{1, -1\}$. Then, for all $s \in T$, $S_{\varsigma,s}$ has the same distribution as $S_{\rho,s}$, i.e., the image measures of Γ under $S_{\varsigma,s}$ and $S_{\rho,s}$ are the same.*

Proof Set $n := H \cdot s$. To apply Proposition 3.1.2, fix an internal sequence $\left(\lambda_j(i)\right)_{i \leq \delta, j \leq n+1}$ in ${}^*\mathbb{R}$. Then

$$\int_{{}^*\mathbb{R}^{\delta(n+1)}} e^{\sum_{i \leq \delta, j \leq n+1} \lambda_j(i) \cdot y_j(i)} d\Gamma_{S^{\mathfrak{E}}_{\varsigma,s}}(y_j(i)) = a \cdot b_\varsigma < \infty \text{ in } {}^*\mathbb{R},$$

where

$$a = \int_{{}^*\mathbb{R}^{\delta \cdot n}} e^{\sum_{i \leq \delta, j \leq n} \lambda_j(i) \cdot \left(\sum_{k \leq j} y_k(i) \right) + \lambda_{n+1}(i) \left(\sum_{j \leq n} y_j(i) \right)} d\gamma^{\delta \cdot n}_{\frac{1}{H}} \left(y_j(i) \right),$$

$$b_\varsigma = \int_{{}^*\mathbb{R}^{\delta \cdot (H^2 - n)}} e^{\sum_{i \leq \delta, j > n} \lambda_{n+1}(i) \cdot \varsigma_j \cdot y_j(i)} d\gamma^{\delta \cdot (H^2 - n)}_{\frac{1}{H}} \left(y_j(i) \right).$$

Now $b_\varsigma = b_\rho$, because of Fubini's theorem and since $\int_{{}^*\mathbb{R}} e^{\lambda y} \cdot e^{-\frac{H}{2} y^2} dy = \int_{{}^*\mathbb{R}} e^{-\lambda y} \cdot e^{-\frac{H}{2} y^2} dy$ for each $\lambda \in {}^*\mathbb{R}$. □

Theorem 11.6.2 (Lévy [63]) *For each $j \in \mathbb{N}$, $E \in \mathcal{E}(\mathbb{H})$, $\sigma \in T$ and for each $r \in {}^*\mathbb{R}$ we have*

$$\Gamma\left\{\max_{s\in T_\sigma} |\pi_{\mathbb{F}E}(B_s)|_j \geq r\right\} \leq 2\Gamma\left\{|\pi_{\mathbb{F}E}(B_\sigma)|_j \geq r\right\}.$$

Proof Note that $\pi_{\mathbb{F}E}(B_s(X)) = \sum_{j\leq s} x_j =: S_s(X)$. First define a stopping time $\tau : \Omega \to T_\sigma \cup \left\{\sigma + \frac{1}{H}\right\}$ by

$$\tau := \inf\left\{s \in T_\sigma \mid |S_s|_j \geq r\right\},$$

where $\inf\emptyset := \sigma + \frac{1}{H}$. We obtain

$$\Gamma\left\{\max_{s\in T_\sigma} |S_s|_j \geq r\right\} = \sum_{s\in T_\sigma} \Gamma\left\{|S_s|_j \geq r \text{ and } \tau = s\right\}.$$

Note that $|S_s|_j \geq r$ implies $|S_\sigma|_j \geq r$ or $|S_s - (S_\sigma - S_s)|_j \geq r$. Assume that this is not true. Then we obtain a contradiction as follows:

$$\begin{aligned}|2S_s|_j &\leq |S_s + (S_\sigma - S_s)|_j + |S_s - (S_\sigma - S_s)|_j \\ &= |S_\sigma|_j + |S_s - (S_\sigma - S_s)|_j < 2r, \text{ thus } |S_s|_j < r.\end{aligned}$$

Therefore,

$$\begin{aligned}\Gamma\left\{|S_s|_j \geq r \text{ and } \tau = s\right\} \leq \Gamma&\left\{|S_\sigma|_j \geq r \text{ and } \tau = s\right\} \\ &+ \Gamma\left\{|S_s - (S_\sigma - S_s)|_j \geq r \text{ and } \tau = s\right\}.\end{aligned}$$

In order to compute the second summand, we set

$$\varsigma : T \to \{1, -1\}, \quad \varsigma_i := \varsigma(i) := \begin{cases} 1 & \text{if } i \leq s \\ -1 & \text{if } s < i. \end{cases}$$

Recall that $x_i := \pi_{\mathbb{F}E}(X_i)$. Now Lemma 11.6.1 shows that

$$\begin{aligned}&\Gamma\left\{|S_s - (S_\sigma - S_s)|_j \geq r \text{ and } \tau = s\right\} \\ &= \Gamma\left\{X \mid \left|\sum_{i\leq s} x_i - \sum_{s<i\leq\sigma} x_i\right|_j \geq r \text{ and } \tau(X) = s\right\} \\ &= \Gamma\left\{X \mid \left|\sum_{i\leq s} x_i + \sum_{s<i\leq\sigma} \varsigma_i x_i\right|_j \geq r, \left|\sum_{i\leq s} x_i\right|_j \geq r, \forall t < s\left(\left|\sum_{i\leq t} x_t\right|_j < r\right)\right\}\end{aligned}$$

$$= \Gamma \left\{ X \mid \left| \sum_{t \in T_\sigma} x_t \right|_j \geq r, \left| \sum_{t \leq s} x_t \right|_j \geq r, \forall t < s \left(\left| \sum_{i \leq t} x_i \right|_j < r \right) \right\}$$

$$= \Gamma \left\{ |S_\sigma|_j \geq r \text{ and } \tau = s \right\}.$$

This proves that

$$\Gamma \left\{ \max_{s \in T_\sigma} |S_s|_j \geq r \right\} \leq 2 \sum_{s \in T_\sigma} \Gamma \left\{ |S_\sigma|_j \geq r \text{ and } \tau = s \right\} = 2\Gamma \left\{ |S_\sigma|_j \geq r \right\}.$$

□

11.7 The final construction

We shall convert B to a $\mathbb{B}$-valued Brownian motion $b_{\mathbb{B}} : \Omega \times [0, \infty[\to \mathbb{B}$ such that each continuous function from $[0, \infty[$ into $\mathbb{B}$ is a path of $b_{\mathbb{B}}$. Let $\mathcal{E}$ be the set of finite-dimensional subspaces of $\mathbb{H}$.

Since $\left(|\cdot|_j\right)_{j \in \mathbb{N}}$ is a sequence of measurable semi-norms, there exists a monotone increasing function $h \colon \mathbb{N} \to \mathcal{E}$ such that, for each $\sigma, j \in \mathbb{N}, j \leq \sigma$, and each $E \in \mathcal{E}$ with $E \perp h(\sigma)$,

$$\gamma_\sigma^E \left\{ x \in E \mid |x|_j \geq \frac{1}{2^\sigma} \right\} \leq \frac{1}{2^{\sigma+1}}.$$

By the transfer principle, for each $\sigma \in {}^*\mathbb{N}$, $j \leq \sigma$ and each $E \perp {}^*h(\sigma)$, $E \in {}^*\mathcal{E}$, $E \subseteq \mathbb{F}$

$$\gamma_\sigma^E \left\{ x \in E \mid |x|_j \geq \frac{1}{2^\sigma} \right\} \leq \frac{1}{2^{\sigma+1}}.$$

By Proposition 11.2.1 (d), we may replace the previous inequality by

$$\Gamma \left\{ |\pi_{\mathbb{F}E}(B_\sigma)|_j \geq \frac{1}{2^\sigma} \right\} \leq \frac{1}{2^{\sigma+1}}.$$

By Theorem 11.6.2,

$$\Gamma \left\{ \max_{s \in T_\sigma} |\pi_{\mathbb{F}E}(B_s)|_j \geq \frac{1}{2^\sigma} \right\} \leq \frac{1}{2^\sigma}.$$

Lemma 11.7.1 *There exists an ONB $(\mathfrak{e}_i)_{i \in \mathbb{N}}$ of $\mathbb{H}$ such that for all $\sigma \in \mathbb{N}$ there exists an $s_\sigma \in \mathbb{N}$ with $h(\sigma) \subseteq \operatorname{span}\left\{\mathfrak{e}_1, \ldots, \mathfrak{e}_{s_\sigma}\right\}$, thus, for all $j \leq \sigma$,*

$$\Gamma \left\{ \max_{t \in T_\sigma} \left| \sum_{s_\sigma < i \in \omega} \langle B_t, \mathfrak{e}_i \rangle \cdot \mathfrak{e}_i \right|_j \geq \frac{1}{2^\sigma} \right\} \leq \frac{1}{2^\sigma},$$

where $(\mathfrak{e}_i)_{i\in\omega}$ *is an extension of* $(\mathfrak{e}_i)_{i\in\mathbb{N}}$ *to an internal ONB of* $\mathbb{F}$.

Proof We construct $(\mathfrak{e}_i)_{i\in\mathbb{N}}$ by recursion. Fix an ONB $(\mathfrak{b}_i)_{i\in\mathbb{N}}$ of $\mathbb{H}$. For $\sigma=1$ let $(\mathfrak{e}_i)_{i\le s_1}$ be an ONB of span $(h_1\cup\{b_1\})$. Assume that s_σ is already defined with the desired property. Let $(\mathfrak{e}_i)_{i\le s_{\sigma+1}}$ be an extension of $(\mathfrak{e}_i)_{i\le s_\sigma}$ to an ONB of span $(h(\sigma+1)\cup\{\mathfrak{b}_1,\ldots,\mathfrak{b}_{\sigma+1}\})$. Finally, we obtain an ONB $(\mathfrak{e}_i)_{i\in\mathbb{N}}$ of $\mathbb{H}$ and the assertion of the lemma is true. By Lemma 11.7.1 and Proposition 9.8.1, there exists an internal extension $(\mathfrak{e}_i)_{i\in\omega}$ of $(\mathfrak{e}_i)_{i\in\mathbb{N}}$ to an ONB of $\mathbb{F}$. □

Define for all $\sigma\in\mathbb{N}$

$$A_\sigma:=\bigcup_{j\le\sigma}\left\{\max_{s\in T_\sigma}\left|\sum_{s_\sigma<i\in\omega}\langle B_s,\mathfrak{e}_i\rangle\cdot\mathfrak{e}_i\right|_j\ge\frac{1}{2^\sigma}\right\},$$

and

$$U_0^-:=\bigcup_{l\in\mathbb{N}}\bigcap_{\sigma\ge l}\Omega\setminus A_\sigma.$$

By Theorem 11.4.2, for all $a\in\mathbb{F}$ of norm 1 there exists a set $U_a\subseteq\Omega$ of $\widehat{\Gamma}$-measure 1 such that $\langle B(X,\cdot),a\rangle$ is S-continuous for all $X\in U_a$. Set

$$U_0:=U_0^-\cap\bigcap_{i\in\mathbb{N}}U_{\mathfrak{e}_i}.$$

Let us write $\alpha\approx\beta$ if $\alpha,\beta\in{}^*\mathbb{B}$ are infinitely close in the topology of $\mathbb{B}$.

Lemma 11.7.2 $\widehat{\Gamma}(U_0)=1$.

Proof Since $\sum_{\sigma=1}^\infty\widehat{\Gamma}(A_\sigma)\le\sum_{\sigma=1}^\infty\frac{\sigma}{2^\sigma}<\infty$, we have $\widehat{\Gamma}\left(U_0^-\right)=1$ by the Borel–Cantelli lemma. It follows that $\widehat{\Gamma}(U_0)=1$. □

Corollary 11.7.3 *Fix* $X\in U_0$. *Then* $B(X,\cdot)$ *is pre-S-continuous in the topology of* $\mathbb{B}$.

Proof Fix limited $s,t\in T$ with $s\approx t$ and $j\in\mathbb{N}$. Since $X\in U_0^-$, there is an $l\in\mathbb{N}$ such that $X\notin A_\sigma$ for all $\sigma\ge l$. We may assume that $s,t,j\le l$. Then for all $\sigma\ge l$

$$|B_s(X)-B_t(X)|_j\le A+B+C$$

with

$$A=\left|B_s(X)-\sum_{i=1}^{s_\sigma}\langle B_s(X),\mathfrak{e}_i\rangle\,\mathfrak{e}_i\right|_j<\frac{1}{2^\sigma},$$

$$B = \left| \sum_{i=1}^{s_\sigma} \langle B_s(X), \mathfrak{e}_i \rangle \mathfrak{e}_i - \sum_{i=1}^{s_\sigma} \langle B_t(X), \mathfrak{e}_i \rangle \mathfrak{e}_i \right|_j \approx 0,$$

$$C = \left| \sum_{i=1}^{s_\sigma} \langle B_t(X), \mathfrak{e}_i \rangle \mathfrak{e}_i - B_t(X) \right|_j < \frac{1}{2^\sigma}.$$

It follows that $B_s(X) \approx B_t(X)$. □

In order to prove that there exists a set U_1 of $\widehat{\Gamma}$-measure 1 such that $B_t(X)$ is nearstandard in $\mathbb{B}$ for all limited $t \in T$ and all $X \in U_1$, we now use the fact that the metric d is measurable (see Corollary 10.2.2). Fix a positive $t \in \mathbb{Q}$. Then $t \in T$. There exists an increasing function $f : \mathbb{N} \to \mathcal{E}$ such that for all $E \in \mathcal{E}$, $E \perp f(m)$,

$$\gamma_t^E \left\{ x \in E \mid d(x,0) \geq \frac{1}{2^m} \right\} \leq \frac{1}{2^m}.$$

By transfer, $\gamma_t^E \left\{ x \in E \mid d(x,0) \geq \frac{1}{2^m} \right\} \leq \frac{1}{2^m}$ for all $m \in {}^*\mathbb{N}$ and $E \in {}^*\mathcal{E}$, $E \perp {}^*f(m)$. The preceding inequality may be replaced by the inequality $\Gamma \left\{ d(\pi_{\mathbb{F}E}(B_t), 0) \geq \frac{1}{2^m} \right\} \leq \frac{1}{2^m}$. According to Lemma 11.7.1, we obtain the following.

Lemma 11.7.4 *There exists an ONB* $(\mathfrak{b}_i)_{i \in \mathbb{N}}$ *of* $\mathbb{H}$ *and a mapping* $r : \mathbb{N} \to \mathbb{N}$ *such that for all* $\sigma \in \mathbb{N}$

$$\Gamma \left\{ d\Big(\sum_{r_\sigma < i \in \omega} \langle B_t, \mathfrak{b}_i \rangle \mathfrak{b}_i, 0 \Big) \geq \frac{1}{2^\sigma} \right\} \leq \frac{1}{2^\sigma},$$

where $(\mathfrak{b}_i)_{i \in \omega}$ *is an extension of* $(\mathfrak{b}_i)_{i \in \mathbb{N}}$ *to an internal ONB of* $\mathbb{F}$.

For all $\sigma \in \mathbb{N}$ set

$$A_\sigma^t := \left\{ d\left(\sum_{r_\sigma < i \in \omega} \langle B_t, \mathfrak{b}_i \rangle \mathfrak{b}_i, 0 \right) \geq \frac{1}{2^\sigma} \right\},$$

and

$$U_1^{t,-} := \bigcup_{l \in \mathbb{N}} \bigcap_{\sigma \geq l} \Omega \setminus A_\sigma^t.$$

Recall the definition of U_a for $a \in \mathbb{F}$ with norm 1. Set

$$U_1^t := U_1^{t,-} \cap \bigcap_{i \in \mathbb{N}} U_{\mathfrak{b}_i}.$$

In analogy to Lemma 11.7.2 we obtain the following.

Lemma 11.7.5 $\widehat{\Gamma}(U_1^t) = 1$.

Moreover, for each $X \in U_1^t$ there exists an $l \in \mathbb{N}$ such that $X \notin A_\sigma^t$ for all $\sigma \geq l$. Therefore, we obtain for all $\sigma, \rho \geq l$

$$d\left(\sum_{i \leq r_\sigma} \langle B_t(X), \mathfrak{b}_i \rangle \mathfrak{b}_i, \sum_{i \leq r_\rho} \langle B_t(X), \mathfrak{b}_i \rangle \mathfrak{b}_i\right)$$

$$\leq d\left(\sum_{i \leq r_\sigma} \langle B_t(X), \mathfrak{b}_i \rangle \mathfrak{b}_i, B_t(X)\right) + d\left(B_t(X), \sum_{i \leq r_\rho} \langle B_t(X), \mathfrak{b}_i \rangle \mathfrak{b}_i\right)$$

$$= d\left(\sum_{r_\sigma < i \in \omega} \langle B_t(X), \mathfrak{b}_i \rangle \mathfrak{b}_i, 0\right) + d\left(0, \sum_{r_\rho < i \in \omega} \langle B_t(X), \mathfrak{b}_i \rangle \mathfrak{b}_i\right) < \frac{1}{2^\sigma} + \frac{1}{2^\rho}.$$

It follows that $\left({}^\circ \sum_{i \leq r_\sigma} \langle B_t(X), \mathfrak{b}_i \rangle \mathfrak{b}_i\right)_{\sigma \in \mathbb{N}}$ is a Cauchy sequence and its limit is infinitely close to $B_t(X)$. Thus $B_t(X)$ is nearstandard. Now set

$$U := \left(\bigcap_{t \in]0,\infty[\cap \mathbb{Q}} U_1^t\right) \cap U_0.$$

Note that $\widehat{\Gamma}(U) = 1$. The following theorems are the main results in this section.

Theorem 11.7.6 *Fix $X \in U$. The $\mathbb{F}$-valued function $B(X, \cdot)$ is S-continuous in the topology of $\mathbb{B}$.*

Proof Fix $X \in U$ and a limited $s \in T$. By Corollary 11.7.3, it remains to prove that $B(X,s)$ is nearstandard: Take a sequence (t_n) of positive rationals with $\lim_{n\to\infty} t_n = {}^\circ s$. Let $(t_n)_{n \in {}^*\mathbb{N}}$ be an internal extension of $(t_n)_{n\in\mathbb{N}}$ in T. Then there exists an unlimited $M_\infty \in {}^*\mathbb{N}$ such that $t_n \approx s$ for all unlimited $n \leq M_\infty$. We will show that $\left({}^\circ B_{t_n}(X)\right)_{n\in\mathbb{N}}$ is a Cauchy sequence. Assume that this is not true. Then there exists an $\varepsilon > 0$ such that $d(B_{t_n}(X), B_{t_m}(X)) \geq \varepsilon$ for arbitrary large n, m. By Proposition 8.4.1, there are unlimited $M, N \leq M_\infty$ with $d(B_{t_N}(X), B_{t_M}(X)) \geq \varepsilon$. Since $t_N \approx t_M$, we have a contradiction to Corollary 11.7.3. In the same way one can prove that $\lim_{n\to\infty} {}^\circ B_{t_n}(X) \approx_{\mathbb{B}} B_s(X)$. This proves that $B_s(X)$ is nearstandard. □

Using this result, B can be converted into a continuous process $b_{\mathbb{B}} : \Omega \times [0,\infty[\to \mathbb{B}$ setting

$$b_{\mathbb{B}}(X, {}^\circ t) := \begin{cases} {}^\circ B_t(X) & \text{if } B(X) \text{ is } S\text{-continuous} \\ 0 & \text{otherwise.} \end{cases}$$

The standard part $^\circ B_t$ of B_t is the standard part with respect to the topology on the Fréchet space $\mathbb{B}$. Finally, we have the following.

Theorem 11.7.7 *The continuous process $b_{\mathbb{B}}$ is a $\mathbb{B}$-valued Brownian motion under $(\mathfrak{b}_t)_{t\in[0,\infty[}$. Moreover, $X \mapsto b_{\mathbb{B}}(X,\cdot)$ is Loeb–Borel-measurable.*

Proof Fix $\varphi,\psi \in \mathbb{B}'$ with $\|\varphi\| = \|\psi\| = 1$ and $\varphi \perp \psi$. Since $^*[\mathbb{H}] \subseteq \mathbb{F}$, we have $^*\varphi, {}^*\psi \in \mathbb{F}$. We identify $^*\varphi$, $^*\psi$ with φ and ψ. Set $b := b_{\mathbb{B}}$. By Theorem 11.5.1, $(\varphi \circ b)$is a one-dimensional Brownian motion under $(\mathfrak{b}_t)_{t\in[0,\infty[}$. To prove the independence of $(\varphi \circ b, \psi \circ b)$, fix $r \in [0,\infty[$ and $t \in T$ with $t \approx r$, and $\lambda, \rho \in \mathbb{R}$. Since $\mathbb{E}e^{\lambda\varphi\circ B_t} = e^{\frac{1}{2}\lambda^2 t}$ is limited, $\mathbb{E}e^{\lambda\varphi\circ B_t} \in SL^p(\Gamma)$for all $p \in [1,\infty[$. Moreover, $e^{\lambda\varphi\circ B_t}$ is a lifting of $e^{\lambda\varphi\circ b_r}$. Therefore,

$$\begin{aligned}
\mathbb{E}\left(e^{\lambda\cdot\varphi\circ b_r}e^{\rho\cdot\psi\circ b_r}\right) &\approx \mathbb{E}\left(e^{\lambda\cdot\varphi\circ B_t}e^{\rho\cdot\psi\circ B_t}\right) = \mathbb{E}e^{\sum_{s\le t}(\lambda\cdot\langle\varphi,X_s\rangle+\rho\cdot\langle\psi,X_s\rangle)} \\
&= \int_{^*\mathbb{R}^{2Ht}} e^{\sum_{s\le t}\left(\lambda\cdot x_s+\rho\cdot y_s-\frac{H}{2}\left(x_s^2+y_s^2\right)\right)} d\,(x_s,y_s)_{s\le t}\left(\sqrt{\frac{H}{2\pi}}\right)^{2Ht} \\
&= \alpha\cdot\beta
\end{aligned}$$

with

$$\alpha = \int_{^*\mathbb{R}^{Ht}} e^{\sum_{s\le t}\left(\lambda\cdot x_s-\frac{H}{2}x_s^2\right)} d\,(x_s)_{s\le t}\sqrt{\frac{H}{2\pi}}^{Ht} = \mathbb{E}e^{\lambda\cdot\varphi\circ B_t} \approx \mathbb{E}e^{\lambda\cdot\varphi\circ b_r},$$

$$\beta = \int_{^*\mathbb{R}^{Ht}} e^{\sum_{s\le t}\left(\rho\cdot x_s-\frac{H}{2}x_s^2\right)} d\,(x_s)_{s\le t}\sqrt{\frac{H}{2\pi}}^{Ht} = \mathbb{E}e^{\rho\cdot\psi\circ B_t} \approx \mathbb{E}e^{\rho\cdot\psi\circ b_r}.$$

By Proposition 3.3.1, $(\varphi \circ b_r, \psi \circ b_r)$ is independent.

The second assertion follows from the fact that $b_{\mathbb{B}}$ is the standard part of the internal function B, and is left to the reader. □

11.8 The Wiener space

Using the Brownian motion $b_{\mathbb{B}}$, we will define the Wiener measure on the space $C_{\mathbb{B}}$ of $\mathbb{B}$-valued continuous functions, defined on $[0,\infty[$.

First of all let us prove in detail that each continuous function is a path of our condinuous Brownian motion $b_{\mathbb{B}}$. Recall the definition of $(\lfloor\cdot\rfloor_\sigma)_{\sigma\in\mathbb{N}}$ in Section 4.5. The following lemma will be used.

Lemma 11.8.1 *Fix $f \in C_{\mathbb{B}}$ and $m \in \mathbb{N}$. Then there exists a continuous function $g := g_m : [0,\infty[\to \mathbb{H}$ such that $\lfloor g - f\rfloor_m < \frac{1}{m}$. It follows that $\lim_{m\to\infty} g_m = f$ in the topology of $C_{\mathbb{B}}$.*

Moreover, $^*g$ is a mapping from $^*[0,\infty[$ into $\mathbb{F}$.

Proof There exists $k := k_m \in \mathbb{N}$ such that $|f(x)-f(y)|_j < \frac{1}{5m}$ for all $x,y \in [0,m]$ with $|x-y| \leq \frac{1}{k}$ and all $j = 1,\ldots,m$. Let $M \in \mathbb{N}$ with $\frac{M-1}{k} < m \leq \frac{M}{k}$. Since $\mathbb{H}$ is dense in $\mathbb{B}$, there exist $a_0,\ldots,a_M \in \mathbb{H}$ such that $\left|f\left(\frac{i}{k}\right)-a_i\right)\Big|_j < \frac{1}{5m}$ for all $i = 0,\ldots,M$ and all $j = 1,\ldots,m$. Define for all $i = 1,\ldots,M$ and all $\lambda \in [0,1]$

$$g\left(\frac{i-1}{k}+\frac{\lambda}{k}\right) := a_{i-1}+\lambda\,(a_i - a_{i-1}).$$

Set $g(x) := a_M$ for all $x > \frac{M}{k}$. Obviously, g is a continuous mapping from $[0,\infty[$ into $\mathbb{H}$. Moreover, since $^*[\mathbb{H}] \subseteq \mathbb{F}$, $^*a_i \in \mathbb{F}$ thus $^*a_{i-1}+\lambda\,(^*a_i - ^*a_{i-1}) \in \mathbb{F}$ for all $\lambda \in {}^*\mathbb{R}$. This proves that the range of $^*g$ is a subset of $\mathbb{F}$. Now fix $x \in [0,m]$. Then there is an $i = 1,\ldots,M$ and $\lambda \in [0,1]$ with $x = \frac{i-1}{k}+\frac{\lambda}{k}$. We obtain for all $j = 1,\ldots,m$

$$\begin{aligned}
&|g(x)-f(x)|_j \\
&\leq |a_{i-1}+\lambda\,(a_i - a_{i-1}) - a_{i-1}|_j + \left|a_{i-1}-f\left(\frac{i-1}{k}\right)\right|_j + \left|f\left(\frac{i-1}{k}\right)-f(x)\right|_j \\
&\leq |a_i - a_{i-1}|_j + \left|a_{i-1}-f\left(\frac{i-1}{k}\right)\right|_j + \left|f\left(\frac{i-1}{k}\right)-f(x)\right|_j.
\end{aligned}$$

Now we estimate

$$|a_i - a_{i-1}|_j \leq \left|a_i - f\left(\frac{i}{k}\right)\right|_j + \left|f\left(\frac{i}{k}\right)-f\left(\frac{i-1}{k}\right)\right|_j + \left|f\left(\frac{i-1}{k}\right)-a_{i-1}\right|_j.$$

It follows that $|g_m(x)-f(x)|_j < \frac{1}{m}$ for all $m \in \mathbb{N}$, $j \leq m$ and $x \in [0,m]$, thus g_m converges to f in the topology of $C_{\mathbb{B}}$. □

Proposition 11.8.2 *Fix a continuous $f : [0,\infty[\to \mathbb{B}$. Then there exists an $X \in \Omega$ such that $B(X)$ is S-continuous and such that for all $r \in [0,\infty[$*

$$f(r) = b_{\mathbb{B}}(X,r).$$

Proof There exists a sequence $(g_m)_{m\in\mathbb{N}}$ of continuous functions $g_m : [0,\infty[\to \mathbb{H}$, converging to f, according to Lemma 11.8.1. Define for $i > 1$

$$X_{m,\frac{i}{H}} := {}^*g_m\left(\frac{i}{H}\right) - {}^*g_m\left(\frac{i-1}{H}\right) \text{ and } X_{m,\frac{1}{H}} := {}^*g_m\left(\frac{1}{H}\right).$$

Then $B_t(X_m) = {}^*g_m(t)$ for all $t \in T$. Since ${}^*g_m$ is S-continuous, $B(X_m)$ is S-continuous. Moreover, we have for all limited $t \in T$

$$b_{\mathbb{B}}(X_m, {}^\circ t) = {}^\circ B(X_m, t) = {}^\circ\left({}^*g_m(t)\right) = g_m({}^\circ t).$$

Therefore and by the construction of g_m, we obtain for all $m \in \mathbb{N}$, $t \in T_m$ and $j \leq m$

$$\left|B(X_m, t) - {}^*f(t)\right|_j < \frac{1}{m}.$$

By Proposition 8.4.1 (b), there exists an unlimited $M \in {}^*\mathbb{N}$ such that the preceding inequality is true for M instead of $m$and for all $j \leq M$ and $t \in T_M$. This proves that $B(X_M, \cdot)$is S-continuous and$f = b_{\mathbb{B}}(X_M, \cdot)$. □

It follows that the mapping $\kappa : \Omega \to C_{\mathbb{B}}, X \mapsto b_{\mathbb{B}}(X, \cdot)$ is surjective. Moreover, it is Loeb–Borel-measurable. Let $\mathcal{W}_{C_{\mathbb{B}},0}$ be the σ-algebra generated by κ, i.e., $\mathcal{W}_{C_{\mathbb{B}},0} = \left\{\kappa^{-1}[B] \mid B \in \mathcal{B}(C_{\mathbb{B}})\right\}$. Set

$$\mathcal{W}_{C_{\mathbb{B}}} := \mathcal{W}_{C_{\mathbb{B}},0} \vee \mathcal{N}_{\widehat{\Gamma}}.$$

The image measure $W_{C_{\mathbb{B}}}$ of $\widehat{\Gamma}$ by κ is called the **Wiener measure** on the σ-algebra

$$\overline{\mathcal{B}(C_{\mathbb{B}})} := \left\{D \subseteq C_{\mathbb{B}} \mid \kappa^{-1}[D] \in \mathcal{W}_{C_{\mathbb{B}}}\right\} \supseteq \mathcal{B}(C_{\mathbb{B}}).$$

It is interesting that, although the Brownian motion $b_{\mathbb{B}}$ starts in 0 $\widehat{\Gamma}$-a.s., all continuous functions, and those that don't start at 0, appear as a path of $b_{\mathbb{B}}$. This proves the following.

Corollary 11.8.3 *The set of continuous functions $f : [0, \infty[\to \mathbb{B}$ such that $f(0) \neq 0$ is a $W_{C_{\mathbb{B}}}$-nullset.*

The following result is also a consequence of Proposition 11.8.2.

Corollary 11.8.4 *For each $r \in [0, \infty[$ the mapping*

$$\kappa_r : \Omega \to \mathbb{B}, X \mapsto b_{\mathbb{B}}(X, r) \in \mathbb{B}$$

is surjective.

Now we obtain Gross' theorem (Theorem 4.3.8). Fix $\sigma \in]0, \infty[$ and let $\mathcal{W}_{\sigma,\mathbb{B}}$ be the σ-algebra on Ω, generated by $b_{\mathbb{B}}(\cdot, \sigma) : \Omega \to \mathbb{B}$, augmented by the $\widehat{\Gamma}$-nullsets.

Corollary 11.8.5 *The image measure of $\widehat{\Gamma}$ by the $\mathbb{B}$-valued random variable*

$$X \mapsto b_{\mathbb{B}}(X, \sigma)$$

is a centred Gaussian measure of variance σ on

$$\overline{\mathcal{B}(\mathbb{B})} := \left\{ D \subseteq \mathbb{B} \mid (b_{\mathbb{B}}(\cdot,\sigma))^{-1}[D] \in \mathcal{W}_{\sigma,\mathbb{B}} \right\} \supseteq \mathcal{B}(\mathbb{B}).$$

Restricted to $\mathcal{B}(\mathbb{B})$, it is identical to the measure γ_σ constructed in Section 4.3 from the Gaussian measure of variance σ on the cylinder sets of $\mathbb{H}$.

A Loeb measure approach to Gross' theorem can be also found in Lindstrøm [65].

We shall now see that $\mathcal{W}_{C_{\mathbb{B}}}$ does not depend on the abstract Wiener-Fréchet space $\mathbb{B}$ over $\mathbb{H}$; $\mathcal{W}_{C_{\mathbb{B}}}$ only depends on $\mathbb{H}$.

Proposition 11.8.6 *$\mathcal{W}_{C_{\mathbb{B}}}$ is generated by the random variables $^\circ\langle a, B_t\rangle$ and the $\widehat{\Gamma}$-nullsets, where $t \in T$ is limited and $a \in \mathbb{H}$. We identify a with $^*a \in \mathbb{F}$.*

Proof By Proposition 4.5.1, $\mathcal{W}_{C_{\mathbb{B}}}$ is generated by the functions $\varphi \circ b_{\mathbb{B}}(\cdot, r)$ and the $\widehat{\Gamma}$-nullsets, where $r \in [0,\infty[$ and $\varphi \in \mathbb{B}'$. By Proposition 4.3.6 (a), for each $a \in \mathbb{H}$ there exists a sequence (φ_k) in $\mathbb{B}'$ with $\lim_{k\to\infty} \|\varphi_k - a\|_{\mathbb{H}} = 0$. Now we obtain for $t \in T$ and $r \in [0,\infty[$ with $r \approx t$

$$\begin{aligned}
&\lim_{k\to\infty} \mathbb{E}\,(^\circ\langle a, B_t\rangle - \varphi_k \circ b_{\mathbb{B}}(\cdot, r))^2 \\
&\quad = \lim_{k\to\infty} \mathbb{E}\,(^\circ\langle a - \varphi_k, B_t\rangle)^2 = \lim_{k\to\infty} {}^\circ\mathbb{E}\langle a - \varphi_k, B_t\rangle)^2 \\
&\quad = \lim_{k\to\infty} {}^\circ\mathbb{E}\left(\sum_{s\le t} (a-\varphi_k)(X_s)\right)^2 = \lim_{k\to\infty} {}^\circ\sum_{s\le t}\sum_{i\in\omega} (a-\varphi_k)^2(\mathfrak{e}_i)\frac{1}{H} \\
&\quad = \lim_{k\to\infty} {}^\circ\left(t \cdot \|a - \varphi_k\|_{\mathbb{F}}^2\right) = \lim_{k\to\infty} {}^\circ t \cdot \|a - \varphi_k\|_{\mathbb{H}}^2 = 0.
\end{aligned}$$

It follows that $^\circ\langle a, B_t\rangle$ is $\mathcal{W}_{C_{\mathbb{B}}}$-measurable for each limited $t \in T$ and $a \in \mathbb{H}$, which proves the result. □

Therefore, we can define, for a fixed separable Hilbert space $\mathbb{H}$ and for each abstract Wiener–Fréchet space $\mathbb{B}$ over $\mathbb{H}$,

$$\mathcal{W}_{C_{\mathbb{H}}} := \mathcal{W}_{C_{\mathbb{B}}}.$$

From the proof of Proposition 11.2.1 part (c) we obtain the following.

Corollary 11.8.7 *Define $N : \mathbb{H} \to L^2\left(\Omega, \mathcal{W}_{C_{\mathbb{H}}}, \widehat{\Gamma}\right)$ by setting*

$$a \mapsto {}^\circ\langle a, B_1\rangle_{\mathbb{F}}.$$

Then $\mathbb{E}\,(N(a)\cdot N(b)) = \langle a,b\rangle_{\mathbb{H}}$ for all $a, b \in \mathbb{H}$ with $\|a\| = \|b\| = 1$.

Fix $p \in [1,\infty[$. Since $\kappa : \Omega \to C_{\mathbb{B}}$ is surjective and $\mathcal{W}_{C_{\mathbb{H}}}$ is – up to $\widehat{\Gamma}$-nullsets – the smallest σ-algebra for which κ is measurable, we may identify the L^p-spaces (see Corollary 5.6.4)

$$L^p(W_{C_{\mathbb{B}}}) := L^p(C_{\mathbb{B}}, \mathcal{B}(C_{\mathbb{B}}), W_{C_{\mathbb{B}}}) \quad \text{and} \quad L^p_{\mathcal{W}_{C_{\mathbb{H}}}}(\widehat{\Gamma}) := L^p(\Omega, \mathcal{W}_{C_{\mathbb{H}}}, \widehat{\Gamma}),$$

because $S(\varphi)(X) := \varphi(b_{\mathbb{B}}(X,\cdot))$ defines a **canonical**, i.e., basis-independent, isometric isomorphism S from $L^p(W_{C_{\mathbb{B}}})$ onto $L^p_{\mathcal{W}_{C_{\mathbb{H}}}}(\widehat{\Gamma})$. In the same way we may identify the following three product L^p-spaces:

$$L^p(W_{C_{\mathbb{B}}} \otimes \lambda) := L^p(C_{\mathbb{B}} \times [0,\infty[, \mathcal{B}(C_{\mathbb{B}}) \otimes \mathrm{Leb}[0,\infty[, W_{C_{\mathbb{B}}} \otimes \lambda),$$

$$L^p_{\mathcal{B}(C_{\mathbb{B}}) \otimes \mathcal{L}^1}(W_{C_{\mathbb{B}}} \otimes \widehat{\nu}) := L^p(C_{\mathbb{B}} \times T, \mathcal{B}(C_{\mathbb{B}}) \otimes \mathcal{L}^1, W_{C_{\mathbb{B}}} \otimes \widehat{\nu})$$

and

$$L^p_{\mathcal{W}_{C_{\mathbb{H}}} \otimes \mathcal{L}^1}(\widehat{\Gamma \otimes \nu}) := L^p(\Omega \times T, (\mathcal{W}_{C_{\mathbb{H}}} \otimes \mathcal{L}^1) \vee \mathcal{N}_{\widehat{\Gamma \otimes \nu}}, \widehat{\Gamma \otimes \nu}).$$

The last space $L^p_{\mathcal{W}_{C_{\mathbb{H}}} \otimes \mathcal{L}^1}(\widehat{\Gamma \otimes \nu})$ is the simplest one of these three spaces, because Ω is a finite-dimensional Euclidean space in the sense of our poly-saturated model and T is a finite set in this sense. It follows that analysis on the Wiener space is similar to finite-dimensional analysis, which is the topic of these notes.

Exercises

11.1 Prove Lemma 11.1.3.

11.2 Prove Lemma 11.7.5.

11.3 Prove the second assertion in Theorem 11.7.7.

Now we construct Anderson's Brownian motion [4]: Let ζ^1 be the internal Borel probability measure on ${}^*\mathbb{R}$, defined by

$$\zeta^1(D) := \frac{\left| D \cap \left\{ -\frac{1}{\sqrt{H}}, \frac{1}{\sqrt{H}} \right\} \right|}{2}$$

and let ζ be its product on ${}^*\mathbb{R}^T$. Use the notation in this chapter, but replace Γ with ζ and $\mathbb{F}$ with ${}^*\mathbb{R}$.

11.4 Prove that $\mathbb{E}_\zeta \max_{t \in T_\sigma} B_t^2$ is limited for all $\sigma \in \mathbb{N}$.

11.5 Prove that for all $\lambda \in \mathbb{R}$, limited $t,s \in T, t < s$, and $E \in \mathcal{B}_t$,

$$\int_E e^{\lambda B_s - \frac{\lambda^2 s}{2}} d\zeta \approx \int_E e^{\lambda B_t - \frac{\lambda^2 t}{2}} d\zeta.$$

11.6 Prove that B is S-continuous $\widehat{\zeta}$-a.s.

11.7 Define $a : \Lambda \times [0, \infty[\to \mathbb{R}$ by setting for all limited $t \in T$,

$$a(X, {}^\circ t) := \begin{cases} {}^\circ B(X, t) & \text{if } B(X, \cdot) \text{ is } S\text{-continuous} \\ 0, & \text{otherwise.} \end{cases}$$

Prove that a is a one-dimensional Brownian motion under the standard part of $(\mathcal{B}_s)_{s \in T}$ for $\widehat{\zeta}$.

12

The Itô integral for infinite-dimensional Brownian motion

Let us fix an abstract Wiener–Fréchet space $\mathbb{B}$ over $\mathbb{H}$ and an internal representation $\mathbb{F}$ of $\mathbb{H}$, according to Theorem 9.8.1. The scalar products in $\mathbb{F}$ and $\mathbb{H}$ are both denoted by $\langle \cdot, \cdot \rangle$ if there is no risk of confusion.

In this chapter we construct, following [84], the Itô integral as a continuous square-integrable martingale. The integrands are $\mathbb{H}$-valued square-integrable adapted processes and the integrator is the Brownian motion $b := b_{\mathbb{B}}$, which was constructed in the previous chapter. We introduce the integral as the standard part of an internal Riemann–Stieltjes integral according to Anderson's [4] construction of the Itô integral for finite-dimensional Brownian motion.

A standard approach to vector-valued stochastic integration can be found in the work of Duncan [36] and Duncan and Varayia [35], who constructed the integral as a continuous version of a certain stochastic process. Here we construct the integral path-by-path as a continuous process. Lindstrøm [68] has also studied stochastic vector integrals, where the integrator is a Brownian motion with values in an abstract Wiener space. This Brownian motion is the standard part of a hyperfinite random walk. However, to be able to differentiate, we use a smooth approach to Brownian motion in analogy to Cutland and Ng's work [24] for the classical Wiener space $C_{\mathbb{R}} := C([0,1], \mathbb{R})$. A disadvantage of our approach will become apparent in the more complicated treatment of the quadratic variation.

12.1 The S-continuity of the internal Itô integral

In order to define the Itô integral with respect to the Brownian motion $b_{\mathbb{B}}$ as a continuous process, we first introduce the internal Riemann–Stieltjes integral with respect to the internal Brownian motion B and then prove that this integral is S-continuous, provided that the integrand is square-integrable and adapted.

Therefore, by Proposition 9.7.1, it can be converted to a continuous process. We use the notation of the preceding chapter.

For the moment let $\Phi : \Omega \times T \to \mathbb{F}$ be internal. The **internal stochastic integral** $\int \Phi \Delta B : \Omega \times T \to {}^*\mathbb{R}$ is defined by setting

$$\left(\int \Phi \Delta B\right)(X,\sigma) := \sum_{s \in T_\sigma} \Phi_s(X)(X_s) = \sum_{s \in T_\sigma} \langle \Phi_s(X), X_s \rangle_{\mathbb{F}}.$$

Note that

$$\int \Phi \Delta B(X,\sigma) = \sum_{(s,i) \in T_\sigma \times \omega} \langle \Phi_s(X), \mathfrak{e}_i \rangle_{\mathbb{F}} \cdot \langle X_s, \mathfrak{e}_i \rangle_{\mathbb{F}},$$

where $(\mathfrak{e}_i)_{i \in \omega}$ is an internal ONB of $\mathbb{F}$. The notion internal 'stochastic integral' is too sophisticated. It is nothing but the Riemann–Stieltjes integral with respect to the path-by-path defined signed measure on the internal power set $\mathcal{P}(T)$ of T, setting

$$B(X)(]s,t]) := B(X,t) - B(X,s) \text{ for } X \in \Omega.$$

Since T is *finite, this measure is of bounded variation.

We shall see that for adapted and S-square-integrable Φ the standard part ${}^\circ\int \Phi \Delta B$ of $\int \Phi \Delta B$ is the Itô integral of the standard part ${}^\circ\Phi$ of Φ with respect to the Fréchet space valued Brownian motion $b_{\mathbb{B}}$. Moreover, in the non-adapted case, ${}^\circ\int \Phi \Delta B$ is the Skorokhod integral and leads to the Skorokhod integral process of ${}^\circ\Phi$.

We shall also see that, if Φ is $(\mathcal{B}_{t^-})_{t \in T}$-adapted and locally in $SL^2(\Gamma \otimes \nu, \mathbb{F})$, then $\int \Phi \Delta B$ can be converted to a continuous process $\int \Phi db_{\mathbb{B}}$, defined on $\Omega \times [0,\infty[$ by setting, for all limited $s \in T$,

$$\int \Phi db_{\mathbb{B}}(\cdot, {}^\circ s) := {}^\circ\int \Phi \Delta B(\cdot, s).$$

If Φ is a lifting of a standard function $g : \Omega \times [0,\infty[\to \mathbb{H}$ (i.e., $g(X, {}^\circ t) \approx \Phi(X,t)$ for $\widehat{\Gamma \otimes \nu}$-almost all (X,t)), then the stochastic integral $\int g db_{\mathbb{B}}$ of g is identical to $\int \Phi db_{\mathbb{B}}$.

Remark 12.1.1 However, there exist many Φ which are not liftings of standard functions: Note that the following function F is locally in $SL^2(\nu, \mathbb{F})$, thus locally in $SL^2(\Gamma \otimes \nu, \mathbb{F})$, and is $(\mathcal{B}_{t^-})$-adapted, but F is not a lifting of a standard function: Let $a \in \mathbb{H}$, $a \neq 0$. Set

$$F(t) := \begin{cases} {}^*a & \text{if } t \cdot H \text{ is odd} \\ 0 & \text{if } t \cdot H \text{ is even.} \end{cases}$$

Therefore, we have here an extension of the standard integration theory.

The following lemma is a simple application of Lemma 11.1.3. We will prove it in detail, because similar arguments will often be used in what follows.

Lemma 12.1.2 *Fix $(\mathcal{B}_{t^-})_{t\in T}$-adapted $\Theta, \Psi : \Omega \times T \to \mathbb{F}$ such that $\Theta_t, \Psi_t \in L^2(\Gamma, \mathbb{F})$ for all $t \in T$. Then*

(a) *for all $\sigma \in T$,*

$$\mathbb{E}\left(\left(\int \Theta \Delta B\right)_\sigma \cdot \left(\int \Psi \Delta B\right)_\sigma\right) = \mathbb{E}\int_{T_\sigma} \langle \Theta_s, \Psi_s \rangle_{\mathbb{F}}\, d\nu(s) \in {}^*\mathbb{R},$$

in particular, $\mathbb{E}\left(\int \Theta \Delta B\right)_\sigma^2 = \mathbb{E}\int_{T_\sigma} \|\Theta_s\|_{\mathbb{F}}^2\, d\nu(s) < \infty$ in ${}^*\mathbb{R}$;

(b) *$\int \Theta \Delta B$ is an internal $(\mathcal{B}_t)_{t\in T}$-martingale.*

Proof (a) Fix an ONB $(\mathfrak{e}_i)_{i\in\omega}$ of $\mathbb{F}$. We obtain, using Lemma 11.1.3,

$$\begin{aligned}
&\mathbb{E}\left(\int \Theta \Delta B\right)_\sigma \cdot \left(\int \Psi \Delta B\right)_\sigma \\
&\quad = \mathbb{E}\sum_{s,t\in T_\sigma,(i,j)\in\omega^2} \langle \Theta_s, \mathfrak{e}_i \rangle_{\mathbb{F}} \left\langle \Psi_t, \mathfrak{e}_j \right\rangle_{\mathbb{F}} \langle X_s, \mathfrak{e}_i \rangle_{\mathbb{F}} \left\langle X_t, \mathfrak{e}_j \right\rangle_{\mathbb{F}} = \alpha + \beta + \gamma,
\end{aligned}$$

where

$$\alpha = \mathbb{E}\sum_{s\neq t\in T_\sigma,(i,j)\in\omega^2} \langle \Theta_s, \mathfrak{e}_i \rangle_{\mathbb{F}} \left\langle \Psi_t, \mathfrak{e}_j \right\rangle_{\mathbb{F}} \langle X_s, \mathfrak{e}_i \rangle_{\mathbb{F}} \left\langle X_t, \mathfrak{e}_j \right\rangle_{\mathbb{F}} = 0,$$

because, for $s < t$,

$$\begin{aligned}
&\mathbb{E}\langle \Theta_s, \mathfrak{e}_i \rangle_{\mathbb{F}} \left\langle \Psi_t, \mathfrak{e}_j \right\rangle_{\mathbb{F}} \langle X_s, \mathfrak{e}_i \rangle_{\mathbb{F}} \left\langle X_t, \mathfrak{e}_j \right\rangle_{\mathbb{F}} \\
&\quad = \mathbb{E}\langle \Theta_s, \mathfrak{e}_i \rangle_{\mathbb{F}} \left\langle \Psi_t, \mathfrak{e}_j \right\rangle_{\mathbb{F}} \langle X_s, \mathfrak{e}_i \rangle_{\mathbb{F}} \mathbb{E}^{\mathcal{B}_{t^-}} \left\langle X_t, \mathfrak{e}_j \right\rangle_{\mathbb{F}} = 0, \\
\beta &= \mathbb{E}\sum_{s\in T_\sigma, i\neq j\in\omega} \langle \Theta_s, \mathfrak{e}_i \rangle_{\mathbb{F}} \left\langle \Psi_s, \mathfrak{e}_j \right\rangle_{\mathbb{F}} \langle X_s, \mathfrak{e}_i \rangle_{\mathbb{F}} \left\langle X_s, \mathfrak{e}_j \right\rangle_{\mathbb{F}} \\
&= \mathbb{E}\sum_{s\in T_\sigma, i\neq j\in\omega} \langle \Theta_s, \mathfrak{e}_i \rangle_{\mathbb{F}} \left\langle \Psi_s, \mathfrak{e}_j \right\rangle_{\mathbb{F}} \mathbb{E}^{\mathcal{B}_{s^-}} \langle X_s, \mathfrak{e}_i \rangle_{\mathbb{F}} \left\langle X_s, \mathfrak{e}_j \right\rangle_{\mathbb{F}} = 0, \\
\gamma &= \mathbb{E}\sum_{s\in T_\sigma, i\in\omega} \langle \Theta_s, \mathfrak{e}_i \rangle_{\mathbb{F}} \langle \Psi_s, \mathfrak{e}_i \rangle_{\mathbb{F}} \langle X_s, \mathfrak{e}_i \rangle_{\mathbb{F}}^2
\end{aligned}$$

$$= \mathbb{E} \sum_{s \in T_\sigma, i \in \omega} \langle \Theta_s, \mathfrak{e}_i \rangle_{\mathbb{F}} \langle \Psi_s, \mathfrak{e}_i \rangle_{\mathbb{F}} \mathbb{E}^{\mathcal{B}_{s^-}} \langle X_s, \mathfrak{e}_i \rangle_{\mathbb{F}}^2$$

$$= \mathbb{E} \sum_{s \in T_\sigma, i \in \omega} \langle \Theta_s, \mathfrak{e}_i \rangle_{\mathbb{F}} \langle \Psi_s, \mathfrak{e}_i \rangle_{\mathbb{F}} \frac{1}{H} = \mathbb{E} \int_{T_\sigma} \langle \Theta_s, \Psi_s \rangle_{\mathbb{F}} \, d\nu(s).$$

This proves part (a). The proof of part (b) is similar. □

We use the following trick, essentially due to Lindstrøm (see [2]). In order to handle the quadratic variation of $\int \Phi \Delta B$ successfully, we modify the timeline T to the timeline $\overline{T} := \{(s,i) \mid s \in T, i \in \{1, \ldots, \omega\}\}$. On $\overline{T}$ we use the lexicographic order, denoted by $<$. Fix an ONB $(\mathfrak{e}_i)_{i \in \omega}$ of $\mathbb{F}$. We can identify Ω with ${}^*\mathbb{R}^{\overline{T}}$ via

$$X \longmapsto (\langle X_s, \mathfrak{e}_i \rangle_{\mathbb{F}})_{i \in \omega, s \in T}.$$

Set $E_i := \operatorname{span}\{\mathfrak{e}_1, \ldots, \mathfrak{e}_i\}$ and define for $s \in T$ and $i \in \omega$

$$\alpha_{s,i} : \Omega \to \mathbb{F}^{T_{s^-}} \times E_i, X \mapsto \left(X_{\frac{1}{H}}, \ldots, X_{s^-}, \pi_{\mathbb{F}E_i}(X_s) \right).$$

Now define a new filtration $\left(\mathcal{B}_{s,i}\right)_{(s,i) \in \overline{T}}$ on the internal Borel algebra $\mathcal{B}$ of Ω setting

$$\mathcal{B}_{(s,i)} := \left\{ \alpha_{s,i}^{-1}[A] \mid A \in \mathcal{B}\left(\mathbb{F}^{T_{s^-}} \times E_i\right) \right\}.$$

Define $\left(\frac{1}{H}, 1\right)^- := 0$ and recall that $\mathcal{B}_0 := \{\Omega, \emptyset\}$. Note that an internal Borel-measurable function $F : \Omega \to \mathbb{F}$ is $\mathcal{B}_{s,i}$-measurable if and only if $F(X) = F(Y)$ for all $X, Y \in \Omega$ with $X =_{(s,i)} Y$, i.e., $\langle X_t, \mathfrak{e}_j \rangle = \langle Y_t, \mathfrak{e}_j \rangle$ for all $(t,j) \leq (s,i)$. Moreover,

$$\mathcal{B}_{s^-} \subseteq \mathcal{B}_{(s,i)} \subseteq \mathcal{B}_{(s,\omega)} = \mathcal{B}_s.$$

We modify each internal $\Phi : \Omega \times T \to \mathbb{F}$ to $\overline{\Phi} : \Omega \times \overline{T} \to {}^*\mathbb{R}$, setting

$$\overline{\Phi}(X, (s,i)) := \langle \Phi(X,s), \mathfrak{e}_i \rangle_{\mathbb{F}}.$$

Suppose that Φ is $(\mathcal{B}_{s^-})_{s \in T}$-adapted. Then $\overline{\Phi}$ is $(\mathcal{B}_{s^-})_{s \in \overline{T}}$-adapted. If, in addition, $\Phi(\cdot, r) \in L^2(\Gamma, \mathbb{F})$ for all $r \in T$, a new internal integral $\int \overline{\Phi} \Delta B : \Omega \times \overline{T} \to {}^*\mathbb{R}$ is defined by

$$\left(\int \overline{\Phi} \Delta B \right)(X, (\sigma, j)) := \sum_{(s,i) \leq (\sigma,j)} \langle \Phi_s(X), \mathfrak{e}_i \rangle_{\mathbb{F}} \cdot \langle X_s, \mathfrak{e}_i \rangle_{\mathbb{F}}.$$

Note that $\int \overline{\Phi} \Delta B$ is a Γ-square-integrable $(\mathcal{B}_s)_{s \in \overline{T}}$-martingale. We call (s,i) **limited** if s is limited, and we call (s,i) **infinitely close** to (t,j) if $s \approx t$.

Lemma 12.1.3 *Fix a $(\mathcal{B}_{t^-})$-adapted Φ locally in $SL^2(\Gamma \otimes \nu, \mathbb{F})$ and limited $\sigma \in T$, $r \in \overline{T}$. Then*

$$\mathbb{E}\max_{t\in T_\sigma}\left(\int \Phi\Delta B\right)_t^2, \quad \mathbb{E}\max_{t\in \overline{T}_r}\left(\int \overline{\Phi}\Delta B\right)_t^2 \text{ are limited.}$$

There exists a set U of $\widehat{\Gamma}$-measure 1 such that, for all $X \in U$ and all limited $t \in T$, $s \in \overline{T}$, $\left(\int \Phi\Delta B(X,t)\right)^2, \left(\int \overline{\Phi}\Delta B(X,s)\right)^2$ are limited.

Proof Fix $\sigma \in \mathbb{N}$. Then $\sigma \in T$ and $(\sigma,\omega) =: \rho \in \overline{T}$. By Lemma 12.1.2 and Theorem 2.4.2,

$$\mathbb{E}\max_{t\in T_\sigma}\left(\int \Phi\Delta B\right)_t^2 \le 4\cdot\mathbb{E}\left(\int \Phi\Delta B\right)_\sigma^2 \text{ is limited.}$$

In the same way we obtain that $\mathbb{E}\max_{t\in\overline{T}_\rho}\left(\int \overline{\Phi}\Delta B\right)_t^2$ is limited. It follows that there exists a set U_σ of $\widehat{\Gamma}$-measure 1 such that $\max_{t\in T_\sigma}\left(\int \Phi\Delta B\right)_t^2(X)$ and $\max_{t\in\overline{T}_\rho}\left(\int \overline{\Phi}\Delta B\right)_t^2(X)$ are limited for all $X \in U_\sigma$. Set $U := \bigcap_{\sigma\in\mathbb{N}} U_\sigma$. Then $\widehat{\Gamma}(U) = 1$. Since for all limited $t \in T$, $s \in \overline{T}$, there exists a $\sigma \in \mathbb{N}$ with $t \le \sigma, s \le (\sigma,\omega)$, the proof is finished. □

In the following we fix a $(\mathcal{B}_{t^-})$-adapted Φ locally in $SL^2(\Gamma \otimes \nu, \mathbb{F})$. The proof of the following result is a consequence of Lemmas 12.1.2 and 12.1.3 and Theorem 10.14.2 and is left to the reader.

Lemma 12.1.4 (a) *If $\int \overline{\Phi}\Delta B(X,\cdot)$ is S-continuous, then $\int \Phi\Delta B(X,\cdot)$ is S-continuous.*

(b) *If the quadratic variation $\left[\int \overline{\Phi}\Delta B\right]$ of $\int \overline{\Phi}\Delta B$ is S-continuous $\widehat{\Gamma}$-a.s., then $\int \overline{\Phi}\Delta B$ is S-continuous $\widehat{\Gamma}$-a.s.*

(c) $\left[\int \overline{\Phi}\Delta B\right]_{(\sigma,j)}(X) = \sum_{(s,i)\le(\sigma,j)} \langle \Phi_s(X), \mathfrak{e}_i\rangle_{\mathbb{F}}^2 \langle X_s, \mathfrak{e}_i\rangle_{\mathbb{F}}^2$.
Define

$$\Psi : \Omega \times T \to {}^*\mathbb{R}, (X,\sigma) \longmapsto \sum_{(s,i)\in T_\sigma\times\omega} \langle \Phi_s(X), \mathfrak{e}_i\rangle_{\mathbb{F}}^2 \langle X_s, \mathfrak{e}_i\rangle_{\mathbb{F}}^2.$$

(d) *If $\Psi(X)$ is S-continuous, then $\left[\int \overline{\Phi}\Delta B\right](X)$ is S-continuous.*

From this lemma it follows that, in order to prove that $\int \Phi\Delta B$ is S-continuous $\widehat{\Gamma}$-a.s., it suffices to show that Ψ is S-continuous $\widehat{\Gamma}$-a.s. To this end we define for each $m \in {}^*\mathbb{N}$ stopping times $\tau_m : \Omega \to T \cup \{H + \frac{1}{H}\}$, setting

$$\tau_m := \inf\left\{t \in T \mid \int_{T_t} \|\Phi_s\|_{\mathbb{F}}^2 d\nu(s) \ge m\right\},$$

where $\inf \emptyset := H + \frac{1}{H}$. Set for $X \in \Omega, t \in T$ and $m \in {}^*\mathbb{N}$

$$\Phi_t^m := \mathbf{1}_{T_{\tau_{m-\frac{1}{H}}}}(t)\Phi_t,$$

$$\Psi_t^m := \sum_{(s,i)\in T_t\times\omega} \langle \Phi_s^m, \mathfrak{e}_i\rangle_{\mathbb{F}}^2 \langle \cdot_s, \mathfrak{e}_i\rangle_{\mathbb{F}}^2.$$

The key for the main results in this and the subsequent section is the following lemma.

Lemma 12.1.5 *Fix $m \in {}^*\mathbb{N}$, $k, \sigma \in \mathbb{N}$ and $r \in T_\sigma$.*

(a) *Φ_r^m is $\mathcal{B}_{r^-}$-measurable.*

(b) *$r \mapsto \Psi_r^m - \int_{T_r} \|\Phi_s^m\|_{\mathbb{F}}^2 d\nu(s) = \sum_{i\in\omega, s\in T_r} \langle \Phi_s^m, \mathfrak{e}_i\rangle_{\mathbb{F}}^2 \left(\langle \cdot_s, \mathfrak{e}_i\rangle_{\mathbb{F}}^2 - \frac{1}{H}\right)$ is a Γ-square-integrable $(B_r)_{r\in T_\sigma}$-martingale and $\mathbb{E}\left(\Psi_r^k - \int_{T_r} \|\Phi_s^k\|_{\mathbb{F}}^2 d\nu(s)\right)^2 \approx 0.$*

(c) *$\mathbb{E}\max_{r\in T_\sigma}\left(\Psi_r^k - \int_{T_r} \|\Phi_s^k\|_{\mathbb{F}}^2 d\nu(s)\right)^2 \approx 0.$*

(d) *$\Psi_\sigma^k \in SL^1(\Gamma)$.*

(e) *$\lim_{k\to\infty} {}^\circ\mathbb{E}\Psi_\sigma^k = {}^\circ\mathbb{E}\Psi_\sigma$.*

Proof (a) is true, since Φ is $(\mathcal{B}_{r^-})_{r\in T_\sigma}$-adapted.

(b) Using Lemma 11.1.3, we obtain (see the proof of Lemma 12.1.2 part (a)),

$$\mathbb{E}\left(\Psi_r^m - \int_{T_r} \|\Phi_s^m\|_{\mathbb{F}}^2 d\nu(s)\right)^2 = \mathbb{E}\left(\sum_{i\in\omega, s\in T_r} \langle \Phi_s^m, \mathfrak{e}_i\rangle_{\mathbb{F}}^2 \left(\langle \cdot_s, \mathfrak{e}_i\rangle_{\mathbb{F}}^2 - \frac{1}{H}\right)\right)^2$$

$$= \mathbb{E}\sum_{i\in\omega, s\in T_r} \langle \Phi_s^m, \mathfrak{e}_i\rangle_{\mathbb{F}}^4 \left(\langle \cdot_s, \mathfrak{e}_i\rangle_{\mathbb{F}}^2 - \frac{1}{H}\right)^2$$

$$= \mathbb{E}\sum_{i\in\omega, s\in T_r} \langle \Phi_s^m, \mathfrak{e}_i\rangle_{\mathbb{F}}^4 \mathbb{E}^{\mathcal{B}_{s^-}}\left(\langle \cdot_s, \mathfrak{e}_i\rangle_{\mathbb{F}}^2 - \frac{1}{H}\right)^2$$

$$= 2\cdot\mathbb{E}\sum_{i\in\omega, s\in T_r} \langle \Phi_s^m, \mathfrak{e}_i\rangle_{\mathbb{F}}^4 \frac{1}{H^2} \le 2\mathbb{E}\alpha,$$

where

$$\alpha := \int_{\{(s,t)\in T_r^2 | s=t\}} \|\Phi_s^m\|_{\mathbb{F}}^2 \cdot \|\Phi_t^m\|_{\mathbb{F}}^2 d\nu^2(s,t).$$

Note that $\mathbb{E}\alpha$ is limited in ${}^*\mathbb{R}$. Now let $m := k$. Since $\int_{T_r} \|\Phi_s^k\|_{\mathbb{F}}^2 d\nu_r(s) < k$, $\alpha \in SL^1(\Gamma)$. Since $\|\Phi^k\|_{\mathbb{F}}^2 \in SL^1(\Gamma \otimes \nu_r)$, from Theorem 10.12.1 it follows that

$t \mapsto \left\| \Phi_t^k \right\|_{\mathbb{F}}^2 \in SL^1(\nu_r)\,\widehat{\Gamma}$-a.s., thus, by Lemma 10.12.2 (b),

$$(s,t) \mapsto \left\| \Phi_s^k \right\|_{\mathbb{F}}^2 \cdot \left\| \Phi_t^k \right\|_{\mathbb{F}}^2 \in SL^1(\nu_r^2)\,\widehat{\Gamma}\text{-a.s.}$$

It follows that $\alpha \approx 0\,\widehat{\Gamma}$-a.s., because $\nu^2\left\{(s,t) \in T_r^2 \mid s = t\right\} \approx 0$. Therefore, $\mathbb{E}\alpha \approx 0$. The martingale properties follow from Lemma 11.1.3.

(c) follows from (b) and Doob's inequality.

(d) By (b), $\mathbb{E}\left|\Psi_\sigma^k - \int_{T_\sigma} \left\| \Phi_s^k \right\|_{\mathbb{F}}^2 d\nu(s)\right| \approx 0$. Therefore, for each $N \in \mathcal{B}$,

$$\int_N \Psi_\sigma^k d\Gamma \leq \mathbb{E}\left|\Psi_\sigma^k - \int_{T_\sigma} \left\| \Phi_s^k \right\|_{\mathbb{F}}^2 d\nu(s)\right| + \int_{N \times T_\sigma} \left\| \Phi^k \right\|_{\mathbb{F}}^2 d\Gamma \otimes \nu$$

$$\begin{cases} \approx 0 & \text{if } \Gamma(N) \approx 0 \\ \text{is limited.} & \end{cases}$$

It follows that $\Psi_\sigma^k \in SL^1(\Gamma)$.

(e) Assume that (e) is not true. By saturation, there exists a standard $\varepsilon > 0$ and an unlimited $M \in {}^*\mathbb{N}$ such that

$$\begin{aligned} \varepsilon \leq A &:= \mathbb{E}\left(\Psi - \Psi^M\right)_\sigma \\ &= \mathbb{E} \sum_{i \in \omega, s \in T_\sigma} \left((\Phi_s)^2(\mathfrak{e}_i) - \left(\Phi_s^M\right)^2(\mathfrak{e}_i)\right) \mathbb{E}^{\mathcal{B}_{s^-}}\left(\langle \cdot_s, \mathfrak{e}_i \rangle_{\mathbb{F}}^2\right) \\ &= \mathbb{E} \int_{T_\sigma} \|\Phi_s\|_{\mathbb{F}}^2 - \left\| \Phi_s^M \right\|_{\mathbb{F}}^2 d\nu(s). \end{aligned}$$

By Theorem 10.12.1, $\int_{T_\sigma} \|\Phi_s\|_{\mathbb{F}}^2 d\nu(s) \in SL^1(\Gamma)$, thus $\int_{T_\sigma} \|\Phi_s\|_{\mathbb{F}}^2 d\nu(s)$ is limited $\widehat{\Gamma}$-a.s. Therefore,

$$\int_{T_\sigma} \|\Phi_s\|_{\mathbb{F}}^2 d\nu(s) = \int_{T_\sigma} \left\| \Phi_s^M \right\|_{\mathbb{F}}^2 d\nu(s)\,\widehat{\Gamma}\text{-a.s.}$$

This proves that $A \approx 0$, which is a contradiction. □

Theorem 12.1.6 *$\int \Phi \Delta B$ is S-continuous $\widehat{\Gamma}$-a.s.*

Proof We use Lemma 10.7.1 and Theorem 10.12.1. There exists a $U \in L_\Gamma(\mathcal{B})$ with $\widehat{\Gamma}(U) = 1$ and such that for all $X \in U$ and all $\sigma \in \mathbb{N}$:

(i) $\Psi_t^k(X) \approx \int_{T_t} \left\| \Phi_s^k(X) \right\|_{\mathbb{F}}^2 d\nu(s)$ is limited for all $k \in \mathbb{N}$ and $t \in T_\sigma$,

(ii) $\int_{T_\sigma} \|\Phi_s(X)\|_{\mathbb{F}}^2 d\nu(s)$ is limited, thus there exists a $k \in \mathbb{N}$ with $\Phi_s(X) = \Phi_s^k(X)$ for all $s \in T_\sigma$,

(iii) $t \mapsto \|\Phi_t^k(X)\|_{\mathbb{F}}^2 \in SL^1(\nu_\sigma)$, therefore $\int_{T_s \setminus T_t} \|\Phi_r^k(X)\|_{\mathbb{F}}^2 d\nu(r) \approx 0$ if $s \approx t$ and $t \leq s \in T_\sigma$.

We obtain for all $X \in U$ and all limited $s,t \in T$ with $s \approx t$ and $s \leq t$

$$\Psi_s(X) = \Psi_s^k(X) \approx \int_{T_s} \|\Phi_r^k(X)\|_{\mathbb{F}}^2 d\nu(r) \approx \int_{T_t} \|\Phi_r^k(X)\|_{\mathbb{F}}^2 d\nu(r) \approx \Psi_t(X),$$

where we may assume that $s,t \in T_\sigma$. Moreover, $\Psi_t(X)$ is nearstandard for all limited $t \in T$ and all $X \in U$. By Lemma 12.1.5, $\int \Phi \Delta B$ is S-continuous $\widehat{\Gamma}$-a.s. □

12.2 On the S-square-integrability of the internal Itô integral

Proposition 12.2.1 *Fix a $(\mathcal{B}_{t^-})_{t\in T}$-adapted $\Phi : \Omega \times T \to \mathbb{F}$ locally in $SL^2(\Gamma \otimes \nu, \mathbb{F})$ and $\sigma \in \mathbb{N}$. Then*

(a)

$$\max_{t \in T_\sigma} \left| \int \Phi \Delta B(\cdot, t) \right| \in SL^2(\Gamma),$$

(b)

$$\int \Phi \Delta B \restriction \Omega \times T_\sigma \in SL^2(\Gamma \otimes \nu_\sigma).$$

Proof (a) We use the notation in the preceding section. Note that it suffices to prove that $\max_{s \in \overline{T}, s \leq (\sigma,\omega)} \left| \int \overline{\Phi} \Delta B_s \right| \in SL^2(\Gamma)$. Since $\int \overline{\Phi} \Delta B$ is a $(\mathcal{B}_t)_{t \in \overline{T}}$-martingale and $\mathbb{E}\left(\int \overline{\Phi} \Delta B\right)^2_{(\sigma,\omega)}$ is limited, by Theorem 10.13.1, it suffices to show that $\left[\int \overline{\Phi} \Delta B\right]_{(\sigma,\omega)} \in SL^1(\Gamma)$. Therefore, it suffices to prove that $\Psi_\sigma \in SL^1(\Gamma)$. Now, by Lemma 12.1.5 parts (d) and (e), $\lim_{k \to \infty} {}^\circ\mathbb{E}(\Psi - \Psi^k)_\sigma = 0$ and $\Psi_\sigma^k \in SL^1(\Gamma)$. It follows that $\Psi_\sigma \in SL^1(\Gamma)$ (see Corollary 10.7.3).

(b) follows from (a). □

Corollary 12.2.2 *Fix a $(\mathcal{B}_{t^-})_{t \in T}$-adapted $\Phi : \Omega \times T \to \mathbb{F}$ now in $SL^2(\Gamma \otimes \nu, \mathbb{F})$ (not only locally). Then*

$$\max_{t \in T} \left(\int \Phi \Delta B \right)_t \in SL^2(\Gamma),$$

thus $\int \Phi \Delta B \in SL^2(\Gamma \otimes \nu)$.

Proof Note that, by Doob's inequality, for $\sigma \in \mathbb{N}$,

$$
{}^\circ\mathbb{E} \max_{t\in T\setminus T_\sigma} \left(\left(\int \Phi\Delta B \right)_t - \left(\int \Phi\Delta B \right)_\sigma \right)^2 \leq 4 \cdot {}^\circ\mathbb{E} \left(\sum_{t\in T\setminus T_\sigma} \Phi_t(X)(X_t) \right)^2
$$

$$
= 4 \cdot {}^\circ\mathbb{E} \sum_{t\in T\setminus T_\sigma} \|\Phi_t\|_{\mathbb{F}}^2 \frac{1}{H} \to_{\sigma\to\infty} 0.
$$

Assume that this convergence fails. Then there exists a standard $\varepsilon > 0$ and infinitely many $\sigma \in \mathbb{N}$ with $\mathbb{E}\sum_{t\in T\setminus T_\sigma} \|\Phi_t\|_{\mathbb{F}}^2 \frac{1}{H} \geq \varepsilon$. Then there exists an unlimited $S \in {}^*\mathbb{N}$ with $\mathbb{E}\sum_{t\in T\setminus T_S} \|\Phi_t\|_{\mathbb{F}}^2 \frac{1}{H} \geq \varepsilon$, which contradicts the S-square-integrability of Φ. Therefore, for all $\sigma \in \mathbb{N}$

$$
\left(\int_A \max_{t\in T} \left(\int \Phi\Delta B \right)_t^2 d\Gamma \right)^{\frac{1}{2}}
$$

$$
\leq \left(\int_A \max_{t\in T_\sigma} \left(\int \Phi\Delta B \right)_t^2 d\Gamma \right)^{\frac{1}{2}} + \left(\int_A \max_{t\in T\setminus T_\sigma} \left(\int \Phi\Delta B \right)_t^2 d\Gamma \right)^{\frac{1}{2}}.
$$

By Proposition 12.2.1 (a), the first summand is limited and infinitesimal if $\Gamma(A) \approx 0$. The second summand equals

$$
\left(\int_A \max_{t\in T\setminus T_\sigma} \left(\left(\int \Phi\Delta B \right)_t - \left(\int \Phi\Delta B \right)_\sigma + \left(\int \Phi\Delta B \right)_\sigma \right)^2 d\Gamma \right)^{\frac{1}{2}}
$$

$$
\leq \left(\int_A \max_{t\in T\setminus T_\sigma} \left(\left(\int \Phi\Delta B \right)_t - \left(\int \Phi\Delta B \right)_\sigma \right)^2 d\Gamma \right)^{\frac{1}{2}}
$$

$$
+ \left(\int_A \left(\int \Phi\Delta B \right)_\sigma^2 d\Gamma \right)^{\frac{1}{2}},
$$

which is limited and can be made arbitrarily small standardly if $\Gamma(A) \approx 0$. This proves the assertion. □

12.3 The standard Itô integral

Fix a $(\mathcal{B}_{t^-})_{t\in T}$-adapted Φ locally in $SL^2(\Gamma \otimes \nu, \mathbb{F})$. By Theorem 12.1.6 and Proposition 9.7.1, we may convert $\int \Phi\Delta B$ to a continuous stochastic process

$\int \Phi db_{\mathbb{B}}$ on the timeline $[0,\infty[$): Define for $\widehat{\Gamma}$-almost all $X \in \Omega$ and all limited $t \in T$,

$$\int \Phi db_{\mathbb{B}} : \Omega \times [0,\infty[\ni (X, {}^\circ t) \mapsto {}^\circ\left(\int \Phi \Delta B(X,t)\right).$$

Now we shall see that non-time-anticipating processes have non-time-anticipating liftings. Here we use the equivalence of 'adapted' and 'predictable' (see Corollary 5.4.2).

Theorem 12.3.1 *Let $\varphi : \Omega \times [0,\infty[\to \mathbb{H}$ be $(\mathfrak{b}_r)_{r\in[0,\infty[}$-adapted and (locally) in $L^2\left(\widehat{\Gamma}\otimes\lambda, \mathbb{H}\right)$. Then φ has a $(\mathcal{B}_{t}\text{-})_{t\in T}$-adapted lifting $\Phi : \Omega \times T \to \mathbb{F}$ (locally) in $SL^2(\Gamma\otimes\nu, \mathbb{F})$.*

Proof We only prove the case of locally L^2. The proof for full L^2 is left to the reader. Fix $\sigma \in \mathbb{N}$ and let L be the Hilbert space of strongly $(\mathfrak{b}_r)_{t\in[0,\sigma]}$-predictable square-integrable functions $f : \Omega \times [0,\sigma] \to \mathbb{H}$ and let M be the set of all these functions having a $(\mathcal{B}_{t}\text{-})_{t\in T_\sigma}$-adapted lifting $F : \Omega \times T_\sigma \to \mathbb{F} \in SL^2(\Gamma\otimes\nu, \mathbb{F})$. Let $B \times]r,u]$ be a predictable rectangle with $B \in \mathfrak{b}_r$ and $r < u \in [0,\sigma]$. Then there exists an $s \approx r, s \in T_\sigma$, and a μ-approximation $A \in \mathcal{B}_{s}\text{-}$ of B. Let $v \approx u, v \in T_\sigma$. Set $Y := A \times (]s,v] \cap T_\sigma)$ and $X := B \times \left(st^{-1}[]r,u]] \cap T_\sigma\right)$. Then

$$X \Delta Y = (A \Delta B \times T_\sigma) \cup (\Omega \times (\widetilde{r} \cup \widetilde{u}))$$

is a $\widehat{\mu\otimes\nu}$- nullset. Recall that $\widetilde{r} = \{t \in T_\sigma \mid t \approx r\}$. It follows that ${}^*a \cdot \mathbf{1}_Y \in SL^2(\Gamma\otimes\nu, \mathbb{F})$ and is a $(\mathcal{B}_{t}\text{-})_{t\in T_\sigma}$-adapted lifting of $a \cdot \mathbf{1}_{B\times]r,u]}$ for $a \in \mathbb{H}$, thus $a \cdot \mathbf{1}_{B\times]r,u]} \in M$. In the same way one can see that $a \cdot \mathbf{1}_{B\times[0,u]} \in M$ if $B \in \mathfrak{b}_0$. Obviously, M is a linear space. The completeness of M can be proved in the same manner as has been done several times before, using Corollary 10.9.3. Therefore, $M = L$, by Proposition 5.6.1.

It follows that φ, restricted to $\Omega \times [0,\sigma]$, has a $(\mathcal{B}_{t}\text{-})_{t\in T_\sigma}$-adapted lifting $\Phi_\sigma : \Omega \times T_\sigma \to \mathbb{F} \in SL^2(\Gamma\otimes\nu, \mathbb{F})$. We may assume that $\Phi_\sigma(x) = \Phi_{\sigma+1}(x)$ for $x \in \Omega \times T_\sigma$. Let $(\Phi_\sigma)_{\sigma\in{}^*\mathbb{N}}$ be an internal extension of $(\Phi_\sigma)_{\sigma\in\mathbb{N}}$ such that $\Phi_\sigma(x) = \Phi_{\sigma+1}(x)$ for $x \in \Omega \times T_\sigma$ and all $\sigma \in {}^*\mathbb{N}$, $\sigma \leq H$. Moreover, we may assume that all Φ_σ are $(\mathcal{B}_{t}\text{-})_{t\in T_\sigma}$-adapted. Then $\Phi := \Phi_H$ is locally in $SL^2(\Gamma\otimes\nu, \mathbb{F})$ and a $(\mathcal{B}_{t}\text{-})_{t\in T}$-adapted lifting of φ. □

Now fix φ and Φ as in the preceding theorem. We define

$$\int \varphi db_{\mathbb{B}} := \int \Phi db_{\mathbb{B}}.$$

This process $\int \varphi db_{\mathbb{B}}$ is called the **Itô integral of** φ. One has to prove that $\int \varphi db_{\mathbb{B}}$ is well defined $\widehat{\Gamma}$-a.s., i.e., it does not depend on the chosen lifting. This follows from:

Lemma 12.3.2 *Suppose that Φ is locally in $SL^2(\Gamma \otimes \nu, \mathbb{F})$ and a lifting of the 0-function. Then $\widehat{\Gamma}$-a.s.*

$$\left(\int \Phi \Delta B\right)_t \approx 0 \text{ for all limited } t \in T.$$

Proof Fix $\sigma \in \mathbb{N}$. By Lemma 11.1.3 and Doob's inequality, we obtain

$$\begin{aligned}
\mathbb{E} \max_{t \in T_\sigma} \left(\int \Phi \Delta B\right)_t^2 &\leq 4 \cdot \mathbb{E}\left(\sum_{s \in T_\sigma} \langle \Phi_s, \cdot_s \rangle_{\mathbb{F}}\right)^2 \\
&= 4\mathbb{E} \sum_{i \in \omega, s \in T_\sigma} \langle \Phi_s, \mathfrak{e}_i \rangle_{\mathbb{F}}^2 \langle \cdot_s, \mathfrak{e}_i \rangle_{\mathbb{F}}^2 \\
&= 4\mathbb{E} \sum_{i \in \omega, s \in T_\sigma} \langle \Phi_s, \mathfrak{e}_i \rangle_{\mathbb{F}}^2 \mathbb{E}^{\mathcal{B}_{s^-}} \langle \cdot_s, \mathfrak{e}_i \rangle_{\mathbb{F}}^2 \\
&= 4\mathbb{E} \int_{T_\sigma} \|\Phi_s\|_{\mathbb{F}}^2 d\nu(s) \approx 0.
\end{aligned}$$

By Lemma 10.7.1 (b), $\max_{t \in T_\sigma} \left(\int \Phi \Delta B\right)_t^2 \approx 0$ $\widehat{\Gamma}$-a.s. Therefore, $\left(\int \Phi \Delta B\right)_t^2 \approx 0$ for all limited $t \in T$ $\widehat{\Gamma}$-a.s. □

We want to define the **Itô integral as a random variable** for $(\mathfrak{b}_r)_{r \in [0,\infty[}$-adapted processes $\varphi : \Omega \times [0, \infty[\to \mathbb{H} \in L^2(\widehat{\Gamma} \otimes \lambda, \mathbb{H})$: If $\Phi : \Omega \times T \to \mathbb{F}$ is a $(\mathcal{B}_{t^-})_{t \in T}$-adapted lifting of φ in $SL^2(\Gamma \otimes \nu, \mathbb{F})$, define

$$\int^V \varphi db_{\mathbb{B}} : \Omega \to \mathbb{R}, X \mapsto {}^\circ \sum_{s \in T} \langle \Phi(X, s), X_s \rangle_{\mathbb{F}}.$$

Note that $\int^V \varphi db_{\mathbb{B}}$ is $\widehat{\Gamma}$-a.s. well defined.

Now suppose that $\varphi : [0, \infty[\to \mathbb{H} \in L^2(\lambda, \mathbb{H})$, i.e., φ is deterministic. We set

$$I(\varphi) := I_1(\varphi) := \int^V \varphi db_{\mathbb{B}}$$

and call $I(\varphi)$ the **Wiener integral** of φ. For internal $\Phi : T \to \mathbb{F}$ define

$$I(\Phi) := I_1(\Phi) := \sum_{s \in T} \langle \Phi(s), X_s \rangle_{\mathbb{F}},$$

called the **internal Wiener integral**.

12.4 On the integrability of the Itô integral

Fix a $(\mathfrak{b}_r)_{r\in[0,\infty[}$-adapted φ locally in $L^2(\widehat{\Gamma}\otimes\lambda,\mathbb{H})$ and a $(\mathcal{B}_{t^-})_{t\in T}$-adapted lifting Φ locally in $SL^2(\Gamma\otimes\nu,\mathbb{F})$, according to Theorem 12.3.1.

Theorem 12.4.1 *Fix $\sigma\in\mathbb{N}$. Set $b:=b_{\mathbb{B}}$.*

(a) *The Itô integral of φ is a continuous $(\mathfrak{b}_r)_{r\in[0,\infty[}$-martingale, with*

$$\mathbb{E}\sup_{r\in[0,\sigma]}\left(\int\varphi db\right)_r^2<\infty.$$

(b) *Suppose that $(\varphi_k)_{k\in\mathbb{N}}$ is a sequence of $(\mathfrak{b}_t)_{t\in[0,\infty[}$-adapted functions $\varphi_k:\Omega\times[0,\sigma]\to\mathbb{H}$, converging to $\varphi:\Omega\times[0,\sigma]\to\mathbb{H}$ in $L^2(\widehat{\Gamma}\otimes\lambda,\mathbb{H})$. Then the Itô integral of φ exists and*

$$\lim_{k\to\infty}\mathbb{E}\sup_{r\in[0,\sigma]}\left(\int(\varphi_k-\varphi)\,db\right)_r^2=0.$$

(c) *Suppose that $(\varphi_k)_{k\in\mathbb{N}}$ is a sequence of $(\mathfrak{b}_t)_{t\in[0,\infty[}$-adapted functions $\varphi_k:\Omega\times[0,\infty[\to\mathbb{H}$, converging to $\varphi:\Omega\times[0,\infty[\to\mathbb{H}$ in $L^2(\widehat{\Gamma}\otimes\lambda,\mathbb{H})$. Then the Itô integral of φ exists and*

$$\lim_{k\to\infty}\mathbb{E}\left(\int^V(\varphi_k-\varphi)\,db\right)^2=0.$$

Proof (a) We have already seen that $\left(\int\varphi db\right)$ is continuous $\widehat{\Gamma}$-a.s. By Proposition 12.2.1,

$$\mathbb{E}\sup_{r\in[0,\sigma]}\left(\int\varphi db\right)_r^2\approx\mathbb{E}\max_{t\in T_\sigma}\left(\int\Phi\Delta B\right)_t^2\text{ is limited.}$$

Since $\left(\int\varphi db\right)_{{}^\circ t}\approx\left(\int\Phi\Delta B\right)_t$ $\widehat{\Gamma}$-a.s for limited $t\in T$ and $\left(\int\Phi\Delta B\right)_t$ is $\mathcal{B}_t$-measurable, $\left(\int\varphi db\right)_{{}^\circ t}$ is $\mathcal{B}_t\vee\mathcal{N}_{\widehat{\Gamma}}$-measurable, thus $\left(\int\varphi db\right)_{{}^\circ t}$ is $\mathfrak{b}_{{}^\circ t}$-measurable. To prove the martingale property, fix $r<u$ in $[0,\infty[$ and $B\in\mathfrak{b}_r$. Then there exists an $s\in T$, $s\approx r$, and a Γ-approximation $A\in\mathcal{B}_s$ of B. Let $t\approx u$. By Propositions 12.2.1, 12.1.2 (b) and Theorem 10.8.1,

$$\begin{aligned}\int_B\left(\int\varphi db\right)_u d\widehat{\Gamma}&=\int_A{}^\circ\left(\int\Phi\Delta B\right)_t d\widehat{\Gamma}\approx\int_A\left(\int\Phi\Delta B\right)_t d\Gamma\\&=\int_A\left(\int\Phi\Delta B\right)_s d\Gamma\approx\int_B\left(\int\varphi db\right)_r d\widehat{\Gamma}.\end{aligned}$$

It follows that $\mathbb{E}^{\mathfrak{b}_r}\left(\int\varphi db\right)_u=\left(\int\varphi db\right)_r$.

(b) Assume that $\Phi_k : \Omega \times T_\sigma \to \mathbb{F} \in SL^2(\Gamma \otimes \nu_\sigma, \mathbb{F})$ is a $(\mathcal{B}_{t^-})_{t \in T}$-adapted lifting of φ_k. Then $\lim_{k,l \to \infty} {}^\circ \int_{\Omega \times T_\sigma} \|\Phi_k - \Phi_l\|_{\mathbb{F}}^2 \, d\Gamma \otimes \nu = 0$ and, as in the proof of Lemma 12.3.2, we see that

$$\lim_{k,l \to \infty} {}^\circ\mathbb{E} \max_{t \in T_\sigma} \left(\int (\Phi_k - \Phi_l) \, \Delta B \right)_t^2 = 0.$$

By Corollary 10.9.3, there is a $(\mathcal{B}_{t^-})_{t \in T}$-adapted function $\Phi : \Omega \times T_\sigma \to \mathbb{F}$ in $SL^2(\Gamma \otimes \nu_\sigma, \mathbb{F})$ such that $\lim_{k \to \infty} {}^\circ \int_{\Omega \times T_\sigma} \|\Phi_k - \Phi\|_{\mathbb{F}}^2 \, d\Gamma \otimes \nu = 0$ and $\lim_{k \to \infty} {}^\circ\mathbb{E} \max_{t \in T_\sigma} \left(\int (\Phi_k - \Phi) \Delta B \right)_t^2 = 0$. It follows that Φ is a lifting of φ and we obtain

$$\mathbb{E} \sup_{r \in [0,\sigma]} \left(\int (\varphi_k - \varphi) \, db \right)_r^2 = {}^\circ\mathbb{E} \max_{t \in T_\sigma} \left(\int (\Phi_k - \Phi) \, \Delta B \right)_t^2 \to_{k \to \infty} 0.$$

The proof of (c) is left to the reader. □

12.5 $\mathcal{W}_{C_{\mathbb{H}}}$ is generated by the Wiener integrals

We take the σ-algebra, generated by the Lévy and Wiener integrals, as a basis for the chaos expansion. In the Wiener case we obtain the following.

Proposition 12.5.1 *Let $\mathfrak{J}$ be the set of Lebesgue-measurable functions $f : [0, \infty[\to \mathbb{H}' \in L^2(\lambda, \mathbb{H}')$ with compact support. Then*

$$\mathcal{W}_{C_{\mathbb{H}}} = \{ I(f) \mid f \in \mathfrak{J} \} \vee \mathcal{N}_{\widehat{\Gamma}} = \sigma \left(\{ I(f) \mid f \in \mathfrak{J} \} \cup \mathcal{N}_{\widehat{\Gamma}} \right).$$

Proof We first prove '$\subseteq$': Recall that the sets $\{ \varphi \circ b_{\mathbb{B}}(\cdot, r) < c \}$ generate $\mathcal{W}_{C_{\mathbb{H}}}$, where $\varphi \in \mathbb{B}' \subseteq \mathbb{H}$, $r \in [0, \infty[$ and $c \in \mathbb{R}$. Set $g(s) := \mathbf{1}_{[0,r]}(s) \cdot \varphi$. Let $t \in T$ with $t \approx r$ and define $G(s) := \mathbf{1}_{T_t}(s) \cdot {}^*\varphi$. Then $G : T \to \mathbb{F}$ is a lifting of g in $SL(\nu_t, \mathbb{F})$ and $g \in \mathfrak{J}$. Now $\widehat{\Gamma}$-a.s.

$$I(g) \approx I(G) = \sum_{s \le t} \left\langle {}^*\varphi, \cdot_s \right\rangle_{\mathbb{F}} = \left\langle {}^*\varphi, B_t \right\rangle_{\mathbb{F}} \approx \varphi \circ b_{\mathbb{B}}(\cdot, r).$$

For the reverse inclusion we prove that each $I(g)$ with $g \in \mathfrak{J}$ and $r \in [0, \infty[$ is $\mathcal{W}_{C_{\mathbb{H}}}$-measurable: Since g is the limit in $L^2(\lambda, \mathbb{H})$ of simple functions and the Borel sets of $[0, \infty[$ are generated by sets of the form $[0, r]$, it suffices to prove that $I(g)$ is $\mathcal{W}_{C_{\mathbb{H}}}$-measurable for each g with $g(s) = \mathbf{1}_{[0,r]}(s) \cdot \varphi$ where $\varphi \in \mathbb{H}$. Since, by Proposition 4.3.6 (a), $\varphi \in \mathbb{H}$ can be approximated by functions in $\mathbb{B}'$ and since, by Corollary 12.4.1 (b), I is continuous, it suffices to prove that $I(g)$

is $\mathcal{W}_{C_{\mathbb{H}}}$-measurable for each g with $g(s) = \mathbf{1}_{[0,r]}(s) \cdot \varphi$ where $\varphi \in \mathbb{B}'$. But we have already seen that $I(g) = \varphi \circ b_{\mathbb{B}}(\cdot, r)$ $\widehat{\Gamma}$-a.s. □

12.6 †The distribution of the Wiener integrals

In this section we will show that for each $\sigma \in [0,\infty[$ and each finite ONS $\{f_1,\dots,f_n\}$ in $L^2(\widehat{\nu_\sigma},\mathbb{H})$ the image measure $\widehat{\Gamma}_{(I(f_1),\dots,I(f_n))}$ of $\widehat{\Gamma}$ by $(I(f_1),\dots,I(f_n)) : \Omega \to \mathbb{R}^n$ is equal to the n-fold centred Gaussian measure γ_1^n of variance 1.

Lemma 12.6.1 *Fix $\sigma \in \mathbb{N}$, $n \in {}^*\mathbb{N}$ and internal n-tuples $(F_1,\dots,F_n)$ in $L^2(\nu_\sigma,\mathbb{F})$ and $(\alpha_1,\dots,\alpha_n)$ in ${}^*\mathbb{R}$. Then*

(a) $\mathbb{E}e^{\sum_{j=1}^n \alpha_j I(F_j)} = e^{\frac{1}{2}\int_{T_\sigma}\left\|\sum_{j=1}^n \alpha_j F_j\right\|_{\mathbb{F}}^2 d\nu}$.

(b) *If $n \in \mathbb{N}$ and the α_j are limited and $\int_{T_\sigma}\|F_j\|_{\mathbb{F}}^2 d\nu$ is limited for each $j = 1,\dots,n$, then $e^{\sum_{j=1}^n \alpha_j I(F_j)} \in SL^p(\Gamma)$ for all $p \in [1,\infty[$.*

Proof (a) Fix an internal ONB $(\mathfrak{e}_i)_{i\in\omega}$ of $\mathbb{F}$. Set $c := \sqrt{\frac{H}{2\pi}}^{H\sigma\omega}$. Then

$$
\begin{aligned}
&\mathbb{E}e^{\sum_{j=1}^n \alpha_j I(F_j)} \\
&= c \cdot \int_\Omega e^{\sum_{t\in T_\sigma, i\in\omega}\left(-\frac{H}{2}\langle X_t,\mathfrak{e}_i\rangle_{\mathbb{F}}^2 + \langle X_t,\mathfrak{e}_i\rangle_{\mathbb{F}} \sum_{j=1}^n \alpha_j F_j(t)(\mathfrak{e}_i)\right)} dX \\
&= c \cdot \int_\Omega e^{\sum_{t\in T_\sigma, i\in\omega}\left(-\frac{H}{2}\left(\langle X_t,\mathfrak{e}_i\rangle_{\mathbb{F}} - \frac{1}{H}\sum_{j=1}^n \alpha_j F_j(t)(\mathfrak{e}_i)\right)^2 + \frac{1}{2H}\left(\sum_{j=1}^n \alpha_j F_j(t)(\mathfrak{e}_i)\right)^2\right)} dX \\
&= e^{\sum_{t\in T_\sigma, i\in\omega}\frac{1}{2H}\left(\sum_{j=1}^n \alpha_j F_j(t)(\mathfrak{e}_i)\right)^2} = e^{\frac{1}{2}\int_{T_\sigma}\left\|\sum_{j=1}^n \alpha_j F_j\right\|_{\mathbb{F}}^2 d\nu}.
\end{aligned}
$$

(b) Under the assumptions of (b) we obtain from (a) that for each $p \in [1,\infty[$

$$
\mathbb{E}e^{p\cdot\sum_{j=1}^n \alpha_j I(F_j)} = e^{\frac{p^2}{2}\int_{T_\sigma}\left\|\sum_{j=1}^n \alpha_j F_j\right\|_{\mathbb{F}}^2 d\nu} \text{ is limited.}
$$

It follows that $e^{\sum_{j=1}^n \alpha_j I(F_j)} \in SL^p(\Gamma)$ for all $p \in [1,\infty[$. □

Corollary 12.6.2 *Suppose that $n,\sigma \in \mathbb{N}$ and $\{f_1,\dots,f_n\}$ is an ONS in $L^2(\widehat{\nu_\sigma},\mathbb{H})$. Then*

$$
\widehat{\Gamma}_{(I(f_1),\dots,I(f_n))} = \gamma_1^n.
$$

Proof Fix $(\alpha_1,\ldots,\alpha_n)\in\mathbb{R}^n$ and liftings $F_i\in SL^2(\nu_\sigma,\mathbb{F})$ of f_i. Then we obtain from Lemma 12.6.1 part (b):

$$\begin{aligned}\int_{\mathbb{R}^n} e^{\sum_{j=1}^n \alpha_j x_j}\,d\widehat{\Gamma}_{(I(f_1),\ldots,I(f_n))} &= \mathbb{E}e^{\sum_{j=1}^n \alpha_j I(f_j)} = {}^{\circ}\mathbb{E}e^{\sum_{j=1}^n \alpha_j I(F_j)}\\ &= {}^{\circ}e^{\frac{1}{2}\int_{T_\sigma}\left\|\sum_{j=1}^n \alpha_j F_j\right\|_{\mathbb{F}}^2 d\nu}\\ &= e^{\frac{1}{2}\int_{T_\sigma}\left\|\sum_{j=1}^n \alpha_j f_j\right\|_{\mathbb{H}}^2 d\widehat{\nu}} = e^{\frac{1}{2}\sum_{j=1}^n \alpha_j^2}\\ &= \int_{\mathbb{R}^n} e^{\sum_{j=1}^n \alpha_j x_j}\,d\gamma_1^n(x_1,\ldots,x_n)<\infty.\end{aligned}$$

By Proposition 3.1.2, $\widehat{\Gamma}_{(I(f_1),\ldots,I(f_n))}=\gamma_1^n$. □

Corollary 12.6.3 *Suppose that $n,\sigma\in\mathbb{N}$ and $\{f_1,\ldots,f_n\}$ is an ONS in $L^2(\widehat{\nu_\sigma},\mathbb{H})$. If g_k converges to g in $L^2(\gamma_1^n)$, then $g_k\circ(I(f_1),\ldots,I(f_n))$ converges to $g\circ(I(f_1),\ldots,I(f_n))$ in $L^2(\widehat{\Gamma})$.*

Exercises

Use the notation in Section 12.1.

12.1 Prove Lemma 12.1.2 (b).
12.2 Prove that $\int\overline{\Phi}\Delta B$ is a $(\mathcal{B}_s)_{s\in\overline{T}}$-martingale.
12.3 Prove Lemma 12.1.4.
12.4 Prove Theorem 12.3.1 for full L^2.
12.5 Prove Theorem 12.4.1 (c).

As a continuation of Section 11.9:
Fix $(\mathcal{B}_{s^-})_{s\in T}$-adapted $G,F:\Lambda\times T\to {}^*\mathbb{R}$.

12.6 Prove that $\mathbb{E}_\zeta\left(\left(\int F\Delta B\right)_\sigma\cdot\left(\int G\Delta B\right)_\sigma\right)=\mathbb{E}_\zeta\int_{T_\sigma}F_s\cdot G_s d\nu(s)$ for all $\sigma\in T$.
12.7 Prove that $\int F\Delta B$ is an internal $(\mathcal{B}_t)_{t\in T}$-martingale.
12.8 Prove that $\int F\Delta B$ is S-continuous $\widehat{\zeta}$-a.s.
12.9 Use Exercise 12.8 to define the standard continuous stochastic integral $\int f da:\Lambda\times[0,\infty[\to\mathbb{R}$ for all locally square-integrable $(\mathfrak{u}_r)_{r\in[0,\infty[}$-adapted processes $f:\Lambda\times[0,\infty[\to\mathbb{R}$, where $(\mathfrak{u}_r)_{r\in[0,\infty[}$ is the standard part of $(\mathcal{B}_s)_{s\in T}$ for $\widehat{\zeta}$ and prove that it is well defined.

13

The iterated integral

Following [84], we study the iterated Itô integral and simplify some of the results and proofs there. The iterated Itô integral is infinitely close to a *finite sum of multilinear forms defined on Ω. We shall later see that each functional in $L^2(W_{C_{\mathbb{B}}})$ can be written as an orthogonal series of iterated Itô integrals. This so-called chaos decomposition result is the basis for the Malliavin calculus on $C_{\mathbb{B}}$. Recall that $L^2(W_{C_{\mathbb{B}}})$ can be identified with $L^2_{W_{C_{\mathbb{H}}}}(\widehat{\Gamma})$ (see the end of Section 11.8).

13.1 The iterated integral with and without parameters

First of all we introduce a mixture of deterministic and random functions in connection with iterated integrals. Fix an internal $F : T^{n+m} \to \mathbb{F}^{\otimes(n+m)}$ with $n, m \in \mathbb{N}_0$. Define $I_{n,m}(F) : \Omega \times T^m \to \mathbb{F}^{\otimes m}$, setting

$$I_{n,m}(F)(X,s) := \sum_{t \in T^n_<} F_{t,s}(X_{t_1}, \ldots, X_{t_n}, \cdot) = \sum_{t \in T^n_<} F_{t,s}(t, s, X_{t_1}, \ldots, X_{t_n}, \cdot).$$

Fix an ONB $\mathfrak{E}$ of $\mathbb{F}$. The following characterization of $I_{n,m}(F)(X,s)$ will often be used:

$$I_{n,m}(F)(X,s) = \sum_{t \in T^n_<} \sum_{\mathfrak{e} \in \mathfrak{E}^n} F_{t,s}(\mathfrak{e}_1, \ldots, \mathfrak{e}_n, \cdot) \cdot \langle X_{t_1}, \mathfrak{e}_1 \rangle_{\mathbb{F}} \cdot \ldots \cdot \langle X_{t_n}, \mathfrak{e}_n \rangle_{\mathbb{F}}.$$

If $n = 0$, then $I_{n,m}(F) = F$. If $n = m = 0$, then $F \in {}^*\mathbb{R}$. The following result is an application of Lemma 11.1.3.

Proposition 13.1.1 *Fix internal functions* $F : T^{n+m} \to \mathbb{F}^{\otimes(n+m)}$ *and* $G : T^{k+m} \to \mathbb{F}^{\otimes(k+m)}$. *Then*

$$\int_{\Omega \times T^m} \langle I_{n,m}(F), I_{k,m}(G)\rangle_{\mathbb{F}^m} \, d\Gamma \otimes \nu^m = \begin{cases} 0 & \text{if } n \neq k \\ \int_{T^n_< \times T^m} \langle F, G\rangle_{\mathbb{F}^{n+m}} d\nu^{n+m} & \text{if } n = k. \end{cases}$$

The function $I_{n,m}(F)$ is called the **internal iterated integral of** F **with** m **parameters**. If $m = 0$, then we write $I_n(F)$ instead of $I_{n,0}(F)$ and call $I_n(F)$ the **internal iterated integral of** F. Recall from Section 12.1 that for $n = 1$ we have $I(F) := I_1(F) = \int F \Delta B(\cdot, H)$.

Now we will show that all standard moments of $I_n(F)$ are limited if $\int_{T^n_<} \|F\|^2_{\mathbb{F}^n} d\nu^n$ is limited. The following simple but interesting lemma is used. The proof is by induction on even s.

Lemma 13.1.2 *Fix an even* $s \in \mathbb{N}$, $k \in \mathbb{N}$ *and* $m_1, \ldots, m_k \in \mathbb{N}$ *with* $m_1 + \ldots + m_k = \frac{s}{2}$. *Recall that* $s!! = (s-1) \cdot (s-3) \cdot \ldots \cdot 3$. *Then*

$$\frac{s!}{(2m_1)! \cdot \ldots \cdot (2m_k)!} \leq \frac{s!! \cdot \frac{s}{2}!}{m_1! \cdot \ldots \cdot m_k!}.$$

Theorem 13.1.3 *Fix* $s, n \in \mathbb{N}$ *and an internal* $F : T^n \to \mathbb{F}$. *Then*

$$\mathbb{E}_\Gamma \, (I_n(F))^s \begin{cases} = 0, & \text{if } s \text{ is odd} \\ \leq (s!!)^{n+1} \left(\sum_{t \in T^n_<} \|F_t\|^2_{\mathbb{F}^n} \frac{1}{H^n} \right)^{\frac{s}{2}} & \text{if } s \text{ is even.} \end{cases}$$

Proof Fix an internal ONB $(\mathfrak{e}_i)_{i \in \omega}$ of $\mathbb{F}$. We use the following shorthand: Let $t \in T^n_<$ and $i \in \omega^n$; set $F_t(\mathfrak{e}_i) = F_{t_1,\ldots,t_n}(\mathfrak{e}_{i_1}, \ldots, \mathfrak{e}_{i_n})$ and $\langle X_t, \mathfrak{e}_i\rangle = \langle X_{t_1}, \mathfrak{e}_{i_1}\rangle \cdot \ldots \cdot \langle X_{t_n}, \mathfrak{e}_{i_n}\rangle$. Let $\sqsubset$ be the lexicographic order on $T^n_< \times \omega^n$. We apply the hyperfinite-dimensional binomial formula (see Heuser [45, page 69] and Lemma 11.1.3, in particular part (d)):

$$\begin{aligned} &\mathbb{E}_\Gamma \, (I_n(F))^s \\ &= \mathbb{E}_\Gamma \Bigg(\sum_{(t,i) \in T^n_< \times \omega^n} \langle X_t, \mathfrak{e}_i\rangle_{\mathbb{F}} F_t(\mathfrak{e}_i) \Bigg)^s \\ &= \sum_{k=1}^{s} \sum_{\substack{m_1, \ldots, m_k \in \mathbb{N}, \\ m_1 + \ldots + m_k = s}} \sum_{(t^1, i^1) \sqsubset \ldots \sqsubset (t^k, i^k) \in T^n_< \times \omega^n} \frac{s!}{m_1! \cdot \ldots \cdot m_k!} \\ &\quad \cdot \left(F_{t^1}(\mathfrak{e}_{i^1})\right)^{m_1} \cdot \ldots \cdot \left(F_{t^k}(\mathfrak{e}_{i^k})\right)^{m_k} \mathbb{E}_\Gamma \langle X_{t^1}, \mathfrak{e}_{i^1}\rangle^{m_1}_{\mathbb{F}} \cdot \ldots \cdot \langle X_{t^k}, \mathfrak{e}_{i^k}\rangle^{m_k}_{\mathbb{F}}. \end{aligned}$$

We compute the expected value in the preceding equality. It is 0 if one of the m_j is odd. So we may assume that all m_j are even, thus s is also even. Using the fact that $\left(\langle X_{t^1}, \mathfrak{e}_{i^1}\rangle_{\mathbb{F}}^{m_1}, \ldots, \langle X_{t^k}, \mathfrak{e}_{i^k}\rangle_{\mathbb{F}}^{m_k}\right)$ is independent if the $(t_j, \mathfrak{e}_j)$ are pairwise different, we obtain

$$\mathbb{E}_\Gamma (I_n(F))^s$$

$$\leq \sum_{k=1}^{s} \sum_{\substack{m_1,\ldots,m_k \in 2\mathbb{N},\\ m_1+\ldots+m_k=s}} \sum_{(t^1,i^1)\sqsubset\ldots\sqsubset(t^k,i^k)\in T^n_< \times \omega^n} \cdot \frac{s!\,(s!!)^n}{m_1!\ldots m_k!}\left(\frac{1}{H^{\frac{m_1}{2}}}\cdot\ldots\cdot\frac{1}{H^{\frac{m_k}{2}}}\right)^n$$

$$\cdot \left(F_{t^1}(\mathfrak{e}_{i^1})\right)^{m_1}\cdot\ldots\cdot\left(F_{t^k}(\mathfrak{e}_{i^k})\right)^{m_k}$$

$$= \left(\frac{1}{H^{\frac{s}{2}}}\right)^n \sum_{k=1}^{\frac{s}{2}} \sum_{\substack{m_1,\ldots,m_k \in \mathbb{N},\\ m_1+\ldots+m_k=\frac{s}{2}}} \sum_{(t^1,i^1)\sqsubset\ldots\sqsubset(t^k,i^k)\in T^n_< \times \omega^n} \frac{s!\,((s!!))^n}{(2m_1)!\ldots(2m_k)}$$

$$\cdot \left(F^2_{t^1}(\mathfrak{e}_{i^1})\right)^{m_1}\ldots\left(F^2_{t^k}(\mathfrak{e}_{i^k})\right)^{m_k}$$

$$\leq \left(\frac{s!!}{H^{\frac{s}{2}}}\right)^n \sum_{k=1}^{\frac{s}{2}} \sum_{\substack{m_1,\ldots,m_k \in \mathbb{N},\\ m_1+\ldots+m_k=\frac{s}{2}}} \sum_{(t^1,i^1)\sqsubset\ldots\sqsubset(t^k,i^k)\in T^n_< \times \omega^n} \frac{s!!\,\frac{s}{2}!}{m_1!\ldots m_k!}$$

$$\cdot \left(F^2_{t^1}(\mathfrak{e}_{i^1})\right)^{m_1}\cdot\ldots\cdot\left(F^2_{t^k}(\mathfrak{e}_{i^k})\right)^{m_k}$$

$$\leq (s!!)^{n+1} \sum_{k=1}^{\frac{s}{2}} \sum_{\substack{m_1,\ldots,m_k \in \mathbb{N},\\ m_1+\ldots+m_k=\frac{s}{2}}} \sum_{(t^1,i^1)\sqsubset\ldots\sqsubset(t^k,i^k)\in T^n_< \times \omega^n} \frac{\frac{s}{2}!}{m_1!\ldots m_k!}\left(\frac{1}{H^{\frac{s}{2}}}\right)^n$$

$$\cdot \left(F^2_{t^1}(\mathfrak{e}_{i^1})\right)^{m_1}\cdot\ldots\cdot\left(F^2_{t^k}(\mathfrak{e}_{i^k})\right)^{m_k}$$

$$= (s!!)^{n+1}\left(\sum_{(t,i)\in T^n_<\times\omega^n} F^2_t(\mathfrak{e}_i)\frac{1}{H^n}\right)^{\frac{s}{2}} = (s!!)^{n+1}\left(\int_{T^n_<}\|F_t\|^2_{\mathbb{F}^n}\,d\nu^n\right)^{\frac{s}{2}}.$$

□

Corollary 13.1.4 *Assume that $\int_{T^n_<}\|F_t\|^2_{\mathbb{F}^n}\,d\nu^n$ is limited. Then, for all $p\in[1,\infty[$, $I_n(F)\in SL^p(\Gamma)$.*

In order to define the iterated integral with parameters, Theorem 13.1.6 below is crucial, where we use the following lemma, which is a straightforward application of Jensen's inequality.

Lemma 13.1.5 *Fix $D:\Omega\times T\to\mathbb{F}\in SL^2(\Gamma\otimes\nu,\mathbb{F})$. Then*

$$A:(\cdot,s)\mapsto\mathbb{E}^{\mathcal{B}_{s^-}}D(\cdot,s)\in SL^2(\Gamma\otimes\nu,\mathbb{F}).$$

Moreover, the close relationship between Loeb and Lebesgue measure is used. The results in Section 10.5 imply that for each $f\in L^2_{\mathcal{L}^m}\left(\widehat{\Gamma\otimes\nu^m},\mathbb{H}^{\otimes m}\right)$ there exists a $g\in L^2\left(\widehat{\Gamma}\otimes\lambda^m,\mathbb{H}^{\otimes m}\right)$ with $f(X,t)=g(X,{}^\circ t)$ for $\widehat{\Gamma\otimes\nu^m}$-almost all (X,t). We can identify f and g.

Theorem 13.1.6 *Fix $f:[0,\infty[^{n+m}\to\mathbb{H}^{\otimes(n+m)}\in L^2\left(\lambda^{n+m},\mathbb{H}^{\otimes(n+m)}\right)$. Let $F:T^{n+m}\to\mathbb{F}^{\otimes(n+m)}\in SL^2\left(\nu^{n+m},\mathbb{F}^{\otimes(n+m)}\right)$ be a lifting of f (see Theorem 10.9.2).*

(a) *Then $I_{n,m}(F)\in SL^2\left(\Gamma\otimes\nu^m,\mathbb{F}^{\otimes m}\right)$ and the standard part ${}^\circ I_{n,m}(F)$ exists in $L^2\left(\widehat{\Gamma}\otimes\lambda^m,\mathbb{H}^{\otimes m}\right)=L^2_{\mathcal{L}^m}\left(\widehat{\Gamma\otimes\nu^m},\mathbb{H}^{\otimes m}\right)$.*

(b) *Let $m=1$. Then $t\mapsto\mathbb{E}^{\mathrm{B}_{t^-}}I_{n,1}(F)(\cdot,t)\in SL^2(\Gamma\otimes\nu,\mathbb{F})$.*

Proof

(a) To save indices, let $n=m=1$. Let M be the set of all $g\in L^2\left(\lambda^2,\mathbb{H}^{\otimes 2}\right)$ such that there exists a lifting $G\in SL^2\left(\nu^2,\mathbb{F}^{\otimes 2}\right)$ of g with $I_{1,1}(G)\in SL^2(\Gamma\otimes\nu,\mathbb{F})$ and ${}^\circ I_{1,1}(G)$ exists in $L^2\left(\widehat{\Gamma}\otimes\lambda,\mathbb{H}\right)$. Then M is a linear space and complete, which is an application of Theorem 10.9.2, Corollary 10.9.3 and Proposition 13.1.1. To prove that $M=L^2\left(\lambda^2,\mathbb{H}^{\otimes 2}\right)$, let B_1,B_2 be bounded Lebesgue-measurable subsets of $[0,\infty[$ and $a\in\mathbb{H}^2$. By Theorems 10.5.2 and 10.1.2, there exist limited internal subsets A_1,A_2 of T with $\widehat{\nu}(A_i\Delta st^{-1}[B_i])=0$. Set $A:=A_1\times A_2$, $B:=B_1\times B_2$, ${}^*a:=({}^*a_1,{}^*a_2)$. Then

$$G:=\mathbf{1}_A\otimes[{}^*a]:T^2\to\mathbb{F}^{\otimes 2},(t,b)\mapsto\mathbf{1}_A(t)\cdot\langle{}^*a_1,b_1\rangle\cdot\langle{}^*a_2,b_2\rangle$$

is in $SL^2(\nu^2,\mathbb{F}^{\otimes 2})$ and a lifting of $g:=\mathbf{1}_B\otimes[a]$. Moreover,

$$I_{1,1}(G)=I_1(\mathbf{1}_{A_1}\otimes[{}^*a_1])\otimes(\mathbf{1}_{A_2}\otimes[{}^*a_2]).$$

By Corollary 13.1.4, $I_1(\mathbf{1}_{A_1}\otimes[{}^*a_1])\in SL^2(\Gamma)$ and, therefore, its standard part exists in $L^2\left(\widehat{\Gamma}\right)$. Moreover, $\mathbf{1}_{A_2}\otimes[{}^*a_2]$ is in $SL^2(\nu,\mathbb{F})$ and is a lifting of $\mathbf{1}_{B_2}\otimes[a_2]$. By Lemma 10.12.2 (b), $I_{1,1}(G)\in SL^2(\Gamma\otimes\nu,\mathbb{F})$ and its standard part exists. By Corollary 5.6.3, $M=L^2\left(\lambda^2,\mathbb{H}^{\otimes 2}\right)$.

Now fix a lifting $G\in SL^2\left(\nu^2,\mathbb{F}^{\otimes 2}\right)$ of f with $I_{1,1}(G)\in SL^2(\Gamma\otimes\nu,\mathbb{F})$ and ${}^\circ I_{1,1}(G)$ exists in $L^2\left(\widehat{\Gamma}\otimes\lambda,\mathbb{H}\right)$. By Proposition 13.1.1,

$$\int_{\Omega\times T}\|I_{1,1}(F-G)\|^2_{\mathbb{F}}\,d\Gamma\otimes\nu=\int_{T^2}\|F-G\|^2_{\mathbb{F}^2}\,d\nu^2\approx 0.$$

Part (a) follows. Part (b) follows from Theorem 13.1.6 and Lemma 13.1.5. □

Now fix $f \in L^2\left(\lambda^{n+m}, \mathbb{H}^{\otimes(n+m)}\right)$ and a lifting $F \in SL^2\left(\nu^{n+m}, \mathbb{F}^{\otimes(n+m)}\right)$ and define

$$I_{n,m}(f) = I_{n,m}({}^\circ F) := {}^\circ I_{n,m}(F) : \Omega \times [0,\infty[^m \to \mathbb{H}^{\otimes m} \in L^2\left(\widehat{\Gamma} \otimes \lambda^m, \mathbb{H}^{\otimes m}\right).$$

By Theorem 13.1.6 and Proposition 13.1.1, $I_{n,m}(f)$ is well defined. From Proposition 13.1.1 we obtain, using S-square-integrable liftings of f and g and the simple fact that $\langle a,b\rangle_{\mathbb{H}^m} = \frac{1}{2}\left(\|a\|^2_{\mathbb{H}^m} + \|b\|^2_{\mathbb{H}^m} - \|a-b\|^2_{\mathbb{H}^m}\right)$, the following result.

Proposition 13.1.7 *Fix functions $f \in L^2\left(\lambda^{n+m}, \mathbb{H}^{\otimes(n+m)}\right)$ and $g \in L^2\left(\lambda^{k+m}, \mathbb{H}^{\otimes(k+m)}\right)$ Then*

$$\int_{\Omega\times[0,\infty[^m} \langle I_{n,m}(f), I_{k,m}(g)\rangle_{\mathbb{H}^m}\, d\widehat{\Gamma}\otimes\lambda^m$$
$$= \begin{cases} 0 & \text{if } n \neq k \\ \int_{[0,\infty[^n_< \times [0,\infty[^m} \langle f,g\rangle_{\mathbb{H}^{n+m}} d\lambda^{n+m} & \text{if } n = k. \end{cases}$$

Corollary 13.1.8 *The iterated integral with m parameters is a continuous operator, i.e., if $(f_j)_{j\in\mathbb{N}}$ is a sequence converging to f in the space $L^2\left(\lambda^{n+m}, \mathbb{H}^{\otimes(n+m)}\right)$, then*

$$\lim_{j\to\infty} I_{n,m}(f_j) = I_{n,m}(f) \text{ in } L^2\left(\widehat{\Gamma}\otimes\lambda^m, \mathbb{H}^{\otimes m}\right).$$

The function $I_{n,m}(f)$ is called the **iterated integral with m parameters.** Recall that this process is equivalent to a process $g : \Omega \times T^m \to \mathbb{H}^{\otimes m} \in L^2_{\mathcal{L}^m}\left(\widehat{\Gamma \otimes \nu^m}, \mathbb{H}^{\otimes m}\right)$, which is also denoted by $I_{n,m}(f)$. Now assume that $m = 0$, i.e., $f \in L^2\left(\lambda^n, \mathbb{H}^{\otimes n}\right)$. By Theorem 10.9.2 f has a lifting $F : T^n \to \mathbb{F}^{\otimes n}$ in $SL^2\left(\nu^n, \mathbb{F}^{\otimes n}\right)$. Then we set

$$I_n(f) = I_n({}^\circ F) := {}^\circ I_n(F)$$

and call $I_n(f)$ the **n-fold iterated Itô integral of** f. Note that it is well defined $\widehat{\Gamma}$-a.s.

13.2 The product of an internal iterated integral and an internal Wiener integral

In this section fix an internal symmetric function $F : T^n_{\neq} \to \mathbb{F}^{\otimes n}$ and $G : T \to \mathbb{F}$. We use the following common notation. If $K : T^{n+1}_{\neq} \to \mathbb{F}^{\otimes(n+1)}$ is internal and symmetric in the first n arguments, define

$$\widetilde{K}_{(t_1,\ldots,t_{n+1})}(a_1,\ldots,a_{n+1})$$
$$:= \sum_{j=1}^{n+1} K_{(t_1,\ldots,t_{j-1},t_{j+1},\ldots,t_{n+1}t_j,)}(a_1,\ldots,a_{j-1},a_{j+1},\ldots,a_{n+1},a_j).$$

Then $\widetilde{K}$ is symmetric. Define $\int_T \langle F_s, G_s \rangle_{\mathbb{F}} d\nu(s) : T^{n-1}_{\neq} \to \mathbb{F}^{\otimes(n-1)}$, setting

$$\left(\int_T \langle F_s, G_s \rangle_{\mathbb{F}} d\nu(s)\right)_{(t_1,\ldots,t_{n-1})} (a_1,\ldots,a_{n-1})$$
$$:= \int_T \left\langle F_{(t_1,\ldots,t_{n-1},s)}(a_1,\ldots,a_{n-1},\cdot), G_s(\cdot)\right\rangle_{\mathbb{F}} d\nu(s)$$
$$= \int_T \sum_{i\in\omega} F_{(t_1,\ldots,t_{n-1},s)}(a_1,\ldots,a_{n-1},\mathfrak{e}_i) \cdot G_s(\mathfrak{e}_i) d\nu(s),$$

where $(\mathfrak{e}_i)_{i\in\omega}$ is an internal ONB of $\mathbb{F}$. Note that $(\widetilde{F \otimes G})$ is symmetric.

Theorem 13.2.1 *Assume that $F \in SL^2(\nu^n, \mathbb{F}^{\otimes n})$ is symmetric and G belongs to $SL^2(\nu, \mathbb{F})$. Then*

$$I_n(F) \cdot I(G) \approx I_{n+1}(\widetilde{F \otimes G}) + I_{n-1}\left(\int_T \langle F_s, G_s \rangle_{\mathbb{F}} d\nu(s)\right) \text{ in } L^2(\Gamma). \quad (27)$$

In particular, for $F := G^{\otimes n} := G \otimes \ldots \otimes G$ (n times) we have:

$$I_n(G^{\otimes n}) \cdot I_1(G) \approx (n+1) I_{n+1}(G^{\otimes(n+1)}) + I_{n-1}(G^{\otimes(n-1)}) \left(\int_T \|G\|^2_{\mathbb{F}} d\nu\right).$$

Proof We may assume that $F : T^n_{\neq} \to \mathbb{F}^{\otimes n}$. Fix $S \in \mathbb{N}$ and set $F^S := F \upharpoonright T^n_S$ and $G^S := G \upharpoonright T_S$. Note that

$$I_n(F^S)(X) \cdot I(G^S)(X) = A + \sum_{j=1}^{n} B_j + \sum_{j=1}^{n} C_j,$$

where

$$A = \sum_{t \in T^n_<, s \notin \{t_1,\ldots,t_n\}} F^S_t(X_{t_1},\ldots,X_{t_n}) G^S_s(X_s)$$

$$= \sum_{j=1}^{n+1} \sum_{t \in T^{n+1}_<} F^S_{t_1,\ldots,[t_j],\ldots,t_{n+1}}(X_{t_1},\ldots,[X_{t_j}],\ldots,X_{t_{n+1}}) G^S_{t_j}(X_{t_j}) = I_{n+1}(\widetilde{F^S \otimes G^S}),$$

$$B_j = \sum_{(t,i) \in T^n_< \times \omega^n} \sum_{k \neq i_j} \langle X_{t_1}, \mathfrak{e}_{i_1}\rangle_{\mathbb{F}} \cdots \langle X_{t_n}, \mathfrak{e}_{i_n}\rangle_{\mathbb{F}} F^S_t(\mathfrak{e}_{i_1},\ldots,\mathfrak{e}_{i_n}) \left\langle X_{t_j}, \mathfrak{e}_k \right\rangle_{\mathbb{F}} G^S_{t_j}(\mathfrak{e}_k).$$

Note that

$$\mathbb{E}_\Gamma B_j^2 \leq \sum_{t \in T^n_<} \left\| F^S_t \right\|^2_{\mathbb{F}^n} \cdot \left\| G^S_{t_j} \right\|^2_{\mathbb{F}} \frac{1}{H^{n+1}}$$

$$= \sum_{\{(t,s) \in T^n_< \times T \mid s = t_j\}} \left\| F^S_t \right\|^2_{\mathbb{F}^n} \cdot \left\| G^S_s \right\|^2_{\mathbb{F}} \frac{1}{H^{n+1}} \approx 0,$$

because $\{(t,s) \in T^n_< \times T \mid s = t_j\}$ is a $\widehat{\nu^{n+1}}$-nullset, and, by Lemma 10.12.2 part (b), $S^{n+1} \ni (t,s) \mapsto \left\| F^S_t \right\|^2_{\mathbb{F}^n} \cdot \left\| G^S_s \right\|^2_{\mathbb{F}} \in SL^1\left(\nu^{n+1}_S\right)$.

$$C_j = \sum_{t \in T^n_<} \sum_{i \in \omega^n} \langle X_{t_1}, \mathfrak{e}_{i_1}\rangle_{\mathbb{F}} \cdots \left\langle X_{t_j}, \mathfrak{e}_{i_j} \right\rangle^2_{\mathbb{F}} \cdots \langle X_{t_n}, \mathfrak{e}_{i_n}\rangle_{\mathbb{F}} F^S_t(\mathfrak{e}_{i_1},\ldots,\mathfrak{e}_{i_n}) G^S_{t_j}(\mathfrak{e}_{i_j}).$$

Define

$$\widetilde{C}_j = \sum_{t \in T^n_<} \sum_{i \in \omega^n} \langle X_{t_1}, \mathfrak{e}_{i_1}\rangle_{\mathbb{F}} \cdots \frac{1}{H} \cdots \langle X_{t_n}, \mathfrak{e}_{i_n}\rangle_{\mathbb{F}} F^S_t(\mathfrak{e}_{i_1},\ldots,\mathfrak{e}_{i_n}) G^S_{t_j}(\mathfrak{e}_{i_j}).$$

Since $\mathbb{E}\left(\left(\langle X_s, \mathfrak{e}_i\rangle^2 - \frac{1}{H}\right)^2\right) = \frac{2}{H^2}$, we have $\mathbb{E}_\Gamma\left(C_j - \widetilde{C}_j\right)^2 \approx 0$. Since F^S is symmetric, we see that in $L^2(\Gamma)$

$$\sum_{j=1}^n C_j \approx \sum_{j=1}^n \widetilde{C}_j = I_{n-1}\left(\int_T \langle F^S_s, G^S_s\rangle_{\mathbb{F}} \, d\nu(s)\right).$$

It follows that equation (27) is true for $F := F^S$ and $G = G^S$ with $S \in \mathbb{N}$. Using Proposition 8.4.1 (b), there exists an unlimited $S \in {}^*\mathbb{N}$ such that equation (27) is true for $F = F^S$ and $G = G^S$. Using the Cauchy–Schwarz inequality,

Theorem 13.1.3, the S-square-integrability of F and G and finite combinatorics, we obtain

$$\mathbb{E}_\Gamma\left(I_n(F)\cdot I(G) - I_n(F^S)\cdot I(G^S)\right)^2 \approx 0,$$

$$\mathbb{E}_\Gamma\left(I_{n+1}(\widetilde{F\otimes G}) - I_{n+1}(\widetilde{F^S\otimes G^S})\right)^2 \approx 0,$$

$$\mathbb{E}_\Gamma\left(I_{n-1}\left(\int_T \langle F_s, G_s\rangle_{\mathbb{F}}\, d\nu(s)\right) - I_{n-1}\left(\int_T \langle F_s^S, G_s^S\rangle_{\mathbb{F}}\, d\nu(s)\right)\right)^2 \approx 0$$

The proof of equation (27) is finished. □

Using Theorem 13.2.1 and the standard part map, one can prove by induction on the degree of polynomials the following crucial result.

Corollary 13.2.2 *Fix a polynomial p in $\mathbb{R}$.*

(a) *Suppose that $G \in SL^2(\nu, \mathbb{F})$. Then $p(I(G))$ is infinitely close in $L^2(\Gamma)$ to a standardly long linear combination of internal iterated integrals of the form $I_n(G^{\otimes n})$ with limited scalars.*
(b) *Suppose that $g \in L^2(\lambda, \mathbb{H})$. Then $p(I(g))$ is a linear combination of iterated Itô integrals, where the integrands are tensor products of g.*

13.3 The continuity of the standard iterated integral process

Now we study the n-fold iterated Itô integral as a continuous stochastic process. Fix an internal $F: T^n \to \mathbb{F}^{\otimes n}$. The internal process $I_n^M(F): \Omega \times T \to {}^*\mathbb{R}$, defined by

$$I_n^M(F)(X,s) = \sum_{t_1<\ldots<t_n\leq s} F_{(t_1,\ldots,t_n)}(X_{t_1},\ldots,X_{t_n})$$

is called the **internal iterated integral process of** F. Note that $I_n^M(F)$ is an internal stochastic integral:

$$I_n^M(F)(X,s) = \left(\int r \mapsto \mathbb{E}^{\mathcal{B}_{r^-}} I_{n-1,1}(F)(\cdot, r)\Delta B\right)(X,s)$$
$$= \sum_{r\leq s}\left(\sum_{t_1<\ldots<t_{n-1}<r} F_{t_1,\ldots,t_{n-1},r}(X_{t_1},\ldots,X_{t_{n-1}},\cdot)\right)(X_r).$$

By Theorem 13.1.6, Proposition 12.2.1 and Theorem 12.1.6, we obtain:

Theorem 13.3.1 *Let f be locally in $L^2\left(\lambda^n, \mathbb{H}^{\otimes n}\right)$ and let F be a lifting of f locally in $SL^2\left(\nu^n, \mathbb{F}^{\otimes n}\right)$. Then $I_{n-1,1}(F)$ and $r \mapsto \mathbb{E}^{\mathcal{B}_{r^-}} I_{n-1,1}(F)(\cdot, r)$ are locally in $SL^2(\Gamma \otimes \nu, \mathbb{F})$ and, for all $\sigma \in \mathbb{N}$,*

$$\max_{s \in T_\sigma} \left| I_n^M(F)(X,s) \right| \in SL^2(\Gamma), I_n^M(F) \upharpoonright \Omega \times T_\sigma \in SL^2(\Gamma \otimes \nu).$$

Moreover, $I_n^M(F)$ is S-continuous $\widehat{\Gamma}$-a.s.

Fix f, F as in Theorem 13.3.1. Then, by Theorem 12.4.1, the process $I_n^M(f)$: $\Omega \times [0, \infty[\to \mathbb{R}$, defined for all limited t by

$$I_n^M(f)(\cdot, {}^\circ t) := {}^\circ\left(I_n^M(F)(\cdot, t)\right),$$

is a continuous square-integrable $(\mathfrak{b}_r)_{r \in [0,\infty[}$-martingale. The result is an extension of a result due to Bouleau and Hirsch [16] for the classical Wiener space. The process $I_n^M(f)$ is called the **continuous iterated integral of** f.

Using suitable S-square-integrable liftings and Proposition 13.1.1 we obtain:

Corollary 13.3.2 *Fix $r \in [0, \infty[$ and f locally in $L^2\left(\lambda^n, \mathbb{H}^{\otimes n}\right)$ and g locally in $L^2\left(\lambda^m, \mathbb{H}^{\otimes m}\right)$. Then*

$$\mathbb{E}\left(I_n^M f\right)(\cdot, r) \cdot \left(I_m^M g\right)(\cdot, r) = \begin{cases} \int_{[0,r]^n} \langle f_s, g_s \rangle_{\mathbb{H}^n}^2 \, d\lambda^n(s) & \text{if } m = n \\ 0 & \text{if } m \neq n. \end{cases}$$

13.4 The $\mathcal{W}_{C_{\mathbb{H}}}$-measurability of the iterated Itô integral

Fix $r \in \mathbb{N}$ and $f \in L^2\left(\lambda_r^n, \mathbb{H}^{\otimes n}\right)$, the space of square-integrable functions with support $[0,r]^n$. Let F be a lifting of f in $SL^2\left(\nu_r^n, \mathbb{F}^{\otimes n}\right)$. We shall now see that $I_n(f)$ is $\mathcal{W}_{C_{\mathbb{H}}}$-measurable. A modified iterated Itô integral $J_n(F): \Omega \to {}^*\mathbb{R}$ is defined by setting:

$$J_n(F)(X) := \sum_{(t_1,\ldots,t_n) \in T_{\neq}^n} F_{(t_1,\ldots,t_n)}(X_{t_1}, \ldots, X_{t_n}).$$

Set $F^\sigma_{(t_1,\ldots,t_n)}(X_{t_1}, \ldots, X_{t_n}) := F_{(t_{\sigma_1},\ldots,t_{\sigma_n})}(X_{t_{\sigma_1}}, \ldots, X_{t_{\sigma_n}})$, where σ is a permutation of $\{1, \ldots, n\}$. Note that $J_n(F)(X) = I_n\left(\sum_\sigma F^\sigma\right)$, where σ runs through all permutations of $\{1, \ldots, n\}$. We obtain results which are similar to the results for $I_n(f)$, in particular, $J_n(F) \in SL^2(\Gamma)$ and $\mathbb{E}_\Gamma\left(J_n(F-G)\right)^2 \approx 0$ if $F \approx_{\mathbb{F}^n} G$ in $L^2(\nu^n)$. Therefore, we may set $J_n(f) := {}^\circ J_n(F)$. Moreover, J_n is a continuous operator and we obtain the following recursion formula for J:

Lemma 13.4.1 *Fix $F \in SL^2(\nu_r^n, \mathbb{F}^{\otimes n})$ and $G \in SL^2(\nu_r, \mathbb{F})$, where r is limited. Then in $L^2(\Gamma)$*

$$J_{n+1}(F \otimes G) \approx J_n(F) \cdot J(G) - \sum_{i=1}^{n} J_{n-1}\left(\int_{T_r} \langle F^i(\cdot,t), G(t)\rangle_{\mathbb{F}} \, d\nu(t) \right),$$

where

$$\begin{aligned} &F^i_{(t_1,\ldots,t_{n-1},t)}(a_1,\ldots,a_{n-1},a) \\ &:= F_{(t_1,\ldots,t_{i-1},t,t_i,\ldots,t_{n-1})}(a_1,\ldots,a_{i-1},a,a_i,\ldots,a_{n-1}). \end{aligned}$$

Using this result and suitable liftings of $f \in L^2(\lambda_r^n, \mathbb{H}^{\otimes n})$ and $g \in L^2(\lambda_r, \mathbb{H})$ we obtain

$$J_{n+1}(f \otimes g) = J_n(f) \cdot J(g) - \sum_{i=1}^{n} J_{n-1} \int_T \langle f^i(\cdot,t), g(t)\rangle_{\mathbb{H}} \, d\widehat{\nu}(t) \text{ in } L^2\left(\widehat{\Gamma}\right).$$

Proposition 13.4.2 *$J_n(f)$ is $\mathcal{W}_{C_{\mathbb{H}}}$-measurable.*

Proof Fix $r \in [0,\infty[$ and let M be the set of functions $f \in L^2(\lambda_r^n, \mathbb{H}^{\otimes n})$ such that $J_n(f)$ is $\mathcal{W}_{C_{\mathbb{H}}}$-measurable. Since J_n is a continuous operator on $L^2(\lambda_r^n, \mathbb{H}^{\otimes n})$, M is a complete linear space. In order to prove that $M = L^2(\lambda_r^n, \mathbb{H}^{\otimes n})$, it is sufficient to prove (see Proposition 5.6.1) that $f \in M$ for functions f of the following form: $f = \mathbf{1}_B \otimes [h_1,\ldots,h_n] : [0,r]^n \to \mathbb{H}^{\otimes n}$ with $f_{(t_1,\ldots,t_n)}(a_1,\ldots,a_n) = \mathbf{1}_B(t_1,\ldots,t_n) \cdot h_1(a_1) \cdot \ldots \cdot h_n(a_n)$, where $B = B_1 \times \ldots \times B_n$ with $B_i \in \text{Leb}\,[0,r]$ and $h_1,\ldots,h_n \in \mathbb{H}$. We prove this result by induction on n. For $n=1$ it is true by Proposition 12.5.1. We set for $i = 1,\ldots,n$

$$\mathbf{1}_{B_1\times\ldots[B_i]\ldots\times B_n} : (t_1,\ldots,t_{n-1}) \mapsto \mathbf{1}_{B_1}(t_1)\ldots\mathbf{1}_{B_{i-1}}(t_{i-1})\mathbf{1}_{B_{i+1}}(t_i)\ldots\mathbf{1}_{B_n}(t_{n-1})$$

and

$$[h_1,\ldots,[h_i],\ldots,h_n] : (a_1,\ldots,a_{n-1}) \mapsto h_1(a_1)\ldots h_{i-1}(a_{i-1})h_{i+1}(a_i)\ldots h_n(a_{n-1}).$$

Then

$$\begin{aligned} J_{n+1}(f) &= J_{n+1}(\mathbf{1}_{B_1\times\ldots\times B_{n+1}} \otimes [h_1,\ldots,h_{n+1}]) = A - B\cdot C \\ \text{with} \quad A &= J_n(\mathbf{1}_{B_1\times\ldots\times B_n} \otimes [h_1,\ldots,h_n]) \cdot J_1(\mathbf{1}_{B_{n+1}} \otimes h_{n+1}) \\ B &= \sum_{i=1}^{n} J_{n-1}\left(\mathbf{1}_{B_1\times\ldots[B_i]\ldots\times B_n} \otimes [h_1,\ldots,[h_i],\ldots,h_n]\right) \\ C &= \int_{[0,r]} \mathbf{1}_{B_i}(t) \cdot \mathbf{1}_{B_{n+1}}(t) \cdot \langle h_i, h_{n+1}\rangle_{\mathbb{H}} \, d\lambda(t) \end{aligned}$$

is $\mathcal{W}_{C_{\mathbb{H}}}$-measurable by the induction hypothesis. It follows that $J_n(f)$ is $\mathcal{W}_{C_{\mathbb{H}}}$-measurable for all $f \in L^2(\lambda_r^n, \mathbb{H}^{\otimes n})$, thus for all $f \in L^2(\lambda^n, \mathbb{H}^{\otimes n})$. □

Corollary 13.4.3 *$I_n(f)$ belongs to $L^2_{\mathcal{W}_{C_{\mathbb{H}}}}(\widehat{\Gamma})$.*

Proof Since in $I_n(f)$ only increasing n-tuples are involved, we may assume that f is symmetric. Now $I_n(f) = \frac{1}{n!} J_n(f) \in L^2_{\mathcal{W}_{C_{\mathbb{H}}}}(\widehat{\Gamma})$. □

Using the proof of Theorem 13.1.6 and Corollary 13.4.3, we obtain a measurability result for iterated integrals with parameters:

Corollary 13.4.4 *Fix $f : [0,\infty[^{n+m} \to \mathbb{H}^{\otimes(n+m)}$ in $L^2\left(\lambda^{n+m}, \mathbb{H}^{\otimes(n+m)}\right)$. Then $I_{n,m}(f)$ is in $L^2_{\mathcal{W}_{C_{\mathbb{H}}} \otimes \mathrm{Leb}^m}\left(\widehat{\Gamma} \otimes \lambda^m, \mathbb{H}^{\otimes m}\right)$. Equivalently, we obtain for the timeline T: if F is a lifting of f, then*

$$^{\circ}I_{n,m}(F) : (X,t) \mapsto I_{n,m}(f)(\cdot, {}^{\circ}t) \in L^2_{\mathcal{W}_{C_{\mathbb{H}}} \otimes \mathcal{L}^m}\left(\widehat{\Gamma \otimes \nu^m}, \mathbb{H}^{\otimes m}\right).$$

13.5 $I_n^M(f)$ is a continuous version of the standard part of $I_n^M(F)$

In the preceding section we constructed a continuous $(\mathfrak{b}_r)_{r \in [0,\infty[}$-martingale $I_n^M(f)$ from $I_n^M(F)$, where F is locally in $SL^2(\nu^n, \mathbb{F}^{\otimes n})$ and is a lifting of an f, locally in $L^2(\lambda^n, \mathbb{H}^{\otimes n})$. Since $I_n^M(F)$ is a $(\mathcal{B}_t)_{t \in T}$-martingale and since, by Theorem 13.3.1, $I_n^M(F)(\cdot,t) \in SL^2(\Gamma)$ for all limited t, we also can take the standard part $i_n^M(f)$ of $I_n^M(F)$, according to Corollary 10.16.3, to obtain a càdlàg $(\mathfrak{b}_r)_{r \in [0,\infty[}$-martingale. Recall that

$$i_n^M(f)(\cdot,r) = \lim_{{}^{\circ}t \downarrow r} {}^{\circ}I_n^M(F)(\cdot,t).$$

Let f, g be processes defined on $\Omega \times [0,\infty[$. Then f is called a **version of** g if, for all $r \in [0,\infty[$, $g_r = f_r$ $\widehat{\Gamma}$-a.s., where the exceptional nullset may depend on r.

Theorem 13.5.1 (a) *$I_n^M(f)$ is a continuous version of $i_n^M(f)$.*
(b) $\mathbb{E}^{\mathfrak{b}_r}({}^{\circ}I_n(F)) = I_n^M(f)(\cdot,r) = i_n^M(f)(\cdot,r)$ *$\widehat{\Gamma}$-a.s. for $r \in [0,\infty[$. The exceptional nullset depends on r.*

Proof Fix $r \in [0,\infty[$.

(a) By Theorem 10.16.2 (b), there exists a $t \in T$ with $t \approx r$ such that $I_n^M(F)(\cdot,t)$ is a lifting of $i_n^M(f)(\cdot,r)$. It follows that $\widehat{\Gamma}$-a.s.

$$I_n^M(f)(\cdot,r) = {}^{\circ}I_n^M(F)(\cdot,t) = i_n^M(f)(\cdot,r).$$

(b) Let $(t_k)_{k\in\mathbb{N}}$ be a decreasing sequence in the limited part of T such that $\lim_{k\to\infty} {}^\circ t_k = r$ and ${}^\circ t_k > r$. By Theorem 10.6.1, $\mathfrak{b}_r = \bigcap_{k\in\mathbb{N}} \left(\mathcal{B}_{t_k} \vee \mathcal{N}_{\widehat{\Gamma}}\right)$. By the martingale convergence theorem (see Billingsley [13] Theorem 35.9),

$$\lim_{k\to\infty} \mathbb{E}^{\mathcal{B}_{t_k}\vee\mathcal{N}_{\widehat{\Gamma}}} \left({}^\circ I_n(F)\right) = \mathbb{E}^{\mathfrak{b}_r} \left({}^\circ I_n(F)\right).$$

On the other side, $\lim_{k\to\infty} {}^\circ I_n^M(F)(\cdot,t_k) = i_n^M(f)(\cdot,r)\ \widehat{\Gamma}$-a.s. Now note that ${}^\circ I_n^M(F)(\cdot,t_k) = {}^\circ\mathbb{E}^{\mathcal{B}_{t_k}}(I_n(F)) = \mathbb{E}^{\mathcal{B}_{t_k}\vee\mathcal{N}_{\widehat{\Gamma}}}({}^\circ I_n(F))\ \widehat{\Gamma}$-a.s. to finish the proof of (b). □

Instead of $I_n^M(F)(\cdot,t)$ we could have taken $I_n^M(F)\left(\cdot,t^-\right)$ and the filtration $(\mathcal{B}_{t^-})_{t\in T}$, leading to a slight modification of Theorem 13.3.1: recall that we have defined $G(\frac{1}{H}^-) := 0$.

Lemma 13.5.2 *Fix a lifting F, locally in $SL^2\left(\nu^n,\mathbb{F}^{\otimes n}\right)$, of some f, locally in $L^2\left(\lambda^n,\mathbb{H}^{\otimes n}\right)$. The mapping $T\ni t\mapsto I_n^M(F)(\cdot,t^-)$ is S-continuous $\widehat{\Gamma}$-a.s.*

Proof We have to prove $I_n^M(F)(\cdot,\frac{1}{H}) \approx 0\ \widehat{\Gamma}$-a.s. Since $I_n^M(F)(\cdot,\frac{1}{H}) = 0$ for all $n>1$, we may assume that $n=1$. Now, since F is locally in $SL^2(\nu,\mathbb{F})$, $\mathbb{E}_\Gamma\left(I_1^M(F)(\cdot,\frac{1}{H})\right)^2 = \int_{\left\{\frac{1}{H}\right\}} \|F\|_{\mathbb{F}}^2 d\nu \approx 0$. This proves that $I_n^M(F)(\cdot,\frac{1}{H}) \approx 0$ $\widehat{\Gamma}$-a.s. □

13.6 Continuous versions of internal iterated integral processes

In the preceding two sections we have extensively used the fact that the integrand F in $I_n^M(F)$ is a lifting of a standard function. Now we take an arbitrary F, locally in $SL^2\left(\nu^n,\mathbb{F}^{\otimes n}\right)$. We will show that $I_n^M(F)$ has a continuous version. The techniques in this section will be used later to obtain continuous versions of Skorokhod integral processes on finite chaos levels. Define for all $t\in T$

$$\alpha(t) := \sum_{t_1<\ldots<t_n\le t} \left\|F_{t_1,\ldots,t_n}\right\|_{\mathbb{F}^n}^2 \frac{1}{H^n}.$$

Since F is locally S-square-integrable, and thus α is S-continuous, we may define a continuous function ${}^\circ\alpha : [0,\infty[\to [0,\infty[$ by setting

$${}^\circ\alpha({}^\circ t) := {}^\circ(\alpha(t)) \quad \text{for limited } t\in T.$$

By Theorem 13.1.3, $\mathbb{E}_\Gamma I_n^M(F)^2(\cdot,r)$ is limited for all limited $r\in T$. Since $I_n^M(F)$ is a $(\mathcal{B}_t)_{t\in T}$-martingale, we may apply Theorem 10.16.2 and Corollary

10.16.3, to obtain the standard part $m = {}^{\circ}M$ of $M := I_n^M(F)$. Recall that m is a càdlàg $(\mathfrak{b}_r)_{r\in[0,\infty[}$-martingale.

Our aim is to prove that m has a continuous version. Let us only sketch the main ideas, because in the next section we describe how to convert internal processes, for example Skorokhod integral processes, to continuous versions.

By Theorem 10.16.2, for each $r \in [0,\infty[$ there exists an $s_r \in T, s_r \approx r$, such that M_{s_r} is a lifting of m_r. Moreover, there exists a $t_r \in T, t_r \approx r$, such that M_{t_r} is a lifting of m_{r^-}. Since $M_{s_r} \in SL^{2p}(\Gamma)$ (see Corollary 13.1.4), we have $\mathbb{E}_{\widehat{\Gamma}}(m_r - m_{\widetilde{r}})^{2p} = {}^{\circ}\mathbb{E}_{\Gamma}\left(M_{s_r} - M_{s_{\widetilde{r}}}\right)^{2p}$ for all $p \in \mathbb{N}$.

Lemma 13.6.1 *Fix limited $r, \widetilde{r} \in T$ with $r < \widetilde{r}$ and $p \in \mathbb{N}$. Then*

$$\mathbb{E}_{\Gamma}(M_{\widetilde{r}} - M_r)^{2p} \leq n^{2p}\left((2p)!!\right)^{n+1}(\alpha_{\widetilde{r}} - \alpha_r)^p.$$

In particular, if $r \approx \widetilde{r}$, then $\mathbb{E}_{\Gamma}(M_{\widetilde{r}} - M_r)^{2p} \approx 0$, thus $M_{\widetilde{r}} \approx M_r$ $\widehat{\Gamma}$-a.s.

Proof By the triangle inequality in $L^{2p}(\Gamma)$,

$$\left(\mathbb{E}_{\Gamma}(M_{\widetilde{r}} - M_r)^{2p}\right)^{\frac{1}{2p}} \leq \sum_{i=1}^{n}\left(\mathbb{E}_{\Gamma}\left(A_i^{2p}\right)\right)^{\frac{1}{2p}},$$

where $A_i := \sum_{t_1<\ldots<t_{i-1}\leq r<t_i<\ldots<t_n\leq\widetilde{r}} F_{t_1,\ldots,t_n}(X_{t_1},\ldots,X_{t_n})$. By Theorem 13.1.3,

$$\mathbb{E}_{\Gamma}\left(A_i^{2p}\right) \leq ((2p)!!)^{n+1}\left(\sum_{t_1<\ldots<t_{i-1}\leq r<t_i<\ldots<t_n\leq\widetilde{r}} \left\|F_{t_1,\ldots,t_n}\right\|_{\mathbb{F}^n}^2 \frac{1}{H^n}\right)^p.$$

This proves that

$$\mathbb{E}_{\Gamma}(M_{\widetilde{r}} - M_r)^{2p} \leq n^{2p}\left((2p)!!\right)^{n+1}(\alpha_{\widetilde{r}} - \alpha_r)^p.$$

□

A process $f : \Omega \times [0,\infty[\to \mathbb{R}$ is called **continuous in probability in** $r \in [0,\infty[$ if for each $\varepsilon > 0$ there exists a $\delta > 0$ such that for all $s \in [0,\infty[$ with $|s - r| < \delta$

$$\widehat{\Gamma}\{|f_s - f_r| \geq \varepsilon\} < \varepsilon.$$

Corollary 13.6.2

(a) *For all $r < \widetilde{r}$ in $[0,\infty[$*

$$\mathbb{E}_{\widehat{\Gamma}}(m_{\widetilde{r}} - m_r)^{2p} \leq n^{2p}\left((2p)!!\right)^{n+1}\left({}^{\circ}\alpha_{\widetilde{r}} - {}^{\circ}\alpha_r\right)^p.$$

(b) *m has a continuous version.*
(c) *m is continuous* $\widehat{\Gamma}$*-a.s. on each countable subset* $D \subseteq [0,\infty[$.
(d) *m is continuous in probability.*

Proof (a) follows from Lemma 13.6.1 and (b) follows from (a) and a slight modification of the Kolmogorov continuity theorem (see Karatzas and Shreve [52]). For the details see Theorem 13.7.2 below.

(c) For $r \in D$ choose $s_r, t_r \in T, s_r, t_r \approx r$ such that M_{s_r} is a lifting of m_r and M_{t_r} is a lifting of m_{r-}. Since $\mathbb{E}_{\widehat{\Gamma}} |m_r - m_{r-}| \approx \mathbb{E}_\Gamma |M_{s_r} - M_{t_r}| \approx 0$, $m_r = m_{r-}$ $\widehat{\Gamma}$-a.s., thus m is continuous in r $\widehat{\Gamma}$-a.s. The result (c) follows.

The proof of part (d) is left to the reader. □

In the following section we are going to extend some of the results in this section to processes that are not martingales.

13.7 Kolmogorov's continuity criterion

In this section we present a slight modification of Kolmogorov's continuity criterion, by which it is possible to convert certain internal processes $M : \Omega \times T \to {}^*\mathbb{R}$ to continuous versions $f : \Omega \times [0,\infty[\to \mathbb{R}$. In contrast to the preceding section, M is not necessarily a martingale. An application will later show that the Skorokhod integral process of the form $(X,r) \mapsto \int_0^r I_{n,1}(f) db_{\mathbb{B}}$ has a continuous version, where $f : [0,\infty[^{n+1} \to \mathbb{H}^{\otimes n+1}$ is locally square-integrable.

A subset D of T, where D may be external, is called S-**dense** if for any standard ε-neighbourhood of each $r \in [0,\infty[$ there exists a $t \in D$ with $|t - r| < \varepsilon$.

Fix an internal process $M : \Omega \times T \to {}^*\mathbb{R}$. We call M S-**continuous on** D if for $\widehat{\Gamma}$-almost all $X \in \Omega$ the following holds: $M_t(X)$ is limited for all $t \in D$, and for each standard $\varepsilon > 0$ there exists a standard $\delta > 0$ such that $|M(X,t) - M(X,s)| < \varepsilon$ for all $s,t \in D$ with $|s - t| < \delta$. If M is S-continuous on D, then M can be converted to a continuous process ${}^D M$ depending on D: Fix $r \in [0,\infty[$ and let $(t_n)_{n\in\mathbb{N}}$ be a sequence in D with $\lim_{n\to\infty} {}^\circ t_n = r$. Define

$${}^D M(r) := \lim_{n\to\infty} {}^\circ M(t_n).$$

It follows from the S-continuity of M on D that ${}^D M : \Omega \times [0,\infty[\to \mathbb{R}$ is well defined and continuous $\widehat{\Gamma}$-a.s.; ${}^D M$ is called a **continuous version of** M **on** D. There are many continuous versions of M in general. We call a continuous version ${}^D M$ of M on D a **uniquely determined (u.d.) continuous version of** M if any other continuous version ${}^E M$ of M on some S-dense subset E of T is a version of ${}^D M$. When do we have u.d. continuous versions?

The function M is **weakly S-continuous on** T if for all limited $s,t\in T$ with $s\approx t$ there exists a set $U_{s,t}\subseteq\Omega$ of $\widehat{\Gamma}$-measure 1 such that $M(X,s)\approx M(X,t)$ for all $X\in U_{s,t}$. In the preceding section we have seen that $I_n^M(F)$ is weakly S-continuous. We have the following result.

Proposition 13.7.1 *Let M be weakly S-continuous on T and let ${}^DM$ be a continuous version of M on an S-dense subset D of T, where we assume that M is S-continuous on D. Then ${}^DM$ is a u.d. continuous version of M.*

Proof Let ${}^DM$ and ${}^EM$ be two continuous versions of M. We have to prove that for all $k\in\mathbb{N}$ and all $r\in[0,\infty[$

$$\widehat{\Gamma}\left\{\left|{}^DM_r-{}^EM_r\right|\geq\frac{1}{k}\right\}=0.$$

To this end let $(r_n)_{n\in\mathbb{N}},(s_n)_{n\in\mathbb{N}}$ be sequences in D, E, respectively, with $\lim{}^\circ r_n=r=\lim{}^\circ s_n$. Then for all $n\in\mathbb{N}$

$$\widehat{\Gamma}\left\{\left|{}^DM_r-{}^EM_r\right|\geq\frac{1}{k}\right\}\leq A+B_n+C$$

with

$$A:=\widehat{\Gamma}\left\{\left|{}^DM_r-{}^\circ M_{r_n}\right|\geq\frac{1}{3k}\right\},$$

$$B_n:=\widehat{\Gamma}\left\{\left|{}^\circ M_{r_n}-{}^\circ M_{s_n}\right|>\frac{1}{3k}\right\},$$

$$C:=\widehat{\Gamma}\left\{\left|{}^\circ M_{s_n}-{}^EM_r\right|\geq\frac{1}{3k}\right\}.$$

The first and third summand can be made arbitrarily small, because $\left({}^\circ M_{r_n}\right)$ and $\left({}^\circ M_{s_n}\right)$ converge to ${}^DM_r$, ${}^EM_r$ in measure, respectively. Assume that $(B_n)_{n\in\mathbb{N}}$ does not converge to 0. Then there exists an $\varepsilon>0$ such that $\Gamma\left\{\left|M_{r_n}-M_{s_n}\right|\geq\frac{1}{3k}\right\}>\varepsilon$ for infinitely many $n\in\mathbb{N}$. By overspill, this inequality is true for some unlimited $n\in{}^*\mathbb{N}$. We may assume that $r_n\approx r\approx s_n$. But then $M_{r_n}\approx M_{s_n}$ $\widehat{\Gamma}$-a.s., by the weak S-continuity of M on T, which is a contradiction. □

The following result is relevant for finding u.d. continuous versions of internal processes. The proof is a slight modification of Kolmogorov's continuity theorem (see Karatzas and Shreve [52]).

Theorem 13.7.2 *Suppose that $M : \Omega \times T \to {}^*\mathbb{R}$ is an internal process. Set $M(\cdot,0) := 0$. Suppose that there exists a continuous monotone increasing function $\alpha : [0,\infty[\to [0,\infty[$ with $\alpha(0) = 0$ and a constant $c \in \mathbb{R}^+$ such that for all limited $t, s \in T \cup \{0\}$*

$$^\circ\mathbb{E}_\Gamma\left((M_t - M_s)^4\right) \le c\left(\alpha(^\circ t) - \alpha(^\circ s)\right)^2.$$

(We may replace 4 by any number $p \ge 1$ and 2 by any number $q > 1$.)

Then M has a u.d. continuous version.

Proof Obviously, M is weakly S-continuous and M_t is limited $\widehat{\Gamma}$-a.s. for all limited t (the exceptional nullset depends on t). Fix limited $s, t \in T$. There exists a $\sigma \in \mathbb{N}$ with $s, t \in T_\sigma$. For simplicity we may assume that $\sigma = 1$ and $\alpha(1) = 1$. For all $n \in \mathbb{N}$ and all $i = 0, \ldots, 2^n$ choose $t_i^n \in T \cup \{0\}$ such that $\alpha(^\circ t_i^n) = \frac{i}{2^n}$. Set $t_0^n = 0$ and $t_{2^n}^n = 1$. We further assume that $t_{2^m \cdot k}^{n+m} = t_k^n$. Set

$$D_n := \left\{t_i^n \mid i = 0, \ldots, 2^n\right\} \cup \left\{\frac{i}{m} \mid m \le n, i = 1, \ldots, m-1\right\}.$$

Since each $n \in \mathbb{N}$ divides H, $D_n \subseteq T \cup \{0\}$. Let $D_n = \left\{b_i^n \mid i = 0, \ldots, \overline{n}\right\}$ with $0 = b_0^n < \ldots < b_{\overline{n}}^n$. Note that $D_n \subseteq D_{n+1}$. Since α is monotone increasing, $0 \le \left(\alpha(^\circ b_i^n) - \alpha(^\circ b_{i-1}^n)\right) \le \frac{1}{2^n}$. Set

$$D := \bigcup_{n \in \mathbb{N}} D_n.$$

There exists a set $U_1 \subseteq \Omega$ of $\widehat{\Gamma}$-measure 1 such that $F_a(X) \approx F_b(X)$ is limited for all $X \in U_1$ and all $a, b \in D$ with $\alpha(^\circ b) = \alpha(^\circ a)$. By Tschebyschev's inequality, we obtain

$$\widehat{\Gamma}\left\{\left|M_{b_i^n} - M_{b_{i-1}^n}\right| \ge 2^{-\frac{1}{8}n}\right\} \le c \cdot 2^{-\frac{3}{2}n}.$$

Since $\overline{n} \le 2^n + n^2$, we obtain

$$\widehat{\Gamma}\left\{\max_{i=1,\ldots,\overline{n}}\left|M_{b_i^n} - M_{b_{i-1}^n}\right| \ge 2^{-\frac{1}{8}n}\right\} \le c\left(2^{-\frac{1}{2}n} + n^2 2^{-\frac{3}{2}n}\right).$$

Since the last term is the general term of a convergent series, by the Borel–Cantelli lemma, there exists a set $U_2 \subseteq \Omega$ of $\widehat{\Gamma}$-measure 1 and a function $g : U_2 \to \mathbb{N}$ such that for all $X \in U_2$ and all $n \ge g(X)$

$$\max_{i=1,\ldots,\overline{n}}\left|M_{b_i^n}(X) - M_{b_{i-1}^n}(X)\right| < 2^{-\frac{1}{8}n}. \tag{28}$$

Set $U := U_1 \cap U_2$. Fix $X \in U$ and $n \geq g(X)$. We will prove by induction that for all $m > n$ and for all $s,t \in D_m$, with $|\alpha(°t) - \alpha(°s)| < \frac{1}{2^n}$,

$$|M_s(X) - M_t(X)| < 2 \sum_{j=n+1}^{m} j^2 2^{-\frac{1}{8}j}. \tag{29}$$

Suppose that $s,t \in D_{n+1}$ with $s < t$ and $|\alpha(°t) - \alpha(°s)| < \frac{1}{2^n}$. Let $t_i^n \leq s < t_{i+1}^n$. Then $t < t_{i+2}^n$. Since between t_i^n and t_{i+2}^n there are less than $2(n+1)^2$ elements of D_{n+1}, we obtain from (28)

$$|M_s(X) - m_t(X)| < 2(n+1)^2 2^{-\frac{1}{8}(n+1)}.$$

Assume that $N > n+1$ and (29) is true for all $m < N$. Suppose that $s,t \in D_N$ with $|\alpha(°t) - \alpha(°s)| < \frac{1}{2^n}$. We may assume $s < t$. Set $s_0 := \min\{a \in D_{N-1} \mid s \leq a\}$ and $t_0 := \max\{a \in D_{N-1} \mid a \leq t\}$. Since between t_0 and t and s and s_0 there are at most N^2 many elements of D_N and $|\alpha(°t_0) - \alpha(°s_0)| < \frac{1}{2^n}$, we obtain from (28) and the induction hypothesis:

$$|M_s(X) - M_t(X)| \leq 2 \sum_{j=n+1}^{M} j^2 2^{-\frac{1}{8}j}.$$

To prove our result, fix $s,t \in D$ with $0 \leq \alpha(°t) - \alpha(°s) < \frac{1}{2^{g(X)}}$. If $\alpha(°t) = \alpha(°s)$, then the inequalities under (30) below become true. Now assume $0 < \alpha(°t) - \alpha(°s)$. We select n such that $\frac{1}{2^{n+1}} \leq \alpha(°t) - \alpha(°s) < \frac{1}{2^n}$. Then $n \geq g(X)$ and $\left(\frac{1}{2^{n+1}}\right)^{\frac{1}{16}} \leq (\alpha(°t) - \alpha(°s))^{\frac{1}{16}}$. We obtain

$$\begin{aligned} |M_t(X) - M_s(X)| &\leq 2 \sum_{j=n+1}^{\infty} j^2 2^{-\frac{1}{16}j} (\alpha(°t) - \alpha(°s))^{\frac{1}{16}} \\ &\leq \left(2 \sum_{j=0}^{\infty} j^2 2^{-\frac{1}{16}j}\right) (\alpha(°t) - \alpha(°s))^{\frac{1}{16}}. \end{aligned} \tag{30}$$

We have proved that there exists a countable dense subset $D \subset T$, a constant d, a set $U \subseteq \Omega$ of $\widehat{\Gamma}$-measure 1 and a function $g : U \to \mathbb{N}$ such that for all $X \in U$ and all $s,t \in D$ with $0 \leq \alpha(°t) - \alpha(°s) < \frac{1}{2^{g(X)}}$

$$|M_t(X) - M_s(X)| \leq d \cdot |\alpha(°t) - \alpha(°s)|^{\frac{1}{16}}.$$

It follows that $M(X,\cdot)$ is S-continuous on D. From the hypothesis it follows that M is weakly S-continuous. By Proposition 13.7.1, ${}^{D}M$ is a u.d. continuous version of M. □

Exercises

In Exercise 13.1 we continue Exercises 3.10–3.11.

13.1 Fix $F \in SL^2(\nu,\mathbb{F})$ with $\int_T \|F\|^2_{\mathbb{F}}\, d\nu \approx 1$. Prove that in $L^2(\widehat{\Gamma})$

$$H_n({}^{\circ}I(F)) = {}^{\circ}I_n(F^{\otimes n}) \text{ for all } n \geq 0.$$

Therefore, by the definition of the standard iterated Itô integral, we have, for all $f \in L^2(\widehat{\nu},\mathbb{H})$ with $\|f\|_{\widehat{\nu}} = 1$,

$$H_n(I(f)) = I_n(f^{\otimes n}).$$

13.2 Prove the three equalities at the end of the proof of Theorem 13.2.1.

13.3 Prove Corollary 13.3.2.

13.4 Prove Corollary 13.6.2 part (d).

14

†Infinite-dimensional Ornstein–Uhlenbeck processes

This chapter contains extensions of results in joint work with Jiang-Lun Wu [87]. We noticed that the proof for the existence of continuous solutions to equation (31) below also works under milder assumptions on S than stated there in the corresponding theorem. Moreover, instead of Banach space valued OU-processes there, here we take OU-processes with values in any abstract Wiener–Fréchet space $(\mathbb{B}, d)$ over a fixed infinite-dimensional separable Hilbert space $\mathbb{H}$ with norm $\|\cdot\|$. Assume that the metric d is generated by a separating sequence $\left(|\cdot|_j\right)_{j\in\mathbb{N}}$ of semi-norms. We will write $\alpha \approx \beta$ if $\alpha, \beta \in {}^*\mathbb{B}$ and ${}^*d(\alpha, \beta) \approx 0$. Let $b := b_{\mathbb{B}}$ be the Brownian motion defined in Section 11.7.

In the first part we construct a continuous solution $\mathfrak{o} : \Omega \times [0, \infty[\to \mathbb{B}$ on Ω of the infinite-dimensional Langevin differential equation

$$d\,\mathfrak{o}_t = db_t + S(\mathfrak{o}_t)dt \tag{31}$$

with initial data $\mathfrak{o}_0 = 0$ for each d-continuous operator $S : \mathbb{B} \to \mathbb{B}$ such that $S \restriction \mathbb{H} : \mathbb{H} \to \mathbb{H}$ is a Hilbert–Schmidt operator. Since the solution $\mathfrak{o}$ is constructed path by path, we are able to construct from any continuous function $f : [0, \infty[\to \mathbb{B}$ an $X \in \Omega$ with $\mathfrak{o}(X, \cdot) = f$.

The equality (31) is equivalent to the integral equation

$$\mathfrak{o}_t = b_t + \int_0^t S(\mathfrak{o}_s)d\lambda(s). \tag{32}$$

We have the following example.

Example 14.0.1 Let $\mathbb{B}$ be the abstract Wiener–Fréchet space over l^2 in Example 4.3.7. Set $S\left((a_i)_{i\in\mathbb{N}}\right) := \left(\frac{1}{i}a_i\right)_{i\in\mathbb{N}}$. Then $S : \mathbb{B} \to \mathbb{B}$ is continuous and $S \restriction \mathbb{H}$ is a Hilbert Schmidt operator on $\mathbb{H}$.

Let us point out that we consider directly equation (32), which is equivalent to equation (31), while, in the literature, the following mild integral form of equation (32) is considered:

$$\mathfrak{o}_t = e^{tS}\mathfrak{o}_0 + \int_0^t e^{(t-s)\cdot S} db_s, \tag{33}$$

where $\left(e^{tS}\right)_{t\in[0,1]}$ stands for the analytic semi-group on $L^2(\gamma_{\mathbb{B}})$, where $\gamma_{\mathbb{B}}$ is the image measure of $\widehat{\Gamma}$ by $b_{\mathbb{B}}(\cdot,1)$. The solution to equation (33) is called a weak solution to equation (31).

In the second part we sketch the construction of a path-by-path continuous solution to equations of the form

$$d\mathfrak{o}_t = \sqrt{2|\alpha|\beta}db_t + \alpha\mathfrak{o}_t dt,$$

with initial data $\mathfrak{o}_0 = 0$. Here $\alpha, \beta \in \mathbb{R}$, $0 < \beta$ and $\alpha \neq 0$. Again, each continuous function $f : [0,\infty[\to \mathbb{B}$ is a path of $\mathfrak{o}$.

Two types of infinite-dimensional Langevin equations have been studied by K. Itô [50]: Ornstein–Uhlenbeck equations in the Malliavin style (see also [75]) and Ornstein–Uhlenbeck equations of Gaveau type (see also [40]). In our present framework the Malliavin type is the preceding equation in the case that the drift operator is constant and hence a bounded operator from $\mathbb{H}$ to $\mathbb{H}$, while the Gaveau type refers to equation (31) with an unbounded drift operator S from $\mathbb{H}$ to $\mathbb{H}$. There are many works devoted to the mild solutions to infinite-dimensional Langevin equations for both types (see e.g. Da Prato, Kwapien and Zabczyk [28] as well as Da Prato and Zabczyk [27] and references therein and more recently Brzezniak and van Neerven [17]). Continuity of the mild solutions to equation (31) has been discussed in a number of papers. Here we mention just a few: Iscoe *et al.* [49], Simao [106] for the case that $\mathbb{B}$ is a separable Hilbert space, and Brzezniak and van Neerven [17], Smoleński [107] and Millet and Smoleński [77] for the case that $\mathbb{B}$ is a separable Banach space. For nonstandard analysis attempts to study infinite-dimensional Ornstein–Uhlenbeck processes, let us also mention Lindstrøm [66] for a hyperfinite construction of infinite-dimensional Ornstein–Uhlenbeck processes via Anderson's random walk and Cutland [25] for another nonstandard construction of infinite-dimensional Ornstein–Uhlenbeck processes via the Wiener sphere. A nonstandard approach to the existence (in Keisler's sense) as well as the support property of the unique solution to equation (31) of the Gaveau type can be found in [118], where a more general, infinite-dimensional stochastic evolution equation in Itô's framework (see [50]) is treated. We would like to point out further that Lindstrøm [68] gives an early attempt to use nonstandard analysis to infinite-dimensional Langevin equations with solutions in the infinite-dimensional Hilbert space $\mathbb{H}$. The interested reader is also referred to the monograph [20] by Capinski and Cutland for the best illustration of the power of the nonstandard approach to infinite-dimensional stochastic differential equations.

14.1 Ornstein–Uhlenbeck processes for shifts given by Hilbert–Schmidt operators

We will prove the following result.

Theorem 14.1.1 *Suppose that $S : \mathbb{B} \to \mathbb{B}$ is a continuous linear operator such that $S \upharpoonright \mathbb{H} : \mathbb{H} \to \mathbb{H}$ is a Hilbert–Schmidt operator. Then there exists a continuous solution $\mathfrak{o} : \Omega \times [0, \infty[\to \mathbb{B}$ of the infinite-dimensional Langevin differential equation*

$$\mathfrak{o}_t = b_t + \int_0^t S(\mathfrak{o}_s)\, d\lambda(s),\ t \in [0, \infty[\tag{34}$$

with initial data $\mathfrak{o}_0 = 0$. Moreover, the set of all paths of $\mathfrak{o}$ is identical to the set $C_{\mathbb{B}}$.

Let us follow Keisler's method to solve this equation.

Step 1: Using the internal lifting B of b (see Sections 11.2 and 11.7), convert Equation (34) into an appropriate internal difference equation

$$\Theta_t = B_t + \sum_{s<t} \widetilde{S}(\Theta_s) \frac{1}{H}. \tag{35}$$

Then Θ is a recursively defined solution to this equation.

Step 2: Using an appropriate internal Girsanov transformation, find a new internal probability measure $D\Gamma$ with density D and show Θ is an internal ω-dimensional Brownian motion with respect to $D\Gamma$.

Step 3: Prove that for each $a \in \mathbb{F}$ with $\|a\| = 1$ the standard part $^\circ\langle \Theta, a \rangle$ of $\langle \Theta, a \rangle$ defines a continuous one-dimensional Brownian motion with respect to the Loeb measure $\widehat{D\Gamma}$.

Step 4: Convert Θ to a continuous $\mathbb{B}$-valued Brownian motion $\mathfrak{o}$ with respect to the measure $\widehat{D\Gamma}$ for any abstract Wiener space $\mathbb{B}$ over $\mathbb{H}$.

Step 5: Show that $\mathfrak{o}$ is a continuous solution of Equation (34) on the original probability space Ω with measure $\widehat{\Gamma}$.

The simplest part is Step 1. However, this part is crucial, because the internal solution generates an appropriate Girsanov transformation.

'Step 1' Define $\widetilde{S}: {}^*\mathbb{H} \to \mathbb{F}$, $a \mapsto \pi_{{}^*\mathbb{H},\mathbb{F}}(a)$, where $\pi_{{}^*\mathbb{H},\mathbb{F}}$ denotes the internal orthogonal projection from ${}^*\mathbb{H}$ onto $\mathbb{F}$. By internal recursion on $t \in T$ define

$$\Theta_t(X) = B_t(X) + \sum_{s<t, s\in T} \widetilde{S}(\Theta_s(X))\frac{1}{H} = \sum_{s\leq t, s\in T} \left(X_s + \frac{\widetilde{S}}{H}(\Theta_{s^-}(X))\right).$$

Recall that $\Theta_{s^-} := \Theta_{s-\frac{1}{H}}$ with $\Theta_0 := 0$. By internal induction on $t \in T$, we obtain the following simple result.

Lemma 14.1.2 *For all $t \in T$ and all $X \in \Omega$*

$$\Theta_t(X) = \sum_{s\leq t, s\in T} \left(\mathrm{id} + \frac{\widetilde{S}}{H}\right)^{H(t-s)} (X_s),$$

where $F^m := F \circ F \circ \ldots \circ F$ m-times and F^0 is the identity map.

Our aim is to prove that Θ is S-continuous $\widehat{\Gamma}$-a.s. and that its continuous standard part, according to Proposition 9.7.1, is a solution to Equation (34). Since it suffices to find a solution on each compact timeline, fix $\sigma \in \mathbb{N}$. We define a mapping $\Phi : \Omega \to \Omega$ via

$$\Phi(X) := \left(\left(X_s + \frac{\widetilde{S}}{H}(\Theta_{s^-}(X))\right)_{s\in T_\sigma}, (X_s)_{s\in T\setminus T_\sigma}\right).$$

By Lemma 14.1.2, Φ is linear. Note that for all $t \in T_\sigma$ and all $X \in \Omega$

$$\Theta_t(X) = B_t \circ \Phi(X).$$

Let us fix an ONB $\mathfrak{E} := (\mathfrak{e}_i)_{i\in\omega}$ of $\mathbb{F}$.

Lemma 14.1.3 *The function Φ is bijective and* $\det \Phi = 1$.

Proof To prove that Φ is injective, let us fix $X, Y \in \Omega$ with $X \neq Y$. Then there exists an $i \in T$ such that $X_i \neq Y_i$ and $X_j = Y_j$ for all $j < i$. If $i > \sigma$, then $\Phi_i(X) \neq \Phi_i(Y)$. Let $i \leq \sigma$. From Lemma 14.1.2 it follows that $\Phi_i(X) \neq \Phi_i(Y)$. To prove that Φ is surjective, fix $Z = (Z_{\frac{1}{H}}, \ldots, Z_H) \in \Omega$. If $s > \sigma$, set $X_s := Z_s$. Otherwise, define recursively

$$X_{\frac{1}{H}} := Z_{\frac{1}{H}}, X_{s+\frac{1}{H}} := Z_{s+\frac{1}{H}} - \frac{\widetilde{S}}{H}\left(\Theta_s(X_{\frac{1}{H}}, \ldots, X_s, 0, \ldots, 0)\right)$$

for $s+\frac{1}{H}\leq\sigma$. Note that $\Phi(X_{\frac{1}{H}},\ldots,X_H)=(Z_{\frac{1}{H}},\ldots,Z_H)$. From $\mathfrak{E}$ define an ONB $(\mathfrak{e}_s(i))_{s\in T,i\in\omega}$ of Ω, setting

$$\mathfrak{e}_s(i)(t):=\begin{cases}\mathfrak{e}_i & \text{if } s=t\\ 0 & \text{if } s\neq t,\quad s,\,t\in T.\end{cases}$$

Note that the representing matrix of Φ with respect to this basis is a triangle matrix with 1 in the diagonal. It follows that $\det\Phi=1$. □

'Step 2' Let $\langle\cdot,\cdot\rangle$, $\|\cdot\|$ denote the internal scalar product and norm on $\mathbb{F}$. Define an internal mapping $D:\Omega\to{}^*\mathbb{R}$ by

$$D(X):=\exp\left[-\int\widetilde{S}\circ\Theta_{s^-}\Delta B(X,\sigma)-\frac{1}{2}\int_{T_\sigma}\left\|\widetilde{S}\circ\Theta_{s^-}(X)\right\|^2 d\nu(s)\right].$$

Corollary 14.1.4 *The internal measure $D\Gamma:A\mapsto\int_A D\,d\Gamma$ on $\mathcal{B}$ is the internal image measure $\Gamma_{\Phi^{-1}}$ of Γ by Φ^{-1}.*

Proof Set $g(X):=e^{-\frac{H}{2}\sum_{i\in\omega,s\in T}\langle X_s,\mathfrak{e}_i\rangle^2}\cdot w$, where $w:=\sqrt{\frac{H}{2\pi}}^{\,\omega\cdot H^2}$. By Lemma 14.1.3 and the transfer principle applied to the finite-dimensional substitution rule, we obtain for all $A\in\mathcal{B}$:

$$\begin{aligned}&\Gamma(\Phi[A])\\ &=\int_{\Phi[A]}g(X)dX=\int_A g\circ\Phi(X)dX\\ &=w\cdot\int_A\exp\left[-\frac{H}{2}\sum_{i\in\omega,s\in T_\sigma}\left\langle X_s+\frac{\widetilde{S}}{H}(\Theta_{s^-}(X)),\mathfrak{e}_i\right\rangle^2-\frac{H}{2}\sum_{i\in\omega,s>\sigma}\langle X_s,\mathfrak{e}_i\rangle^2\right]dX\\ &=\int_A D\,d\Gamma=D\Gamma(A).\end{aligned}$$

□

The following results can also be obtained from Lemma 14.1.3. We first prove that $(\Theta_t)_{t\in T_\sigma}$ is an internal ω-dimensional Brownian motion under $D\Gamma$.

Corollary 14.1.5 *Fix $a,b\in\mathbb{F}$ with $\|a\|=1=\|b\|$ and $\lambda\in{}^*\mathbb{R}$. Then the following hold.*

(a) *The internal process $\left(e^{\lambda\langle a,\Theta_t\rangle-\frac{1}{2}\lambda^2 t}\right)_{t\in T_\sigma}$ is an internal martingale with respect to the measure $D\Gamma$.*
(b) *For all $t\in T_\sigma$, $\int_\Omega\langle a,\Theta_t\rangle\langle b,\Theta_t\rangle\,dD\Gamma=\langle a,b\rangle\cdot t$.*

Proof (a) Using Lemma 11.1.2, it is easy to see that $e^{\lambda\langle a,\Theta_t\rangle-\frac{1}{2}\lambda^2 t}$ is $\mathcal{B}_t$-measurable. Note that $\Phi[A]\in\mathcal{B}_s$ for any $A\in\mathcal{B}_s$. By Lemma 14.1.3, we obtain for all $s,t\in T_\sigma$ with $s<t$ and all $A\in\mathcal{B}_s$

$$\begin{aligned}
&\int_A \exp\left[\lambda\langle a,\Theta_t(X)\rangle-\frac{1}{2}\lambda^2 t\right]dD\Gamma\\
&=\int_A \exp\left[\lambda\langle a,B_t\circ\Phi(X)\rangle-\frac{1}{2}\lambda^2 t\right]d\Gamma_{\Phi^{-1}}\\
&=\int_{\Phi[A]} \exp\left[\lambda\langle a,B_t\rangle-\frac{1}{2}\lambda^2 t\right]d\Gamma \quad (11.2.1)\\
&=\int_{\Phi[A]} \exp\left[\lambda\langle a,B_s\rangle-\frac{1}{2}\lambda^2 s\right]d\Gamma\\
&=\int_A \exp\left[\lambda\langle a,\Theta_s(X)\rangle-\frac{1}{2}\lambda^2 s\right]dD\Gamma.
\end{aligned}$$

(b) The proof is similar to the proof of (a). □

'Step 3' Next fix an $a\in\mathbb{F}$ with $\|a\|=1$. Our aim is to show that $\langle\Theta,a\rangle$ is S-continuous $\widehat{D\Gamma}$-a.s. on T_σ. Using Lemma 14.1.3, the substitution rule and the facts that $(\langle B_t,a\rangle)_{t\in T_\sigma}$ is a $(\mathcal{B}_t)_{t\in T_\sigma}$-martingale for Γ and $\int_\Omega\langle B_t,a\rangle^2 d\Gamma$ is limited for all $t\in T_\sigma$, we obtain the following.

Lemma 14.1.6 *The internal process $(\langle\Theta_t,a\rangle)_{t\in T_\sigma}$ is a $(\mathcal{B}_t)_{t\in T_\sigma}$-martingale under $D\Gamma$ and $\int_\Omega\langle\Theta_t,a\rangle^2 dD\Gamma$ is limited for all $t\in T_\sigma$.*

Now we are able to apply Theorem 10.14.2 and show the following.

Proposition 14.1.7 *The ${}^*\mathbb{R}$-valued function $t\mapsto\langle\Theta_t,a\rangle$ defined on T_σ is S-continuous $\widehat{D\Gamma}$-a.s.*

Proof By Lemma 14.1.6 and Theorem 10.14.2, it suffices to show that the quadratic variation $[\langle\Theta,a\rangle]$ of $\langle\Theta,a\rangle$ is S-continuous $\widehat{D\Gamma}$-a.s. Fix $t\in T_\sigma$. Note that

$$[\langle\Theta(X),a\rangle]_t=\sum_{s\le t}\langle(\Phi(X))_s,a\rangle^2,$$

and $\int_\Omega[\langle\Theta,a\rangle]_t dD\Gamma=\int_\Omega[\langle B,a\rangle]_t d\Gamma=t$ for all $\sigma\in\mathbb{N}$. Therefore, $[\langle\Theta,a\rangle]_t$ is limited $\widehat{D\Gamma}$-a.s.. Now we apply Josef Berger's trick [7] and set $F(X,t):=H\langle\pi_t\circ\Phi,a\rangle^2$ for all $X\in\Omega$ and all $t\in T_\sigma$. Since $\int_{\Omega\times T_\sigma}F^2 dD\Gamma\otimes\nu=3\sigma$, we have $F\in SL^1(D\Gamma\otimes\nu_\sigma)$. By Keisler's Fubini theorem, Theorem 10.12.1, $F(X,\cdot)\in SL^1(\nu_\sigma)$ for $\widehat{\Gamma}$-almost all $X\in\Omega$. Fix such an X and limited $s,t\in T_\sigma$ with $s<t$

and $s \approx t$. Set $N := \{r \in T \mid s < r \leq t\}$. Since $\nu(N) \approx 0$, we obtain

$$[\langle\Theta, a\rangle]_t(X) - [\langle\Theta, a\rangle]_s(X) = \int_N F(X, \cdot) d\nu_\sigma \approx 0.$$

Moreover, $[\langle\Theta, a\rangle]_\sigma(X)$ is limited. □

We define for all X such that $\langle\Theta(X), a\rangle$ is S-continuous on T_σ and for all $r \in [0, \sigma]$ and all $t \in T$ with $t \approx r$

$$o_a(r, X) := {}^\circ\langle\Theta_t(X), a\rangle .$$

By Proposition 9.7.1, $o_a(\cdot, X)$ is continuous.

'Step 4' Using the techniques in Section 11.7 we can prove the following result.

Theorem 14.1.8 *The $\mathbb{F}$-valued process Θ is S-continuous $\widehat{D\Gamma}$-a.s. on T_σ with respect to the topology on $\mathbb{B}$.*

'Step 5' In order to prove that Θ is S-continuous on T_σ $\widehat{\Gamma}$-a.s. and to finish the proof of Theorem 14.1.1, we have to prove the following.

Lemma 14.1.9 *The original measure $\widehat{\Gamma}$ is absolutely continuous with respect to the new measure $\widehat{D\Gamma}$.*

Proof Fix $A \in \mathcal{B}$ and assume that $D\Gamma(A) \approx 0$. Then

$$\int_A \exp\left[-\left(\int \widetilde{S} \circ \Theta_{s^-} \Delta B\right)_\sigma - \frac{1}{2}\int_{T_\sigma} \|\widetilde{S} \circ \Theta_{s^-}\|^2 d\nu(s)\right] d\Gamma \approx 0,$$

thus $\mathbf{1}_A \cdot \exp\left[-\left(\int \widetilde{S} \circ \Theta_{s^-} \Delta B\right)_\sigma - \frac{1}{2}\int_{T_\sigma} \|\widetilde{S} \circ \Theta_{s^-}\|^2 d\nu(s)\right] \approx 0$ $\widehat{\Gamma}$-a.s.

In order to prove that $\mathbf{1}_A \approx 0$ $\widehat{\Gamma}$-a.s., thus $\Gamma(A) \approx 0$, it suffices to prove that the internal random variable $\left(\int \widetilde{S} \circ \Theta_{s^-} \Delta B\right)_\sigma + \frac{1}{2}\int_{T_\sigma} \|\widetilde{S} \circ \Theta_{s^-}\|^2 d\nu$ is limited $\widehat{\Gamma}$-a.s. Now, by the properties of Gaussian measures, for each ONB $(\mathfrak{e}_i)_{i \in \omega}$ of $\mathbb{F}$, we have

$$\mathbb{E}_\Gamma \left(\int \widetilde{S} \circ \Theta_{\cdot^-} \Delta B\right)_\sigma^2 = \int_\Omega \left(\sum_{t \leq \sigma} \widetilde{S} \circ \Theta_{t^-}(X)(X_t)\right)^2 d\Gamma(X)$$

$$= \int_\Omega \left(\sum_{t \leq \sigma} \sum_{i \in \omega} \langle X_t, \mathfrak{e}_i\rangle \cdot \widetilde{S} \circ \Theta_{t^-}(X)(\mathfrak{e}_i)\right)^2 d\Gamma(X)$$

$$= \int_{\Omega} \sum_{t \le \sigma} \sum_{i \in \omega} \langle X_t, \mathfrak{e}_i \rangle^2 \cdot \left(\widetilde{S} \circ \Theta_{t^-}(X)(\mathfrak{e}_i) \right)^2 d\Gamma(X)$$

$$= \int_{\Omega} \sum_{t \le \sigma} \sum_{i \in \omega} \frac{1}{H} \cdot \left(\widetilde{S} \circ \Theta_{t^-}(X)(\mathfrak{e}_i) \right)^2 d\Gamma(X)$$

$$= \int_{\Omega \times T_\sigma} \left\| \widetilde{S} \circ \Theta_{t^-}(X) \right\|^2 d(\Gamma \otimes \nu)(X, t).$$

Therefore, it suffices to prove that $\int_{\Omega \times T_\sigma} \left\| \widetilde{S} \circ \Theta_{t^-} \right\|^2 d\Gamma \otimes \nu(t)$ is limited. First we will prove by internal induction that for all $n \in H \cdot T \cup \{0\}$ the following inequality holds

$$\left(\sum_{i \in \omega} \left\| \widetilde{S} \circ \left(\mathrm{id} + \frac{\widetilde{S}}{H} \right)^n (\mathfrak{e}_i) \right\|^2 \right)^{\frac{1}{2}} \le \left(1 + \frac{\|\widetilde{S}\|}{H} \right)^n \left(\sum_{i \in \omega} \|\widetilde{S}(\mathfrak{e}_i)\|^2 \right)^{\frac{1}{2}}.$$

For $n = 0$ the result is obvious. Let us assume that the previous inequality holds for n. By Hölder's inequality, we have

$$\left(\sum_{i \in \omega} \left\| \widetilde{S} \circ \left(\mathrm{id} + \frac{\widetilde{S}}{H} \right)^{n+1} (\mathfrak{e}_i) \right\|^2 \right)^{\frac{1}{2}}$$

$$= \left(\sum_{i \in \omega} \left\| \left(\mathrm{id} + \frac{\widetilde{S}}{H} \right) \circ \widetilde{S} \circ \left(\mathrm{id} + \frac{\widetilde{S}}{H} \right)^n (\mathfrak{e}_i) \right\|^2 \right)^{\frac{1}{2}}$$

$$\le \left(\sum_{i \in \omega} \left\| \widetilde{S} \circ \left(\mathrm{id} + \frac{\widetilde{S}}{H} \right)^n (\mathfrak{e}_i) \right\|^2 \right)^{\frac{1}{2}} + \left(\sum_{i \in \omega} \left\| \frac{\widetilde{S}}{H} \circ \widetilde{S} \circ \left(\mathrm{id} + \frac{\widetilde{S}}{H} \right)^n (\mathfrak{e}_i) \right\|^2 \right)^{\frac{1}{2}}$$

$$\le \left(1 + \frac{\|\widetilde{S}\|}{H} \right)^n \left(\sum_{i \in \omega} \|\widetilde{S}(\mathfrak{e}_i)\|^2 \right)^{\frac{1}{2}} + \frac{\|\widetilde{S}\|}{H} \left(1 + \frac{\|\widetilde{S}\|}{H} \right)^n \left(\sum_{i \in \omega} \|\widetilde{S}(\mathfrak{e}_i)\|^2 \right)^{\frac{1}{2}}$$

$$\le \left(1 + \frac{\|\widetilde{S}\|}{H} \right)^{n+1} \left(\sum_{i \in \omega} \|\widetilde{S}(\mathfrak{e}_i)\|^2 \right)^{\frac{1}{2}}.$$

Therefore,

$$\left(\int_{\Omega \times T_\sigma} \|\widetilde{S} \circ \Theta_{t^-}\|^2 d\Gamma \otimes \nu(t) \right)^{\frac{1}{2}}$$

$$= \left(\int_\Omega \sum_{t \in T_\sigma} \left\| \sum_{s<t} \widetilde{S} \left(\mathrm{id} + \frac{\widetilde{S}}{H} \right)^{Ht-1-Hs} (X_s) \right\|^2 \frac{1}{H} d\Gamma(X) \right)^{\frac{1}{2}}$$

$$= \left(\int_\Omega \sum_{t \in T_\sigma} \sum_{i \in \omega} \left(\sum_{s<t} \left\langle \widetilde{S} \left(\mathrm{id} + \frac{\widetilde{S}}{H} \right)^{Ht-1-Hs} (X_s), \mathfrak{e}_i \right\rangle \right)^2 \frac{1}{H} d\Gamma(X) \right)^{\frac{1}{2}}$$

$$= \left(\int_\Omega \sum_{t \in T_\sigma} \sum_{i \in \omega} \left(\sum_{s<t} \sum_{j \in \omega} \langle X_s, \mathfrak{e}_j \rangle \left\langle \widetilde{S} \left(\mathrm{id} + \frac{\widetilde{S}}{H} \right)^{Ht-1-Hs} (\mathfrak{e}_j), \mathfrak{e}_i \right\rangle \right)^2 \frac{1}{H} d\Gamma(X) \right)^{\frac{1}{2}}$$

$$= \left(\int_\Omega \sum_{t \in T_\sigma} \sum_{i \in \omega} \sum_{s<t} \sum_{j \in \omega} \langle X_s, \mathfrak{e}_j \rangle^2 \left\langle \widetilde{S} \left(\mathrm{id} + \frac{\widetilde{S}}{H} \right)^{Ht-1-Hs} (\mathfrak{e}_j), \mathfrak{e}_i \right\rangle^2 \frac{1}{H} d\Gamma(X) \right)^{\frac{1}{2}}$$

$$= \left(\sum_{s<t \in T_\sigma} \sum_{i \in \omega} \sum_{j \in \omega} \left\langle \widetilde{S} \left(\mathrm{id} + \frac{\widetilde{S}}{H} \right)^{Ht-1-Hs} (\mathfrak{e}_j), \mathfrak{e}_i \right\rangle^2 \frac{1}{H^2} \right)^{\frac{1}{2}}$$

$$= \left(\sum_{s<t \in T_\sigma} \sum_{j \in \omega} \left\| \widetilde{S} \left(\mathrm{id} + \frac{\widetilde{S}}{H} \right)^{Ht-1-Hs} (\mathfrak{e}_j) \right\|^2 \frac{1}{H^2} \right)^{\frac{1}{2}}$$

$$\leq \left(\sum_{s<t \in T_\sigma} \left(\left(1 + \frac{\|\widetilde{S}\|}{H} \right)^{2Ht-2-2Hs} \sum_{j \in \omega} \|\widetilde{S}(\mathfrak{e}_j)\|^2 \right) \frac{1}{H^2} \right)^{\frac{1}{2}}$$

$$\leq \sigma \left(\left(1 + \frac{\|\widetilde{S}\|}{H} \right)^{2H\sigma} \sum_{j \in \omega} \|\widetilde{S}(\mathfrak{e}_j)\|^2 \right)^{\frac{1}{2}}$$

$$= \sigma \left(1 + \frac{\|\widetilde{S}\|}{H} \right)^{H\sigma} \left(\sum_{j \in \omega} \|\widetilde{S}(\mathfrak{e}_j)\|^2 \right)^{\frac{1}{2}} \text{ is limited,}$$

because $S \restriction \mathbb{H}$ is a Hilbert–Schmidt operator. □

It follows that Θ is S-continuous $\widehat{\Gamma}$-a.s. on T_σ Since σ is any positive integer, Θ is S-continuous on T. Therefore, Θ can be converted to a $\widehat{\Gamma}$-a.s. continuous process $\mathfrak{o} : \Omega \times [0, \infty[\to \mathbb{B}$ by setting

$$\mathfrak{o}_{\circ t}(X) := {}^{\circ}\Theta_t(X)$$

for all X such that $\Theta(X)$ is S-continuous. For the other X we may just take $\mathfrak{o}_{\circ t}(X) := 0$.

Now fix an X such that $\Theta(X)$ is S-continuous and $B(X)$ is S-continuous. This happens on a set of $\widehat{\Gamma}$-measure 1. We see that for all limited $t \in T$

$$S(\mathfrak{o}_{\circ t}(X)) \approx_{\mathbb{B}} \widetilde{S}(\Theta_t(X)).$$

Fix $r \in [0, \infty[$ and $t \in T$ with $t \approx r$. Since $b_{\mathbb{B}}(X, r) = {}^{\circ}B(X, t)$, we obtain

$$\begin{aligned} \mathfrak{o}_r(X) \approx_{\mathbb{B}} \Theta_t(X) &= B_t(X) + \sum_{s<t} \widetilde{S}(\Theta_s(X)) \frac{1}{H} \\ &\approx_{\mathbb{B}} b_r(X) + \int_0^r S(\mathfrak{o}_s(X)) d\lambda(s). \end{aligned}$$

This proves that $\mathfrak{o}$ is a path-by-path solution of the original stochastic differential equation (SDE).

Here we have used the Loeb–Anderson lifting result for functions with values in $\mathbb{B}$. If $f : [0, r] \to \mathbb{B}$ is bounded and measurable and $F : T_t \to \mathbb{F}$ is a limited **lifting of** f, i.e., $F(s) \approx_{\mathbb{B}} f({}^{\circ}s)$ for $\widehat{\nu}$-almost all $s \in T_t$, then

$$\sum_{s \leq t} F(s) \frac{1}{H} \approx_{\mathbb{B}} \int_0^r f(s) d\lambda(s).$$

The proof is similar to the proof of Theorem 10.8.1.

We know that almost every path of the constructed Ornstein–Uhlenbeck process $\mathfrak{o}$ is an element of $C_{\mathbb{B}}$, but, conversely, also each continuous function is a path of $\mathfrak{o}$.

Proposition 14.1.10 *For each continuous $f : [0, \infty[\to \mathbb{B}$ there exists an $X \in \Omega$ such that $\Theta(X)$ is S-continuous and for all $r \in [0, \infty[$ and, for all $t \in T$ with $t \approx r$, we have*

$$f(r) = {}^{\circ}\Theta_t(X) = \mathfrak{o}_r(X).$$

Proof By Proposition 11.8.2, there exists a $Y \in \Omega$ such that $B_t(Y) \approx f({}^{\circ}t)$ for all limited $t \in T$. We may assume that $t \in T_{\sigma}$. It follows that for $X := \Phi^{-1}(Y)$

$$\Theta_t(X) := B_t \circ \Phi(X) \approx f({}^{\circ}t).$$

Therefore, $\Theta(X)$ is S-continuous and

$$\mathfrak{o}(X) = f.$$

□

The image measure of $\widehat{\Gamma}$ by the $C_{\mathbb{B}}$-valued random variable $X \mapsto \mathfrak{o}(X,\cdot)$ can be viewed as an **Ornstein–Uhlenbeck measure** on $C_{\mathbb{B}}$.

14.2 Ornstein–Uhlenbeck processes for shifts by scalars

Our aim is now to solve the simple infinite-dimensional stochastic differential equation

$$do_t = \sqrt{2|\alpha|\beta}db_t + \alpha o_t dt \quad \text{with } 0 < \beta, \alpha \neq 0 \text{ in } \mathbb{R},$$

and initial data $\mathfrak{o}_0 = 0$. Equivalent to this equality is:

$$o_t = \sqrt{2|\alpha|\beta}b_t + \alpha \int_0^t o_s ds. \tag{36}$$

The integral $\int_0^t o_s ds$ is the Bochner integral of o_s. We shall solve equation (36) path by path by a continuous process $o : \Omega \times [0,\infty[\to \mathbb{B}$. It follows again that each continuous function $f : [0,\infty[\to \mathbb{B}$ is a path of o.

Following again Keisler's approach [53] to stochastic differential equations, we convert equation (36) to an internal difference equation and solve this equation by recursion: The result is given by

$$\Xi_t = \sqrt{2|\alpha|\beta}B_t + \alpha \sum_{s \le t} \Xi_{s^-} \frac{1}{H}.$$

Here $\Xi_{s^-} := \Xi_{s-\frac{1}{H}}$ with $\Xi_0 = 0$. The process Ξ is $\mathbb{F}$-valued, defined on $\Omega \times T$. Our aim is to convert this process Ξ to a solution of the original differential equation. Set $\eta := 1 + \frac{\alpha}{H}$ and $w := \sqrt{2|\alpha|\beta}$.

Lemma 14.2.1 *Fix $t \in T$. Then*

(a) $\Xi_t(X) = w \cdot \sum_{s \le t} \eta^{H(t-s)} X_s$,
(b) $\sum_{s \le t} \eta\left(^{2H(t-s)}\right) = \sum_{s \le t} \eta^{2Hs} + 1 - \eta^{2Ht}$

Proof By internal induction on t. □

We define for $\sigma \in T$

$$c_\sigma := \frac{w^2}{H} \sum_{s \le \sigma} \eta^{2H(\sigma - s)},$$

and will prove that Ξ_σ is $\mathcal{N}(0,c_\sigma)$-distributed. Let us first prove the following lemma.

Lemma 14.2.2 *Fix a limited $\sigma \in T$. Then c_σ is limited. If $\sigma \not\approx 0$, then $c_\sigma \not\approx 0$.*

Proof By Lemma 14.2.1,

$$\frac{1}{w^2}c_\sigma \approx \frac{1}{H}\sum_{s\leq\sigma}\eta^{2Hs} \approx \frac{1}{H}\frac{1-\eta^{2H\sigma+2}}{1-\eta^2} = \frac{1-\eta^{2H\sigma+2}}{-2\alpha-\frac{\alpha^2}{H}} \approx \frac{1-\eta^{2H\sigma}}{-2\alpha}.$$

Since $\eta^{2H\sigma}$ is limited, c_σ is limited. Suppose $\sigma \not\approx 0$. Then

$$\eta^{2H\sigma} \approx e^{2\alpha\sigma} \not\approx 1,$$

thus $c_\sigma \not\approx 0$. □

Proposition 14.2.3 *For each $\sigma \in T$ and each $a \in \mathbb{F}$ with $\|a\| = 1$, $\langle \Xi_\sigma, a\rangle$ is $\mathcal{N}(0,c_\sigma)$-distributed, thus $\frac{1}{\sqrt{c_\sigma}}\langle \Xi_\sigma, a\rangle$ is $\mathcal{N}(0,1)$-distributed, with respect to Γ (see Corollary 3.2.4).*

Proof In the first equality of the computation below we use Fubini's theorem and the fact that a can be extended to an internal ONB of $\mathbb{F}$. Fix $\lambda \in {}^*\mathbb{R}$. By Lemma 14.2.1, we obtain for all $\sigma \in T$,

$$\int_\Omega \exp[\lambda\langle \Xi_\sigma, a\rangle]\,d\Gamma$$

$$= \int_{{}^*\mathbb{R}^T} \exp\left[\sum_{s\leq\sigma}\left(\lambda w\eta^{H(\sigma-s)}X_s - \frac{H}{2}X_s^2\right) + \sum_{\sigma<s} -\frac{H}{2}X_s^2\right]\sqrt{\frac{H}{2\pi}}^{H^2} d(X_s)_{s\in T}$$

$$= \int_{{}^*\mathbb{R}^{H\sigma}} \exp\left[\sum_{s\leq\sigma}\left(\lambda w\eta^{H(\sigma-s)}X_s - \frac{H}{2}X_s^2\right)\right]\sqrt{\frac{H}{2\pi}}^{H\sigma} d(X_s)_{s\leq H\sigma}$$

$$= \prod_{s\leq\sigma}\int_{{}^*\mathbb{R}} \exp\left[\lambda w\eta^{H(\sigma-s)}X_s - \frac{H}{2}X_s^2\right]\sqrt{\frac{H}{2\pi}}dX_s$$

$$= \prod_{s\leq\sigma}\int_{{}^*\mathbb{R}} \exp\left[-\frac{H}{2}\left(X_s^2 - \frac{2\lambda}{H}w\eta^{H(\sigma-s)}X_s\right)\right]\sqrt{\frac{H}{2\pi}}dX_s$$

$$= \prod_{s\leq\sigma}\int_{{}^*\mathbb{R}} \exp\left[-\frac{H}{2}\left(X_s - \frac{\lambda}{H}w\eta^{H(\sigma-s)}\right)^2 + \frac{\lambda^2 w^2}{2H}\eta^{2H(\sigma-s)}\right]\sqrt{\frac{H}{2\pi}}dX_s$$

$$= \prod_{s\leq\sigma} \int_{^*\mathbb{R}} \exp\left[-\frac{H}{2}X_s^2 + \frac{\lambda^2 w^2}{2H}\eta^{2H(\sigma-s)}\right]\sqrt{\frac{H}{2\pi}}dX_s$$

$$= \exp\left[\frac{\lambda^2 w^2}{2}\sum_{s\leq t}\eta^{2H(\sigma-s)}\frac{1}{H}\right].$$

By Proposition 3.2.3, $\langle \Xi_t, a\rangle$ is $\mathcal{N}(0,c_t)$-distributed. □

From Proposition 14.2.3 it follows in a similar way to Proposition 11.2.1 (d) that the following holds.

Corollary 14.2.4 *Set* $\widetilde{\Xi}_\sigma := \frac{1}{\sqrt{c_\sigma}}\Xi_\sigma$. *For each* $E \in {}^*\mathcal{E}$, γ_1^E *is the image measure of* Γ *by* $\pi_{\mathbb{F}E}\left(\widetilde{\Xi}_\sigma\right)$.

Since d is a measurable metric, there exists a function $g : \mathbb{N} \to \mathcal{E}(\mathbb{H})$ with ${}^*g(m) \subseteq {}^*g(m+1)$ for all $m \in {}^*\mathbb{N}$ and such that for each $E \in {}^*\mathcal{E}$ with $E \perp {}^*g(m)$ and $E \subseteq \mathbb{F}$

$$\gamma_1^E\left\{x \in E \mid d(x,0) \geq \frac{1}{2^m}\right\} \leq \frac{1}{2^m}. \tag{37}$$

By Corollary 14.2.4, we may replace (37) by

$$\Gamma\left\{d\left(\pi_{\mathbb{F}E} \circ \widetilde{\Xi}_\sigma, 0\right) \geq \frac{1}{2^m}\right\} \leq \frac{1}{2^m}. \tag{38}$$

Now we proceed in a similar way to the construction in Section 11.7: There exists an ONB $(\mathfrak{c}_i)_{i\in\mathbb{N}}$ of $\mathbb{H}$, an internal extension $(\mathfrak{c}_i)_{i\in\omega}$ to an ONB of $\mathbb{F}$ and an increasing function $f : \mathbb{N} \to \mathbb{N}$ such that for all $m \in \mathbb{N}$

$$\Gamma\left\{d\left(\sum_{f(m)<i\in\omega}\langle\widetilde{\Xi}_\sigma, \mathfrak{c}_i\rangle \cdot \mathfrak{c}_i, 0\right) \geq \frac{1}{2^m}\right\} \leq \frac{1}{2^m}.$$

Since $\langle \Xi_\sigma, a\rangle$ is $\mathcal{N}(0,c_\sigma)$-distributed for each $\sigma \in T$ and each $a \in \mathbb{F}$ with $\|a\| = 1$, the image measures of Γ by $\langle \Xi_\sigma, a\rangle$ and $\langle B_{c_\sigma}, a\rangle$ are the same. Since $\mathbb{E}_\Gamma \langle B_{c_\sigma}, a\rangle^2 = c_\sigma$, we obtain, using the transformation rule:

Corollary 14.2.5 *For each limited* $\sigma \in T$ *and each* $a \in \mathbb{F}$ *with* $\|a\| = 1$, $\mathbb{E}\langle \Xi_\sigma, a\rangle^2 = c_\sigma$ *is limited. It follows that* $\langle \Xi_\sigma, a\rangle^2$ *is limited* $\widehat{\Gamma}$*-a.s.*

Define for limited $\sigma \in T$

$$U_\sigma := \bigcap_{n\in\mathbb{N}}\bigcup_{n\leq m\in\mathbb{N}}\left\{d\left(\sum_{f(m)<i\in\omega}\langle\widetilde{\Xi}_\sigma, \mathfrak{c}_i\rangle \cdot \mathfrak{c}_i, 0\right) < \frac{1}{2^m}\right\} \cap U_0,$$

where $U_0 = \bigcap_{i \in \mathbb{N}} \left\{ \langle \Xi_\sigma, \mathfrak{c}_i \rangle^2 \text{ and } \langle \widetilde{\Xi}_\sigma, \mathfrak{c}_i \rangle^2 \text{ are limited} \right\}$. Then $\widehat{\Gamma}(U_\sigma) = 1$, by the Borel–Cantelli lemma. The preceding results yield the following corollary in analogy to the corresponding result in Section 11.7.

Corollary 14.2.6 *Fix a limited $\sigma \in T$. For all $X \in U_\sigma$, $\widetilde{\Xi}_\sigma(X)$ and $\Xi_\sigma(X)$ are nearstandard in $\mathbb{B}$, and*

$$^\circ\widetilde{\Xi}_\sigma(X) = \lim_{m \to \infty} \sum_{i=1}^{f(m)} {}^\circ\left\langle \widetilde{\Xi}_t(X), \mathfrak{c}_i \right\rangle \mathfrak{c}_i.$$

Moreover. we obtain in a similar way,

$$^\circ\Xi_\sigma(X) = \lim_{m \to \infty} \sum_{i=1}^{f(m)} {}^\circ\langle \Xi_t(X), \mathfrak{c}_i \rangle \mathfrak{c}_i.$$

It follows that the image measure of $\widehat{\Gamma}$ by $^\circ\widetilde{\Xi}_\sigma$ on the Borel σ-algebra of $(\mathbb{B}, |\cdot|)$ is the Gaussian measure γ_1 of variance 1 on $\mathbb{B}$.

Unfortunately, in contrast to Lemma 11.7.2, U_σ depends on $\sigma \in T$. To overcome this difficulty, we use the following lemma and the internal version of a celebrated result due to Fernique [39]. The proof is transferred from Kuo's proof in [56] to our setting.

Lemma 14.2.7 *Fix $\sigma \in {}^*\mathbb{N}$, $\sigma \leq H$. Then $\mathbb{E}\max_{t \in T_\sigma} |\Xi_t|^2$ is less or equal or infinitely close to $4e^{2|\alpha|}\mathbb{E}\,|\Xi_\sigma|^2$.*

Proof Fix $t \in T_\sigma$. Note that $\eta^{2Ht - 2H\sigma} \leq 1$ for $\alpha > 0$ and that for $\alpha < 0$, by Proposition 9.5.1 (c),

$$\eta^{2Ht-2H\sigma} \leq \frac{1}{\left(1+\frac{\alpha}{H}\right)^{2H\sigma-2Ht}} \leq \frac{1}{\left(1+\frac{\alpha}{H}\right)^{2H\sigma}} \approx e^{-2\alpha\sigma}.$$

Therefore, using Doob's and Jensen's inequality, we obtain:

$$\frac{1}{2|\alpha|\beta}\mathbb{E}\max_{t \in T_\sigma} |\Xi_t|^2 = \mathbb{E}\max_{t \in T_\sigma}\left(\left|\mathbb{E}^{\mathcal{B}_t}\sum_{s \in T_\sigma}\eta^{H\sigma - Hs}X_s\right|^2 \cdot \eta^{2Ht-2H\sigma}\right)$$

$$\leq \text{ or } \approx e^{2|\alpha|\sigma}\mathbb{E}\max_{t \in T_\sigma}\left|\mathbb{E}^{\mathcal{B}_t}\sum_{s \in T_\sigma}\eta^{H\sigma-Hs}X_s\right|^2 \leq 4e^{2|\alpha|\sigma}\mathbb{E}\left|\sum_{s \in T_\sigma}\eta^{H\sigma-Hs}X_s\right|^2.$$

□

Theorem 14.2.8 (Fernique [39]) *Fix $\sigma, j \in \mathbb{N}$. There exists a constant $\rho > 0$ in $\mathbb{R}$ such that $\mathbb{E}e^{\rho|\Xi_\sigma|_j}$ is limited. It follows that $\mathbb{E}|\Xi_\sigma|_j^p$ is limited for all $p \in [1,\infty[$ and, by Lemma 14.2.7, $\mathbb{E}\max_{t\in T_\sigma}|\Xi_t|_j^2$ is limited.*

Proof Set $c := c_\sigma$ and $T_\sigma^2 := \left\{\frac{1}{H},\ldots,\sigma,\sigma+\frac{1}{H},\ldots,2\sigma\right\}$. We introduce the internal centred Gaussian measure Γ^2 on $\mathbb{F}^{T_\sigma^2}$ of variance $\frac{1}{H}$, i.e., for all internal Borel sets B in $\mathbb{F}^{T_\sigma^2}$

$$\Gamma^2(B) := \int_{B^{\mathfrak{E}}} e^{-\frac{H}{2}\sum_{s\in T_\sigma^2, i\in\omega} x_s(i)^2}\, d\,(x_s(i))_{s\in T_\sigma^2, i\in\omega}\cdot\sqrt{\frac{H}{2\pi}}^{\,2H\sigma\omega},$$

where $\mathfrak{E}$ is an internal ONB of $\mathbb{F}$. Define mappings $A, B : \mathbb{F}^{T_\sigma^2} \to \mathbb{F}$ by setting

$$A(X) := \frac{\sqrt{2|\alpha|\beta}}{\sqrt{c}}\sum_{s\in T_\sigma}\eta^{H(\sigma-s)}X_s$$

and

$$B(X) := \frac{\sqrt{2|\alpha|\beta}}{\sqrt{c}}\sum_{s\in T_\sigma}\eta^{H(\sigma-s)}X_{\sigma+s}.$$

Proposition 3.2.3 tells us that $A, B, \frac{1}{\sqrt{2}}(A+B)$ and $\frac{1}{\sqrt{2}}(A-B)$ are $\mathcal{N}(0,1)$-distributed. Obviously, A and B are independent. Using a vector-valued version of Proposition 3.3.1 (a) and the fact that $\int_{\mathbb{R}} e^{i\lambda x}e^{-\frac{1}{2}x^2}dy\sqrt{\frac{1}{2\pi}} = e^{-\frac{1}{2}\lambda^2}$, one can prove: If internal random variables F, G are $\mathcal{N}(0,1)$-distributed and (F,G) is independent, then $\left(\frac{F-G}{\sqrt{2}}, \frac{F+G}{\sqrt{2}}\right)$ is also independent. It follows that $\left(\frac{1}{\sqrt{2}}(A+B), \frac{1}{\sqrt{2}}(A-B)\right)$ is independent. We obtain for all $s, t \in {}^*\mathbb{R}$ with $0 < s \le t$,

$$\begin{aligned}
\Gamma\left\{\left|\widetilde{\Xi}_\sigma\right|_j \le s\right\}\cdot\Gamma\left\{\left|\widetilde{\Xi}_\sigma\right|_j > t\right\} &= \Gamma^2\left\{\left|\frac{A+B}{\sqrt{2}}\right|_j \le s\right\}\cdot\Gamma^2\left\{\left|\frac{A-B}{\sqrt{2}}\right|_j > t\right\}\\
&= \Gamma^2\left\{|A+B|_j \le \sqrt{2}s \text{ and } |A-B|_j > \sqrt{2}t\right\}\\
&\le \Gamma^2\left\{\left||A|_j - |B|_j\right| \le \sqrt{2}s \text{ and } |A|_j + |B|_j > \sqrt{2}t\right\}\\
&\le \Gamma^2\left\{|A|_j > \frac{t-s}{\sqrt{2}} \text{ and } |B|_j > \frac{t-s}{\sqrt{2}}\right\}\\
&= \Gamma^2\left\{|A|_j > \frac{t-s}{\sqrt{2}}\right\}\cdot\Gamma^2\left\{|B|_j > \frac{t-s}{\sqrt{2}}\right\}\\
&= \Gamma\left\{\left|\widetilde{\Xi}_\sigma\right|_j > \frac{t-s}{\sqrt{2}}\right\}^2.
\end{aligned}$$

This proves that

$$\Gamma\left\{\left|\widetilde{\Xi}_\sigma\right|_j \leq s\right\} \cdot \Gamma\left\{\left|\widetilde{\Xi}_\sigma\right|_j > t\right\} \leq \Gamma\left\{\left|\widetilde{\Xi}_\sigma\right|_j > \frac{t-s}{\sqrt{2}}\right\}^2. \tag{39}$$

Now set $t_0 := s$ and $t_{n+1} := s + \sqrt{2}t_n$. Then $t_n = \sum_{i=0}^n \sqrt{2}^i s$. For $s \in \mathbb{N}$ define $\delta_0(s) = \frac{\Gamma\left\{\left|\widetilde{\Xi}_\sigma\right|_j > s\right\}}{\Gamma\left\{\left|\widetilde{\Xi}_\sigma\right|_j \leq s\right\}}$ with $\delta_0(s) = \frac{1}{0} := \infty$ if $\Gamma\left\{\left|\widetilde{\Xi}_\sigma\right|_j \leq s\right\} = 0$. We will prove that there exists a standard $s \in \mathbb{N}$ such that ${}^\circ\delta_0(s) < 1$. Assume that this is not true. Then, for all $k \in \mathbb{N}$, ${}^\circ\Gamma\left\{\left|\widetilde{\Xi}_\sigma\right|_j > k\right\} \geq {}^\circ\Gamma\left\{\left|\widetilde{\Xi}_\sigma\right|_j \leq k\right\}$. It follows that

$$\begin{aligned}\widehat{\Gamma}\bigcap_{k\in\mathbb{N}}\left\{\left|\widetilde{\Xi}_\sigma\right|_j > k\right\} &= \lim_{k\to\infty} {}^\circ\Gamma\left\{\left|\widetilde{\Xi}_\sigma\right|_j > k\right\} \\ &\geq \lim_{k\to\infty} {}^\circ\Gamma\left\{\left|\widetilde{\Xi}_\sigma\right|_j \leq k\right\} = \widehat{\Gamma}\bigcup_{k\in\mathbb{N}}\left\{\left|\widetilde{\Xi}_\sigma\right|_j \leq k\right\}.\end{aligned}$$

By Corollary 14.2.6, the right-hand side equals 1, thus, the left-hand side is 1, which is a contradiction. Now choose $s \in \mathbb{N}$ such that the standard part of $\delta_0(s)$ is strictly smaller than 1. We define for each $n \in {}^*\mathbb{N}_0$

$$\delta_n(s) = \frac{\Gamma\left\{\left|\widetilde{\Xi}_\sigma\right|_j > t_n\right\}}{\Gamma\left\{\left|\widetilde{\Xi}_\sigma\right|_j \leq s\right\}}.$$

We will show that $\delta_{n+1}(s) \leq \delta_n(s)^2$. By (39) we see that

$$\begin{aligned}\frac{\Gamma\left\{\left|\widetilde{\Xi}_\sigma\right|_j > t_{n+1}\right\}}{\Gamma\left\{\left|\widetilde{\Xi}_\sigma\right|_j \leq s\right\}} &= \frac{\Gamma\left\{\left|\widetilde{\Xi}_\sigma\right|_j > s + \sqrt{2}t_n\right\}}{\Gamma\left\{\left|\widetilde{\Xi}_\sigma\right|_j \leq s\right\}} \\ &= \frac{\Gamma\left\{\left|\widetilde{\Xi}_\sigma\right|_j > s + \sqrt{2}t_n\right\} \cdot \Gamma\left\{\left|\widetilde{\Xi}_\sigma\right|_j \leq s\right\}}{\Gamma\left\{\left|\widetilde{\Xi}_\sigma\right|_j \leq s\right\}^2} \\ &\leq \frac{\Gamma\left\{\left|\widetilde{\Xi}_\sigma\right|_j > t_n\right\}^2}{\Gamma\left\{\left|\widetilde{\Xi}_\sigma\right|_j \leq s\right\}^2} = \delta_n(s)^2.\end{aligned}$$

Therefore, $\delta_n(s) \leq \delta_0(s)^{2^n} = \exp(2^n \ln \delta_0(s))$ and with $v_n := 2^{n+1}s \geq t_n$:

$$\Gamma\left\{\left|\widetilde{\Xi}_\sigma\right|_j > v_n\right\} \leq \Gamma\left\{\left|\widetilde{\Xi}_\sigma\right|_j > t_n\right\} = \delta_n(s) \cdot \Gamma\left\{\left|\widetilde{\Xi}_\sigma\right|_j \leq s\right\}$$
$$\leq \Gamma\left\{\left|\widetilde{\Xi}_\sigma\right|_j \leq s\right\} \cdot \exp\left(2^n \ln \delta_0(s)\right) = \Gamma\left\{\left|\widetilde{\Xi}_\sigma\right|_j \leq s\right\} \cdot \exp\left(\frac{v_n}{2s} \ln \delta_0(s)\right).$$

Set $a := -\frac{\ln \delta_0(s)}{2s}$. Then $0 < {}^\circ a$. Choose a standard $b \in]0, \frac{{}^\circ a}{2}[$. Then we obtain for all standard ρ with $0 < \rho < b$

$$\int_{\left\{\left|\widetilde{\Xi}_\sigma\right|_j > v_0\right\}} \exp\left(\rho \cdot \left|\widetilde{\Xi}_\sigma\right|_j\right) d\Gamma = \sum_{n \in {}^*\mathbb{N}_0} \int_{\left\{v_n < \left|\widetilde{\Xi}_\sigma\right|_j \leq v_{n+1}\right\}} \exp\left(\rho \cdot \left|\widetilde{\Xi}_\sigma\right|_j\right) d\Gamma$$
$$\leq \sum_{n \in {}^*\mathbb{N}_0} \exp\left(\rho \cdot v_{n+1}\right) \Gamma\left\{\left|\widetilde{\Xi}_\sigma\right|_j > v_n\right\} \leq \sum_{n \in {}^*\mathbb{N}_0} \exp\left(\rho \cdot v_{n+1} - 2bv_n\right)$$
$$= \sum_{n \in \mathbb{N}_0} \exp\left((\rho - b) \cdot v_{n+1}\right) < \infty \text{ in } \mathbb{R}.$$

Since $\Xi_\sigma = \frac{1}{\sqrt{c}} \widetilde{\Xi}_\sigma$ and $\frac{1}{\sqrt{c}} < \infty$ in $\mathbb{R}$, the proof is finished. □

Now we prove, using Fernique's Theorem, the main result in this section.

Theorem 14.2.9 *There exists a set U with $\widehat{\Gamma}(U) = 1$ such that, for all $X \in U$, $\Xi(X)$ is S-continuous with respect to the topology of $\mathbb{B}$.*

Proof By Theorem 14.2.8, for all $\sigma, j \in \mathbb{N}$ there exists an event $U_{\sigma,j}$ of $\widehat{\Gamma}$-measure 1 and a mapping $g_{\sigma,j} : U_{\sigma,j} \to \mathbb{N}$ such that, for all $X \in U_{\sigma,j}$, $\max_{t \in T_\sigma} |\Xi_t(X)|_j \leq g_{\sigma,j}(X)$. Set $U := \bigcap_{\sigma,j \in \mathbb{N}} U_{\sigma,j}$. Then $\widehat{\Gamma}(U) = 1$. Let $s, t \in T$ be limited with $s \approx t$ and $s \leq t$. Then there exists a $\sigma \in \mathbb{N}$ with $s, t \leq \sigma$. We obtain for all $j \in \mathbb{N}$ and all $X \in U$

$$\left|\int_{T_t} \Xi_{r^-}(X) d\nu(r) - \int_{T_s} \Xi_{r^-}(X) d\nu(r)\right|_j \leq \int_{T_t \setminus T_s} |\Xi_{r^-}(X)|_j d\nu(r)$$
$$\leq g_{\sigma,j}(X) \cdot \nu(T_t \setminus T_s) \approx 0,$$

thus $t \mapsto \int_{T_t} \Xi_{r^-}(X) d\nu(r)$ is pre-S-continuous. Since $t \mapsto B_t$ is S-continuous $\widehat{\Gamma}$-a.s., we may assume that, for all $X \in U$, $t \mapsto \Xi_t(X)$ is pre-S-continuous. Moreover, we may assume that $\Xi_t(X)$ is nearstandard for all $X \in U$ and all $t \in]0, \infty[\cap \mathbb{Q} \subseteq T$. Using the proof of Theorem 11.7.6, one can see that $t \mapsto \Xi_t(X)$ is S-continuous for all $X \in U$. □

Using this result, the internal $\mathbb{F}$-valued process Ξ can be converted into a $\widehat{\Gamma}$-a.s. continuous process $o : \Omega \times [0, \infty[\to \mathbb{B}$ setting $o_{{}^\circ t}(X) := {}^\circ\Xi_t(X)$ for all

$X \in U$, where U is defined in the proof of Theorem 14.2.9. Because of the well-known close relationship between the Loeb and Lebesgue measure, and since $o(X) \upharpoonright [0,n]$ is Bochner integrable for all $X \in U$ and all $n \in \mathbb{N}$ (in fact, $d(o(X),0)$ is bounded on $[0,n]$), we obtain for all $X \in U$ and all limited $t \in T$

$$o_{^\circ t}(X) \approx \Xi_t(X) = \sqrt{2\alpha\beta} B_t(X) + \alpha \sum_{s \le t} \Xi_{s^-}(X) \frac{1}{H}$$

$$\approx \sqrt{2|\alpha|\beta} b_{^\circ t}(X) + \alpha \int_0^{^\circ t} o_s(X) d\lambda(s).$$

Since the Brownian motion b was constructed path by path, o is a path-by-path continuous solution to equation (36).

An interesting and simple result arises as a consequence:

Corollary 14.2.10 *For each continuous function $f : [0,\infty[\to \mathbb{B}$ there exists an $X \in \Omega$ with $f = o(X)$.*

Proof Using the proof of Proposition 11.8.2, we may assume that f is a polygon with values in $\mathbb{H}$. Then ${}^*f : {}^*[0,\infty[\to \mathbb{F}$. Define, by internal recursion,

$$X_{\frac{1}{H}} := \frac{1}{\sqrt{2|\alpha|\beta}} {}^*f\left(\frac{1}{H}\right),$$

$$X_{t+\frac{1}{H}} = \frac{1}{\sqrt{2|\alpha|\beta}} {}^*f\left(t+\frac{1}{H}\right) - \eta {}^*f(t).$$

Then, for all limited $t \in T$,

$$f(^\circ t) \approx {}^*f(t) = \Xi_t(X) \approx o_{^\circ t}(X),$$

where the proof of the equation in the middle uses internal induction. □

Exercises

14.1 Prove Lemma 14.1.2.
14.2 Prove Corollary 14.1.5 part (b).
14.3 Prove Lemma 14.1.6.
14.4 Prove Theorem 14.1.8.
14.5 Prove Lemma 14.2.1.
14.6 Prove Corollary 14.2.4.

15

Lindstrøm's construction of standard Lévy processes from discrete ones

In this chapter and the next chapter we turn to Lévy processes. Following Lindstrøm's [67] approach to Lévy processes, we will show that the standard part of each internally square-integrable Lévy processes is a standard Lévy process and each standard Lévy process can be obtained in this way. Here we have identified two Lévy processes if they satisfy the same Lévy triplet. Moreover, we will use Lindstrøm's work to prove the well-known fact that each Lévy process can be divided into its continuous and pure jump part. The standard theory of Lévy processes can be found in Bertoin [8] or Sato [103]. I also refer to Ng [78] for a nonstandard approach to Lévy processes.

The results of Lindstrøm are transferred to the setting in [89]. While Lindstrøm uses internal probability measures concentrated on a *finite set, we study arbitrary Borel probability measures on $^*\mathbb{R}$ and their products on $^*\mathbb{R}^T$, where $T = \{\frac{1}{H}, \ldots, H\}$. The proofs are straightforward modifications of the proofs in Lindstrøm's article. For simplicity we assume that Lévy processes are one-dimensional.

Recall that $T_t = \left\{\frac{1}{H}, \ldots, t\right\}$ for all $t \in T$, and $T^n_< := \{t \in T^n \mid t_1 < \ldots < t_n\}$. Set $T_0 := T \cup \{0\}$. Moreover, recall that $\Omega = {}^*\mathbb{R}^T$ and μ is the internal product measure of some internal Borel probability measure μ^1 on $^*\mathbb{R}$ and that

$$\mathcal{B}_t = \left\{D \times {}^*\mathbb{R}^{T \setminus T_t} \mid D \text{ is an internal Borel set in } {}^*\mathbb{R}^{T_t}\right\}.$$

By means of the transfer principle, applied to the results in Section 3.4, we may assume that each internal (one-dimensional) Lévy process defined on T is of the form $B : \Omega \times T \to {}^*\mathbb{R}$, $(X, t) \mapsto \sum_{s \in T_t} X_s$. Note that, by transfer of Condition (L 3) in Section 3.4, $\mathbb{E}_{\mu^1} x^2 = \mathbb{E}_\mu X^2_{\frac{1}{H}} < \infty$ in $^*\mathbb{R}$.

The measure μ^1 can be regarded as a measure on the increments of B, and $H\mu^1$ leads to the Lévy measure of the standard part $^\circ B$ of B.

15.1 Exponential moments for processes with limited increments

An internal $\mathcal{C}$-measurable process $W : \Lambda \times T \to {}^*\mathbb{R}$ has **ρ-limited increments** if there exists a standard $\delta > 0$ such that $\rho\{\Delta W_t > \delta\} = 0$ for all $t \in T$. The aim in this section is to prove that Lévy processes with limited increments have exponential moments.

In [89] there is a proof of this result for Lévy processes with weighted increments, using concentration inequalities, due to Boucheron, Lugosi and Massart [15]. In many situations weighted Lévy processes lead to better models. For example, if L_t counts the number of cars crossing a street up to time t, then weighted Lévy processes provide a more accurate description. Moreover, we will see in Section 16.3 that Wiener–Lévy integrals are standard parts of a weighted Lévy process. The result in [89] can be used to develop Malliavin calculus for weighted Lévy processes

Theorem 15.1.1 (Lindstrøm [67], Protter [97]) *Fix a standard $\sigma > 0$. Assume that* $\max_{t \in T_\sigma} |B(\cdot, t)|$ *is limited $\widehat{\mu}$-a.s. and B has μ-limited increments, i.e., there is a standard $\delta > 0$ with $\mu^1\{|x| > \delta\} = 0$. Then there exists a standard $\varepsilon > 0$ such that*

$$\mathbb{E}_\mu e^{\varepsilon |B_\sigma|} \text{ is limited.}$$

Proof If $\mathbb{E}_\mu B_\sigma^2 = 0$ or $\sigma = 0$, the result is obvious. So we may assume that $\mathbb{E}_\mu B_\sigma^2 = \eta \neq 0$ and $\sigma > 0$. Extend Ω to $\Omega_\infty := {}^*\mathbb{R}^{T^\infty}$ with $T_\infty := \frac{{}^*\mathbb{N}}{H}$ and let μ^∞ be the internal product of μ^1 on Ω_∞. Fix $k \in {}^*\mathbb{N}$. We need to know that there exists an $n_k \in {}^*\mathbb{N}$ such that

$$\mu^\infty \left\{ \left| \sum_{s \in T_{n_k}} X_s \right| < 2k \right\} < 1. \tag{40}$$

Assume that $\mu^\infty \left\{ \left| \sum_{s \in T_n} X_s \right| < 2k \right\} = 1$ for all $n \in {}^*\mathbb{N}$. Then

$$\begin{aligned} n \cdot \eta = n \cdot \mathbb{E}_{\mu^\infty} B_\sigma^2 &= n \cdot \mathbb{E}_{\mu^\infty} \left(\sum_{s \in T_\sigma} X_s \right)^2 \\ &= Hn\sigma \mathbb{E}_{\mu^1} x^2 + Hn\sigma (H\sigma - 1) \left(\mathbb{E}_{\mu^1} x\right)^2 \\ &\leq Hn\sigma \mathbb{E}_{\mu^1} x^2 + Hn\sigma (Hn\sigma - 1) \left(\mathbb{E}_{\mu^1} x\right)^2 \\ &= \mathbb{E}_{\mu^\infty} \left(\sum_{s \in T_{n\sigma}} X_s \right)^2 \leq 4k^2. \end{aligned}$$

Since this is true for all $n \in {}^*\mathbb{N}$, we have a contradiction. Define a sequence $\left(\tau_m^k\right)_{m\in {}^*\mathbb{N}}$ of $(\mathcal{B}_t)_{t\in T_\infty}$-stopping times by internal recursion:

$$\tau_0^k := 0$$

and for $m \geq 1$

$$\tau_m^k := \inf\left\{t \in T_\infty \mid \left|\sum\nolimits_{\tau_{m-1}<s\leq t} X_s\right| \geq k\right\} \text{ with } \inf\emptyset = \infty.$$

We prove first that $\mu^\infty\left\{\tau_1^k = \infty\right\} = 0$:

$$\begin{aligned}
&\mu^\infty\left\{\tau_1^k = \infty\right\} \\
&= \mu^\infty \bigcap_{t\in T_\infty}\left\{\left|\sum_{s\leq t} X_s\right| < k\right\} \\
&\leq \mu^\infty \bigcap_{n\in {}^*\mathbb{N}}\left\{\left|\sum_{s\leq n\cdot n_k} X_s\right| < k\right\} \leq \mu^\infty \bigcap_{n\in {}^*\mathbb{N}}\left\{\left|\sum_{(n-1)n_k<s\leq n\cdot n_k} X_s\right| < 2k\right\} \\
&= \lim_{l\to\infty, l\in {}^*\mathbb{N}} \mu^\infty \bigcap_{n\leq l}\left\{\left|\sum_{(n-1)n_k<s\leq n\cdot n_k} X_s\right| < 2k\right\} \\
&= \lim_{l\to\infty}\left(\mu^\infty\left\{\left|\sum_{s\leq n_k} X_s\right| < 2k\right\}\right)^l = 0,
\end{aligned}$$

because of (40). Now we will prove that $\mu^\infty\left\{\tau_m^k = \infty\right\} = 0$ for $m > 1$ under the assumption that $\mu^\infty\left\{\tau_{m-1}^k = \infty\right\} = 0$:

$$\begin{aligned}
&\mu^\infty\left\{\tau_m^k = \infty\right\} \\
&= \sum_{r\in T_\infty} \mu^\infty\left(\left\{\tau_{m-1}^k = r\right\} \cap \bigcap_{t\in T_\infty}\left\{\left|\sum_{r<s\leq r+t} X_s\right| < k\right\}\right) \\
&\leq \sum_{r\in T_\infty} \mu^\infty\left(\left\{\tau_{m-1}^k = r\right\} \cap \bigcap_{n\in {}^*\mathbb{N}}\left\{\left|\sum_{r+(n-1)n_k<s\leq r+n\cdot n_k} X_s\right| < 2k\right\}\right) \\
&= \sum_{r\in T_\infty} \mu^\infty\left\{\tau_{m-1}^k = r\right\} \cdot \mu_\infty\left(\bigcap_{n\in {}^*\mathbb{N}}\left\{\left|\sum_{r+(n-1)n_k<s\leq r+n\cdot n_k} X_s\right| < 2k\right\}\right) \\
&= \lim_{l\to\infty} \mu_\infty\left\{\left|\sum_{s\leq n_k} X_s\right| < 2k\right\}^l = 0.
\end{aligned}$$

This proves that $\mu^\infty\left\{\tau_m^k = \infty\right\} = 0$ for all $k, m \in {}^*\mathbb{N}_0$, $k \geq 1$. We will prove that, for all unlimited $k \in {}^*\mathbb{N}$, $\widehat{\mu^\infty}\left\{\tau_m^k - \tau_{m-1}^k < \sigma\right\} = 0$:

$$\begin{aligned}
\mu^\infty\left\{\tau_m^k - \tau_{m-1}^k < \sigma\right\} &= \sum_{r \in T_\infty} \mu^\infty\left\{\tau_{m-1}^k = r \wedge \max_{t \in T_\sigma}\left|\sum_{r<s\leq r+t} X_s\right| \geq k\right\} \\
&= \sum_{r \in T_\infty}\left(\mu^\infty\left\{\tau_{m-1}^k = r\right\} \cdot \mu^\infty\left\{\max_{t \in T_\sigma}\left|\sum_{r<s\leq r+t} X_s\right| \geq k\right\}\right) \\
&= \sum_{r \in T_\infty}\left(\mu^\infty\left\{\tau_{m-1}^k = r\right\} \cdot \mu^\infty\left\{\max_{t \in T_\sigma}\left|\sum_{s\leq t} X_s\right| \geq k\right\}\right) \\
&\leq \mu^\infty\left\{\max_{t \in T_\sigma}\left|\sum_{s\leq t} X_s\right| \geq k\right\} \approx 0.
\end{aligned}$$

It follows that $\mu^\infty\left\{\tau_m^k - \tau_{m-1}^k < \sigma\right\} < i := \frac{1}{2}\left(1 - \frac{1}{e^\sigma}\right) < \frac{1}{2}$ for all unlimited $k \in {}^*\mathbb{N}$. By Proposition 8.4.1 (a), there exists a $k \in \mathbb{N}$ with $\delta \leq k$ and $\mu^\infty\left\{\tau_m^k - \tau_{m-1}^k < \sigma\right\} < i$. We will write τ_m instead of τ_m^k. Now

$$\mathbb{E}_{\mu^\infty} e^{-(\tau_m - \tau_{m-1})} \leq \int_{\{\tau_m - \tau_{m-1} \geq \sigma\}} \frac{1}{e^\sigma} d\mu^\infty + \int_{\{\tau_m - \tau_{m-1} < \sigma\}} 1 d\mu \leq \frac{1}{e^\sigma} + i < 1.$$

This is true for all $m \in {}^*\mathbb{N}$. Set $\alpha := \frac{1}{e^\sigma} + i$. Since the internal m-tuple $\left(e^{-\tau_1}, e^{-(\tau_2 - \tau_1)}, \ldots, e^{-(\tau_m - \tau_{m-1})}\right)$ is independent, by Proposition 3.3.1 (d),

$$\mathbb{E}_{\mu^\infty} e^{-\tau_m} = \mathbb{E}_{\mu^\infty} e^{-\tau_1} \mathbb{E}_{\mu^\infty} e^{-(\tau_2 - \tau_1)} \mathbb{E}_{\mu^\infty} e^{-(\tau_m - \tau_{m-1})} \leq \alpha^m \to_{m \to \infty} 0 \text{ in } {}^*\mathbb{N}.$$

We will now show that for all $n \in {}^*\mathbb{N}_0$ and all $t \in T_\infty$

$$\mu^\infty\left\{\left|\sum_{s\leq t} X_s\right| \geq 2nk\right\} \leq \mu^\infty\left\{\tau_n \leq t\right\}.$$

For $n = 0$ the result is clear. Set

$$U := \{X \in \Omega \mid |X_t| \leq \delta \text{ for all } t \in T\}.$$

Then $\mu(U) = 1$. Fix $X \in U$ and let $\tau_{n+1}(X) > t$. Then $\left|\sum_{\tau_n < s \leq t} X_s\right| < k$. It follows that

$$\begin{aligned}
\left|\sum_{s\leq t} X_s\right| &\leq \sum_{i<n, i \in \mathbb{N}_0}\left(\left|\sum_{\tau_i(X)<s<\tau_{i+1}(X)} X_s\right| + \left|X_{\tau_{i+1}(X)}\right|\right) + \left|\sum_{\tau_n(X)<s\leq t} X_s\right| \\
&< kn + nk + k = (2n+1)k < 2(n+1)k.
\end{aligned}$$

Moreover,

$$\mu^{\infty}\{\tau_n \leq t\} = \mu^{\infty}\left\{e^{-t} \leq e^{-\tau_n}\right\} = \mu^{\infty}\left\{1 \leq e^{t-\tau_n}\right\} \leq e^t \mathbb{E}_{\mu^{\infty}} e^{-\tau_n} \leq e^t \alpha^n.$$

Choose $\varepsilon > 0$ such that $\alpha \cdot e^{2k\cdot\varepsilon} < 1$. Then we obtain, for all limited $\rho \in T^{\infty}$,

$$\begin{aligned}\mathbb{E}_{\mu^{\infty}} e^{\varepsilon\left|\sum_{s\leq\rho} X_s\right|} &\leq \sum_{n\in {}^*\mathbb{N}} \int_{\left\{2(n-1)k \leq \left|\sum_{s\leq\rho} X_s\right| < 2nk\right\}} e^{\varepsilon 2nk} d\mu^{\infty} \\ &\leq \sum_{n\in {}^*\mathbb{N}} e^{\rho}\alpha^{n-1} e^{\varepsilon 2nk} = e^{\rho+\varepsilon 2k} \sum_{n\in {}^*\mathbb{N}} \left(\alpha e^{\varepsilon 2k}\right)^{n-1}\end{aligned}$$

is limited. The proof of Theorem 15.1.1 is finished. □

15.2 Limited Lévy processes

Using the work of Lindstrøm [67], we will study limited Lévy processes in detail. Fix an internal probability space $(\Lambda, \mathcal{C}, \rho)$. An internal $\mathcal{C}$-measurable process $F : \Lambda \times T \to {}^*\mathbb{R}$ is called ρ-**limited** if for all standard $n \in \mathbb{N}$

$$\widehat{\rho}\left\{\max_{t\in T_n} |F(\cdot,t)| \text{ is limited}\right\} = 1.$$

By transfer of the results in Section 3.4, $B : \Omega \times T \to {}^*\mathbb{R}$ with measure μ can be used instead of any internal Lévy process on T and we will do so. The notion 'limited' for the process B depends on μ, actually on μ^1. The main result in this section is a useful characterization of μ-limited Lévy processes, which can be applied to the following important examples.

Examples 15.2.1
(BM) Set

$$\mu^1(A) = \int_A e^{-\frac{H}{2}x^2} dx \sqrt{\frac{H}{2\pi}}.$$

Cutland [23] proved that the standard part of B under μ is a Brownian motion (see Sections 11.4 and 11.5).

(PP) Set

$$\mu^1(A) := \sum_{j\in {}^*\mathbb{N}_0 \cap A} e^{-\frac{\beta}{H}} \left(\frac{\beta}{H}\right)^i \frac{1}{i!}.$$

The $\widehat{\mu}$-a.s. defined standard part of B is a Poisson process with rate β. (See Section 15.5.)

In order to work with Example BM, the following result is useful.

Lemma 15.2.2 *For all standard* $\varepsilon > 0$,

$$H\mu^1\left\{x \in {}^*\mathbb{R} \,|\, x \geq \varepsilon\right\} \leq \frac{H}{2} e^{-\frac{H}{2}\varepsilon^2} \approx 0.$$

Proof We first prove that, for $x > 0$, $\int_x^\infty e^{-\frac{1}{2}t^2} dt \frac{1}{\sqrt{2\pi}} \leq \frac{1}{2} e^{-\frac{x^2}{2}}$. Set $f(x) := \frac{1}{2} e^{-\frac{x^2}{2}} + \int_0^x e^{-\frac{1}{2}t^2} dt \frac{1}{\sqrt{2\pi}}$. Since $\int_0^\infty e^{-\frac{1}{2}t^2} dt \frac{1}{\sqrt{2\pi}} = \frac{1}{2}$, it suffices to prove that $\frac{1}{2} \leq f$ on $[0, \infty[$. Note that $f'(x) = 0$ if and only if $x = \frac{2}{\sqrt{2\pi}}$, and $f' > 0$ on $[0, \frac{2}{\sqrt{2\pi}}[$ and $f' < 0$ on $]\frac{2}{\sqrt{2\pi}}, \infty[$, so f is strictly monotone increasing on $[0, \frac{2}{\sqrt{2\pi}}[$ and strictly monotone decreasing on $]\frac{2}{\sqrt{2\pi}}, \infty[$. Moreover, $\lim_{x \downarrow 0} f(x) = \frac{1}{2}$ and $\lim_{x \to \infty} f(x) = \frac{1}{2}$. Therefore, $\frac{1}{2} \leq f$. By the substitution rule, we obtain

$$\begin{aligned} H \cdot \mu^1 \{x \geq \varepsilon\} &= H \int_\varepsilon^\infty e^{-\frac{H}{2}x^2} \sqrt{\frac{H}{2\pi}} dx = H \int_{\sqrt{H}\varepsilon}^\infty e^{-\frac{1}{2}x^2} \sqrt{\frac{1}{2\pi}} dx \\ &= H \cdot \frac{1}{2} e^{-\frac{H\varepsilon^2}{2}} \approx \lim_{n \to \infty} n \cdot \frac{1}{2} e^{-\frac{n\varepsilon^2}{2}} = 0. \end{aligned}$$

This proves the lemma. □

For internal Lévy processes with μ-limited increments we obtain a first result:

Proposition 15.2.3 *Suppose that B has μ-limited increments. Then B is μ-limited if and only if $H\mathbb{E}_{\mu^1} x$ and $H\mathbb{E}_{\mu^1} x^2$ are both limited.*

Proof Assume first that B is μ-limited. Then $H\mathbb{E}_{\mu^1} x = \mathbb{E}_\mu B(\cdot, 1)$ and $\mathbb{E}_\mu B^2(\cdot, 1)$ are limited by Theorem 15.1.1. By Lemma 3.4.2 (e) with $\square = \frac{1}{H}$,

$$H\mathbb{E}_{\mu^1} x^2 = \mathbb{E}_\mu B^2(\cdot, 1) - H(H-1)\left(\mathbb{E}_{\mu^1} x\right)^2$$

is limited. To prove the converse, we assume that $H\mathbb{E}_{\mu^1} x$ and $H\mathbb{E}_{\mu^1} x^2$ are both limited. Since

$$\mathbb{E}_\mu\left(B_n{}^2\right) = Hn\mathbb{E}_{\mu^1} x^2 + Hn(Hn-1)\left(\mathbb{E}_{\mu^1} x\right)^2$$

is limited, we can apply Lemma 3.4.5 to see that $\mathbb{E}_\mu\left(\max_{t \in T_n} B_t{}^2\right)$ is limited. By Lemma 10.7.1 (a), $\max_{t \leq n} B_t^2$ is limited $\widehat{\mu}$-a.s., thus $\max_{t \leq n} |B_t|$ is limited $\widehat{\mu}$-a.s. □

Fix an internal Borel set $A \subseteq {}^*\mathbb{R}$. Recall that B is the representation of $B^A : (X,t) \mapsto \sum_{s \le t} \mathbf{1}_A(X_s) \cdot X_s$ by (μ_A, μ) if μ_A^1 is the image measure of μ by $B^A_{\frac{1}{H}}$. The following lemma is the key to the main results in this section.

Lemma 15.2.4 (a) *Fix $n \in \mathbb{N}$ and an internal Borel set $A \subseteq {}^*\mathbb{R}$ such that $H \cdot \mu^1(A)$ is limited. Then the set of all $X \in \Omega$ such that infinitely many X_s with $s \in T_n$ are elements of A is a $\widehat{\mu}$-nullset, i.e., $\widehat{\mu}\left(\bigcap_{k \in \mathbb{N}} N_k\right) = 0$, where*

$$N_k := \{X \in \Omega \mid |\{s \in T_n \mid X_s \in A\}| > k\}.$$

(b) *Suppose that B is μ-limited. Then $\lim_{k \to \infty} {}^\circ H \mu^1 \{k < |x|\} = 0$.*

(c) *Suppose that $\lim_{k \to \infty} {}^\circ H \mu^1 \{k < |x|\} = 0$ and $H \mu^1 \{\delta < |x|\}$ is limited. Then $B^{\{\delta < |x|\}}$ is μ-limited.*

(d) *Assume that $0 \not\approx \delta > 0$ and $H \int_{\{|x| \le k\}} x^2 d\mu^1$ and $H \mu^1 \{k < |x|\}$ are both limited for some $k \ge 0$. Then $H \mu^1 \{\delta < |x|\}$ is limited.*

Proof (a) By Proposition 8.4.1 (b), it suffices to show that $\mu(N_{K_0}) \approx 0$ for all unlimited $K_0 \in {}^*\mathbb{N}$. Note that $\mu(N_{K_0}) = \sum_{K_0 < K \le H \cdot n} \mu(G_K)$ with

$$G_K := \left\{\exists s_1 < \ldots < s_K \in T_n(D_{s_1,\ldots,s_K})\right\},$$

where

$$D_{s_1,\ldots,s_K} := \left\{X_{s_1} \in A, \ldots, X_{s_K} \in A \wedge \forall s \in T_n \setminus \{s_1, \ldots, s_K\} (X_s \notin A)\right\}.$$

Fix $r \in \mathbb{N}$ with $H \mu^1(A) \le r$. Since the components of X are independent and identically distributed, we obtain

$$\begin{aligned}
\mu(G_K) &= \binom{H \cdot n}{K} \cdot \left(\mu^1(A)\right)^K \cdot \left(\mu^1(\Omega \backslash A)\right)^{Hn-K} \\
&= \frac{1}{K!} \frac{Hn(Hn-1)\ldots(Hn-K+1)}{H^K} \cdot \left(H \cdot \mu^1(A)\right)^K \\
&\quad \cdot \left(1 - \mu^1(A)\right)^{Hn-K} \\
&\le \frac{1}{K!} \cdot \left(H \cdot n \cdot \mu^1(A)\right)^K \le \frac{1}{K!} (n \cdot r)^K.
\end{aligned}$$

Then $\mu(N_{K_0}) = \sum_{K_0 < K \le H \cdot n} \mu(G_K) \le \sum_{K_0 < K \le H \cdot n} \frac{1}{K!} (n \cdot r)^K \approx 0$ by Proposition 9.5.1 (c).

(b) Assume that (b) is not true. By Proposition 8.4.1 (b), there exists a standard $\delta > 0$ and an unlimited $N \in {}^*\mathbb{N}$ with $H \mu^1 \{N < |x|\} \ge \delta$.

By the assumption,

$$1 \approx \mu \bigcap_{t \leq 1} \left\{ |B(\cdot,t)| \leq \frac{N}{2} \right\} \leq \mu \bigcap_{t \leq 1} \{|X_t| \leq N\} = \mu^1 \{|x| \leq N\}^H$$

$$= \left(1 - \mu^1 \{N < |x|\}\right)^H \leq \left(1 - \frac{\delta}{H}\right)^H \approx (\,9.5.1\,(\mathrm{c}))\; e^{-\delta} < 1,$$

which is a contradiction.

(c) Assume that $B^{\{\delta<|x|\}}$ is not μ-limited. Then there exist an $n \in \mathbb{N}$, a standard $\varepsilon > 0$ and an unlimited $N \in {}^*\mathbb{N}$ such that the set $J := \left\{ \max_{t \in T_n} \left| B_t^{\{\delta<|x|\}} \right| \geq N \right\}$ has μ-measure greater than ε. First fix $m \in \mathbb{N}$ with $m > \delta$. Note that the set $I := \left\{ \exists t \in T_n (B_t^{\{\delta<|x|\leq m\}} \text{ is not limited}) \right\}$ is a subset of the set of all $X \in \Omega$ such that the set of all $s \in T_n$ with $\delta < |X_s| \leq m$ is infinite. By (a), $\widehat{\mu}(I) = 0$. This proves that $B^{\{\delta<|x|\leq m\}}$ is μ-limited. To obtain a contradiction to the assumption, we first compute

$$\mu \left\{ \forall t \in T_n \left(B_t^{\{\delta<|x|\}} = B_t^{\{\delta<|x|\leq m\}} \right) \right\}$$
$$= \mu \{ \forall t \in T_n (|X_t| \leq m) \} = \mu^1 \{|x| \leq m\}^{Hn}$$
$$= \left(1 - \mu^1 \{m < |x|\}\right)^{Hn} = \left(1 - \frac{q_m}{H}\right)^{Hn} \approx e^{-q_m} \quad (\text{by } 9.5.1\,(\mathrm{d}))$$

with limited $q_m := H\mu^1 \{m < |x|\}$. Since $\lim_{k\to\infty} {}^\circ H\mu^1 \{k < |x|\} = 0$, we can choose $m > \delta$ such that $1 - \mu(J) < \left(1 - \frac{q_m}{H}\right)^{Hn} - \frac{\varepsilon}{2}$. On the other hand,

$${}^\circ\mu(J) \leq {}^\circ\mu \left\{ \exists t \in T_n \left(B_t^{\{\delta<|x|\}} \neq B_t^{\{\delta<|x|\leq m\}} \right) \right\} \approx 1 - \left(1 - \frac{q_m}{H}\right)^{Hn},$$

thus $1 - {}^\circ\mu(J) \geq {}^\circ\left(1 - \frac{q_m}{H}\right)^{Hn}$, which is a contradiction.

(d) We may assume $\delta < k$. Then

$$H\mu^1 \{\delta < |x| \leq k\} = H \int_{\left\{1 < \frac{1}{\delta^2} x^2 \text{ and } |x| \leq k\right\}} 1 d\mu^1 \leq \frac{1}{\delta^2} H \int_{\{|x| \leq k\}} x^2 d\mu^1$$

is limited. Since $H\mu^1 \{k < |x|\}$ is limited, $H\mu^1 \{\delta < |x|\}$ is limited. □

Theorem 15.2.5 *Assume that the process B is μ-limited. If $\delta > 0$ is non-infinitesimal, then $B^{\{\delta<|x|\}}$ and $B^{\{|x|\leq\delta\}}$ are both μ-limited.*

Proof By Lemma 15.2.4 (b), $H\mu^1 \{k < |x|\}$ is limited for certain $k \in \mathbb{N}$. By (c), $B^{\{k<|x|\}}$ is μ-limited. Since B is μ-limited,

$$B^{\{|x|\leq k\}} = B - B^{\{k<|x|\}} \text{ is } \mu\text{-limited}$$

with μ-limited increments. Therefore, $H\int_{\{|x|\le k\}} x^2 d\mu^1$ is limited (Proposition 15.2.3). By Lemma 15.2.4 (d), $H\mu^1\{\delta<|x|\}$ is limited. By (c) and (b), $B^{\{\delta<|x|\}}$ is μ-limited, and thus $B^{\{|x|\le\delta\}}=B-B^{\{\delta<|x|\}}$ is μ-limited. □

Theorem 15.2.6 *The process B is μ-limited if and only if the following conditions are fulfilled.*

(i) $H\mathbb{E}_{\mu^1}\mathbf{1}_{\{|x|\le\delta\}}(x)\cdot x$ *is limited for all limited non-infinitesimal* $\delta>0$.
(ii) $H\mathbb{E}_{\mu^1}\mathbf{1}_{\{|x|\le\delta\}}(x)\cdot x^2$ *is limited for all limited* $\delta>0$.
(iii) $\lim_{k\to\infty}{}^\circ H\mu^1\{k<|x|\}=0$.

Proof Suppose that B is μ-limited. Then (iii) follows from Lemma 15.2.4 (b). Fix a non-infinitesimal $\delta>0$. By Theorem 15.2.5, $B^{\{|x|\le\delta\}}$ is μ-limited. Proposition 15.2.3 tells us that (i) and (ii) are true. Of course, (ii) is true for all limited $\delta>0$.

For the converse, assume that (i), (ii) and (iii) are true. Fix a non-infinitesimal $\delta>0$. From (i), (ii) and Proposition 15.2.3 it follows that $B^{\{|x|\le\delta\}}$ is μ-limited. By (iii) and Lemma 15.2.4 (d) and (c), $B^{\{\delta<|x|\}}$ is μ-limited. It follows that

$$B=B^{\{|x|\le\delta\}}+B^{\{\delta<|x|\}}$$

is μ-limited. □

From Theorem 15.2.6 (ii) and (iii) and Lemma 15.2.4 (d) we obtain the following Lemma.

Corollary 15.2.7 *Assume that B is μ-limited. Then $H\mu^1\{\varepsilon<|x|\}$ is limited for all $\varepsilon>0$, $\varepsilon\not\approx 0$. In particular,*

$$\widehat{\mu}\left\{B_{\frac{1}{H}}\not\approx 0\right\}=\widehat{\mu^1}\{x\not\approx 0\}=0.$$

The following example, also due to Lindstrøm, shows that there is an internal measure μ^1 such that B is μ-limited, but (i) fails for certain infinitesimal $\varepsilon>0$.

Example 15.2.8 By Proposition 8.4.1 (b), there exists an unlimited $N\in{}^*\mathbb{N}$ with $\frac{N}{H}<\frac{1}{N}$ and $\frac{1}{H-N}\left(\sum_{i=1}^N\frac{1}{i}\right)^2<\frac{1}{N}$. Set $\varepsilon:=\frac{1}{H-N}\sum_{i=1}^N\frac{1}{i}$. Set $\mu^1\left\{\frac{1}{i}\right\}:=\frac{1}{H}$ and $\mu^1\{-\varepsilon\}=\frac{H-N}{H}$, thus μ^1 is concentrated on $\{-\varepsilon\}\cup\left\{\frac{1}{i}\mid i\in{}^*\mathbb{N}\text{ and }i\le N\right\}$. Note that $H\mathbb{E}_{\mu^1}x^2$ is limited and $H\mathbb{E}_{\mu^1}x=0$. Since $\mu^1\{1<|x|\}=0$, B is μ-limited by Proposition 15.2.3. On the other hand we have

$$H\mathbb{E}_{\mu^1}\mathbf{1}_{\{|x|\le\varepsilon\}}(x)\cdot x=-\sum_{i=1}^N\frac{1}{i}\text{ is unlimited.}$$

Example 15.2.9 Use Theorem 15.2.6 to prove that the process B is μ-limited if μ is one of the measures in Examples 15.2.1.

15.3 Approximation of limited processes by processes with limited increments

Theorem 15.3.1 *Let B be μ-limited. Then for each $n \in \mathbb{N}$*

$$\lim_{r\to\infty} {}^{\circ}\mu\left\{\forall t \in T_n\left(B_t = B_t^{\{|x|\leq r\}}\right)\right\} = 1.$$

By Theorem 15.2.5, $B^{\{|x|\leq r\}}$ is μ-limited for all $r > 0$, $r \not\approx 0$.

Proof Define $q_r := H\mu^1\{r < |x|\}$. Then

$$\mu\left\{\forall t \in T_n\left(B_t = B_t^{\{|x|\leq r\}}\right)\right\}$$
$$= \mu\{\forall t \in T_n(|X_t| \leq r)\} = \mu^1\{|x| \leq r\}^{Hn} = \left(1 - \frac{q_r}{H}\right)^{Hn}.$$

By Lemma 15.2.4 (b), we can choose r such that $\left(1 - \frac{q_r}{H}\right)^{Hn}$ is arbitrarily close to 1 in the standard sense. □

15.4 Splitting infinitesimals

Following Lindstrøm [67], we will use so-called splitting infinitesimals to derive the standard Lévy–Khintchine formula from the internal one. Moreover, they are used to divide each Lévy process into its continuous and pure jump parts.

Proposition 15.4.1 *Fix μ^1 such that B is μ-limited. There exists an unlimited $N_\infty \in {}^*\mathbb{N}$ such that (SI 1), (SI 2) and (SI 3) are true for all unlimited $N \in {}^*\mathbb{N}$ with $N \leq N_\infty$ and for all standard $\delta > 0$, where*

(SI 1) $H\left|\mathbb{E}_{\mu^1}\mathbf{1}_{\left\{|x|\leq\frac{1}{N}\right\}}(x)\cdot x\right| \leq \sqrt[4]{H}$,

(SI 2) $H\left|\mathbb{E}_{\mu^1}\mathbf{1}_{\left\{\frac{1}{N}<|x|\leq\delta\right\}}(x)\cdot x\right| \leq \sqrt[4]{H}$.

It follows that $\sqrt{H}\left|\mathbb{E}_{\mu^1}\mathbf{1}_{\left\{|x|\leq\frac{1}{N}\right\}}(x)\cdot x\right| \approx 0 \approx H\left(\mathbb{E}_{\mu^1}\mathbf{1}_{\left\{|x|\leq\frac{1}{N}\right\}}(x)\cdot x\right)^2$ *and that, for all standard* $\delta > 0$, $\sqrt{H}\left|\mathbb{E}_{\mu^1}\mathbf{1}_{\left\{\frac{1}{N}<|x|\leq\delta\right\}}(x)\cdot x\right| \approx 0.$

(SI 3) $\lim_{n\to\infty} {}^\circ H\mathbb{E}_{\mu^1}\mathbf{1}_{\left\{\frac{1}{N}<|x|\leq\frac{1}{n}\right\}}(x)\cdot x^2 = 0.$

Proof Define $\iota := \lim_{n\to\infty} {}^\circ H\mathbb{E}_{\mu^1}\mathbf{1}_{\{|x|\leq\frac{1}{n}\}}(x)\cdot x^2$. By Theorem 15.2.6 (ii) and (i), $i < \infty$ and for all standard $\delta > 0$,

$$H\left|\mathbb{E}_{\mu^1}\mathbf{1}_{\{\frac{1}{n}<|x|\leq\delta\}}(x)\cdot x\right| = H\left|\mathbb{E}_{\mu^1}\mathbf{1}_{\{|x|\leq\delta\}}(x)\cdot x - \mathbb{E}_{\mu^1}\mathbf{1}_{\{|x|\leq\frac{1}{n}\}}(x)\cdot x\right| \leq \sqrt[4]{H}.$$

By Propositions 8.4.1 (b) and 8.2.1 (f), there exists an unlimited $N_\infty \in {}^*\mathbb{N}$ such that the preceding equalities hold for all standard $\delta > 0$ and all unlimited $N \in {}^*\mathbb{N}$ with $N \leq N_\infty$ if we replace n by N. We may also assume that $\iota \approx H\mathbb{E}_{\mu^1}\mathbf{1}_{\{|x|\leq\frac{1}{N}\}}(x)\cdot x^2$ for all these N, whence

$$\lim_{n\to\infty} {}^\circ H\mathbb{E}_{\mu^1}\mathbf{1}_{\{\frac{1}{N}<|x|\leq\frac{1}{n}\}}(x)\cdot x^2 \approx \iota - H\mathbb{E}_{\mu^1}\mathbf{1}_{\left\{|x|\leq\frac{1}{N}\right\}}(x)\cdot x^2 \approx 0.$$

This proves (SI 3) and also (SI 1), (SI 2) for all unlimited $N \in {}^*\mathbb{N}$ with $N \leq N_\infty$ and all standard $\delta > 0$. □

Lindstrøm has called an infinitesimal η such that (SI 1), (SI 2) and (SI 3) are true with η instead of $\frac{1}{N}$ a **splitting infinitesimal for** μ.

Example 15.4.2 Note that $H^{-\frac{1}{4}}$ and all positive infinitesimals are splitting infinitesimals for the measures μ in the Examples 15.2.1 (BM), (PP), respectively.

15.5 Standard Lévy processes

Fix a μ-limited Lévy process B. We construct a standard Lévy process by taking the standard part ${}^\circ B : \Omega \times [0,\infty[\to \mathbb{R}$ of B, setting

$${}^\circ B(X,r) := \lim_{{}^\circ s\downarrow r} {}^\circ B(X,r).$$

We will prove that ${}^\circ B$ exists and is a Lévy process.

Proposition 15.5.1 *${}^\circ B$ exists $\widehat{\mu}$-a.s. and is a càdlàg process.*

Proof First assume that B has μ-limited increments. Then, by Proposition 15.2.3, $\mathbb{E}_\mu\left(B_t - Ht\mathbb{E}_{\mu^1}x\right)^2$ is limited for all limited t. By Theorem 10.16.2, ${}^\circ\left(B - \left(t \mapsto Ht\mathbb{E}_{\mu^1}x\right)\right)$ exists and therefore ${}^\circ B$ exists $\widehat{\mu}$-a.s. and is a càdlàg process. For general B there exists a sequence $(B^n)_{n\in\mathbb{N}}$ of μ-limited processes with

μ-limited increments approximating B, according to Theorem 15.3.1. Fix $\sigma \in \mathbb{N}$. Then

$$\widehat{\mu}\{\forall r \leq \sigma\, ((^\circ B)_r \text{ exists})\} = \widehat{\mu}\left\{\forall r \leq \sigma \left(^\circ\left(B - B^n\right)_r \text{ and } \left(^\circ B^n\right)_r \text{ exist}\right)\right\}$$
$$= \widehat{\mu}\left\{\forall r \leq \sigma \left(^\circ\left(B - B^n\right)_r \text{ exists}\right)\right\} \geq \widehat{\mu}\left\{\forall s \in T_\sigma \left(B_s = B_s^n\right)\right\},$$

which can be made arbitrarily close to 1. It follows that

$$\widehat{\mu}\{\forall r \in [0,\infty[\, ((^\circ B)_r \text{ exists})\} = 1.$$

□

In order to prove that $^\circ B$ is a Lévy process, the following lemma is useful.

Lemma 15.5.2 *Fix $n \in \mathbb{N}$ and $\varepsilon > 0$ in $\mathbb{R}$. Then there exists a $\delta > 0$ in $\mathbb{R}$ such that, for all $r < t$ in T_n with $t - r < \delta$,*

$$\mu\{|B_t - B_r| \geq \varepsilon\} < \varepsilon.$$

It follows that for all $r \in [0,\infty[$ and all $t \in T$, $t \approx r$, $(^\circ B)_r \approx B_t\, \widehat{\mu}$-a.s., thus B_t is a lifting of $(^\circ B)_r$.

Proof First assume that B has μ-limited increments. Then

$$\mu\{|B_t - B_r| \geq \varepsilon\} = \mathbb{E}_\mu \mathbf{1}_{\{|B_t - B_r| \geq \varepsilon\}} \leq \frac{1}{\varepsilon^2} \mathbb{E}_\mu |B_t - B_r|^2$$
$$= \frac{1}{\varepsilon^2} H(t-r) \left(\mathbb{E}_{\mu^1} x^2 - (H(t-r) - 1)\left(\mathbb{E}_{\mu^1} x\right)^2\right) < \varepsilon$$

for small $t - r$. For arbitrary μ-limited B fix a μ-limited $B^S := B^{\{|x| \leq S\}}$ with $S \in \mathbb{N}$. Then, for $r < t$ in T_n with $n \in \mathbb{N}$,

$$\mu\{|B_t - B_r| \geq \varepsilon\}$$
$$\leq \mu\left\{|B_t - B_t^S| \geq \frac{\varepsilon}{3}\right\} + \mu\left\{|B_t^S - B_r^S| \geq \frac{\varepsilon}{3}\right\} + \mu\left\{|B_r - B_r^S| \geq \frac{\varepsilon}{3}\right\} < \varepsilon$$

for large S and small $t - r$ (see Theorem 15.3.1).

To prove the second assertion, fix $r \in [0,\infty[$ and $t \approx r$. Let $(t_k)_{k \in \mathbb{N}}$ be a sequence in T with $\lim_{k\to\infty} {^\circ t_k} \downarrow r$. Then $\lim_{k\to\infty} {^\circ(B_{t_k})} = (^\circ B)_r$ $\widehat{\mu}$-a.s. By the preceding computation, $\lim_{k\to\infty} {^\circ(B_{t_k})} = {^\circ B_t}$ in measure, thus $(^\circ B)_r \approx B_t$ $\widehat{\mu}$-a.s. □

A process $L : \Omega \times [0,\infty[\to \mathbb{R}$ is called a (one-dimensional) **Lévy process** if

(L 1) $L_0 = 0$ $\widehat{\mu}$-a.s.,
(L 2) L has independent increments, which means that the $n-1$-tuple $\left(L_{r_i} - L_{r_{i-1}}\right)_{i=2,\ldots,n}$ is independent for all $r_1 < \ldots < r_n$ in $[0,\infty[$,
(L 3) for all $r,h \in [0,\infty[$, $L_{r+h} - L_r$ and L_h have the same distribution.

Theorem 15.5.3 *$L := {}^\circ B$ is a càdlàg Lévy process and L is continuous in probability.*

Proof Suppose that $r_1 < \ldots < r_n$ in $[0,\infty[$. Fix $s_i \approx r_i, s_i \in T$. By Lemma 15.5.2, B_{s_i} is a lifting of L_{r_i}. We obtain

$$\begin{aligned}
&\widehat{\mu}\left\{L_{r_2} - L_{r_1} \le c_2, \ldots, L_{r_n} - L_{r_{n-1}} \le c_n\right\} \\
&= \lim_{k\to\infty} {}^\circ\mu\left\{B_{s_2} - B_{s_1} \le c_2 + \frac{1}{k}, \ldots, B_{s_n} - B_{s_{n-1}} \le c_n + \frac{1}{k}\right\} \\
&= \lim_{k\to\infty}\left({}^\circ\mu\left\{B_{s_2} - B_{s_1} \le c_2 + \frac{1}{k}\right\} \cdot \ldots \cdot {}^\circ\mu\left\{B_{s_n} - B_{s_{n-1}} \le c_n + \frac{1}{k}\right\}\right) \\
&= \widehat{\mu}\left\{L_{r_2} - L_{r_1} \le c_2\right\} \cdot \ldots \cdot \widehat{\mu}\left\{L_{r_n} - L_{r_{n-1}} \le c_n\right\}.
\end{aligned}$$

Thus, the increments of L are independent.

Fix $r,h \in [0,\infty[$ and $\rho,\theta \in T$ with $\rho \approx r, \theta \approx h$. By Lemma 15.5.2 again, B_ρ, B_θ, $B_{\rho+\theta}$ are liftings of L_r, L_h, L_{r+h}, respectively. We obtain for $c \in \mathbb{R}$

$$\begin{aligned}
\widehat{\mu}\left\{L_{r+h} - L_r \le c\right\} &= \lim_{k\to\infty} {}^\circ\mu\left\{B_{\rho+\theta} - B_\rho \le c + \frac{1}{k}\right\} \\
&= \lim_{k\to\infty} {}^\circ\mu\left\{B_\theta \le c + \frac{1}{k}\right\} = \widehat{\mu}\left\{L_h \le c\right\}.
\end{aligned}$$

It follows that the increments of L are identically distributed. By Corollary 15.2.7, $L_0 = 0$ $\widehat{\mu}$-a.s. This proves that L is a Lévy process. That L is continuous in probability follows from Lemma 15.5.2. □

The following surprising, but very simple, observation shows how rich Loeb spaces are, in particular the space Ω.

Proposition 15.5.4 *Suppose $f : [0,\infty[\to \mathbb{R}$ is right continuous. Then there exists an $X \in \Omega$ such that $r \mapsto ({}^\circ B)_r(X) = f(r)$, i.e., f is a path of ${}^\circ B$.*

Proof Define $X_{\frac{1}{H}} := {}^*f\left(\frac{1}{H}\right)$ and $X_{\frac{i}{H}} := {}^*f\left(\frac{i}{H}\right) - {}^*f\left(\frac{i-1}{H}\right)$ if $i > 1$ and $\frac{i}{H} \in T$. Then $B(X,t) = {}^*f(t)$ for all $t \in T$. Suppose that f is right continuous and $r \in [0,\infty[$.

Since ${}^*f\,(a) \approx f\,(r)$ if $r \leq a \approx r$ (see Proposition 9.9.1), we obtain

$$f\,(r) = \lim_{{}^\circ t \downarrow r} {}^\circ\left({}^*f\,(t)\right) = \lim_{{}^\circ t \downarrow r} {}^\circ\left(B(X,t)\right) = {}^\circ B_r(X).$$

□

Now it is time to define Poisson processes. A Lévy process P with values in the non-negative integers is called a **Poisson process** with **rate** $\beta \in \mathbb{R}^+$ if for all $h \geq 0$

$$\widehat{\mu}\{P_h = n\} = e^{-\beta h} \cdot \frac{(\beta h)^n}{n!}.$$

15.6 Lévy measure

Still following Lindstrøm [67], the internal Borel measure μ^1 is used to define the Lévy measure ${}^\infty\mu^1$ for the standard Lévy process ${}^\circ B$, where B is a fixed μ-limited process. In analogy to the construction of the Lebesgue measure λ on $[0,\infty[$ as the image measure of the Loeb counting measure $\widehat{\nu}$ on T by the standard part map, we obtain the Lévy measure ${}^\infty\mu^1$ on $\mathbb{R}$ for ${}^\circ B$ under μ^1 as the image measure of the Loeb measure $\widehat{H\mu^1}$ on ${}^*\mathbb{R}$ by the standard part map. The main difference is the following: $\lambda(U) = \infty$ if U is any neighbourhood of ∞ and finite outside. If ${}^\infty\mu^1(\mathbb{R}) = \infty$, then ${}^\infty\mu^1(U) = \infty$ if U is any neighbourhood of 0 and finite outside (see Corollary 15.2.7). For all Borel sets $A \subseteq \mathbb{R}$ set

$${}^\infty\mu^1(A) := \lim_{\varepsilon \downarrow 0} \widehat{H\mu^1}\left(\mathrm{st}^{-1}\left[A \cap \{x \in \mathbb{R} \mid \varepsilon < |x|\}\right]\right).$$

This measure ${}^\infty\mu^1$ is called the **Lévy measure** of ${}^\circ B$. Note that, for all $a > 0$ in $\mathbb{R}$,

$${}^\infty\mu^1(a) := {}^\infty\mu^1(\{a\}) = \widehat{H\mu^1}\,(\widetilde{a}).$$

Recall that $\widetilde{a} := \{x \approx a\}$ is the monad of a. The number a is called an **atom** of ${}^\infty\mu^1$ if ${}^\infty\mu^1(a) > 0$; otherwise a is called a **natom**. The next result shows that ${}^\infty\mu^1$ satisfies the usual **properties of Lévy measures**.

Theorem 15.6.1

(LM 1) ${}^\infty\mu^1\{0\} = 0$.
(LM 2) $\int_{\mathbb{R}} \left(x^2 \wedge 1\right) d\,{}^\infty\mu^1 < \infty$.

Proof (LM 1) is obvious. To prove (LM 2), note that, for standard $\varepsilon > 0$,

$$\int_{\{|x|\leq 1\}} x^2 d\, {}^{\infty}\mu^1 = \lim_{\varepsilon\to 0}\int_{\{\varepsilon<|x|\leq 1\}} x^2 d\, {}^{\infty}\mu^1 = \lim_{\varepsilon\to 0}\int_{\{\varepsilon<|x|\leq 1\}} x^2 d\widehat{H\mu^1}_{st}$$

$$\leq \lim_{\varepsilon\to 0}\int_{^*\{\varepsilon<|x|\leq 1+\varepsilon\}} x^2 d\, \widehat{H\mu^1} \leq {}^{\circ}\int_{\{x\in {}^*\mathbb{R}\,|\,|x|\leq 2\}} x^2 d\, H\mu^1 < \infty,$$

because $B^{\{|x|\leq 2\}}$ is μ-limited with μ-limited increments (see Proposition 15.2.3). By Theorem 15.2.6 and Lemma 15.2.4 (d)

$$\int_{\{1<|x|\}} 1 d\, {}^{\infty}\mu^1 = {}^{\infty}\mu^1\{1<|x|\} \leq {}^{\circ}H\mu^1\left({}^*\{1<|x|\}\right) < \infty.$$

This proves (LM 2). □

(LM 1) and (LM 2) yield the following properties of Lévy measures.

Proposition 15.6.2 *Let ρ be a Lévy measure on $\mathbb{R}$, i.e., ρ has the properties (LM 1) and (LM 2) with ${}^{\infty}\mu^1 = \rho$. Then the following hold.*

(a) *$\rho\{\varepsilon < |x|\} < \infty$ for all standard $\varepsilon > 0$.*
(b) *$\int_{\{|x|\leq n\}} x^2 d\rho(x) < \infty$ for all $n \in \mathbb{N}$.*
(c) *$\int_{\left\{\frac{1}{n}<|x|\leq n\right\}} |x|\, d\rho(x) < \infty$ for all $n \in \mathbb{N}$.*
(d) *$\lim_{k\to\infty} \rho(\{k < |x|\}) = 0$.*
(e) *The set A of atoms in $\mathbb{R}$ is at most countable.*

Proof (a)

$$\rho\{\varepsilon < |x|\} \leq \frac{1}{\varepsilon^2}\int_{\{|x|\leq 1\}} x^2 d\rho + \int_{\{1<|x|\}} 1 d\rho < \infty.$$

(b) By (a),

$$\int_{\{|x|\leq n\}} x^2 d\rho(x) = \int_{\{|x|\leq 1\}} x^2 d\rho(x) + \int_{\{1<|x|\leq n\}} x^2 d\rho(x)$$

$$\leq \int_{\{|x|\leq 1\}} x^2 d\rho(x) + n^2 \cdot \rho\{1<|x|\} < \infty.$$

(c) By (a),

$$\int_{\left\{\frac{1}{n}<|x|\leq n\right\}} |x|\, d\rho(x) \leq n\rho\left\{\frac{1}{n} < |x|\right\} < \infty.$$

(d) By (a),

$$0 = \lim_{k\to\infty}\left(\rho\{1<|x|\} - \rho\{1<|x|\leq k\}\right) = \lim_{k\to\infty} \rho\{k<|x|\}.$$

(e) Note that $A = \bigcup_{n\in\mathbb{N}} A_n$ with $A_n := \left\{|x| > \frac{1}{n} \text{ and } \rho(\{x\}) > \frac{1}{n}\right\}$. By (a), A_n is finite, whence A is countable. □

Examples 15.6.3 Note that 0 and $A \mapsto \beta \cdot \mathbf{1}_A(1)$ are the Lévy measures of the measure μ in Examples 15.2.1 (BM), (PP), respectively.

Condition (e) implies the following useful result.

Corollary 15.6.4 *Fix a countable set $C \subseteq \mathbb{R}$. There is a sequence $(n_k)_{k\in\mathbb{N}}$ in $]0,1[$ converging to 0 such that $\{x \pm n_k \mid x \in C, k \in \mathbb{N}\}$ is a set of natoms.*

Here is an important example for the relationship between the $H\mu^1$-integral of an internal function $F: {}^*\mathbb{R} \to {}^*\mathbb{C}$ and the ${}^\infty\mu^1$-integral of the standard part ${}^\circ F$ of F. To this end define, for a fixed $y \in \mathbb{R}$,

$$f: \mathbb{R} \to \mathbb{C}, x \mapsto e^{iyx} - 1 - iy\mathbf{1}_{\{x\in\mathbb{R} \mid |x| \le 1\}}(x) \cdot x,$$

and note that $\|f\|(x) \le \sqrt{2}x^2 \cdot y^2$ for $|x| \le 1$. Together with $\|f\|(x) \le 2$ for $1 < |x|$ we obtain

$$\|f\|(x) \le 2\max\{|y|, 1\}^2 \left(x^2 \wedge 1\right). \tag{41}$$

Lemma 15.6.5 (Lindstrøm [67], Rehle [99]) *There exists a sequence $(n_m)_{m\in\mathbb{N}}$ in $]0,1[$ and an infinitesimal $\eta_1 \ge 0$ such that*

$$\rho(\{|x| = 1\}) = {}^\circ H\mu^1\{1 - \eta_1 < |x| \le 1 + \eta_1\}$$

and such that, for all $m \in \mathbb{N}$, the function

$$F_{\eta_1}: {}^*\mathbb{R} \to {}^*\mathbb{C}, x \mapsto e^{iyx} - 1 - iy\mathbf{1}_{\{x\in {}^*\mathbb{R} \mid |x| \le 1+\eta_1\}}(x) \cdot x$$

is an $S_{H\mu^1}$-integrable ***lifting*** *of f on ${}^*\{n_m < |x|\}$, i.e.,*

$$I_m := \left\{x \in {}^*\mathbb{R} \mid n_m < |x| \text{ and } F_{\eta_1}(x) \not\approx f({}^\circ x)\right\}$$

is a $\widehat{H\mu^1}$-nullset.

Proof By Corollary 15.6.4, we may choose $(n_m)_{m\in\mathbb{N}}$ in $]0,1[$ converging to 0 such that all $\pm n_m, \pm 1 \pm n_m$ are natoms. Therefore,

$$\begin{aligned}
\widehat{H\mu^1}(\{|x| \approx 1\}) &= {}^\infty\mu^1(\{1,-1\}) = \lim_{k\to\infty} {}^\infty\mu^1(\{1 - n_k < |x| \le 1 + n_k\}) \\
&= \lim_{k\to\infty} {}^\circ H\mu^1\left({}^*\{1 - n_k < |x| \le 1 + n_k\}\right) \\
&\approx H\mu^1\left({}^*\{1 - n_K < |x| \le 1 + n_k\}\right)
\end{aligned}$$

for some unlimited $K \in {}^*\mathbb{N}$ in an internal extension $(n_k)_{k\in {}^*\mathbb{N}}$ of $(n_k)_{k\in\mathbb{N}}$. Set $\eta_1 := n_K$. To prove that F_{η_1} is a lifting of f on each ${}^*\{n_m < |x|\}$, note that $I_m \subseteq \alpha \cup \beta$ with $\beta := \{1+n_K < |x| \approx 1\}$ and $\alpha := \{|x| \text{ is unlimited}\}$. By Lemma 15.2.4 (b), α is a $\widehat{H\mu^1}$-nullset and β is also a $\widehat{H\mu^1}$-nullset, because

$$\beta = \{|x| \approx 1\} \setminus \{1 \approx |x| \leq 1+n_K\}.$$

Since $\|F_{\eta_1}\|$ is limited, F_{η_1} is $S_{H\mu^1}$-integrable on ${}^*\{n_m < |x|\}$. □

Theorem 15.6.6 (Lindstrøm [67], Rehle [99]) *Let η be a splitting infinitesimal for μ. Then*

$$\int_{\mathbb{R}} fd\,{}^\infty\mu^1 = {}^\circ\int_{\{x\in\,{}^*\mathbb{R}\mid\eta<|x|\}} F_{\eta_1} dH\mu^1, \tag{42}$$

Proof We obtain, from Lemma 15.6.5,

$$\begin{aligned}
\int_{\mathbb{R}} fd\,{}^\infty\mu^1 &= \lim_{k\to\infty}\int_{\{n_k<|x|\}} fd\,{}^\infty\mu^1 = \lim_{k\to\infty}\int_{\{n_k<|x|\}} fd\,\widehat{H\mu^1}_{st} \\
&= \lim_{k\to\infty}\int_{{}^*\{n_k<|x|\}} f({}^\circ x)\,d\,\widehat{H\mu^1}(x) \\
&= \lim_{k\to\infty}\int_{{}^*\{n_k<|x|\}} {}^\circ F_{\eta_1}(x)\,dH\mu^1(x) \\
&\approx \lim_{k\to\infty}{}^\circ\int_{{}^*\{n_k<|x|\}} F_{\eta_1}(x)\,dH\mu^1(x) \\
&\approx \int_{{}^*\{n_M<|x|\}} F_{\eta_1}(x)\,dH\mu^1(x)
\end{aligned}$$

for some unlimited $M \leq \frac{1}{\eta}$. Since

$$\left\|{}^\circ\int_{\{\eta<|x|\leq n_M\}} F_{\eta_1} d\,H\mu^1\right\| \leq 2\max\{|y|,1\}^2\,{}^\circ\int_{\left\{\eta<|x|\leq\frac{1}{n}\right\}} x^2 d\,H\mu^1 \to_{n\to\infty} 0,$$

we obtain the desired result. □

15.7 The Lévy–Khintchine formula

We study the Fourier transformation Φ of a fixed μ-limited Lévy process B. The standard part of Φ is the so-called Lévy–Khintchine formula of $^{\circ}B$. Fix a splitting infinitesimal η for μ and use the terminology in the preceding section. Define

$$C := H\mathbb{E}_{\mu^1}\left(\mathbf{1}_{\{|x|\leq\eta\}}(x)\cdot x^2\right) = \int_{\{|x|\leq\eta\}} x^2 dH\mu^1, \tag{43}$$

$$\gamma := H\mathbb{E}_{\mu^1}\left(\mathbf{1}_{\{|x|\leq 1+\eta_1\}}(x)\cdot x\right) = \int_{\{|x|\leq 1+\eta_1\}} x dH\mu^1. \tag{44}$$

By Theorem 15.2.6 (i) and (ii), γ and C are limited. Fix a limited $t\in T$. Then, for all $y\in\mathbb{R}$,

$$\mathbb{E}_{\mu}e^{iyB_t} = \left(\mathbb{E}_{\mu^1}e^{iyx}\right)^{Ht}$$

with

$$\begin{aligned}\mathbb{E}_{\mu^1}e^{iyx} &= 1+\mathbb{E}_{\mu^1}\left(e^{iyx}-1+iy\frac{\gamma}{H}-iy\cdot\mathbf{1}_{\{|x|\leq 1+\eta_1\}}(x)\cdot x\right)\\ &= 1+iy\frac{\gamma}{H}+\mathbb{E}_{\mu^1}\left(e^{iyx}-1-iy\mathbf{1}_{\{|x|\leq 1+\eta_1\}}(x)\cdot x\right) = u+v+w+\alpha\end{aligned}$$

where

$$\begin{aligned}u &:= 1+iy\frac{\gamma}{H},\\ v &:= \int_{\{|x|\leq\eta\}}\left(e^{iyx}-1-iyx\right)d\mu^1,\\ w &:= \int_{\{\eta<|x|\leq 1+\eta_1\}}\left(e^{iyx}-1-iyx\right)d\mu^1,\\ \alpha &:= \int_{\{1+\eta_1<|x|\}}\left(e^{iyx}-1\right)d\mu^1.\end{aligned}$$

We expand the integrand of v into its Taylor series and obtain

$$e^{iyx}-1-iyx = -\frac{y^2}{2}x^2+j(x),$$

where $\int_{\{|x|\leq\eta\}} jdH\mu^1\approx 0$, by using Proposition 15.2.3. The preceding computations and Proposition 9.5.1 (d) imply the following result.

Proposition 15.7.1 *Let B be μ-limited and $y\in\mathbb{R}$. Then, for all limited $t\in T$,*

$$\mathbb{E}_{\mu}e^{iyB_t}\approx\left(1+\frac{1}{H}\left[iy\gamma-\frac{1}{2}Cy^2+\int_{\{\eta<|x|\}}F_{\eta_1}dH\mu^1\right]\right)^{Ht} \tag{45}$$

$$\approx \exp\left(t\left(iy\gamma - \frac{1}{2}Cy^2 + \int_{\{\eta<|x|\}} F_{\eta_1} dH\mu^1\right)\right). \tag{46}$$

Let $^\circ B$ be the standard part of a μ-limited B. Fix $s \in [0,\infty[$ and $t \in T$ with $t \approx s$. By Lemma 15.5.2, B_t is a lifting of $(^\circ B)_s$. Since e^{iyB_t} is limited, this function is an S_μ-integrable lifting of $e^{iy^\circ B_s}$ for all limited y. We obtain from Proposition 15.7.1 the Lévy–Khintchine formula:

Theorem 15.7.2

$$\mathbb{E}_{\widehat{\mu}} e^{iy^\circ B_s} = \exp\left(s\left(iy\,^\circ\gamma - \frac{1}{2}\,^\circ Cy^2 + \int_{\mathbb{R}} fd\,^\infty\mu^1\right)\right).$$

This triplet $(^\circ\gamma,\ ^\circ C,\ ^\infty\mu^1)$ is called the **Lévy triplet**, generating the Lévy process $^\circ B$. In the following section it will be shown that 'the Levy triplet generating the Lévy process' is an appropriate phrase.

Example 15.7.3 Note that $(0,1,0)$ and $(\beta,0,A \mapsto \beta \cdot \mathbf{1}_A(1))$ are the Lévy triplets for $^\circ B$ under the measures μ in Examples 15.2.1 (BM), (PP), respectively.

15.8 Lévy triplets generate Lévy processes

Theorem 15.8.1 *Fix a **Lévy triplet** (γ, C, ρ), i.e., $\gamma, C \in \mathbb{R}$, $C \geq 0$, and ρ is a Lévy measure. There exists an internal Borel measure μ^1 on $^*\mathbb{R}$ such that, B is μ-limited and such that, for all $s \in [0,\infty[$,*

$$\mathbb{E}_{\widehat{\mu}} e^{iy\,^\circ B_s} = \exp\left(s\left[iy\gamma - \frac{1}{2}Cy^2 + \int_{\mathbb{R}} \left(e^{iyx} - 1 - iy \cdot 1_{\{|x|\leq 1\}}(x) \cdot x\right) d\rho\right]\right).$$

Proof We use the *extension $^*\rho$ of ρ. By Proposition 15.6.2 and the transfer principle, for all $n \in \mathbb{N}$,

$$^*\rho\left\{\frac{1}{n} < |x| \leq n\right\}, \int_{\left\{\frac{1}{n}<|x|\leq n\right\}} |x|\, d^*\rho, \int_{\{|x|\leq n\}} x^2 d^*\rho \leq \sqrt[4]{H}. \tag{47}$$

Therefore, by Proposition 8.4.1 (b), there exists an infinitesimal $\eta \geq \frac{1}{\sqrt[4]{H}}$ such that the inequalities (47) are true with n replaced by $\frac{1}{\eta}$. This η will play the role of a splitting infinitesimal. For each internal Borel set $A \subseteq \left\{\eta < |x| \leq \frac{1}{\eta}\right\}$ define

$$\mu^1(A) := \frac{^*\rho(A)}{H}.$$

Then $\mu^1(A) \approx 0$. Since $\rho\{\pm 1\} = H\mu^1\{\pm 1\}$, we can choose $\eta_1 = 0$ in Lemma 15.6.5. Set

$$d := 1 - \mu^1\left(\left\{\eta < |x| \leq \frac{1}{\eta}\right\}\right),$$

$$a^+ = +\sqrt{\frac{C}{Hd} + \frac{\gamma}{Hd} - \frac{1}{Hd}\int_{\{\eta<|x|\leq 1\}} xd^*\rho},$$

$$a^- = -\sqrt{\frac{C}{Hd} + \frac{\gamma}{Hd} - \frac{1}{Hd}\int_{\{\eta<|x|\leq 1\}} xd^*\rho}.$$

Note that $|a^+|, |a^-| < \eta$. Set

$$\mu^1(\{a^+\}) := \mu^1(\{a^-\}) := \frac{d}{2}.$$

Note that μ^1 is concentrated on $\left\{\eta < |x| \leq \frac{1}{\eta}\right\} \cup \{a^+, a^-\}$. Straightforward computations show that

$$\int_{\{|x|\leq 1\}} xdH\mu^1 = \gamma \quad \text{and} \quad \int_{\{|x|\leq \eta\}} x^2 dH\mu^1 \approx C.$$

Now, using Theorem 15.2.6 and the properties of Lévy measures, it is easy to see that for our measure μ^1 the process B is μ-limited. Moreover, B satisfies equations (45) and (46). Altogether, $^\circ B$ exists and, by Theorem 15.7.2, $^\circ B$ is the desired Lévy process. □

Example 15.8.2 It is left to the reader to construct from the Lévy triplets $(0,1,0)$ and $(\beta, 0, A \mapsto \beta \cdot \mathbf{1}_A(1))$ their associated measures μ^1.

15.9 Each Lévy process can be divided into its continuous and pure jump part

Let L be a ρ-limited internal Lévy process, defined on an internal probability space with measure ρ. Let B be the representation of L by (μ, ρ). We say that L has ρ-**infinitesimal increments** if there exists an infinitesimal ε such that $\mu^1\{|x| > \varepsilon\} = 0$. The process L is called a ρ-**pure jump process** if

$$\lim_{n\to\infty} {}^\circ H\mathbb{E}_{\mu^1} \mathbf{1}_{\left\{|x|\leq \frac{1}{n}\right\}}(x) \cdot x^2 = 0.$$

The following question arises: What is the relationship between this formula and L? Here is an answer.

Proposition 15.9.1 *Suppose that L is a ρ-pure jump process. Then for all $k \in \mathbb{N}$*

$$\lim_{n\to\infty} {}^{\circ}\mathbb{E}_{\rho} \max_{t\in T_k} \left(L_t^{\left\{|x|\le\frac{1}{n}\right\}} - \mathbb{E}_{\rho} L_t^{\left\{|x|\le\frac{1}{n}\right\}} \right)^2 = 0.$$

Proof Instead of L we can take its representation B. Since the process $t \mapsto B_t^{\left\{|x|\le\frac{1}{n}\right\}} - \mathbb{E}_{\mu} B_t^{\left\{|x|\le\frac{1}{n}\right\}}$ is a μ-martingale, we have, by Doob's inequality:

$$
\begin{aligned}
&{}^{\circ}\frac{1}{4}\mathbb{E}_{\mu} \max_{t\in T_k} \left(B_t^{\left\{|x|\le\frac{1}{n}\right\}} - \mathbb{E}_{\mu} B_t^{\left\{|x|\le\frac{1}{n}\right\}} \right)^2 \\
&\le {}^{\circ}\mathbb{E}_{\mu} \left(B_k^{\left\{|x|\le\frac{1}{n}\right\}} - \mathbb{E}_{\mu} B_k^{\left\{|x|\le\frac{1}{n}\right\}} \right)^2 \\
&= {}^{\circ}\mathbb{E}_{\mu} \left(\sum_{t\in T_k} \left(\mathbf{1}_{\left\{|x|\le\frac{1}{n}\right\}}(X_t)\cdot X_t - \mathbb{E}_{\mu^1} \mathbf{1}_{\left\{|x|\le\frac{1}{n}\right\}}(x)\cdot x \right) \right)^2 \\
&= {}^{\circ}\mathbb{E}_{\mu} \sum_{t\in T_k} \left(\mathbf{1}_{\left\{|x|\le\frac{1}{n}\right\}}(X_t)\cdot X_t - \mathbb{E}_{\mu^1} \mathbf{1}_{\left\{|x|\le\frac{1}{n}\right\}}(x)\cdot x \right)^2 \\
&= {}^{\circ} \left(Hk\mathbb{E}_{\mu^1} \mathbf{1}_{\left\{|x|\le\frac{1}{n}\right\}}(x)\cdot x^2 - Hk \left(\mathbb{E}_{\mu^1} \mathbf{1}_{\left\{|x|\le\frac{1}{n}\right\}}(x)\cdot x \right)^2 \right) \\
&= {}^{\circ}Hk\mathbb{E}_{\mu^1} \mathbf{1}_{\left\{|x|\le\frac{1}{n}\right\}}(x)\cdot x^2 \to_{n\to\infty} 0. \quad (15.2.6\,(\mathrm{i}))
\end{aligned}
$$

□

Now it is time to present some examples of μ-pure jump processes.

Examples 15.9.2

(i) Let μ^1 be the measure defined in Example 15.2.1 (BM). Then B is not a μ-pure jump process.
(ii) Let μ^1 be the measure defined in Example 15.2.1 (PP). Then B is a μ-pure jump process.
(iii) Let μ^1 be the measure defined in Example 16.1.2 (iii) or (iv) below. Then B is a μ-pure jump process.
(iv) Check the other examples under 16.1.2 below.

(v) Let B be μ-limited and fix a standard $k > 0$. Then $B^{\{k<|x|\}}$ is a μ-pure jump process, which is intuitively obvious.

Proof We only prove (v). Let μ_k^1 be the image measure of μ by $B_{\frac{1}{H}}^{\{k<|x|\}}$. Then we obtain, for all $n \in \mathbb{N}$ with $\frac{1}{n} < k$,

$$H\mathbb{E}_{\mu_k^1}\mathbf{1}_{\left\{|x|\leq\frac{1}{n}\right\}}(x)\cdot x^2 = H\mathbb{E}_{\mu}\mathbf{1}_{\left\{X\left|\,\left|\mathbf{1}_{\{k<|x|\}}(X_{\frac{1}{H}})\cdot X_{\frac{1}{H}}\right|\leq\frac{1}{n}\right.\right\}}(X)\cdot\mathbf{1}_{\{k<|x|\}}(X_{\frac{1}{H}})\cdot X_{\frac{1}{H}}^2$$
$$= H\mathbb{E}_{\mu^1}\mathbf{1}_{\left\{x\left|\,\left|\mathbf{1}_{\{k<|x|\}}(x)\cdot x\right|\leq\frac{1}{n}\right.\right\}}(x)\cdot\mathbf{1}_{\{k<|x|\}}(x)\cdot x^2 = 0.$$

□

Fix a μ-limited Lévy process B and a standard $k > 0$. From Example 15.9.2 (v) and Theorem 15.2.5, B can be split into the sum of two internal Lévy processes,

$$B = B^{\{|x|\leq k\}} + B^{\{k<|x|\}},$$

where the first summand has μ-limited increments and the second one is a μ-pure jump process. It follows that for all limited $t \in T$

$$B_t = Ht\mathbb{E}_{\mu^1}\mathbf{1}_{\{k<|x|\}}(x)\cdot x + B_t^{\{|x|\leq k\}} + \left(B_t^{\{k<|x|\}} - \mathbb{E}_{\mu}B_t^{\{k<|x|\}}\right),$$

where $B^{\{k<|x|\}} - \mathbb{E}_{\mu}B^{\{k<|x|\}}$ is a μ-pure jump Lévy martingale. It is much more delicate to split $B^{\{|x|\leq k\}}$ into two Lévy processes, where the first one has μ-infinitesimal increments and the second one is again a μ-pure jump process:

Theorem 15.9.3 *Let B be a μ-limited Lévy process with μ-limited increments (for example $B^{\{|x|\leq k\}}$ above) and let η be a splitting infinitesimal for μ. Fix a limited $t \in T$. Then*

$$B_t = c_t + I_t + J_t,$$

where

$$c_t = Ht\mathbb{E}_{\mu^1}x \text{ is limited,}$$

and

$$I_t = B_t^{\{|x|\leq\eta\}} - \mathbb{E}_{\mu}B_t^{\{|x|\leq\eta\}} \quad \text{and} \quad J_t = B_t^{\{\eta<|x|\}} - \mathbb{E}_{\mu}B_t^{\{\eta<|x|\}}$$

are μ-limited Lévy martingales. Moreover, I has μ-infinitesimal increments and J is a μ-pure jump process.

Proof Since t is limited, c_t is limited (see Proposition 15.2.3). Since $\mathbb{E}_{\mu}B_t^{\{|x|\leq\eta\}} = Ht\mathbb{E}_{\mu^1}\mathbf{1}_{\{|x|\leq\eta\}}(x)\cdot x$ and $\mathbb{E}_{\mu}B_t^{\{\eta<|x|\}} = Ht\mathbb{E}_{\mu^1}\mathbf{1}_{\{\eta<|x|\}}(x)\cdot x$, we get $B_t = c_t + I_t + J_t$. Obviously, I has μ-infinitesimal increments. In order to prove that I is μ-limited, we estimate for all $n \in \mathbb{N}$ in analogy to the proof of Proposition 15.9.1,

$$\frac{1}{4}\mathbb{E}_\mu \max_{t\in T_n} I_t^2 \leq \mathbb{E}_\mu I_n^2 = Hn\mathbb{E}_{\mu^1}\mathbf{1}_{\{|x|\leq\eta\}}(x)\cdot x^2 - Hn\left(\mathbb{E}_{\mu^1}\mathbf{1}_{\{|x|\leq\eta\}}(x)\cdot x\right)^2.$$

Since B is μ-limited, the first summand is limited (see Theorem 15.2.6 (ii)). The second summand is infinitesimal, since η is a splitting infinitesimal for μ (see Proposition 15.4.1). It follows that $\mathbb{E}_\mu \max_{t\in T_n} I_t^2$ is limited, thus, by Lemma 10.7.1 (a), $\max_{t\in T_n} I_t^2$ is limited and $\max_{t\in T_n}|I_t|$ is also limited $\widehat{\mu}$-a.s. This proves that I is μ-limited. Since $J_t = B_t - I_t - c_t$, B and I are μ-limited and $\max_{t\leq n}|c_t|$ is limited, J is μ-limited.

It remains to prove that J is a μ-pure jump process. To this end let μ_η^1 be the image measure of μ by $J_{\frac{1}{H}}$, let $n\in\mathbb{N}$ and set $D := \mathbb{E}_{\mu^1}\mathbf{1}_{\{\eta<|x|\}}(x)\cdot x$. Then

$$\begin{aligned}
&{}^\circ H\mathbb{E}_{\mu_\eta^1}\mathbf{1}_{\left\{|x|\leq\frac{1}{n}\right\}}(x)\cdot x^2 \\
&= {}^\circ H\mathbb{E}_\mu \mathbf{1}_{\left\{X\left|\mathbf{1}_{\{\eta<|x|\}}(X_{\frac{1}{H}})\cdot X_{\frac{1}{H}}-D\right|\leq\frac{1}{n}\right\}}(X)\cdot\left(\mathbf{1}_{\{\eta<|x|\}}(X_{\frac{1}{H}})\cdot X_{\frac{1}{H}}-D\right)^2 \\
&= {}^\circ H\mathbb{E}_{\mu^1}\mathbf{1}_{\left\{x\left|\mathbf{1}_{\{\eta<|x|\}}(x)\cdot x-D\right|\leq\frac{1}{n}\right\}}(x)\cdot\left(\mathbf{1}_{\{\eta<|x|\}}(x)\cdot x-D\right)^2 = u+v
\end{aligned}$$

with

$$u = {}^\circ H\int_{\{\eta<|x|\}}\mathbf{1}_{\left\{x\,|\,|x-D|\leq\frac{1}{n}\right\}}(x)\circ(x-D)^2\,d\mu^1,$$

$$v = {}^\circ H\int_{\{|x|\leq\eta\}}\mathbf{1}_{\left\{x\,|\,|D|\leq\frac{1}{n}\right\}}D^2\,d\mu^1.$$

Since, by Proposition 15.4.1, $D\approx 0\approx HD^2$, we see that $v=0$. For u we obtain

$$u \leq {}^\circ H\int_{\{\eta<|x|\}}\mathbf{1}_{\left\{x\,|\,|x|\leq\frac{2}{n}\right\}}(x)\cdot(x-D)^2\,d\mu^1 \to_{n\to\infty} 0,$$

because, again by Proposition 15.4.1,

$${}^\circ H\int_{\left\{\eta<|x|\leq\frac{2}{n}\right\}} x^2\,d\mu^1 \to_{n\to\infty} 0,$$

$$2\,{}^\circ HD\int_{\left\{\eta<|x|\leq\frac{2}{n}\right\}} x\,d\mu^1 = 0 = {}^\circ H\int_{\left\{\eta<|x|\leq\frac{2}{n}\right\}} D^2\,d\mu^1.$$

This proves that $\lim_{n\to\infty} {}^{\circ}H\mathbb{E}_{\mu^1_\eta}\mathbf{1}_{\left\{|x|\leq\frac{1}{n}\right\}}(x)\cdot x^2 = 0$. □

Now we will prove that I is S-continuous $\widehat{\mu}$-a.s. By Proposition 9.7.1, I can be converted to a continuous process defined on $\Omega\times[0,\infty[$.

Theorem 15.9.4 *Suppose that B is a μ-limited Lévy process with μ-infinitesimal increments. Then B is S-continuous $\widehat{\mu}$-a.s.*

Proof It can be assumed that B is a martingale. Since $\mathbb{E}_\mu B_t^2$ is limited for limited $t\in T$, by Theorem 10.14.2, it suffices to show that the quadratic variation $[B]$ of B, is S-continuous. Note that $[B]_t(X) = \sum_{s\leq t} X_s^2$. Define the $(\mathcal{B}_t)_{t\in T}$-martingale $N:(X,t)\mapsto \sum_{s\leq t}\left(X_s^2 - \mathbb{E}_{\mu^1}x^2\right)$ and note that for all $n\in\mathbb{N}$

$$\begin{aligned}\mathbb{E}_\mu \max_{t\in T_n} N_t^2 = \mathbb{E}_\mu N_n^2 = \mathbb{E}_\mu \sum_{s\leq n}\left(X_s^4 - \left(\mathbb{E}_{\mu^1}x^2\right)^2\right)\\ \leq i^2 Hn\mathbb{E}_{\mu^1}x^2 - Hn\left(\mathbb{E}_{\mu^1}x^2\right)^2 \approx 0,\end{aligned}$$

where i is an infinitesimal with $\mu^1\{i<|x|\}=0$. It follows that $\max_{t\in T_n} N_t^2\approx 0$ $\widehat{\mu}$-a.s., thus N is S-continuous $\widehat{\mu}$-a.s. and $[B]$ is S-continuous $\widehat{\mu}$-a.s. □

Let us identify two standard Lévy processes $L,\widetilde{L}:\Lambda\times[0,\infty[\to\mathbb{R}$ if L and $\widetilde{L}$ satisfy the same Lévy triplet. Then we have proved the following well-known standard result.

Corollary 15.9.5 *Let L be a standard Lévy process. Then L can be written in the form*

$$L = c + L_S + L_J,$$

where c is a constant, L is a continuous Lévy martingale and L_J is the standard part of a pure jump Lévy martingale.

Exercises

15.1 Fix an internal Lévy process L and the representation B of L by (μ,ρ). Prove that L is ρ-limited if and only if B is μ-limited.

15.2 Prove the assertions in Examples 15.2.9, 15.4.2, 15.6.3, 15.7.3, 15.8.2.

15.3 Prove the assertions in Examples 15.9.2.

15.4 Finish the proof of Theorem 15.8.1; in particular, prove that by the definition of μ^1 the process B is μ-limited.

16

Stochastic integration for Lévy processes

Stochastic integration will be developed now for a large class of, for simplicity, one-dimensional Lévy processes. Our approach includes Brownian motion and Poisson processes. We use the smooth approach in [89]. Compare the techniques and results in this chapter with the corresponding techniques and results in Chapter 6. Recall the notation from the beginning of Chapter 15. Let μ^1 be an internal Borel probability measure on ${}^*\mathbb{R}$ such that the fixed internal process B is μ-limited. It is now necessary to demand that $H\mathbb{E}_{\mu^1}x^n$ is limited for all $n\in\mathbb{N}$. By Theorem 15.1.1, this condition is true in case B has μ-limited increments. Moreover, each μ-limited Lévy process can be approximated by processes with μ-limited increments (see Theorem 15.3.1).

Theorem 10.9.2 and Corollary 10.9.3 are important, with probability measures different from Γ and with $d=0$. Recall that $\mathbb{H}^{\otimes 0}=\mathbb{R}$ and $\mathbb{F}^{\otimes 0}={}^*\mathbb{R}$.

Our model-theoretic approach to Lévy processes provides a setting where it is possible to orthogonalize the increments of a Lévy process and not only the process itself.

16.1 Orthogonalization of the increments

Our final aim is to establish Malliavin calculus for Lévy processes studied in this chapter. The assumption that $H\mathbb{E}_{\mu^1}x^n$ is limited for all $n\in\mathbb{N}$ implies that $\mathbb{E}_\mu(B_t)^n$ is limited for all limited $t\in T$ and all $n\in\mathbb{N}$ (see Theorem 13.1.3). In order to have Lévy processes different from 0, we assume that $H\mathbb{E}_{\mu^1}x^2\not\approx 0$. Therefore,

$$H\mathbb{E}_{\mu^1}\left(x-\mathbb{E}_{\mu^1}x\right)^2=H\mathbb{E}_{\mu^1}x^2-H\left(\mathbb{E}_{\mu^1}x\right)^2\approx H\mathbb{E}_{\mu^1}x^2\not\approx 0 \tag{48}$$

and

$$\mathbb{E}_\mu\left(B_1(X)\right)^2=H\mathbb{E}_{\mu^1}x^2+H\left(H-1\right)\left(\mathbb{E}_{\mu^1}x\right)^2\not\approx 0.$$

Using a slight modification of the Gram–Schmidt orthonormalization procedure, we construct recursively from $(x^n)_{n\in\mathbb{N}_0}$ a sequence $(\dot{p}_n)_{n\in\mathbb{N}_0}$ of orthogonal polynomials $\dot{p}_n$. Set $\dot{p}_0(x) := 1$. The number 0 is called an **uncritical** exponent. Define $\dot{p}_1(x) := x - \mathbb{E}_{\mu^1}x$. By (48), $H\mathbb{E}_{\mu^1}\dot{p}_1^2 \not\approx 0$. The number 1 is called an **uncritical** exponent.

Assume that $\dot{p}_j$ is defined for $j \leq n-1$. Let $0 = u_0 < \ldots < u_l \leq n-1$ be the uncritical exponents below n. Define

$$\dot{p}_n(x) := x^n - \sum_{i=0}^{l} \frac{\mathbb{E}_{\mu^1}\left(x^n \cdot \dot{p}_{u_i}\right)}{\mathbb{E}_{\mu^1}\dot{p}_{u_i}^2}\dot{p}_{u_i}(x).$$

Since $H\mathbb{E}_{\mu^1}\left(x^n \cdot \dot{p}_{u_i}\right)$ is limited and $H\mathbb{E}_{\mu^1}\dot{p}_{u_i}^2 \not\approx 0$, we see that $\frac{\mathbb{E}_{\mu^1}(x^n\cdot\dot{p}_{u_i})}{\mathbb{E}_{\mu^1}\dot{p}_{u_i}^2}$ is limited. If $H\mathbb{E}_{\mu^1}\dot{p}_n^2 \approx 0$, then n is called **critical**; otherwise n is called **uncritical**.

Set

$$\mathbb{N}_\mu := \{n \in \mathbb{N} \mid n \text{ is uncritical}\}.$$

Note that we do not put the uncritical number 0 into $\mathbb{N}_\mu$. Now we change the polynomials $\dot{p}_n$ slightly. Fix $n \in \mathbb{N}_\mu$. Define

$$p_n := \frac{1}{\sqrt{H\mathbb{E}_{\mu^1}\dot{p}_n^2}}\dot{p}_n.$$

Note that

$$\mathbb{E}_{\mu^1}p_n = 0 = \mathbb{E}_\mu p_k(X_t)p_n(X_s) \text{ if } k \neq n \text{ or } s \neq t, \quad \mathbb{E}_\mu p_n^2 = \frac{1}{H}, \tag{49}$$

and that for all $n, k \in \mathbb{N}$

$$H\mathbb{E}_{\mu^1}p_n^k \text{ is limited.} \tag{50}$$

For each $k \in \mathbb{N}_\mu$,

$$M^k : \Omega \times T \to {}^*\mathbb{R}, (X,t) \mapsto \sum_{s\leq t} p_k(X_s) \tag{51}$$

is a $(\mathcal{B}_t)_{t\in T}$-martingale. Fix $r \in [0,\infty[$ and $t \in T$ with $t \approx r$. Then we obtain

$$\mathbb{E}_\mu\left(M^k\right)^2(t,\cdot) = \mathbb{E}_\mu\sum_{s\leq t}p_k^2(X_s) = t \approx r.$$

This equality follows from the fact that $\frac{1}{H}$ can be seen as the Lebesgue unit volume. We shall see that it yields the continuity in measure of the integrals and makes them independent of $k \in \mathbb{N}_\mu$. By Proposition 8.2.2, we obtain the following.

Proposition 16.1.1 *There exists an internal extension $(p_k)_{k\in {}^*\mathbb{N}_\mu}$ of $(p_k)_{k\in\mathbb{N}_\mu}$ and an unlimited $M \in {}^*\mathbb{N}$ such that, for all $n,k \in {}^*(\mathbb{N}_\mu) \cap M$, and all $s,t \in T$, (49) is true. In the following we fix*

$$M_\mu := {}^*(\mathbb{N}_\mu) \cap M,$$

where we identify M with the set $\{1,\ldots,M\}$.

Examples 16.1.2 Recall that the Lévy measure, generated by μ^1, is denoted by ${}^\infty\mu^1$.

(i) Brownian motion: Let μ^1 be the measure defined in Example 15.2.1 (BM). Note that $H\mathbb{E}_{\mu^1}(x^n)$ is limited for all $n \in \mathbb{N}$, that $\mathbb{N}_\mu = \{1\}$ and that ${}^\infty\mu^1 = 0$.

An alternative way of constructing Brownian motion is Anderson's [4] hyperfinite random walk. Let the measure μ^1 be concentrated on $\left\{-\frac{1}{\sqrt{H}}, \frac{1}{\sqrt{H}}\right\}$, defining $\mu^1\left\{-\frac{1}{\sqrt{H}}\right\} := \frac{1}{2} =: \mu^1\left\{\frac{1}{\sqrt{H}}\right\}$.

(ii) Poisson processes: Let μ^1 be the measure defined in Example 15.2.1 (PP). For the measure μ^1, the standard part of B is a Poisson process with rate β. Note that ${}^\infty\mu^1(C) = \beta \cdot \mathbf{1}_C(1)$. We obtain $\mathbb{N}_\mu = \{1\}$, $H\mathbb{E}_{\mu^1}\dot{p}_1^2 \approx \beta \approx H\mathbb{E}_{\mu^1}\dot{p}_1^3$ and $H\mathbb{E}_{\mu^1}p_1^3 \approx \frac{1}{\sqrt{\beta}}$.

A simpler alternative way of constructing Poisson processes is due to Loeb [69]. Let μ^1 be concentrated on $\{0,1\}$ defining $\mu^1\{1\} = \frac{\beta}{H}$ and $\mu^1\{0\} := 1 - \frac{\beta}{H}$.

(iii) Symmetric Poisson processes: Here is a simple example, where $\mathbb{N}_\mu \neq \{1\}$. Set $\mu^1(B) := \frac{1}{2-e^{-\frac{\beta}{H}}} \sum_{i\in {}^*\mathbb{Z}\cap B} e^{-\frac{\beta}{H}} \left(\frac{\beta}{H}\right)^{|i|} \frac{1}{|i|!}$, where $\beta > 0$ is standard. We obtain

$$\dot{p}_1(x) = x,\ \dot{p}_1^3(x) = 0,\ \dot{p}_2(x) = x^2 - \mathbb{E}_{\mu^1}x^2.$$

Note that $\mathbb{N}_\mu = \{1,2\}$ and ${}^\infty\mu^1(C) = \beta \cdot (\mathbf{1}_C(1) + \mathbf{1}_C(-1))$

Following Loeb [69], we obtain symmetric Poisson processes as follows. Let μ^1 be concentrated on $\{-1,0,1\}$ defining $\mu^1\{-1\} = \frac{\beta}{H} = \mu^1\{1\}$ and $\mu^1\{0\} := 1 - \frac{2\beta}{H}$.

(iv) Here is a simple example where $\mathbb{N}_\mu = \mathbb{N}$. Let $a := (a_i)_{i\in\mathbb{N}}$ be a standard sequence of positive different real numbers a_i such that $\sum_{i=1}^\infty a_i$ converges. The *-extension of a is also denoted by a. Let μ^1 be concentrated on $\{a_i \mid i \leq H\}$, setting $\mu^1(\{a_i\}) := \frac{1}{H}$ for all $i \leq H$. Note that

${}^{\infty}\mu^1(A) = |\{i \in \mathbb{N} \mid a_i \in A\}|$. It is left to the reader to prove that $H\mathbb{E}_{\mu^1}\dot{p}_k^2 \not\approx 0$ for all $k \in \mathbb{N}$.

(v) Here is an example where the Lévy measure is very different from the generating measure μ^1, which is similar to a truncated Cauchy distribution. Set $d\mu^1 := \mathbf{1}_{{}^*[-1,1]}(x)\frac{H}{1+H^2x^2}c$ with c such that $\mu^1({}^*\mathbb{R}) = 1$. Note that $c \approx \frac{1}{\pi}$. Then ${}^{\infty}\mu^1$ has the following form: Let I be an interval in $[-1,1]$ with endpoints $a < b$. If 0 does not belong to the closure of I, then ${}^{\infty}\mu^1(I) = \frac{1}{\pi}\left(\frac{1}{a} - \frac{1}{b}\right)$; otherwise ${}^{\infty}\mu^1(I) = \infty$. Since $\int_{-1}^1 x d\mu^1 = 0$ and

$$H\int_{-1}^1 x^2 d\mu^1 \approx \frac{1}{H}\int_{-1}^1 \frac{(Hx)^2 \cdot H}{1+(Hx)^2}dx \cdot \frac{1}{\pi} = \frac{1}{H}\int_{-H}^H \frac{x^2}{1+x^2}dx \cdot \frac{1}{\pi}$$
$$\approx \lim_{n\to\infty} \frac{1}{n}\int_{-n}^n \frac{x^2}{1+x^2}dx\frac{1}{\pi} = \frac{2}{\pi},$$

B is μ-limited.

(vi) Set $d\mu^1 := \frac{\sqrt{H}c}{1+H^2x4}$ with $c \approx \frac{\sqrt{2}}{\pi}$. Note that $\mathbb{N}_\mu = \{1\}$ and $\dot{p}_1 = x$ as in the case of Brownian motion. Note that ${}^{\infty}\mu^1 = 0$, because for standard $\varepsilon > 0$

$$\frac{1}{2}\,{}^{\infty}\mu^1\left([2\varepsilon,\infty[\right) \leq {}^{\circ}H\int_{\varepsilon}^{\infty} \frac{\sqrt{H}}{1+H^2x^4}dx = \lim_{n\to\infty} n\int_{\sqrt{n}\varepsilon}^{\infty} \frac{1}{1+x^4}dx = 0.$$

(vii) This example is similar to Example 15.2.8. Fix an unlimited $K \in {}^*\mathbb{N}$ with $\frac{K}{H} \approx 0$ and let μ^1 be concentrated on $A := \left\{-\varepsilon, 1, \frac{1}{2}, \ldots, \frac{1}{K}, H\right\}$, where ε is defined such that

$$\sum_{i=1}^K \frac{1}{i}\frac{1}{H} + \frac{1}{H^3} - \varepsilon\left(1 - \frac{K}{H} - \frac{1}{H^4}\right) = 0.$$

Note that $\varepsilon \approx 0$. Define $\mu^1\left\{\frac{1}{i}\right\} := \frac{1}{H}, \mu^1\{H\} := \frac{1}{H^4}, \mu^1\{-\varepsilon\} := 1 - \frac{K}{H} - \frac{1}{H^4}$. Then ${}^{\infty}\mu^1$ is concentrated on $\left\{\frac{1}{n} \mid n \in \mathbb{N}\right\}$ and ${}^{\infty}\mu^1(A) = \left|\left\{n \in \mathbb{N} \mid \frac{1}{n} \in A\right\}\right|$ for all Borel sets $A \subseteq \mathbb{R}$.

(viii) We have seen in Section 15.8 that for each Lévy measure ρ there exists a measure μ^1 such that ρ is the associated measure to μ^1. In the case ρ is absolutely continuous with respect to Lebesgue measure with integrable density $f \geq 0$, then we can obtain μ^1 such that ρ is generated by μ^1 as follows. Let $F : T \cup -T \to {}^*[0,\infty[$ be a lifting of $f :]-\infty,\infty[\to [0,\infty[$, where $-T := \{-t \mid t \in T\}$. We may also assume that $\sum_{i\in T\cup -T} F(\frac{i}{H})\frac{1}{H}$ is limited. Then for $i \neq 0$ define

$$\mu^1\left\{\frac{i}{H}\right\} = F\left(\frac{i}{H}\right) \cdot \frac{1}{H^2} \quad \text{and} \quad \mu^1(\{0\}) = 1 - \sum_{i\neq 0} F\left(\frac{i}{H}\right) \cdot \frac{1}{H^2}.$$

Here is an example where we obtain from an internal measure μ^1 a truncated Cauchy distribution. Set $F(\frac{i}{H}) := \mathbf{1}_{{}^*[-1,1]}(\frac{i}{H})\frac{1}{1+\left(\frac{i}{H}\right)^2}\frac{2}{\pi}$ for $i \in {}^*\mathbb{Z}$ and set $\mu^1\left\{\frac{i}{H}\right\} := F(\frac{i}{H}) \cdot \frac{1}{H^2}$ for $i \neq 0$ and $\mu^1(\{0\}) := 1 - \sum_{i\neq 0} F(\frac{i}{H}) \cdot \frac{1}{H^2}$. Note that $0 \not\approx H\int_{-1}^{1} x^2 d\mu^1 = \frac{2}{\pi}\sum_{\{i||i|\leq H\}} \frac{i^2}{H^2+i^2}\frac{1}{H}$ is limited. The associated Lévy measure ${}^\infty\mu^1$ is the truncated Cauchy distribution,

$$d\,{}^\infty\mu^1(x) = \mathbf{1}_{[-1,1]}(x)\frac{1}{1+x^2}\frac{2}{\pi}dx.$$

16.2 From internal random walks to the standard Lévy integral

The integral will be defined as the orthogonal series $\sum_{k\in\mathbb{N}_\mu}\int f_k p_k$ of stochastic integrals $\int f_k p_k$. Recall that, in the cases of Brownian motion and Poisson processes, $\mathbb{N}_\mu = \{1\}$, so we have only one summand $\int fp$, there with $p = p_1$. In general, the integrand $f = (f_k)_{k\in\mathbb{N}_\mu}$ is a square summable adapted process and the integral $\int f_k p_k$ is the standard part of a suitable internal random walk. We start with the internal integral.

Fix $k \in M_\mu$ (see Section 16.1) and a $(\mathcal{B}_{t^-})_{t\in T}$-adapted internal process $F : \Omega \times T \to {}^*\mathbb{R}$. Define

$$\int Fp_k(X,t) := \sum_{s\leq t, s\in T} F(X,s)\cdot p_k(X_s).$$

By (51), $p_k(X_t) = \Delta M_t^k(X)$. By Proposition 16.1.1, the proof of the following lemma is straightforward and is left to the reader.

Lemma 16.2.1 *Assume that $F(\cdot,s) \in L^2(\mu)$ for all $s \in T$. Then for $k \in M_\mu$*

$$\mathbb{E}_\mu\left(\left(\int Fp_k\right)(\cdot,t)\right)^2 = \mathbb{E}_\mu \sum_{s\leq t} F^2(X,s)\frac{1}{H}.$$

Moreover, $\int Fp_k$ is a $(\mathcal{B}_t)_{t\in T}$-martingale.

In order to use this internal integral to define the standard integral $\int fp_k$ for $(\mathfrak{b}_r)_{r\in[0,\infty[}$-adapted integrands $f : \Omega \times [0,\infty[\to \mathbb{R}$ locally in $L^2(\widehat{\mu}\otimes\lambda)$, we take a $(\mathcal{B}_{t^-})_{t\in T}$-adapted lifting $F : \Omega \times T \to {}^*\mathbb{R}$ locally in $SL^2(\mu\otimes\nu)$, according to Theorem 12.3.1, now with μ instead of Γ, $\mathbb{R}$ instead of $\mathbb{H}$, ${}^*\mathbb{R}$ instead of $\mathbb{F}$. Recall that $\Omega = {}^*\mathbb{R}^T$. It follows that $\int Fp_k(\cdot,t) \in SL^1(\mu)$ for all limited $t \in T$.

Then the k**th integral** $\int fp_k : \Omega \times [0,\infty[\to \mathbb{R}$ **of** f is defined by setting for each $r \in [0,\infty[$

$$\int fp_k(\cdot,r) := \lim_{{}^\circ t \downarrow r, t \in T} {}^\circ\left(\int F \cdot p_k\right)(\cdot,t) \ \widehat{\mu}\text{-a.s.}$$

By Theorem 10.16.2 and Corollary 10.16.3, the right-hand side of the preceding equality exists and is a càdlàg $(\mathfrak{b}_r)_{r\in[0,\infty[}$-martingale. The following lemma and again Theorem 10.16.2 and Corollary 10.16.3 are the keys to Theorem 16.2.3, where the continuity in measure follows from part (b).

Lemma 16.2.2 *Fix $n \in \mathbb{N}$ and a $(\mathcal{B}_{t^-})_{t\in T}$-adapted $F : \Omega \times T \to {}^*\mathbb{R}$ locally in $SL^2(\mu \otimes \nu)$. If $k \in \mathbb{N}_\mu$, then*

(a)

$$\max_{t\in T_n}\left|\int Fp_k(\cdot,t)\right| \in SL^2(\mu).$$

(b) *Fix limited $t \approx \widetilde{t} \in T$. Then $\mathbb{E}_\mu\left(\int Fp_k(\cdot,t) - \int Fp_k(\cdot,\widetilde{t})\right)^2 \approx 0$, thus, $\int Fp_k(\cdot,t) \approx \int Fp_k(\cdot,\widetilde{t})$ $\widehat{\mu}$-a.s.*

Proof (a) By Theorem 10.13.1, it suffices to prove that the quadratic variation $\left[\int Fp_k\right]_n$ at n is in $SL^1(\mu)$. Note that

$$\left[\int Fp_k\right]_n(X) = \sum_{t\in T_n} F_t^2(X)p_k^2(X_t).$$

Fix $m \in {}^*\mathbb{N}$. Define $F \wedge m(t) = F(t)$ if $|F(t)| \leq m$, $F_m(t) := 0$ otherwise. Using the limitedness of $H\mathbb{E}_{\mu^1}p_k^4$ (see (50)) and of $F \wedge m$ for $m \in \mathbb{N}$ and the following equality, we see that $X \mapsto \sum_{t\in T_n}(F\wedge m)_t^2\, p_k^2(X_t) \in SL^1(\mu)$:

$$\mathbb{E}_\mu\left(\sum_{t\in T_n}(F\wedge m)_t^2\, p_k^2(X_t)\right)^2 \leq m^4\mathbb{E}_\mu\left(\sum_{t\in T_n} p_k^2(X_t)\right)^2 \text{ is limited.}$$

It remains to prove that $\lim_{m\to\infty} {}^\circ\mathbb{E}_\mu \sum_{t\in T_n}\left(F_t^2 - (F\wedge m)_t^2\right)p_k^2(X_t) = 0$. Assume that this is not true. Then, by Proposition 8.4.1 (b), there exists a standard $\varepsilon > 0$ and an unlimited number $M \in {}^*\mathbb{N}$ such that

$$\varepsilon \leq \mathbb{E}_\mu \sum_{t\in T_n}\left(F_t^2 - (F\wedge M)_t^2\right)p_k^2(X_t) = \mathbb{E}_\mu \sum_{t\in T_n}\left(F_t^2 - (F\wedge M)_t^2\right)\frac{1}{H}.$$

However, since $F, F\wedge M$, restricted to $\Omega \times T_n$, belong to $SL^2(\mu\otimes\nu_n)$ and thus $F = F\wedge M$ $\widehat{\mu\otimes\nu_n}$-a.s., $\mathbb{E}_\mu \sum_{t\in T_n}\left(F_t^2 - (F\wedge M)_t^2\right)\frac{1}{H} \approx 0$. This contradiction proves that $\left[\int Fp_k\right]_n \in SL^1(\mu)$, thus $\max_{t\in T_n}\left|\left(\int Fp_k\right)_t\right| \in SL^2(\mu)$.

(b) Note that for $\widetilde{t} < t$

$$\mathbb{E}_\mu \left(\int F p_k(\cdot, t) - \int F p_k(\cdot, \widetilde{t}) \right)^2 = \mathbb{E}_\mu \left(\sum_{\widetilde{t} < s \leq t} F_s p_k(X_s) \right)^2$$

$$= \mathbb{E}_\mu \sum_{\widetilde{t} < s \leq t} F_s^2 p_k^2(X_s) = \mathbb{E}_\mu \sum_{\widetilde{t} < s \leq t} F_s^2 \frac{1}{H} \approx 0,$$

since F is locally in $SL^2(\mu \otimes \nu)$. □

Theorem 16.2.3 (a) *The integral $\int f p_k$ is well defined $\widehat{\mu}$-a.s., i.e., the definition does not depend on the chosen S-square-integrable lifting of f, and is a càdlàg $(\mathfrak{b}_r)_{r \in [0,\infty[}$-martingale with $\mathbb{E}_{\widehat{\mu}} \left(\sup_{s \in [0,n]} \int f p_k(\cdot, s) \right)^2 < \infty$ for all $n \in \mathbb{N}$.*

(b) *Fix $(\mathfrak{b}_r)_{r \in [0,\infty[}$-adapted processes $f, g : \Omega \times [0, \infty[\to \mathbb{R}$ locally in $L^2(\widehat{\mu} \otimes \lambda)$ and $k, l \in \mathbb{N}_\mu$. Then for all $r \in [0, \infty[$*

$$\mathbb{E}_{\widehat{\mu}} \left(\int f p_k(\cdot, r) \int g p_l(\cdot, r) \right) = \begin{cases} \int_{\Omega \times [0,r]} f \cdot g \, d\widehat{\mu} \otimes \lambda & \text{if } l = k \\ 0 & \text{if } l \neq k. \end{cases}$$

(c) *$\int \cdot p_k$ is a continuous operator on the space of $(\mathfrak{b}_r)_{r \in [0,\infty[}$-adapted processes in $L^2(\widehat{\mu} \otimes \lambda)$. Moreover, $\int f p_k$ is continuous in measure.*

In order to obtain the integral independent of $k \in \mathbb{N}_\mu$, we integrate processes $g : \mathbb{N}_\mu \times \Omega \times [0, \infty[\to \mathbb{R}$, **locally for** λ in $L^2(c \otimes \widehat{\mu} \otimes \lambda)$, i.e., $g \upharpoonright \mathbb{N}_\mu \times \Omega \times [0, n] \in L^2(c \otimes \widehat{\mu} \otimes \lambda)$ for all $n \in \mathbb{N}$. It is necessary to assume each g_k is $(\mathfrak{b}_r)_{r \in [0,\infty[}$-adapted. Define:

$$\int g \, p := \sum_{k \in \mathbb{N}_\mu} \int g_k p_k : \Omega \times [0, \infty[\to \mathbb{R}.$$

By Theorem 16.2.3, this integral is an orthogonal series of square-integrable $(\mathfrak{b}_r)_{r \in [0,\infty[}$-martingales. The operator $\int \cdot p$ is linear and continuous and $\int g \, p$ is continuous in measure. This process $\int g \, p$ is called the **integral** of g.

It is simpler to introduce the integrals not as processes, but as random variables: Fix a $(\mathfrak{b}_r)_{r \in [0,\infty[}$-adapted process $f : \Omega \times [0, \infty[\to \mathbb{R}$ in $L^2(\widehat{\mu} \otimes \lambda)$ and a $(\mathcal{B}_{t^-})_{t \in T}$-adapted lifting $F : \Omega \times T \to {}^*\mathbb{R}$ in $SL^2(\mu \otimes \nu)$. Define

$$\int^V f p_k(X) := {}^\circ \sum_{s \in T} F(s, X) p_k(X_s).$$

If $f \in L^2(c \otimes \widehat{\mu} \otimes \lambda)$ and f_k is $(\mathfrak{b}_t)_{t\in[0,\infty[}$-adapted for all $k \in \mathbb{N}_\mu$, set

$$\int^V fp := \sum_{k\in \mathbb{N}_\mu} \int^V f_k p_k.$$

Often it is not possible to interpret the standard part of an internal integral as a process defined on the continuous timeline. In this more general case we proceed as follows. The algebra $\mathcal{A}$ of **adapted sets** is the set of all $C \subseteq \Omega \times T$ such that $C(\cdot,t)$ is $\mathcal{B}_{t^-}$-measurable for all $t \in T$.

Proposition 16.2.4 *Fix an $\mathcal{A} \vee \mathcal{N}_{\widehat{\mu\otimes\nu}}$-measurable f locally in $L^2\left(\widehat{\mu\otimes\nu}\right)$. Then there exists a $(\mathcal{B}_{t^-})_{t\in T}$-adapted lifting $F : \Omega \times T \to {}^*\mathbb{R}$ of f locally in $SL^2(\mu\otimes\nu)$.*

Proof Use Loeb theory and the fact that internal Borel-measurable processes $F : \Omega \times T \to {}^*\mathbb{R}$ are $\mathcal{A}$-measurable if and only if they are $(\mathcal{B}_{n^-})_{n\in T}$-adapted. □

Fix f and F according to Proposition 16.2.4. Then we define for limited $t \in T$

$$\left(\int fp_k\right)(X,t) := {}^\circ\sum_{s\le t} F(X,s)p_k(X_s).$$

Note that $\left(\int fp_k\right)(\cdot,t)$ is $\widehat{\mu}$-a.s. well defined and $\left(\int fp_k\right)$ is a $\widehat{\mu}$-square-integrable $\left(\mathcal{B}_t \vee \mathcal{N}_{\widehat{\mu}}\right)_{t\in\widetilde{T}}$-martingale, where $\widetilde{T}$ denotes the set of limited elements of T. Moreover, define for $f \in L^2\left(\widehat{\mu\otimes\nu}\right)$ and liftings $F \in SL^2(\mu\otimes\nu)$ of f,

$$\left(\int^V fp_k\right)(X) := {}^\circ\sum_{s\in T} F(X,s)p_k(X_s).$$

Note that $\left(\int^V fp_k\right)$ is $\widehat{\mu}$-a.s. well defined. It is now clear how to define the process $\int f\,p$ and the random variable $\int^V f\,p$.

16.3 Iterated integrals

Let us start with iterated integrals with parameters. Fix an internal $F : T^{n+m} \to {}^*\mathbb{R}$ and $k := (k_1,\dots,k_n) \in M_\mu^n$. Define $I_{k,m}(F) : \Omega \times T^m \to {}^*\mathbb{R}$, setting

$$I_{k,m}(F)(X,s) := \sum_{t\in T^n_<} F_{t,s}\cdot p_{k_1}(X_{t_1})\cdot\ldots\cdot p_{k_n}(X_{t_n}).$$

Note that, by Proposition 16.1.1, we have the following lemma.

Lemma 16.3.1 *For all $k \in M_\mu^n$, $\widetilde{k} \in M_\mu^{\widetilde{n}}$ and internal $F : T^{n+m} \to {}^*\mathbb{R}$, $G : T^{\widetilde{n}+m} \to {}^*\mathbb{R}$,*

$$\int_{\Omega \times T^m} I_{k,m}(F) \cdot I_{\widetilde{k},m}(G) d\mu \otimes \nu^m = \begin{cases} 0 & \text{if } k \neq \widetilde{k} \\ \int_{T^n_< \times T^m} F \cdot G d\nu^{n+m} & \text{if } k = \widetilde{k}. \end{cases}$$

In order to construct from $I_{k,m}(F)$ the standard iterated integral for Lebesgue square-integrable processes, the following result is crucial.

Lemma 16.3.2 *Fix F (locally) in $SL^2\left(\nu^{n+m}\right)$ and $k \in \mathbb{N}_\mu^n$. Then $I_{k,m}(F)$ is (locally) in $SL^2\left(\mu \otimes \nu^m\right)$.*

Proof Fix $\sigma, S, M \in {}^*\mathbb{N}$. Set $F^{\sigma,S} := F \upharpoonright T_\sigma^n \times T_S^m$ and $F^{\cdot,S} := F \upharpoonright T^n \times T_S^m$. Define $F_M^{\sigma,S}(t) := F^{\sigma,S}(t)$ if $\left|F^{\sigma,S}(t)\right| \leq M$ and set $F_M^{\sigma,S}(t) := 0$ otherwise. First assume that $\sigma, S, M \in \mathbb{N}$. Using (50), we see that $\int_{\Omega \times T^m}\left(I_{k,m}(F_M^{\sigma,S})\right)^4 d\mu \otimes \nu^m$ is limited, thus $I_{k,m}(F_M^{\sigma,S}) \in SL^2\left(\mu \otimes \nu^m\right)$. We will prove that

$$\lim_{M \to \infty} {}^\circ\!\int_{\Omega \times T^m}\left(I_{k,m}(F^{\sigma,S} - F_M^{\sigma,S})\right)^2 d\mu \otimes \nu^m = 0.$$

Assume that this is not true. Set $a_M := \int_{T^n_< \times T^m}\left(F^{\sigma,S} - F_M^{\sigma,S}\right)^2 d\nu^{n+m}$. By Proposition 8.4.1 (b), there exists a standard $\varepsilon > 0$ and an unlimited M with

$$a_M = \int_{\Omega \times T^m}\left(I_{k,m}(F^{\sigma,S} - F_M^{\sigma,S}\right)^2 d\mu \otimes \nu^m \geq \varepsilon.$$

By Lemma 10.7.1 (a), $F^{\sigma,S}$ is limited $\widehat{\nu^{n+m}}$-a.s., thus $F^{\sigma,S} - F_M^{\sigma,S}$ is a lifting of the null-function. It follows that $a_M \approx 0$, which is a contradiction. Therefore, $I_{k,m}(F^{\sigma,S}) \in SL^2\left(\mu \otimes \nu^m\right)$. Note that, using the S-square-integrability of F,

$$\lim_{\sigma \to \infty} {}^\circ\!\int_{\Omega \times T^m}\left(I_{k,m}(F^{\sigma,S} - F^{\cdot,S})\right)^2 d\mu \otimes \nu^m = 0.$$

Therefore, $I_{k,m}(F^{\cdot,S}) \in SL^2\left(\mu \otimes \nu^m\right)$. Since for all unlimited S

$$\int_{\Omega \times (T^m \setminus T_S^m)} (I_{k,m}(F))^2 d\mu \otimes \nu^m = \int_{T^n_< \times (T^m \setminus T_S^m)} F^2 d\nu^{n+m} \approx 0,$$

$I_{k,m}(F) \in SL^2\left(\mu \otimes \nu^m\right)$. □

Therefore, we may define for $f \in L^2\left(\widehat{\nu^{n+m}}\right)$ and liftings $F \in SL^2\left(\nu^{n+m}\right)$ and $k \in \mathbb{N}_\mu^n$

$$I_{k,m}(f) = I_{k,m}({}^\circ F) := {}^\circ I_{k,m}(F).$$

Note that $I_{k,m}(f)$ is well defined $\widehat{\mu \otimes \nu^m}$ a.e. and belongs to $L^2\left(\widehat{\mu \otimes \nu^m}\right)$. Using S-square-integrable liftings of f, g, we obtain the following.

Proposition 16.3.3 *Fix $f \in L^2\left(\widehat{\nu^{n+m}}\right)$, $g \in L^2\left(\widehat{\nu^{\widetilde{n}+m}}\right)$, $k \in \mathbb{N}^n_\mu$ and $\widetilde{k} \in \mathbb{N}^{\widetilde{n}}_\mu$. Then*

$$\int_{\Omega \times T^m} I_{k,m}(f) \cdot I_{\widetilde{k},m}(g) d\widehat{\mu \otimes \nu^m} = \begin{cases} 0 & \text{if } k \neq \widetilde{k} \\ \int_{T^n_< \times T^m} f \cdot g d\widehat{\nu^{n+m}} & \text{if } k = \widetilde{k}. \end{cases}$$

For functions $f \in L^2\left(\lambda^{n+m}\right)$, $g \in L^2\left(\lambda^{\widetilde{n}+m}\right)$ we obtain analogous results later, by an application of Theorem 16.5.2 below.

The function $I_{k,m}(f)$ with $k \in \mathbb{N}^n_\mu$ is called the **iterated integral of f with m parameters of order** k. If $m = 0$, then we write $I_k(f)$ instead of $I_{k,0}(f)$ and call $I_k(f)$ the **iterated integral of f of order** k. If $k \in \mathbb{N}_\mu$, then $I_{(k)}(f)$ is called the **Wiener–Lévy integral** of $f \in L^2(\lambda)$ **of order** k.

Now we define martingales from the iterated integral of order k. Fix $n \in \mathbb{N}$, an internal $F : T^n_< \to {}^*\mathbb{R}$ and $k = (k_1, \ldots, k_n) \in M^n_\mu$. Define:

$$I^M_k(F) : \Omega \times T \to {}^*\mathbb{R}, (X, t) \mapsto \sum_{s \in T^n_<, s \leq t} F_s \cdot p_{k_1}(X_{s_1}) \cdot \ldots \cdot p_{k_n}(X_{s_n}).$$

If $n = 0$, set $I^M_\emptyset(F) := F$ for $F \in {}^*\mathbb{R}$. Note that $I^M_k(F)(\cdot, t)$ is a smooth function in the sense of the model $\mathfrak{W}$. From the following lemmas it follows that $I^M_k(F)$ can be converted to a martingale, defined on the continuous timeline $[0, \infty[$ if F is locally S-square-integrable. The proof of the following lemma is straightforward.

Lemma 16.3.4 *Fix internal $F : T^n_< \to {}^*\mathbb{R}$, $G : T^m_< \to {}^*\mathbb{R}$ and $(\circ) \in M^n_\mu$, $(\cdot) \in M^m_\mu$. Then the following hold.*

(a) *$I^M_{(\circ)}(F)$ is a μ-square-integrable $(\mathcal{B}_t)_{t \in T}$-martingale.*
(b) *For all $t \in T$*

$$\mathbb{E}_\mu\left(I^M_{(\circ)}(F)(\cdot, t) I^M_{(\cdot)}(G)(\cdot, t)\right) = \begin{cases} \int_{s \in T^n_<, s \leq t} F(s) \cdot G(s) \frac{1}{H^n} & \text{if } (\circ) = (\cdot) \\ 0 & \text{if } (\circ) \neq (\cdot). \end{cases}$$

Lemma 16.3.5 *Fix $F : T^n_< \to {}^*\mathbb{R}$ locally in $SL^2(\nu^n)$. Then for all $\sigma \in \mathbb{N}$*

$$\max_{t \in T_\sigma} \left| I^M_{(\circ)}(F)(\cdot, t) \right| \in SL^2(\mu) \text{ for } (\circ) \in \mathbb{N}^n_\mu.$$

Proof Assume first that F is limited. We prove by induction on $m \leq n$ that for all $s_{m+2} < \ldots < s_n \in T_\sigma$

$$\max_{t \in T_\sigma} \left| \sum_{s \in T^m_{\sigma,<}, s<t} F(s,t,s_{m+2},\ldots,s_n) p_{k_1}(X_{s_1}) \ldots p_{k_m}(X_{s_m}) \right| \in SL^2(\mu). \tag{52}$$

For $m=0$ the result is true, because F is limited. In the induction step, in order to prove (52), note that the term in $|\cdot|$ equals $\int G p_{k_m}(t,X)$ with

$$G(X,r) := \sum_{s \in T^{m-1}_{\sigma,<}, s<r} F(s,r,t,s_{m+2},\ldots,s_n) p_{k_1}(X_{s_1}) \ldots p_{k_{m-1}}(X_{s_{m-1}}).$$

By the induction hypothesis, $\max_{r \in T_\sigma} |G(\cdot,r)| \in SL^2(\mu)$, thus G belongs to $SL^2(\mu \otimes \nu_\sigma)$. Since $G(\cdot,r)$ is $\mathcal{B}_{r^-}$-measurable, we obtain statement (52) by Lemma 16.2.2. Now $m := n$ gives the result for limited F. Let F be unlimited. Set $F_k(t) := F(t)$ if $|F(t)| \leq k$, and $F_k(t) = 0$ otherwise. We have already seen that $\max_{t \in T_\sigma} \left| I^M_{(\circ)}(F_k)(\cdot,t) \right| \in SL^2(\mu)$. By Corollary 10.7.3 ($\gamma$), it suffices to show that

$$\lim_{k \to \infty} {}^\circ\mathbb{E}_\mu \max_{t \in T_\sigma} \left(I^M_{(\circ)}(F - F_k)(\cdot,t) \right)^2 = 0. \tag{53}$$

Otherwise, there exists a standard $\varepsilon > 0$ and an unlimited number $K \in {}^*\mathbb{N}$ such that, by Doob's inequality and Lemma 16.3.4,

$$\begin{aligned} \varepsilon &\leq \mathbb{E}_\mu \max_{t \in T_\sigma} \left(I^M_{(\circ)}(F - F_K)(\cdot,t) \right)^2 \leq 4\mathbb{E}_\mu \left(I^M_{(\circ)}(F - F_K)(\cdot,\sigma) \right)^2 \\ &= 4 \sum_{t_1 < \ldots < t_n \in T_\sigma} (F - F_K)^2 (t_1,\ldots,t_n) \frac{1}{H^n} \approx 0, \end{aligned}$$

because F, F_K, restricted to T_σ, belong to $SL^2(\nu^n_\sigma)$ and $F = F_K$ ν^n_σ-a.e. Equation (53) proves that $\max_{t \in T_\sigma} \left| I^M_{(\circ)} F(\cdot,t) \right| \in SL^2(\mu)$. □

Fix $f : [0,\infty[^n_{\leq} \to \mathbb{R}$ locally in $L^2(\lambda^n)$ and a lifting F locally in $SL^2(\nu^n)$ of f. We may assume that F is defined on $T^n_<$. Using Lemmas 16.3.4 and 16.3.5, we can define the **iterated integral process** $i^M_k(f)$ **of order** $k = (k_1,\ldots,k_n)$, by setting for $r \in [0,\infty[$,

$$i^M_k(f)(\cdot,r) := \lim_{{}^\circ t \downarrow r, t \in T} {}^\circ I^M_k(F)(\cdot,t) \ \widehat{\mu}\text{-a.e.}$$

(See Theorem 10.16.2 and Corollary 10.16.3.) In analogy to Theorem 16.2.3 we have the following.

Theorem 16.3.6 *Fix $f : [0,\infty[^n_{\leq} \to \mathbb{R}$ locally in $L^2(\lambda^n)$ and $(\circ) \in \mathbb{N}^n_{\mu}$. Then $i^M_{(\circ)}(f)$ is a $\widehat{\mu}$-a.s. well-defined càdlàg $(\mathfrak{b}_t)_{t\in[0,\infty[}$-martingale and, for each $r \in [0,\infty[$, $\sup_{s\leq r} i^M_{(\circ)}(f)(\cdot,s)$ belongs to $L^2(\widehat{\mu})$. Moreover, $i^M_{(\circ)}$ is a continuous operator on $L^2(\lambda^n)$ and $i^M_{(\circ)}(f)$ is continuous in measure. If $g : [0,\infty[^{\widetilde{n}}_{\leq} \to \mathbb{R}$ locally in $L^2(\lambda^{\widetilde{n}})$ and $(\cdot) \in \mathbb{N}^{\widetilde{n}}_{\mu}$, then*

$$\mathbb{E}_{\widehat{\mu}}\left(i^M_{(\circ)}(f)(\cdot,r)\, i^M_{(\cdot)}(g)(\cdot,r)\right) = \begin{cases} \int_{[0,r]^n_{\leq}} f \cdot g\, d\lambda^n & \text{if } (\circ) = (\cdot) \\ 0 & \text{if } (\circ) \neq (\cdot). \end{cases}$$

16.4 Multiple integrals

In order to obtain the integrals with and without parameters independent of $k \in \mathbb{N}^n_{\mu}$, we define for functions $f : \mathbb{N}^n_{\mu} \times T^{n+m} \to \mathbb{R} \in L^2\left(c^n \otimes \widehat{\nu^{n+m}}\right)$:

$$I_{n,m}(f) := \sum_{k\in\mathbb{N}^n_{\mu}} I_{k,m}(f_k),$$

and call it the **multiple integral of f with m parameters**. Obviously, $I_{n,m}(f)$ is also defined for $f : \mathbb{N}^n_{\mu} \times [0,\infty[^{n+m}_{\leq} \to \mathbb{R} \in L^2\left(c^n \otimes \lambda^{n+m}\right)$. Now let $m = 0$. We set $I_n(f) := I_{n,0}(f)$ and call $I_n(f)$ the **multiple integral** of f. In analogy to Theorem 13.1.6 we obtain the following.

Theorem 16.4.1 *Fix $f \in L^2\left(c^n \otimes \widehat{\nu^{n+m}}\right)$. If $F \in SL^2\left(c^n \otimes \nu^{n+m}\right)$ is a lifting of f, then $I_{n,m}(f) = {}^{\circ}I_{n,m}(F)$ in $L^2(\widehat{\mu \otimes \nu^m})$ and thus $\widehat{\mu \otimes \nu^m}$-a.e., where*

$$I_{n,m}(F) : (X,s) \mapsto \sum_{k\in M^n_{\mu}} \sum_{t\in T^n_{<}} F(k,t,s) \prod_{i=1}^{n} p_{k_i}(X_{t_i}) \in SL^2\left(\mu \otimes \nu^m\right).$$

We see that $I_{n,m}(f)$ is a real polynomial in several variables up to an infinitesimal error and a $\widehat{\mu \otimes \nu^m}$-nullset.

The process $i^M_n(f) : \Omega \times [0,\infty[\to \mathbb{R}$, defined for f locally in $L^2(c^n \otimes \lambda^n)$ with respect to λ^n by setting

$$i^M_n(f) := \sum_{k\in\mathbb{N}^n_{\mu}} i^M_k(f_k),$$

is called the **multiple integral process** of f. From Theorem 16.3.6 we obtain the following.

Theorem 16.4.2 *Fix $f : \mathbb{N}_\mu^n \times [0,\infty[^n_{\leq} \to \mathbb{R}$ locally in $L^2(c^n \otimes \lambda^n)$ with respect to λ^n. Then $i_n^M(f)$ is a $\widehat{\mu}$-a.s. well-defined càdlàg $(\mathfrak{b}_r)_{r\in[0,\infty[}$-martingale and, for each $r \in [0,\infty[$, $\sup_{s\leq r} i_n^M(f)(\cdot,s)$ belongs to $L^2(\widehat{\mu})$. Moreover, i_n^M is a continuous operator and $i_n^M(f)$ is continuous in measure. If $g : \mathbb{N}_\mu^{\widetilde{n}} \times [0,\infty[^{\widetilde{n}}_{\leq} \to \mathbb{R}$ locally in $L^2\left(c^{\widetilde{n}} \otimes \lambda^{\widetilde{n}}\right)$, then*

$$\mathbb{E}_{\widehat{\mu}}\left(i_n^M(f)(\cdot,r)\, i_{\widetilde{n}}^M(g)(\cdot,r)\right) = \begin{cases} \sum_{n\in\mathbb{N}_\mu^n} \int_{[0,r]^n_{\leq}} f_n \cdot g_n \, d\lambda^n & \text{if } n=\widetilde{n} \\ 0 & \text{if } n\neq\widetilde{n}. \end{cases}$$

16.5 The σ-algebra generated by the Wiener–Lévy integrals

Recall that the σ-algebra $\mathcal{W}$ generated by the infinite-dimensional Brownian motion is also generated by the Wiener integrals $I(f)$ with $f \in L^2(\lambda, \mathbb{H})$, where f is bounded and has compact support. There $\mathbb{N}_\mu = \{1\}$. Analogously, let $\mathcal{D}$ be the σ-subalgebra of the Loeb σ-algebra $L_\mu(\mathcal{B})$ generated by the Wiener–Lévy integrals $I_{(k)}(f)$ with $k \in \mathbb{N}_\mu$, $f \in L^2(\lambda)$, augmented by the $\widehat{\mu}$-nullsets. Since $I_{(k)}$ is a linear and continuous operator, we may assume that its integrands f are simple functions.

Following [89], in Section 23.3 we will characterize square-integrable Lévy random variables $\varphi \in L^2_{\mathcal{D}}(\widehat{\mu})$ by a sequence $(f_n)_{n\in\mathbb{N}_0}$ of symmetric square-summable deterministic functions $f_n : \mathbb{N}_\mu^n \times [0,\infty[^n \to \mathbb{R}$. In the case $\mathbb{N}_\mu^n = \{1\}$ (see Examples 16.1.2 (i) and (ii)), we can identify $f : \mathbb{N}_\mu^n \times [0,\infty[^n \to \mathbb{R}$ with $g : [0,\infty[^n \to \mathbb{R}$, where $g(t) := f(1,1,\ldots,1,t)$.

Fix an internal $F : M_\mu^n \times T^n \to {}^*\mathbb{R}$. This F is called **restricted by** m if $|F| \leq m$ and if $F(k,t) = 0$ for all $(k,t) \notin m^n \times T_m^n$, where m is identified with $\{1,\ldots,m\}$. For internal $F_1,\ldots,F_n : M_\mu \times T \to {}^*\mathbb{R}$ we define the **tensor product** $F_1 \otimes \ldots \otimes F_n : M_\mu^n \times T^n \to {}^*\mathbb{R}$ by setting

$$(F_1 \otimes \ldots \otimes F_n)(k,t) := F_1(k_1,t_1) \cdot \ldots \cdot F_n(k_n,t_n).$$

It is clear how to define the notion 'restricted' for $f : \mathbb{N}_\mu^n \times [0,\infty[^n \to \mathbb{R}$ and the **tensor product** of functions $f_1,\ldots,f_n : \mathbb{N}_\mu \times [0,\infty[\to \mathbb{R}$.

We have introduced the polynomials p_k in order to be able to write any polynomial of a Lévy–Wiener integral as a linear combination of iterated integrals.

Theorem 16.5.1 *Fix a linear combination C of multiple integrals of the form $I_{(k)}(f)$ with restricted f, and fix $n \in \mathbb{N}$. Then the nth-power C^n of C is in $L^2(\widehat{\mu})$ a linear combination of iterated integrals with integrands $f_1 \otimes \ldots \otimes f_m$ for certain restricted $f_i \in L^2(\lambda)$.*

Proof We use induction over n. In the induction step we have to show that $I_{(k_1,\ldots,k_m)}(g)\cdot I_{(k)}(f)$ has the desired property if the first factor has the desired property. We may assume that the supports of f,g belong to $[0,N]$ with $N\in\mathbb{N}$. Set $R:=T_N$. There exist limited liftings $F:R\to {}^*\mathbb{R}$ of f and $G:R^m\to {}^*\mathbb{R}$ of g. We obtain for $\widehat{\mu}$-almost all X:

$$I_{(k_1,\ldots,k_m)}(g)\cdot I_{(k)}(f)(X)$$
$$\approx\sum_{t\in R^m_<}G(t)\prod_{i=1}^m p_{k_i}(X_{t_i})\cdot\sum_{t\in R}F(t)p_k(X_t)=A+\sum_{i=1}^m B_i,$$

where

$$A(X)=\sum_{i=1}^{m+1}\sum_{t\in R^{m+1}_<}G\left((t_j)_{j\neq i,j\leq m+1}\right)F(t_i)$$
$$\cdot\prod_{j<i}p_{k_j}(X_{t_j})\prod_{i\leq j\leq m}p_{k_j}(X_{t_{j+1}})p_k(X_{t_i})$$

and

$$B_i(X)=\sum_{t\in R^m_<}G(t)F(t_i)\cdot\prod_{j\neq i}p_{k_j}(X_{t_j})\cdot\left(p_k(X_{t_i})p_{k_i}(X_{t_i})\right).$$

The standard part of A has the desired property. Let us study B_i. Here $p_{k_i}\cdot p_k$ is a linear combination $\alpha_{k_i+k}\dot{p}_{k_i+k}+\ldots+\alpha_0\dot{p}_0$ of polynomials $\dot{p}_l$, such that all the α_l are limited and $H\alpha_0$ is limited. It follows that each summand of B_i has the form

$$D_l=\alpha_l\sum_{t\in R^m_<}G(t)F(t_i)\cdot\prod_{j\neq i}p_{k_j}(X_{t_j})\dot{p}_l(X_{t_i}),$$

whose standard part has the desired property if $l\in\mathbb{N}_\mu$.

If $l\notin\mathbb{N}_\mu, l\neq 0$, then ${}^\circ D_l=0$ in $L^2(\widehat{\mu})$, because $H\mathbb{E}_{\mu^1}\dot{p}_l^2\approx 0$.

It remains for us to study the case $l=0$. There exists a limited number a_0 with $a_0=H\cdot\alpha_0$. Then

$$D_0=a_0\sum_{t\in R^{m-1}_<}\left(\sum_{t_{i-1}<r<t_i}G(t_1,\ldots,t_{i-1},r,t_i,\ldots,t_{m-1})F(r)\frac{1}{H}\right)$$
$$\cdot p_{k_1}(X_{t_1})\cdot\ldots\cdot p_{k_{i-1}}(X_{t_{i-1}})\cdot p_{k_{i+1}}(X_{t_i})\cdot\ldots\cdot p_{k_m}(X_{t_{m-1}}).$$

We may assume that

$$G=G_1\otimes\ldots\otimes G_{i-1}\otimes J\otimes G_i\otimes\ldots\otimes G_{m-1},$$

where all the G_i and also J are 1-ary restricted functions defined on $\mathbb{R}$. Thus, D_0 has the following form for suitable l_j:

$$D_0 = a_0 \sum_{t \in R_<^{m-1}} \left(\sum_{t_{i-1} < r < t_i} J(r) F(r) \frac{1}{H} \right) \prod_{j=1}^{m-1} G_j(t_j) p_{l_j}(X_{t_j}) = \alpha - \beta$$

with

$$\alpha = a_0 \sum_{t \in R_<^{m-1}} \sum_{r < t_i} J \cdot F(r) \frac{1}{H} \prod_{j=1}^{m-1} G_j(t_j) p_{l_j}(X_{t_j}),$$

$$\beta = a_0 \sum_{t \in R_<^{m-1}} \sum_{r \le t_{i-1}} J \cdot F(r) \frac{1}{H} \prod_{j=1}^{m-1} G_j(t_j) p_{l_j}(X_{t_j}).$$

Therefore, the standard part of D_0 has the desired property. □

In order to prove that iterated integrals are $\mathcal{D}$-measurable, we use a slightly modified notion of multiple integral. Fix $f \in L^2(\lambda^n)$ and a lifting $F \in SL^2(\nu^n)$ of f. We define for $k = (k_1, \ldots, k_n) \in \mathbb{N}_\mu^n$

$$J_k(F) := \sum_{t \in T_{\neq}^n} F(t) p_{k_1}(X_{t_1}) \cdot \ldots \cdot p_{k_n}(X_{t_n}), \quad J_{(\cdot)}(f) := {}^\circ J_{(\cdot)}(F).$$

The random variable $J_{(\cdot)}(f)$ is well defined and belongs to $L^2(\widehat{\mu})$ (see Section 16.3). Moreover, $J_{(\cdot)}$ is a continuous operator and $J_{(\cdot)}(F) \in SL^2(\mu)$. If functions in $L^2(\lambda^m)$ are again identified with their equivalent functions in $L^2_{\mathcal{L}^m}\left(\widehat{\nu^m}\right)$, we obtain the following.

Theorem 16.5.2 *Fix $k \in \mathbb{N}_\mu^n$.*

(a) *If $g \in L^2(\lambda^n)$, then $I_k(g) \in L^2_{\mathcal{D}}(\widehat{\mu})$.*

(b) *If $g \in L^2\left(\lambda^{n+m}\right)$, then $I_{k,m}(g) \in L^2_{\mathcal{D}\otimes\mathcal{L}^m}\left(\widehat{\mu \otimes \nu^m}\right)$. It follows that there is an $f \in L^2_{\mathcal{D}}(\widehat{\mu} \otimes \lambda^m)$ which is equivalent to $I_{k,m}(g)$. Let us identify f and $I_{k,m}(g)$.*

Proof (a) Set $f := \mathbf{1}_{[0,\infty[_\le} \cdot g$. It suffices to prove by induction on n that $J_{(k_1,\ldots,k_n)}(f)$ is $\mathcal{D}$-measurable. Assume that the result is true for all $m < n$. Since $J_{(k_1,\ldots,k_n)}$ is linear and continuous, it suffices to prove the result for functions f which are tensor products $f_1 \otimes \ldots \otimes f_n$ of restricted functions $f_1, \ldots, f_n \in L^2(\lambda)$. Let $F_1, \ldots, F_n$ be restricted liftings of $f_1, \ldots, f_n$. We obtain

$$J_{(k_1,\ldots,k_{n-1})}(f_1 \otimes \ldots \otimes f_{n-1}) \cdot J_{(k_n)}(f_n) = J_{(k_1,\ldots,k_n)}(f_1 \otimes \ldots \otimes f_n) + B$$

with $B = \sum_{i=1}^{n-1} B_i$ and

$$B_i = {}^\circ \sum_{t \in T_{\neq}^{n-1}} \left(\prod_{j=1, j\neq i}^{n-1} F_j(t_j) p_{k_j}(X_{t_j}) \right) F_i(t_i) F_n(t_i) p_{k_i}(X_{t_i}) p_{k_n}(X_{t_i}).$$

According to the proof of Theorem 16.5.1, we see that B_i is a finite linear combination of functions of the form $J_{(\circ)}(g)$ with $(\circ) \in \mathbb{N}_\mu^{n-1}$ and standard parts of functions of the form

$$\begin{aligned} C_i &= \sum_{t \in T_{\neq}^{n-1}} \left(\prod_{j=1, j\neq i}^{n-1} F_j(t_j) p_{k_j}(X_{t_j}) \right) F_n(t_i) F_i(t_i) \frac{1}{H} \\ &= \sum_{t \in T_{\neq}^{n-2}} \left(\sum_{s \neq t_1, \ldots, t_{n-2}} F_i(s) F_n(s) \frac{1}{H} \right) \prod_{j=1}^{i-1} F_j(t_j) p_{k_j}(X_{t_j}) \\ &\quad \times \prod_{j=i+1}^{n-1} F_j(t_{j-1}) p_{k_j}(X_{t_{j-1}}) \\ &\approx \sum_{t \in T_{\neq}^{n-2}} \left(\sum_{s \in T} F_i(s) F_n(s) \frac{1}{H} \right) \prod_{j=1}^{i-1} F_j(t_j) p_{k_j}(X_{t_j}) \prod_{j=i+1}^{n-1} F_j(t_{j-1}) p_{k_j}(X_{t_{j-1}}). \end{aligned}$$

By the induction hypothesis, B_i and thus B are $\mathcal{D}$-measurable. It follows that $J_{(k_1,\ldots,k_n)}(f_1 \otimes \ldots \otimes f_n)$ is $\mathcal{D}$-measurable.

(b) Let M be the set of all $g \in L^2(\lambda^{n+m})$ such that $I_{k,m}(g)$ is $\mathcal{D} \otimes \mathcal{L}^m$-measurable. Then M is linear and complete. Since, by (a), $I_{k,m}(\mathbf{1}_{A\times B}) = I_k(\mathbf{1}_A) \otimes \mathbf{1}_B$ is $\mathcal{D} \otimes \mathcal{L}^m$-measurable for all Lebesgue-measurable $A \subseteq [0,\infty[^n$ and $B \subseteq [0,\infty[^m$, we have $M = L^2(\lambda^{n+m})$. □

Exercises

16.1 Prove that, under the assumption that $H\mathbb{E}_{\mu^1} x^n$ is limited for all $n \in \mathbb{N}$, $\mathbb{E}_\mu B_t^n$ is limited for all limited $t \in T$.

16.2 Prove Proposition 16.1.1.

16.3 Prove the assertions in Examples 16.1.2.

16.4 Let $F : \Omega \times T \to {}^*\mathbb{R}$ be $(\mathcal{B}_{t^-})_{t\in T}$-adapted and let $F^2(\cdot, t)$ be μ-integrable for all $t \in T$. Then $\int F p_k$ is a μ-square-integrable $(\mathcal{B}_t)_{t\in T}$-martingale.

16.5 Prove Theorem 16.2.3.

16.6 Prove Lemma 16.3.4.

16.7 Prove Theorem 16.3.6.

16.8 Prove Theorem 16.4.1.

16.9 Compute the Lévy triplets for the Lévy processes defined in Examples 16.1.2.

PART III

Malliavin calculus

Introduction

Following [84], we start Malliavin calculus for infinite-dimensional Brownian motion. Fix an abstract Wiener–Fréchet space $\mathbb{B}$, given by a separating sequence $\left(|\cdot|_j\right)_{j\in\mathbb{N}}$ of semi-norms over an infinite-dimensional but separable Hilbert space $\mathbb{H}$. Fix a finite-dimensional representation $\mathbb{F}$ of $\mathbb{H}$, according to Theorem 9.8.1. Let ω be the dimension of $\mathbb{F}$. The probability space we are working with is $\Omega = \mathbb{F}^T$, endowed with the Loeb measure $\widehat{\Gamma}$ over the $\omega \cdot H^2$ fold product Γ of the internal centred Gaussian distribution of variance $\frac{1}{H}$ on $^*\mathbb{R}$. Recall that $\mathcal{W} := \mathcal{W}_{C_{\mathbb{H}}}$ denotes the σ-algebra on Ω generated by the $\mathbb{B}$-valued Brownian motion $b_{\mathbb{B}} = \Omega \to C_{\mathbb{B}}$, where $C_{\mathbb{B}}$ is the space of continuous functions from $[0,\infty[$ into $\mathbb{B}$. By Proposition 11.8.6, $\mathcal{W}$ only depends on $\mathbb{H}$. Let W be the **Wiener measure** on the space $C_{\mathbb{B}}$ of continuous functions from $[0,\infty[$ into $\mathbb{B}$; it is the image measure of $\widehat{\Gamma}$ by $b_{\mathbb{B}}$.

I refer to Üstünel and Zakai [117] and Nualart [79] for the standard theory; see also [120], [116], [115].

At the end of Section 11.8 we saw that $L^p(W,\mathbb{H}^{\otimes k})$ can be identified with $L^p_{\mathcal{W}}(\widehat{\Gamma},\mathbb{H}^{\otimes k})$ via the canonical isometric isomorphism S from $L^p(W,\mathbb{H}^{\otimes k})$ onto $L^p_{\mathcal{W}}(\widehat{\Gamma},\mathbb{H}^{\otimes k})$, given by $S(\varphi)(X) = \varphi(b_{\mathbb{B}}(X,\cdot))$. So we obtain Malliavin calculus for the standard space $L^2(W)$. Recall that the spaces $L^p(\lambda^n,\mathbb{H}^{\otimes k})$ and $L^p_{\mathcal{L}^n}(\widehat{\nu^n},\mathbb{H}^{\otimes k})$ can also be identified via $S(\varphi)(t) = \varphi(^\circ t)$. The standard part $^\circ t$ of t is $\widehat{\nu^n}$-a.e. well defined. Moreover, we can identify $L^p(W\otimes\lambda^n,\mathbb{H}^{\otimes k})$ with $L^p_{\mathcal{W}\otimes\mathcal{L}^n}(\widehat{\Gamma\otimes\nu^n},\mathbb{H}^{\otimes k})$ via $S(\varphi)(X,t) = \varphi(b_{\mathbb{B}}(X,\cdot),^\circ t)$ and $L^p_{\mathcal{W}\otimes\mathrm{Leb}^n}(\widehat{\Gamma}\otimes\lambda^n,\mathbb{H}^{\otimes k})$ with $L^p_{\mathcal{W}\otimes\mathcal{L}^n}(\widehat{\Gamma\otimes\nu^n},\mathbb{H}^{\otimes k})$ via $S(\varphi)(X,t) = \varphi(X,^\circ t)$. We tacitly identify φ with the corresponding $S(\varphi)$. It is easier to work with the elements of $L^p_{\mathcal{W}}(\widehat{\Gamma},\mathbb{H}^{\otimes k}), L^p_{\mathcal{L}^n}(\widehat{\nu^n},\mathbb{H}^{\otimes k})$ and $L^p_{\mathcal{W}\otimes\mathcal{L}^n}(\widehat{\Gamma\otimes\nu^n},\mathbb{H}^{\otimes k})$ rather than to work with the elements in the equivalent standard spaces, because $\widehat{\Gamma}$ is a measure on a (in the sense of $\mathfrak{W}$) finite-dimensional space and $\widehat{\nu}$ is equivalent to a counting measure (in the sense of $\mathfrak{W}$).

We will use the following short notation: Let Λ be an internal measure space with internal measure ρ, and let $F, G : \Lambda \to \mathbb{F}^{\otimes n}$ be internally measurable. Then we set

$$\langle F, G \rangle_\rho = \int_\Lambda \langle F, G \rangle_{\mathbb{F}^n} \, d\rho, \quad \|F\|_\rho = \sqrt{\int_\Lambda \|F\|_{\mathbb{F}^n}^2 \, d\rho}.$$

Similar notation is used for standard spaces and measurable functions f, g with values in $\mathbb{H}^{\otimes n}$.

At the end of Part III we present Malliavin calculus for more general, but finite-dimensional, Lévy processes. For simplicity, we only study the one-dimensional case.

17

Chaos decomposition

One way to establish Malliavin calculus is the chaos representation property. It says that we can characterize square-integrable Wiener random variables $\varphi \in L^2(W)$ by a sequence $(f_n)_{n\in\mathbb{N}_0}$ of square-summable deterministic functions $f_n : [0,\infty[^n_{\leq} \to \mathbb{H}^{\otimes n}$. Therefore, following the terminology and the intention of Wiener (see 'The homogeneous chaos' [119]), one should rather call this property the 'chaos homogenization property'. Wiener has called measures with values in random variables 'chaos'. The chaos representation property is in fact a chaos avoiding property. Our aim is to enrich Malliavin calculus for Lévy processes by the following aspects:

(1) All Wiener functionals, and those with no moments, can be represented by deterministic functions via chaos expansions.
(2) All Lévy functionals can be represented by smooth functions, in fact, by polynomials of several variables. In an application we study anticipative Girsanov transformations, using the substitution rule in finite-dimensional analysis.
(3) In analogy to Cutland and Ng's [24] approach to Malliavin calculus for the classical Wiener space, there exists a nice method for the computation of the kernels, using a simple and mathematically rigorous notion of Wiener white noise. It is used to prove product and chain rules for the Malliavin derivative.

By the identification of $L^2(W)$ with $L^2_{\mathcal{W}}(\widehat{\Gamma})$, it suffices to prove this expansion for elements in the simpler space $L^2_{\mathcal{W}}(\widehat{\Gamma})$.

17.1 Admissible sequences

Our aim is to expand functionals $\varphi \in L^2_{\mathcal{W}}\left(\widehat{\Gamma}\right)$ into an orthogonal series of iterated integrals $I_n({}^\circ F_n)$. Following [90], admissible sequences are introduced, in order to find minimal closure properties of the deterministic kernels F_n under which such an expansion is possible. A sequence $(\mathfrak{H}_n)_{n\in\mathbb{N}_0}$, where

$$\mathfrak{H}_n \subseteq \left\{F : T^n_< \to \mathbb{F}^{\otimes n} \mid F \text{ is internal}\right\},$$

is called an **admissible sequence** if the following conditions are fulfilled:

(W 1) $\mathfrak{H}_0 = \mathbb{R}$.
(W 2) $\mathfrak{H}_1 = \left\{F \in SL^2(\nu, \mathbb{F}) \mid {}^\circ F \in L^2_{\mathcal{L}^1}(\widehat{\nu}, \mathbb{H})\right\}$.
(W 3) $\mathfrak{H}_n$ is a linear space over $\mathbb{R}$.
(W 4) $\mathfrak{H}_n \subseteq \left\{F \in SL^2(\nu^n, \mathbb{F}^{\otimes n}) \mid {}^\circ F \in L^2_{\mathcal{L}^n}(\widehat{\nu^n}, \mathbb{H}^{\otimes n})\right\}$.
(W 5) ${}^\circ\mathfrak{H}_n := \{{}^\circ F \mid F \in \mathfrak{H}_n\}$ is a closed subspace of $L^2_{\mathcal{L}^n}(\widehat{\nu^n}, \mathbb{H}^{\otimes n})$.
(W 6) If $F \in \mathfrak{H}_1$ is limited and the support of F is a subset of T_σ for some $\sigma \in \mathbb{N}$, then $\mathbf{1}_{T^n_<} \cdot F^{\otimes n} \in \mathfrak{H}_n$.

Before giving examples of admissible sequences, we make the following remark.

Remark 17.1.1

(a) It is required that for each $F \in \mathfrak{H}_n$ the standard part ${}^\circ F$ of F exists almost everywhere in $L^2_{\mathcal{L}^n}\left(\widehat{\nu^n}, \mathbb{H}^{\otimes n}\right)$ or, equivalently, in $L^2\left(\lambda^n, \mathbb{H}^{\otimes n}\right)$. Therefore, ${}^\circ F({}^\circ t) \approx_{\mathbb{F}^n} F(t)$, i.e., $\sum_{\mathfrak{e}\in\mathfrak{E}^n} \left({}^*\left({}^\circ F({}^\circ t)\right)(\mathfrak{e}) - F(t)(\mathfrak{e})\right)^2 \approx 0$, for $\widehat{\nu^n}$-almost all $t \in T$, where $\mathfrak{E}$ is an internal ONB of $\mathbb{F}$.
(b) Fix $F \in \mathfrak{H}_n$ and $k \leq n$. Then $I_{n-k,k}(F) \in SL^2\left(\Gamma \otimes \nu^k, \mathbb{F}^{\otimes k}\right)$ by Theorem 13.1.6, and ${}^\circ I_{n-k,k}(F) = I_{n-k,k}({}^\circ F) \in L^2_{\mathcal{W}\otimes\mathcal{L}^k}(\widehat{\Gamma \otimes \nu^k}, \mathbb{H}^{\otimes k})$ by Corollary 13.4.4, or equivalently, ${}^\circ I_{n-k,k}(F) \in L^2_{\mathcal{W}\otimes \mathrm{Leb}[0,\infty[^k}\left(\widehat{\Gamma} \otimes \lambda^k, \mathbb{H}^{\otimes k}\right)$.
(c) Each function $F : T^n_< \to \mathbb{F}^{\otimes n}$ can be extended to a symmetric function $\overleftrightarrow{F} : T^n_{\neq} \to \mathbb{F}^{\otimes n}$ setting

$$\overleftrightarrow{F}(t_1,\ldots,t_n)(a_1,\ldots,a_n) := F(t_{\sigma_1},\ldots,t_{\sigma_n})(a_{\sigma_1},\ldots,a_{\sigma_n}),$$

where σ is a permutation on $\{1,\ldots,n\}$ which arranges the $t_1,\ldots,t_n$ into the natural order, i.e., $t_{\sigma_1} < \ldots < t_{\sigma_n}$.

Examples 17.1.2

(i) Set $\mathfrak{H}_0 := \mathbb{R}$ and for $n \geq 1$ set

$$\mathfrak{H}_n := \left\{F \in SL^2(\nu^n \upharpoonright T^n_<, \mathbb{F}^{\otimes n}) \mid {}^\circ F \in L^2_{\mathcal{L}^n}(\widehat{\nu^n}, \mathbb{H}^{\otimes n})\right\}.$$

Then $(\mathfrak{H}_n)_{n\in\mathbb{N}_0}$ is an admissible sequence.

Proof Obviously, (W 1), (W 2), (W 3) and (W 4) are true. To prove (W 5), let $({}^\circ F_k)_{k\in\mathbb{N}}$ be a Cauchy sequence in ${}^\circ\mathfrak{H}_n$. Using Corollary 10.9.3, there exists an $F: T^n_< \to \mathbb{F}^{\otimes n} \in SL^2(\nu^n, \mathbb{F}^{\otimes n})$ such that $\lim_{k\to\infty} {}^\circ F_k = {}^\circ F$ in $L^2_{\mathcal{L}^n}(\widehat{\nu^n}, \mathbb{H}^{\otimes n})$. Therefore, $F \in \mathfrak{H}_n$. (W 6) follows from Lemma 10.12.2 (b), Proposition 9.8.3 and the fact that $\mathbf{1}_{T^n_<}$ is $\mathcal{L}^n$-measurable. □

The following example is the key to proving the Clark–Ocone formula.

(ii) Fix an admissible sequence $(\mathfrak{H}_n)_{n\in\mathbb{N}_0}$. Set $\mathfrak{G}_0 := \mathbb{R}$. For $n \geq 1$, let $\mathfrak{G}_n$ be the set of all $F \in \mathfrak{H}_n$ such that in $L^2_{\mathcal{W}\otimes\mathcal{L}}\left(\widehat{\Gamma\otimes\nu}, \mathbb{H}\right)$

$$^\circ I_{n-1,1}(F) = \left((X,t) \mapsto \mathbb{E}^{\mathfrak{b}_{\circ t}}\, {}^\circ I_{n-1,1}(\overleftrightarrow{F})(X,t)\right).$$

Then $(\mathfrak{G}_n)_{n\in\mathbb{N}_0}$ is admissible. It follows that $t \mapsto \mathbb{E}^{\mathfrak{b}_{\circ t}}\, {}^\circ I_{n-1,1}(\overleftrightarrow{F})(\cdot,t)$ is $\mathcal{W}$-measurable for $\widehat{\nu}$-almost all $t \in T$. Moreover, $I_{n-1,1}(F)$ is a $(\mathcal{B}_{t^-})_{t\in T}$-adapted lifting in $SL^2(\Gamma\otimes\nu, \mathbb{F})$ of $r \mapsto \mathbb{E}^{\mathfrak{b}_r}\, {}^\circ I_{n-1,1}(\overleftrightarrow{F})(\cdot,r)$.

Proof Obviously, (W 1),..., (W 4) are true. To prove (W 5), fix a Cauchy sequence $({}^\circ F_m)_{m\in\mathbb{N}}$ in ${}^\circ\mathfrak{G}_n$ with limit ${}^\circ F$, where $F \in \mathfrak{H}_n$. Then, by Corollary 13.1.8,

$$\lim_{m\to\infty} {}^\circ I_{n-1,1}(F_m) = {}^\circ I_{n-1,1}(F) \text{ in } L^2_{\mathcal{W}\otimes\mathcal{L}}\left(\widehat{\Gamma\otimes\nu}, \mathbb{H}\right).$$

Moreover, by Jensen's inequality and Theorems 10.9.2 and 13.1.6,

$$\begin{aligned}
&\int_{\Omega\times T} \left\| \mathbb{E}^{\mathfrak{b}_{\circ r}}\, {}^\circ I_{n-1,1}(\overleftrightarrow{F})(\cdot,r) - \mathbb{E}^{\mathfrak{b}_{\circ r}}\, {}^\circ I_{n-1,1}(\overleftrightarrow{F_m})(\cdot,r) \right\|^2_{\mathbb{H}} d\widehat{\Gamma\otimes\nu}(\cdot,r) \\
&\leq \int_{\Omega\times T} \left\| {}^\circ I_{n-1,1}(\overleftrightarrow{F})(\cdot,r) - {}^\circ I_{n-1,1}(\overleftrightarrow{F_m})(\cdot,r) \right\|^2_{\mathbb{H}} d\widehat{\Gamma\otimes\nu} \\
&= {}^\circ \int_{\Omega\times T} \left\| I_{n-1,1}(\overleftrightarrow{F})(\cdot,r) - I_{n-1,1}(\overleftrightarrow{F_m})(\cdot,r) \right\|^2_{\mathbb{F}} d\Gamma\otimes\nu \\
&= {}^\circ \int_{T^{n-1}_<\times T} \left\| \overleftrightarrow{F} - \overleftrightarrow{F_m} \right\|^2_{\mathbb{F}^n} d\nu^n \\
&= \int_{T^{n-1}_<\times T} \left\| {}^\circ\overleftrightarrow{F} - {}^\circ\overleftrightarrow{F_m} \right\|^2_{\mathbb{H}^n} d\widehat{\nu^n} \to_{m\to\infty} 0.
\end{aligned}$$

Since ${}^\circ I_{n-1,1}(F_m)(\cdot,r) = \mathbb{E}^{\mathfrak{b}_{\circ r}}\, {}^\circ I_{n-1,1}(\overleftrightarrow{F_m})(\cdot,r)$ in $L^2_{\mathcal{W}\otimes\mathcal{L}}\left(\widehat{\Gamma\otimes\nu}, \mathbb{H}\right)$, we also have ${}^\circ I_{n-1,1}(F)(\cdot,r) = \mathbb{E}^{\mathfrak{b}_{\circ r}}\, {}^\circ I_{n-1,1}(\overleftrightarrow{F})(\cdot,r)$ in $L^2_{\mathcal{W}\otimes\mathcal{L}}\left(\widehat{\Gamma\otimes\nu}, \mathbb{H}\right)$.

To prove (W 6), fix a limited $F \in \mathfrak{G}_1$ with support in T_σ for some $\sigma \in \mathbb{N}$. Note that $\overleftrightarrow{\mathbf{1}_{T^n_<} \cdot F^{\otimes n}} = \mathbf{1}_{T^n_{\neq}} \cdot F^{\otimes n} \approx F^{\otimes n}$ in $L^2\left(\nu^n, \mathbb{F}^{\otimes n}\right)$. For $r \in T_\sigma, i = 1,\ldots,n$

and $t := (t_1, \ldots, t_{n-1})$ set

$$\beta_i(X,r) := \sum_{t_1 < \ldots < t_{i-1} < r < t_i < \ldots < t_{n-1}} F^{\otimes(n-1)}(t, X_{t_1}, \ldots, X_{t_{n-1}}).$$

Then

$$\mathbb{E}^{\mathfrak{b}_{\circ r}} {}^{\circ}I_{n-1,1}(F^{\otimes n})(\cdot, r) = \sum_{i=1}^{n} {}^{\circ}F(r) \cdot \mathbb{E}^{\mathfrak{b}_{\circ r}} {}^{\circ}\beta_i(\cdot, r).$$

We will prove that $\mathbb{E}^{\mathfrak{b}_{\circ r}} {}^{\circ}\beta_i(\cdot, r) = 0$ for $i < n$. To this end fix $D \in \mathfrak{b}_{\circ r}$. Then there exists an $s \approx r$ with $r < s^-$ and a $C \in \mathcal{B}_{s^-}$ with $\widehat{\Gamma}(C \Delta D) = 0$. Recall that $C = A \times \mathbb{F}^{T \setminus T_{s^-}}$ for a certain Borel set A in $\mathbb{F}^{T_{s^-}}$. Set

$$\alpha_i(X, r, s) := \sum_{t_1 < \ldots < t_{i-1} < r < t_i < \ldots < t_{n-1} < s} F^{\otimes(n-1)}(t, X_{t_1}, \ldots, X_{t_{n-1}}).$$

Then

$$\begin{aligned} \left| \int_D {}^{\circ}\beta_i(\cdot, r) d\widehat{\Gamma} \right|^2 &\approx \left| \int_C \beta_i(\cdot, r) d\Gamma \right|^2 = \left| \int_A \alpha_i(r, s, \cdot) d\Gamma \right|^2 \\ &\leq \mathbb{E}_{\Gamma} \alpha_i^2(r, s, \cdot) \leq \int_{T^{n-2} \times (]r,s] \cap T)} \left\| F^{\otimes(n-1)} \right\|^2_{\mathbb{F}^{n-1}} d\nu^{n-1} \approx 0, \end{aligned}$$

because $T^{n-2} \times (]r,s] \cap T)$ is a ν_σ^{n-1}-nullset and $F^{\otimes(n-1)}$ is $S_{\nu_\sigma^{n-1}}$-square-integrable. We obtain, in $L^2_{\mathcal{W} \otimes \mathcal{L}}\left(\widehat{\Gamma \otimes \nu}, \mathbb{H}\right)$,

$$\mathbb{E}^{\mathfrak{b}_{\circ r}} {}^{\circ}I_{n-1,1}(F^{\otimes n})(\cdot, r) = {}^{\circ}I_{n-1,1}(\mathbf{1}_{T^n_<} \cdot F^{\otimes n})(\cdot, r).$$

(iii) Note that, if $(\mathfrak{H}_n)_{n \in \mathbb{N}_0}$ and $\left(\mathfrak{H}'_n\right)_{n \in \mathbb{N}_0}$ are admissible sequences, then $\left(\mathfrak{H}_n \cap \mathfrak{H}'_n\right)_{n \in \mathbb{N}_0}$ is an admissible sequence too. □

17.2 Chaos expansion

The following theorem is the key for Malliavin calculus on abstract Wiener spaces.

Theorem 17.2.1 *Fix an admissible sequence $(\mathfrak{H}_n)_{n \in \mathbb{N}_0}$. For each $\varphi \in \mathcal{L}^2_{\mathcal{W}}(\widehat{\Gamma})$ there exists a sequence (F_n) with $F_n \in \mathfrak{H}_n$ such that*

$$\varphi = \sum_{n=0}^{\infty} {}^{\circ}I_n(F_n) \text{ in } L^2(\widehat{\Gamma}) \text{ with } {}^{\circ}I_n(F_n) \in L^2_{\mathcal{W}}(\widehat{\Gamma}). \tag{54}$$

Moreover, $^{\circ}I_0(F_0) = {}^{\circ}F_0 = \mathbb{E}_{\widehat{\Gamma}}(\varphi)$. *Recall that* $^{\circ}I_n(F_n) = I_n(^{\circ}F_n)$. *By Remark 17.1.1 (c), we may assume that* $F_n : T^n_{\neq} \to \mathbb{F}^{\otimes n}$ *is symmetric.*

Proof Let M be the set of all $\varphi \in \mathcal{L}^2_{\mathcal{W}}(\widehat{\Gamma})$ such that φ has an expansion according to equation (54). Since $^{\circ}\mathfrak{H}_n$ is complete, $\{^{\circ}I_n(F) \mid F \in \mathfrak{H}_n\}$ is complete (see Corollary 13.1.8). By Proposition 5.6.5, M is complete. Obviously, M is a linear space. In order to prove that $M = L^2_{\mathcal{W}}(\widehat{\Gamma})$, fix $\varphi \in \mathcal{L}^2_{\mathcal{W}}(\widehat{\Gamma})$ with $\varphi \perp M$. It suffices to prove that $\varphi = 0$ $\widehat{\Gamma}$-a.s., thus we have to prove that for all $B \in \mathcal{W}$

$$\widehat{\Gamma}^+(B) := \int_B \varphi^+ d\widehat{\Gamma} = \int_B \varphi^- d\widehat{\Gamma} =: \widehat{\Gamma}^-(B),$$

where $\varphi^+ = \varphi \vee 0$ and $\varphi^- = (-\varphi) \vee 0$. Therefore, we have to prove the equality of the finite measures μ^+ and μ^- defined on $\mathcal{W}$. By Proposition 12.5.1, $\mathcal{W}$ is generated by the linear space of functions $I(f)$, where $f \in L^2(\lambda, \mathbb{H})$ is bounded with compact support. Therefore, we have to prove that the image measures $\widehat{\Gamma}^+_{I(f)}$ and $\widehat{\Gamma}^-_{I(f)}$ of $\widehat{\Gamma}^+$ and $\widehat{\Gamma}^-$ by all $I(f)$ coincide. By Corollary 13.2.2, each polynomial $p(I(f))$ is a linear combination of iterated integrals with integrands of the form $f \otimes \ldots \otimes f$, n times. Let $F \in SL^2(\nu, \mathbb{F})$ be a bounded lifting of f. Since $F \in \mathfrak{H}_1$ and therefore, by (W 6), $\mathbf{1}_{T^n_<} \cdot F \otimes \ldots \otimes F \in \mathfrak{H}_n$, we have $\varphi \perp p(^{\circ}I(F)) = p(I(f))$. Therefore,

$$\int_\Omega \varphi^+ \cdot p(I(f)) d\widehat{\Gamma} = \int_\Omega \varphi^- \cdot p(I(f)) d\widehat{\Gamma}. \tag{55}$$

We obtain

$$\begin{aligned}
\int_{\mathbb{R}} e^{i \cdot x} d\widehat{\Gamma}^+_{I(f)}(x) &= \lim_{n\to\infty} \int_{\mathbb{R}} \sum_{k=0}^{n} \frac{i^k \cdot x^k}{k!} d\widehat{\Gamma}^+_{I(f)}(x) \\
&= \lim_{n\to\infty} \int_{\Omega} \sum_{k=0}^{n} \frac{i^k \cdot I(f)^k}{k!} \varphi^+ d\widehat{\Gamma} \\
&= \lim_{n\to\infty} \int_{\Omega} \sum_{k=0}^{n} \frac{i^k \cdot I(f)^k}{k!} \varphi^- d\widehat{\Gamma} \quad \text{(by (55))} \\
&= \int_{\mathbb{R}} e^{i \cdot x} d\widehat{\Gamma}^-_{I(f)}(x).
\end{aligned}$$

This proves that $\widehat{\Gamma}^+_{I(f)} = \widehat{\Gamma}^-_{I(f)}$, thus $\varphi = 0$ $\widehat{\Gamma}$-a.s.

Let $\varphi = \sum_{n=0}^{\infty} {}^{\circ}I_n(F_n) \in L^2_{\mathcal{W}}(\widehat{\Gamma})$. Since $\mathbb{E}_{\widehat{\Gamma}} {}^{\circ}I_n(F_n) = {}^{\circ}\mathbb{E}_{\Gamma} I_n(F_n) = 0$ for each $n \geq 1$, we have

$$\mathbb{E}_{\widehat{\Gamma}}(\varphi) = {}^{\circ}I_0(F_0) = {}^{\circ}F_0 \in \mathbb{R}.$$

□

Since the space $L^2(W)$ can be identified with the space $L^2_{\mathcal{W}}(\widehat{\Gamma})$ via $\varphi \mapsto (X \mapsto \varphi(b_{\mathbb{B}}(X,\cdot)))$, we also have a chaos decomposition for square-integrable standard functionals $\varphi : C_{\mathbb{B}} \to \mathbb{R}$.

Fix $\varphi \in L^2_{\mathcal{W}}(\widehat{\Gamma})$ with chaos decomposition $\varphi = \sum_{n=0}^{\infty} I_n({}^\circ F_n)$. Then we say that φ belongs to **chaos level** m if ${}^\circ F_n = 0$ for all $n > m$, in which case we write $\varphi \in L^2_{\mathcal{W},m}(\widehat{\Gamma})$. Let us say that φ belongs to **finite chaos levels** if

$$\varphi \in FinL^2_{\mathcal{W}}(\widehat{\Gamma}) = \bigcup_{m \in \mathbb{N}} L^2_{\mathcal{W},m}(\widehat{\Gamma}).$$

The functions F_n and ${}^\circ F_n$ in the chaos decomposition (54) of φ are called the **kernels** of φ ${}^\circ F_n$ j is uniquely determined by φ, as shown by the following result.

Proposition 17.2.2 *If in* $L^2(\widehat{\Gamma})$

$$\varphi = \sum_{n=0}^{\infty} {}^\circ I_n(F_n) \text{ with symmetric } F_n \in SL^2(\nu^n, \mathbb{F}^{\otimes n})$$

and

$$\varphi = \sum_{n=0}^{\infty} {}^\circ I_n(G_n) \text{ with symmetric } G_n \in SL^2(\nu^n, \mathbb{F}^{\otimes n}),$$

then $\|G_n - F_n\|_{\mathbb{F}^n} \approx 0$ *in* $L^2(\nu^n)$ *and* $\widehat{\nu^n}$*-a.e. Therefore,* $\|{}^\circ G_n - {}^\circ F_n\|_{\mathbb{H}^n} = 0$ *in* $L^2(\lambda^n)$ *if* F *or* G *are nearstandard.*

Proof By Proposition 13.1.1 and Corollary 13.1.4, we have

$$\begin{aligned} 0 &= \mathbb{E}\left(\sum_{n=0}^{\infty} {}^\circ I_n(F_n) - \sum_{n=0}^{\infty} {}^\circ I_n(G_n)\right)^2 = \mathbb{E}\left(\sum_{n=0}^{\infty} {}^\circ I_n(F_n - G_n)\right)^2 \\ &= \sum_{n=0}^{\infty} \mathbb{E}\left({}^\circ I_n(F_n - G_n)\right)^2 = \sum_{n=0}^{\infty} {}^\circ \mathbb{E}\left(I_n(F_n - G_n)\right)^2 \\ &= \sum_{n=0}^{\infty} {}^\circ \int_{T^n_<} \|F_n - G_n\|_{\mathbb{F}^n}{}^2 d\nu^n. \end{aligned}$$

By the symmetry of F_n and G_n, $\|G_n - F_n\|_{\mathbb{F}^n} \approx 0$ in $L^2(\nu^n)$ for each $n \in \mathbb{N}_0$, which proves the uniqueness. □

Proposition 17.2.2 shows that an admissible sequence $(\mathfrak{H}_n)_{n \in \mathbb{N}_0}$ is uniquely determined in the following sense.

Corollary 17.2.3 *Suppose that $(\mathfrak{H}_n)$ and $(\mathfrak{H}'_n)$ are admissible. Then for each $F \in \mathfrak{H}_n$ there exists an $F' \in \mathfrak{H}'_n$ with $\|F - F'\|_{\mathbb{F}^n} \approx 0$ in $L^2(\nu^n)$, thus $\widehat{\nu^n}$-a.e.*

Proof Fix $F \in \mathfrak{H}_n$. Since $^\circ I_n(F) \in L^2_{\mathcal{W}}(\widehat{\Gamma})$, by Theorem 17.2.1, $^\circ I_n(F)$ has a chaos decomposition $^\circ I_n(F) = \sum_{n=0}^{\infty} {}^\circ I_n(F'_n)$ with $F'_n \in \mathfrak{H}'_n$. By Proposition 17.2.2 $\|F - F'_n\|_{\mathbb{F}^n} \approx 0$ in $L^2(\nu^n)$ and therefore $\widehat{\nu^n}$-a.e. □

We end this section with the following remark.

Remark 17.2.4 Fix $\sigma \in \mathbb{N}$ and let $\mathcal{W}_\sigma$ be the σ-algebra on the set of continuous functions $f : [0,\sigma] \to \mathbb{B}$ generated by the set of $I(f)$, where f runs through $L^2(\lambda)$ with support in $[0,\sigma]$. We obtain the following chaos decomposition result. Each $\varphi \in L^2_{\mathcal{W}_\sigma}(\widehat{\Gamma})$ has the decomposition

$$\varphi = \sum_{n=0}^{\infty} {}^\circ I_n(F_n),$$

where now $F \in SL^2(\nu^n, \mathbb{F}^{\otimes n})$ with support in T^n_σ. It is left to the reader to modify the following investigation, using $\mathcal{W}_\sigma$ and Brownian functionals resticted to T^n_σ in the internal seeting and to $[0,\sigma]$ in the standard setting.

17.3 A lifting theorem for functionals in $L^2_{\mathcal{W}}(\widehat{\Gamma})$

We now apply the chaos decomposition theorem to prove that each $\varphi \in L^2_{\mathcal{W}}(\widehat{\Gamma})$ is infinitely close to an internally smooth function.

Theorem 17.3.1 *Suppose that $\varphi \in L^2_{\mathcal{W}}(\widehat{\Gamma})$ and $\varphi = \sum_{n=0}^{\infty} {}^\circ I_n(F_n)$ according to Theorem 17.2.1. Choose an internal extension $(F_n)_{n \in {}^*\mathbb{N}_0}$ of $(F_n)_{n \in \mathbb{N}_0}$ such that $F_n : T^n_{\neq} \to \mathbb{F}^{\otimes n}$ is symmetric for all $n \in {}^*\mathbb{N}$. Then there exists an unlimited number $K \in {}^*\mathbb{N}$ such that for each unlimited $M \in {}^*\mathbb{N}$, $M \leq K$,*

$$\Phi := \sum_{n=0}^{M} I_n(F_n) : \Omega \to {}^*\mathbb{R} \in SL^2(\Gamma) \quad \textit{and} \quad \Phi \approx \varphi\, \widehat{\Gamma}\text{-a.s.}$$

We may assume that all the $F_n : T^n_{\neq} \to \mathbb{F}^{\otimes n}$ are symmetric and that $\int_{T^n} \|F_n\|^2_{\mathbb{F}^n} d\nu^n \approx 0$ for all unlimited $n \leq M$.

Proof Use Corollary 10.9.3. Since $^\circ\mathbb{E}\left(\sum_{n=l+1}^{k} I_n(F_n)\right)^2 \to_{n,k\to\infty} 0$ and $\sum_{n=0}^{k} I_n(F_n) \in SL^2(\Gamma)$, there exists an unlimited $K \in {}^*\mathbb{N}$ such that, for all unlimited $M \in {}^*\mathbb{N}$ with $M \leq K$, $\sum_{n=0}^{M} I_n(F_n)$ is nearstandard and $\sum_{n=0}^{M} I_n(F_n) \in SL^2(\Gamma)$

and

$$\lim_{m\to\infty}\sum_{n=0}^{m} I_n({}^\circ F_n) = \lim_{m\to\infty}{}^\circ\sum_{n=0}^{m} I_n(F_n) = {}^\circ\sum_{n=0}^{M} I_n(F_n) \text{ in } L^2(\widehat{\Gamma}).$$

It follows that $\sum_{n=0}^{M} I_n(F_n) \approx \varphi\ \widehat{\Gamma}$-a.s. □

17.4 Chaos for functions without moments

It will be shown that each $\mathcal{W}$-measurable $\varphi : \Omega \to \mathbb{R}$ is infinitely close $\widehat{\Gamma}$-a.s. to a multilinear form $\sum_{n=0}^{M} I_n(F_n)$. We apply the fact that there exists a sequence $(\varphi_k)_{k\in\mathbb{N}}$ of functions $\varphi_k \in L^2_{\mathcal{W}}(\widehat{\Gamma})$, converging to φ in measure. Let $\Phi_k = \sum_{n=0}^{M} I_n(F_{k,n})$ be a lifting of φ_k according to Theorem 17.3.1. By Proposition 8.2.1 (f), we may assume that M works for all $k \in \mathbb{N}$. By Proposition 8.2.2, there exists an internal extension $\left((F_{k,n})_{n\le M}\right)_{k\in {}^*\mathbb{N}}$ of $\left((F_{k,n})_{n\le M}\right)_{k\in\mathbb{N}}$ with symmetric $F_{k,n} : T^n_{\neq} \to \mathbb{F}^{\otimes n}$ for all $n \le M, k \in {}^*\mathbb{N}$. Set $\Phi_k := \sum_{n=0}^{M} I_n(F_{k,n})$ for all $k \in {}^*\mathbb{N}$. Let Ψ be a $\mathcal{B}$-measurable lifting of φ. Then, for all $m \in \mathbb{N}$, $\lim_{k\to\infty} {}^\circ\mu\left\{|\Phi_k - \Psi| \ge \frac{1}{m}\right\} = 0$. By Propositions 9.5.1 (b) and 8.2.1 (f), there exists an unlimited K such that $\mu\left\{|\Phi_K - \Psi| \ge \frac{1}{m}\right\} < \frac{1}{m}$ for each $m \in \mathbb{N}$. Therefore, $\Phi := \Phi_K$ is a lifting of φ. This proves the following.

Theorem 17.4.1 *Let $\varphi : \Omega \to \mathbb{R}$ be $\mathcal{W}$-measurable. Then φ has a lifting Φ of the form $\Phi = \sum_{n=0}^{M} I_n(F_n)$ with internal and symmetric $F_n : T^n_{\neq} \to \mathbb{F}^{\otimes n}$.*

17.5 Computation of the kernels

Let us now provide a method for computing the kernels of the chaos decomposition. We shall see that the kernels F_n of a Wiener functional φ at the point $(t_1,\ldots,t_n)$ can be computed by computing the internal expected value of an appropriate lifting of φ multiplied by white noise $\dot{B}_{t_1}\ldots\dot{B}_{t_n}$ at $(t_1,\ldots,t_n)$. Our results are straightforward extensions of corresponding results due to Cutland and Ng [24].

Fix an internal mapping $\Phi : \Omega \to {}^*\mathbb{R} \in L^2(\Gamma)$. Then we call the function $\Phi_n : T^n_< \to \mathbb{F}^{\otimes n}$, defined by

$$\Phi_n(t_1,\ldots,t_n)(a_1,\ldots,a_n) := \mathbb{E}\left(\Phi\cdot\langle X_{t_1},a_1\rangle\cdot\ldots\cdot\langle X_{t_n},a_n\rangle\right)H^n, \tag{56}$$

the *nth* **white noise function** of Φ.

Equation (56) has the following intuitive meaning. Set $\Delta t := \frac{1}{H}$ and recall from Section 11.2 that X_t is the increment ΔB_t to time t of the internal Brownian motion B. Then equation (56) can be written in the form

$$\Phi_n(t_1,\ldots,t_n)(a_1,\ldots,a_n)$$
$$= \mathbb{E}\left(\Phi\cdot\left\langle\frac{\Delta B_{t_1}}{\Delta t_1},a_1\right\rangle\cdot\ldots\cdot\left\langle\frac{\Delta B_{t_n}}{\Delta t_n},a_n\right\rangle\right) = \mathbb{E}\left(\Phi\cdot\langle\dot{B}_{t_1},a_1\rangle\cdot\ldots\cdot\langle\dot{B}_{t_n},a_n\rangle\right),$$

where $\dot{B}_t$ may be understood as the 'derivative' of B at time t. Cutland and Ng [24] have pointed out that it was the intention of Wiener [119], to think of the kernels f_n of $\varphi = \sum_{n=0}^{\infty} I_n(f_n)$ as being given by

$$f_n(t_1,\ldots,t_n) = \mathbb{E}(\varphi\dot{b}_{t_1}\ldots\dot{b}_{t_n}), \tag{57}$$

although Brownian motion is not differentiable. By using the hyperfinite timeline T, we have a way (without using Schwartz' distributions), to give a correct mathematical meaning to equation (57). Wiener has used chaotic representations to find better models for telecommunications under noise. (I refer the reader to the marvellous book by Masani [76].)

Theorem 17.5.1 *Suppose that* $\varphi \in L^2_{\mathcal{W}}(\widehat{\Gamma})$. *Let* $\Phi \in SL^2(\Gamma)$ *be a lifting of* φ. *Then*

(a) $\Phi_n \in SL^2(\nu^n, \mathbb{F}^{\otimes n})$ *and* Φ_n *is nearstandard,*
(b) $\varphi = \mathbb{E}_{\widehat{\Gamma}}\varphi + \sum_{n=1}^{\infty} {}^\circ I_n(\Phi_n)$.

Proof By Theorem 17.3.1, $\varphi = \mathbb{E}_{\widehat{\Gamma}}\varphi + \sum_{n=1}^{\infty} {}^\circ I_n(F_n)$ has a lifting $\Xi \in SL^2(\Gamma)$ of the form

$$\Xi = \sum_{n=0}^{K} I_n(F_n) \quad \text{with } I_0(F_0) = \mathbb{E}\Xi,$$

where $F_n : T^n_{\neq} \to \mathbb{F}^{\otimes n} \in SL^2(\nu^n, \mathbb{F}^{\otimes n})$ is symmetric. Since

$$U := \left\{ \sum_{n=0}^{K} I_n(G_n) \mid G_n : T^n_< \to \mathbb{F}^{\otimes n} \text{ internal} \right\}$$

is an internally closed subspace of $L^2(\Gamma)$, there exists an internal $K+1$-tuple $(G_n)_{0\le n\le K}$ of internal $G_n : T^n_< \to \mathbb{F}^{\otimes n}$ and a function $\Psi \in L^2(\Gamma)$ with $\Psi \perp U$ such that

$$\Phi = \sum_{n=0}^{K} I_n(G_n) + \Psi.$$

We will now prove that $G_n = \Phi_n$ for all $n = 0,\ldots,K$. Fix an internal ONB $\mathfrak{E}$ of $\mathbb{F}$ and internal n-tuples $(\mathfrak{e}_1,..,\mathfrak{e}_n) \in \mathfrak{E}^n$ and $(t_1,\ldots,t_n) \in T^n_<$. Then, by Lemma 11.1.3, we obtain

$$\frac{1}{H^n}\Phi_n(t_1,\ldots,t_n)(\mathfrak{e}_1,\ldots,\mathfrak{e}_n) = \mathbb{E}\left(\Phi\cdot\langle X_{t_1},\mathfrak{e}_1\rangle\cdot\ldots\cdot\langle X_{t_n},\mathfrak{e}_n\rangle\right) = \alpha+\beta,$$

where

$$\begin{aligned}\alpha &= \mathbb{E}_\Gamma\left(\sum_{n=0}^{K} I_n(G_n)\cdot\langle X_{t_1},\mathfrak{e}_1\rangle\cdot\ldots\cdot\langle X_{t_n},\mathfrak{e}_n\rangle\right)\\ &= \mathbb{E}_\Gamma\left(G_n(t_1,\ldots,t_n)(\mathfrak{e}_1,\ldots,\mathfrak{e}_n)\cdot\langle X_{t_1},\mathfrak{e}_1\rangle^2\cdot\ldots\cdot\langle X_{t_n},\mathfrak{e}_n\rangle^2\right)\\ &= G_n(t_1,\ldots,t_n)(\mathfrak{e}_1,\ldots,\mathfrak{e}_n)\frac{1}{H^n},\end{aligned}$$

and

$$\beta = \mathbb{E}_\Gamma\left(\Psi\cdot\langle X_{t_1},\mathfrak{e}_1\rangle\cdot\ldots\cdot\langle X_{t_n},\mathfrak{e}_n\rangle\right) = 0.$$

This proves that $G_n = \Phi_n$. Now

$$\begin{aligned}0 &\approx \mathbb{E}_\Gamma(\Phi-\Xi)^2 = \mathbb{E}_\Gamma\left(\sum_{n=0}^{K} I_n(G_n)+\Psi-\sum_{n=0}^{K} I_n(F_n)\right)^2\\ &= \mathbb{E}_\Gamma\left(\sum_{n=0}^{K} I_n(\Phi_n)+\Psi-\sum_{n=0}^{K} I_n(F_n)\right)^2 = \mathbb{E}_\Gamma\sum_{n=0}^{K}(I_n(\Phi_n-F_n))^2\,d\Gamma+\mathbb{E}\Psi^2\\ &\geq \mathbb{E}_\Gamma(I_n(\Phi_n-F_n))^2\,d\Gamma = \int_{T^n_<}\|\Phi_n-F_n\|^2_{\mathbb{F}^n}\,d\nu^n.\end{aligned}$$

It follows that $F_n \approx_{\mathbb{F}^n} \Phi_n$ in $L^2(\nu^n)$. Since F_n belongs to $SL^2\left(\nu^n,\mathbb{F}^{\otimes n}\right)$ and is nearstandard, Φ_n is a nearstandard element of $SL^2\left(\nu^n,\mathbb{F}^{\otimes n}\right)$. This proves (a). Moreover, $\mathbb{E}_\Gamma(I_n(\Phi_n-F_n))^2 \approx 0$. Therefore, ${}^\circ I_n(F_n) = {}^\circ I_n(\Phi_n)$ in $L^2(\widehat{\Gamma})$. This proves (b). □

Example 17.5.2 Let $\varphi = e^{I(f)-\frac{1}{2}\int_T\|f\|^2_{\mathbb{H}}d\widehat{\nu}}$ with $f\in L^2_{\mathcal{L}^1}(\widehat{\nu},\mathbb{H})$. Fix a lifting $F\in SL^2(\nu,\mathbb{F})$ of f. Then $\Phi := e^{I(F)-\frac{1}{2}\int_T\|F\|^2_{\mathbb{F}}d\nu}\in SL^2(\Gamma)$ is a lifting of φ. Let $\mathfrak{E}$ be an ONB of $\mathbb{F}$. Transfer of the substitution rule tells us that, for all $m\in\mathbb{N}$, for all $t\in T^m_<$ and all $\mathfrak{e}\in\mathfrak{E}^m$

$$\mathbb{E}_\Gamma\left(\Phi\cdot\langle X_{t_1},\mathfrak{e}_1\rangle\cdot\ldots\cdot\langle X_{t_m},\mathfrak{e}_m\rangle\right)H^m = F^{\otimes m}(t)(\mathfrak{e}),$$

thus

$$e^{I(f)-\frac{1}{2}\int_T\|f\|^2_{\mathbb{H}}d\widehat{\nu}} = \sum_{n=0}^{\infty} I_n(f^{\otimes n}) \quad \text{with } I_0(f^{\otimes 0}) = 1.$$

In the following section the reader can find a second example. Later we will prove the product and chain rule in quite a constructive way, using the method for the computation of the kernels.

17.6 The kernels of the product of Wiener functionals

Our aim is to compute the kernels of the product $\varphi \cdot \psi$ with $\varphi, \psi \in L^2_W(\widehat{\Gamma})$. Since $\varphi \cdot \psi$ is not square-integrable, in general, we assume that φ, ψ belong to finite chaos levels. By Theorem 17.3.1 φ, ψ have liftings Φ, Ψ of the form

$$\Phi = \sum_{n \in M \cup \{0\}} I_n(F_n), \quad \Psi = \sum_{n \in M \cup \{0\}} I_n(G_n) \quad \text{with } M \in \mathbb{N}, \tag{58}$$

where the $F_n, G_n : T^n_{\neq} \to \mathbb{F}^{\otimes n}$ are symmetric and belong to $SL^2\left(\nu^n, \mathbb{F}^{\otimes n}\right)$. By Corollary 13.1.4, $\Phi \cdot \Psi \in SL^2(\Gamma)$, thus $\varphi \cdot \psi \in L^2_W(\widehat{\Gamma})$. By Theorem 17.5.1, the kernels of $\varphi \cdot \psi$ are the standard parts of internal functions $K_m : T^m_< \to \mathbb{F}^{\otimes m}$, which can be computed by the method

$$K_m(r_1, \ldots, r_m)(a_1, \ldots, a_m) = H^m \cdot \mathbb{E}_\Gamma\left(\Phi \cdot \Psi \cdot \langle X_{r_1}, a_1 \rangle \cdot \ldots \cdot \langle X_{r_m}, a_m \rangle\right).$$

We may assume that the $a_1, \ldots, a_m$ are elements of an ONB of $\mathbb{F}$. To gain control of all these combinations of products, we introduce strictly monotone increasing functions σ, defined on subsets of $m = \{1, \ldots, m\}$, by identification of $m \in \mathbb{N}_0$ with the set $\{1, \ldots, m\}$. It follows that $0 = \emptyset$.

Let $m \in \mathbb{N}_0$ with $k \leq m$. If σ is a strictly monotone increasing function from k into m, then we will write $\sigma : k \uparrow m$. For $\sigma : k \uparrow m$, let $\overline{\sigma} : m - k \uparrow m \setminus \text{range}(\sigma)$. For example, if $k = 0$, then $\sigma : k \uparrow m = \emptyset$ and $\overline{\sigma} : m \uparrow m = \text{id}_m$. Since the F_i, G_i are symmetric, we obtain from Theorem 17.5.1, using somehow technical but elementary finite combinatorics, the following.

Theorem 17.6.1 *Fix an ONB $\mathfrak{E}$ of $\mathbb{F}$. Assume that $\varphi, \psi \in L^2_W(\widehat{\Gamma})$ belong to finite chaos levels and let Φ, Ψ be liftings of φ, ψ, according to equation (58). For all $(r_1, \ldots, r_m) \in T^m_<$ and $(a_1, \ldots, a_m) \in \mathfrak{E}^m$ set*

$$\begin{aligned} K_m(r_1, \ldots r_m, a_1, \ldots, a_m) = & \sum_{k=0}^{m} \sum_{\sigma : k \uparrow m} \sum_{n \in {}^*\mathbb{N}_0} \sum_{t \in T^n_<} \sum_{\mathfrak{b} \in \mathfrak{E}^n} \frac{1}{H^n} \\ & \cdot F_{n+k}(t, r_{\sigma_1}, \ldots, r_{\sigma_k}, \mathfrak{b}, a_{\sigma_1}, \ldots, a_{\sigma_k}) \\ & \cdot G_{n+m-k}(t, r_{\overline{\sigma}_1}, \ldots, r_{\overline{\sigma}_{m-k}}, \mathfrak{b}, a_{\overline{\sigma}_1}, \ldots, a_{\overline{\sigma}_{m-k}}). \end{aligned}$$

Then K_m can be extended to a uniquely determined mapping from $T^m_<$ into $\mathbb{F}^{\otimes m}$, also denoted by K_m, and we have

$$\varphi \cdot \psi = \sum_{m \in {}^*\mathbb{N}_0} {}^\circ I_m(K_m) = \sum_{m \in \mathbb{N}_0} I_m({}^\circ K_m).$$

Recall that these series reduce to finite sums.

Exercises

17.1 Prove that all the F_n in Theorem 17.3.1 belong to $SL^2\left(\nu^n, \mathbb{F}^{\otimes n}\right)$.

17.2 Prove the assertion in Example 17.5.2.

17.3 Recall that H_n denotes the nth Hermite polynomial. Prove

$$\begin{aligned}&\operatorname{span}\left\{H_n(I(f)) \mid n \in \mathbb{N}_0 \text{ and } f \in L^2_{\mathcal{L}^1}(\widehat{\nu}, \mathbb{H}) \text{ and } \|f\|_{\widehat{\nu}} = 1\right\}\\ &\quad = \operatorname{span}\left\{I_n(f^{\otimes n}) \mid n \in \mathbb{N}_0 \text{ and } f \in L^2_{\mathcal{L}^1}(\widehat{\nu}, \mathbb{H}) \text{ and } \|f\|_{\widehat{\nu}} = 1\right\}\end{aligned}$$

is a dense subspace of $L^2_{\mathcal{W}}(\widehat{\Gamma})$.

17.4 Prove that

$$\operatorname{span}\left\{e^{I(f) - \frac{1}{2}\|f\|^2_{\widehat{\nu}}} \mid f \in L^2_{\mathcal{L}^1}(\widehat{\nu}, \mathbb{H})\right\} = \operatorname{span}\left\{e^{I(f)} \mid f \in L^2_{\mathcal{L}^1}(\widehat{\nu}, \mathbb{H})\right\}$$

is a dense subspace of $L^2_{\mathcal{W}}(\widehat{\Gamma})$.

Now we provide a method for the computation of the white noise functions of $\mathbb{H}$-valued processes. In Section 19.1 we will show that each $\varphi \in L^2_{\mathcal{W}\otimes\mathcal{L}^1}(\widehat{\Gamma \otimes \nu}, \mathbb{H})$ has the decomposition

$$\varphi = \sum_{n=0}^{\infty} {}^\circ I_{n,1}(F_n),$$

where $F_n \in SL^2(\nu^{n+1}, \mathbb{F}^{\otimes(n+1)})$ is nearstandard and symmetric in the first n variables. Fix $\varphi \in L^2_{\mathcal{W}\otimes\mathcal{L}^1}(\widehat{\Gamma \otimes \nu}, \mathbb{H})$. Let $\Phi \in SL^2(\Gamma \otimes \nu, \mathbb{F})$ be a lifting of φ. Define $\Phi_n : T^n_< \times T \to \mathbb{F}^{\otimes(n+1)}$ by setting

$$\Phi_n(t_1, \ldots, t_n, s)\,(a_1, \ldots, a_n, c) = \mathbb{E}\left(\langle \Phi(\cdot, s), c\rangle \cdot \langle X_{t_1}, a_1\rangle \cdot \ldots \cdot \langle X_{t_n}, a_n\rangle\right) H^n.$$

17.5 Prove that $\Phi_n \in SL^2(\nu^{n+1}, \mathbb{F}^{\otimes n+1})$ and

$$\varphi = \sum_{n=0}^{\infty} {}^\circ I_{n,1}(\Phi_n) \text{ in } \mathcal{L}^2_{\mathcal{W}\otimes\mathcal{L}^1}(\widehat{\Gamma \otimes \nu}, \mathbb{H}).$$

Since each $\varphi \in L^2_{\mathcal{W}}(\widehat{\Gamma})$ has the decomposition $\varphi = \sum_{n=0}^{\infty} I_n(f_n)$ according to Theorem 17.2.1, $\mathrm{Fin}L^2_{\mathcal{W}}(\widehat{\Gamma}) = \bigcup_{m\in\mathbb{N}} L^2_{m,\mathcal{W}}(\widehat{\Gamma})$ is a dense subspace of $L^2_{\mathcal{W}}(\widehat{\Gamma})$. Fix an ONB $\mathfrak{B}_{\mathcal{L},\mathbb{H}}$ of $L^2_{\mathcal{L}^1}(\widehat{\nu},\mathbb{H})$. Set

$$\begin{aligned}\mathbb{D} &:= \operatorname{span}\left\{H_n(I(f)) \mid n\in\mathbb{N}_0, f\in\operatorname{span}\mathfrak{B}_{\mathcal{L},\mathbb{H}} \text{ and } \|f\|_{\widehat{\nu}} = 1\right\}\\ &= \operatorname{span}\left\{I_n(f^{\otimes n})) \mid n\in\mathbb{N}_0, f\in\operatorname{span}\mathfrak{B}_{\mathcal{L},\mathbb{H}} \text{ and } \|f\|_{\widehat{\nu}} = 1\right\}.\end{aligned}$$

17.6 Prove that $\mathbb{D}$ is a dense subspace of $L^2_{\mathcal{W}}(\widehat{\Gamma})$.

18

The Malliavin derivative

Since our sample space Ω is an internal finite-dimensional Euclidean space, the Malliavin derivative is infinitely close to the elementary derivative in finite-dimensional Euclidean spaces. Here are the details:

18.1 The domain of the derivative

In order to define the Malliavin derivative, we look at the derivative of the function $I_n(F)$ which is defined on a *finite-dimensional space Ω. In analogy to standard analysis, an internal function G, defined on Ω with values in $^*\mathbb{R}$ or $\mathbb{F}$ is called **differentiable at** $(X,t) \in \Omega \times T$ if there exists an internal linear function $L_{X,t} : \mathbb{F} \to {}^*\mathbb{R}$ or $L_{X,t} : \mathbb{F} \to \mathbb{F}$, respectively, such that

$$\lim_{h \to 0} \frac{G(X_1,\ldots,X_{t-1},X_t+h,X_{t+1},\ldots,X_H) - G(X) - L_{X,t}(h)}{\|h\|} = 0.$$

If G is differentiable at (X,t) for each $(X,t) \in \Omega \times T$ then $DG : (X,t) \mapsto L_{X,t}$ is called the **derivative of** G **at** (X,t). We see that

$$DI_n(F) = I_{n-1,1}(F) : \Omega \times T \to \mathbb{F}\,(=\mathbb{F}') \text{ for symmetric } F : T^n_{\neq} \to \mathbb{F}^{\otimes n}.$$

According to Theorem 17.2.1, fix $\varphi \in L^2_{\mathcal{W}}(\widehat{\Gamma})$ with chaos decomposition

$$\varphi = \mathbb{E}\varphi + \sum_{n=0}^{\infty} {}^\circ I_n(F_n) = \mathbb{E}\varphi + \sum_{n=0}^{\infty} I_n({}^\circ F_n).$$

By Theorem 13.1.6 and Corollary 13.4.4, we may assume that $I_{n-1,1}(F_n) \in SL^2(\Gamma \otimes \nu, \mathbb{F})$ and ${}^\circ I_{n-1,1}(F_n) \in L^2_{\mathcal{W} \otimes \mathcal{L}^1}(\widehat{\Gamma \otimes \nu}, \mathbb{H})$. Now the **Malliavin derivative** D of φ is nothing but finite-dimensional differentiation under the standard part map and the sum, i.e., D is defined on a dense subspace of $L^2_{\mathcal{W}}(\widehat{\Gamma})$ by setting

$$D(\varphi) := \sum_{n=0}^{\infty} {}^{\circ}I_{n-1,1}(F_n)$$

for those $\varphi \in L^2_{\mathcal{W}}(\widehat{\Gamma})$ such that $\sum_{n=0}^{\infty} {}^{\circ}I_{n-1,1}(F_n)$ converges in the Hilbert space $L^2_{\mathcal{W}\otimes\mathcal{L}^1}(\Gamma \otimes \nu, \mathbb{H})$. If $D(\varphi)$ converges, φ is called **Malliavin differentiable**. Note that we may assume that F_n is symmetric and $F_n(t_1,\ldots,t_n) \neq 0$ implies $(t_1,\ldots t_n) \in T^n_{\neq}$.

In order to characterize the domain of the Malliavin operator, we need the following lemma.

Lemma 18.1.1 *Suppose that $F = F_n$. Then*

$$\int_{\Omega\times T} \|{}^{\circ}I_{n-1,1}(F)\|^2_{\mathbb{H}} \, d\widehat{\Gamma\otimes\nu} = n \cdot \int_{\Omega} ({}^{\circ}I_n(F))^2 \, d\widehat{\Gamma}.$$

Proof Using the fact that $I_{n-1,1}(F)$ belongs to $SL^2(\Gamma \otimes \nu, \mathbb{F})$ and $I_n(F)$ to $SL^2(\Gamma)$, we obtain from Corollary 9.8.2 and the symmetry of F

$$\begin{aligned}\int_{\Omega\times T} \|{}^{\circ}I_{n-1,1}(F)\|^2_{\mathbb{H}} \, d\widehat{\Gamma\otimes\nu} &= \|{}^{\circ}I_{n-1,1}(F)\|^2_{\widehat{\Gamma\otimes\nu}} = {}^{\circ}\|I_{n-1,1}(F)\|^2_{\Gamma\otimes\nu} \\ &= {}^{\circ}\int_{T^{n-1}_{<}\times T} \|F\|^2_{\mathbb{F}^n} \, d\nu^n = n \cdot {}^{\circ}\int_{T^n_{<}} \|F\|^2_{\mathbb{F}^n} \, d\nu^n \\ &= n \cdot {}^{\circ}\int_{\Omega} (I_n(F))^2 d\Gamma = n \cdot \int_{\Omega} {}^{\circ}(I_n(F))^2 d\widehat{\Gamma}.\end{aligned}$$

□

Now thanks to the orthogonality of ${}^{\circ}I_{n-1,1}(F)$ and ${}^{\circ}I_{m-1,1}(G)$ for $m \neq n$ (see Proposition 13.1.7), we obtain:

Proposition 18.1.2 *A function $\varphi \in L^2_{\mathcal{W}}(\widehat{\Gamma})$,is Malliavin differentiable if and only if $\sum_{n=1}^{\infty} \sqrt{n} \cdot {}^{\circ}I_n(F_n)$ converges in $L^2_{\mathcal{W}}(\widehat{\Gamma})$.*

Therefore, D is densely defined. Since the spaces $L^2_{\mathcal{W}\otimes\mathcal{L}^1}(\widehat{\Gamma\otimes\nu}, \mathbb{H})$, $L^2_{\mathcal{W}\otimes\mathrm{Leb}[0,\infty[}(\widehat{\Gamma}\otimes\lambda, \mathbb{H})$ and $L^2(W \otimes \lambda, \mathbb{H})$ can be identified, and also $L^2(W)$ and $L^2_{\mathcal{W}}(\widehat{\Gamma})$, the Malliavin derivative is now also densely defined for functionals in $L^2(W)$ and takes its values in $L^2(W \otimes \lambda, \mathbb{H})$.

It is common practice to denote the **domain of the Malliavin derivative** by $\mathbb{D}^{1,2}$.

18.2 The Clark–Ocone formula

The Clark–Ocone formula is a martingale representation of a large class of random variables: Each Malliavin differentiable function φ is the Itô integral of the conditional expectation of its derivative. More generally, if φ is not Malliavin differentiable, then φ can be written as a stochastic integral plus a constant. This formula has been proved by Clark [22] and more generally by Ocone [81] for the classical Wiener space $C_{\mathbb{R}}$. In this case, a simple proof, using saturation, can be found in the work of Cutland and Ng [24]. Berger [8] proved the Clark–Ocone formula for the abstract Wiener space using the Üstünel–Zakai–Itô integral, based on a resolution of the identity on $\mathbb{H}$. In [85] there is a proof of this formula for the space $C_{\mathbb{B}}$. In analogy to Cutland and Ng's approach to the Clark–Ocone formula for the classical Wiener space, we will now prove this formula for our general setting in a simple way.

Indeed, in the internal setting the Clark–Ocone formula is obvious, because for all internal functions $F: T^n_{\neq} \to \mathbb{F}^{\otimes n}$

$$I_n(F) = \left(\int^V t \mapsto \mathbb{E}^{\mathcal{B}_{t^-}} I_{n-1,1}(F)(\cdot,t) \Delta B \right),$$

where $\int^V$ was defined in Section 12.2. Note that $\mathbb{E}^{\mathcal{B}_{t^-}} I_{n-1,1}(F)(\cdot,t) = I_{n-1,1}(\mathbf{1}_{T^n_<} \cdot F)(\cdot,t)$. Our aim now is to convert this internal equality to a result in the standard setting.

Theorem 18.2.1 [85] *Fix $\varphi = \mathbb{E}_{\widehat{\Gamma}}\varphi + \sum_{n=1}^{\infty} {}^{\circ}I_n(F_n) \in L^2_{\mathcal{W}}(\widehat{\Gamma})$, according to Theorem 17.2.1, with $\mathbf{1}_{T^n_<} \cdot F_n \in \mathfrak{G}_n$, where $(\mathfrak{G}_n)_{n\in\mathbb{N}_0}$ is the admissible sequence in Example 17.1.2 (ii). Then*

(a)

$$\varphi = \mathbb{E}_{\widehat{\Gamma}}\varphi + \int^V \left(r \mapsto \mathbb{E}^{\mathfrak{b}_r} \sum_{n=1}^{\infty} {}^{\circ}I_{n-1,1}(F_n)(\cdot,r) \right) db_{\mathbb{B}},$$

(b) *if φ is Malliavin differentiable, then*

$$\varphi = \mathbb{E}_{\widehat{\Gamma}}\varphi + \int^V \mathbb{E}^{\mathfrak{b}_r} D\varphi(\cdot,r) db_{\mathbb{B}}.$$

Since $\mathbb{E}^{\mathfrak{b}_r}\left({}^{\circ}I_{n-1,1}(F_n)\right)(\cdot,r)$ is $\mathcal{W}$-measurable, we may replace $\mathfrak{b}_r$ by $\mathfrak{b}_r \cap \mathcal{W}$.

Proof Using Theorem 13.1.6, we obtain, in $L^2(\widehat{\Gamma})$,

$$\begin{aligned}\sum_{n=1}^{\infty} {}^\circ I_n(F_n) &= \sum_{n=1}^{\infty} {}^\circ\left(\int^V t \mapsto I_{n-1,1}(\mathbf{1}_{T^n_<} \cdot F_n)(\cdot,t)\Delta B\right) \\ &= \sum_{n=1}^{\infty} \int^V \left(r \mapsto \mathbb{E}^{\mathfrak{b}_r}\, {}^\circ I_{n-1,1}(F_n)(\cdot,r)\right) db_{\mathbb{B}} \\ &= \int^V \left(r \mapsto \mathbb{E}^{\mathfrak{b}_r} \sum_{n=1}^{\infty} {}^\circ I_{n-1,1}(F_n)(\cdot,r)\right) db_{\mathbb{B}}.\end{aligned}$$

If φ is Malliavin differentiable, then

$$\int^V \left(r \mapsto \mathbb{E}^{\mathfrak{b}_r} \sum_{n=1}^{\infty} {}^\circ I_{n-1,1}(F_n)(\cdot,r)\right) db_{\mathbb{B}} = \int^V \left(r \mapsto \mathbb{E}^{\mathfrak{b}_r} D\varphi(\cdot,r)\right) db_{\mathbb{B}}.$$

□

We see that $r \mapsto \mathbb{E}^{\mathfrak{b}_r} \sum_{n=1}^{\infty} {}^\circ I_{n-1,1}(F_n)(\cdot,r)$ exists in $L^2\left(\widehat{\Gamma} \otimes \lambda, \mathbb{H}\right)$, although $r \mapsto \sum_{n=1}^{\infty} {}^\circ I_{n-1,1}(F_n)(\cdot,r)$ need not be $\widehat{\Gamma} \otimes \lambda$-square-integrable.

18.3 A lifting theorem for the derivative

Using saturation, we obtain the following lifting theorem for the Malliavin derivative. The proof is similar to the proof of Theorem 17.3.1.

Theorem 18.3.1 *Suppose that $\varphi = \sum_{n=0}^{\infty} {}^\circ I_n(F_n) \in \mathbb{D}^{1,2}$. Choose an internal extension $(F_n)_{n \in {}^*\mathbb{N}_0}$ of $(F_n)_{n \in \mathbb{N}_0}$. Then there exists an unlimited number $K \in {}^*\mathbb{N}$ such that for each unlimited $M \in {}^*\mathbb{N}, M \leq K$,*

$$\Phi := \sum_{n=0}^{M} I_n(F_n) \in SL^2(\Gamma) \quad and \quad \Phi \approx \varphi\, \widehat{\Gamma}\text{-a.s.},$$

$$D\Phi : (X,t) \mapsto \sum_{n=1}^{M} I_{n-1,1}(F_n)(X,t) \in SL^2(\Gamma \otimes \nu, \mathbb{F}),$$

and

$$D\Phi \approx_{\mathbb{F}} D\varphi\ \widehat{\Gamma \otimes \nu}\text{-a.e.}$$

*We may assume that all the **kernels** F_n of Φ are symmetric and F_n is defined on $T^n_{\neq}$.*

Here is an example in which we compare the usual derivative with the Malliavin derivative.

Example 18.3.2 Recall the notation and results in Example 17.5.2. Then $\Phi := e^{\sum_{t\in T,\mathfrak{e}\in\mathfrak{E}} F_t(\mathfrak{e})\cdot\langle X_t,\mathfrak{e}\rangle - \frac{1}{2}\int_T \|F\|_{\mathbb{F}}^2 d\widehat{\nu}}$ is a lifting of $\varphi := e^{I(f)-\frac{1}{2}\int_T \|f\|_{\mathbb{H}}^2 d\widehat{\nu}}$. We obtain for the usual derivative

$$\frac{d(\Phi)}{d\,\langle X_s,\mathfrak{e}\rangle} = e^{\sum_{t\in T,\mathfrak{e}\in\mathfrak{E}} F_t(\mathfrak{e})\cdot\langle X_t,\mathfrak{e}\rangle - \frac{1}{2}\int_T \|f\|_{\mathbb{H}}^2 d\widehat{\nu}} \cdot F_s(\mathfrak{e}) = \Phi(X)\cdot F_s(\mathfrak{e}).$$

By the chaos decomposition theorem, φ and $D\varphi$ have liftings of the form

$$\Psi : X \mapsto \sum_{n\leq M}\sum_{t\in T^n_<} F_{t_1}(X_{t_1})\cdot\ldots\cdot F_{t_n}(X_{t_n}),$$

$$\begin{aligned} D\Psi : (X,s,\mathfrak{e}) &\mapsto \sum_{n\leq M}\sum_{t\in(T\setminus\{s\})^n_<} F_{t_1}(X_{t_1})\cdot\ldots\cdot F_{t_n}(X_{t_n})\cdot F_s(\mathfrak{e}) \\ &\approx \Psi(X)\cdot F_s(\mathfrak{e}). \end{aligned}$$

It follows that $\frac{d(\Phi)}{d\langle X_s,\mathfrak{e}\rangle} \approx D\Psi(X,s,\mathfrak{e})$. Thus, the standard parts of both derivatives are the same.

18.4 The directional derivative

A natural question is: What is the derivative into a direction g? In analogy to finite-dimensional analysis, a **direction** g is a function $g \in L^2(\lambda,\mathbb{H})$ with $\int_0^\infty \|g\|_{\mathbb{H}}^2 d\lambda = 1$. Let us study first the internal *finite-dimensional case, which does not differ from the standard finite-dimensional situation. Let $F \in SL^2_{\neq}(\nu^n,\mathbb{F}^{\otimes n})$ be symmetric and $G \in SL^2(\nu,\mathbb{F})$ be a lifting of $g_T : t \mapsto g({}^\circ t)$. We may identify g and g_T. By the transfer principle from standard finite-dimensional analysis, the derivative of $I_n(F)$ at $X \in \Omega$ into the direction G is the scalar product

$$\sum_{t\in T}\langle DI_n(F)(X,t), G(t)\rangle_{\mathbb{F}} = \sum_{t\in T} I_{n-1,1}(X,t)(G(t)).$$

In general, the standard part of this sum does not exist, because it may belong to the unlimited part of ${}^*\mathbb{R}$. Therefore, we multiply this sum by the small normalization factor $\frac{1}{H}$, the internal Lebesgue unit volume. We shall now see that the standard part of this 'normalized' sum exists and becomes a standard integral.

Fix a Malliavin differentiable $\varphi = \mathbb{E}\varphi + \sum_{n=1}^\infty I_n(F_n) \in L^2_{\mathcal{W}}(\widehat{\Gamma})$, according to Theorem 17.2.1. Then $D\varphi = \sum_{n=1}^\infty {}^\circ I_{n-1,1}(F_n) \in L^2_{\mathcal{W}\otimes\mathcal{L}_1}(\widehat{\Gamma\otimes\nu},\mathbb{H})$.

The **derivative of φ into the direction** g, denoted by $D\varphi_{\to g}$, is defined by setting

$$D\varphi_{\to g} := \sum_{n=1}^{\infty} {}^\circ \sum_{t\in T} I_{n-1,1}(F_n)(\cdot,t)(G(t))\frac{1}{H} = \sum_{n=1}^{\infty} {}^\circ \langle I_{n-1,1}(F_n), G\rangle_\nu$$

if this series converges in $L^2_{\mathcal{W}}(\widehat{\Gamma})$. By Theorem 13.2.1, we have in $L^2(\Gamma)$:

$$\langle I_{n-1,1}(F_n), G\rangle_\nu \approx I_n(F_n)\cdot I_1(G) - I_{n+1}\left(\widetilde{F_n \otimes G}\right).$$

By Corollary 13.1.4, $I_n(F_n)\cdot I_1(G) \in SL^2(\Gamma)$ and, by an application of Lemma 10.12.2 (b), $I_{n+1}\left(\widetilde{F_n \otimes G}\right) \in SL^2(\Gamma)$. Therefore, $\langle I_{n-1,1}(F_n), G\rangle_\nu \in SL^2(\Gamma)$. Since ${}^\circ I_n(F_n)\cdot {}^\circ I_1(G)$ and ${}^\circ I_{n+1}\left(\widetilde{F_n \otimes G}\right)$ are $\mathcal{W}$-measurable,

$$D({}^\circ I_n(F_n))_{\to g} = \langle {}^\circ I_{n-1,1}(F_n), g\rangle_{\widehat{\nu}} \in L^2_{\mathcal{W}}\left(\widehat{\Gamma}\right).$$

Therefore, in order to prove that $D\varphi_{\to g}$ exists in $L^2_{\mathcal{W}}\left(\widehat{\Gamma}\right)$, we have to prove the following lemma.

Lemma 18.4.1 *$D\varphi_{\to g}$ converges in $L^2(\widehat{\Gamma})$.*

Proof Note that $\langle {}^\circ I_{n-1,1}(F_n), g\rangle_{\widehat{\nu}} \perp \langle {}^\circ I_{m-1,1}(F_m), g\rangle_{\widehat{\nu}}$ if $m \neq n$, by Proposition 13.1.7. To see this, fix an ONB $\mathfrak{E}$ of $\mathbb{F}$. Then,

$$\begin{aligned}
&\mathbb{E}_{\widehat{\Gamma}}\left(\langle {}^\circ I_{n-1,1}(F_n), g\rangle_{\widehat{\nu}}\cdot \langle {}^\circ I_{m-1,1}(F_m), g\rangle_{\widehat{\nu}}\right)\\
&\approx \mathbb{E}_{\Gamma}\left(\langle I_{n-1,1}(F_n), G\rangle_\nu \cdot \langle I_{m-1,1}(F_n), G\rangle_\nu\right)\\
&= \mathbb{E}_{\Gamma}\int_{T^2} \langle I_{n-1}(F_n(\cdot,t)), G_t\rangle_{\mathbb{F}}\cdot \langle I_{m-1}(F_m(\cdot,s)), G_s\rangle_{\mathbb{F}}\, d\nu^2(t,s)\\
&= \sum_{s,t\in T, \mathfrak{e}\in\mathfrak{E}^2} \mathbb{E}_{\Gamma}\left(I_{n-1}(F_n(\cdot,t)(\mathfrak{e}_1))\cdot I_{m-1}(F_m(\cdot,s)(\mathfrak{e}_2))\right) G_t(\mathfrak{e}_1)G_s(\mathfrak{e}_2)\frac{1}{H^2} = 0.
\end{aligned}$$

We obtain:

$$\begin{aligned}
\mathbb{E}_{\widehat{\Gamma}}\left(D\varphi_{\to g}\right)^2 &= \sum_{n=1}^{\infty}\mathbb{E}_{\widehat{\Gamma}}\left(\langle {}^\circ I_{n-1,1}(F_n), g\rangle_{\widehat{\nu}}\right)^2 \leq \sum_{n=1}^{\infty}\mathbb{E}_{\widehat{\Gamma}}\left\| {}^\circ I_{n-1,1}(F_n)\right\|^2_{\widehat{\nu}}\cdot \|g\|^2_{\widehat{\nu}}\\
&= \sum_{n=1}^{\infty}\left\| {}^\circ I_{n-1,1}(F_n)\right\|^2_{\widehat{\Gamma\otimes\nu}}\cdot \int_T \|g\|^2 d\widehat{\nu} < \infty,
\end{aligned}$$

because φ is Malliavin differentiable. Therefore, $D\varphi_{\to g}$ converges in $L^2_{\mathcal{W}}(\widehat{\Gamma})$.

It follows that for all Malliavin differentiable $\varphi \in \mathbb{D}^{1,2}$ and all $g \in L^2_{\mathcal{L}^1}(\widehat{\nu})$

$$D\varphi_{\to g} = \langle D\varphi, g \rangle_{\widehat{\nu}} \in L^2_{\mathcal{W}}\left(\widehat{\Gamma}\right).$$

□

18.5 A commutation rule for derivative and limit

The next result also reminds us of a result in elementary analysis, which says that we can interchange derivative and limit if the sequence of derivatives converges uniformly and the original sequence converges in at least one point. We will give the proof in full detail.

Theorem 18.5.1

(a) *Suppose that* (φ_k) *is a sequence in* $\mathbb{D}^{1,2}$ *such that* $(D\varphi_k)$ *converges in* $L^2_{\mathcal{W}\otimes\mathcal{L}^1}(\widehat{\Gamma\otimes\nu}, \mathbb{H})$ *and suppose that* $(\mathbb{E}\varphi_k)$ *converges in the real numbers. Then* (φ_k) *converges in* $\mathbb{D}^{1,2}$ *and*

$$D(\lim_{k\to\infty} \varphi_k) = \lim_{k\to\infty} D\varphi_k \text{ in } L^2_{\mathcal{W}\otimes\mathcal{L}^1}(\widehat{\Gamma\otimes\nu}, \mathbb{H}).$$

(b) *Fix* $m \in \mathbb{N}_0$ *and a sequence* (φ_k) *in* $L^2_{\mathcal{W},m}(\widehat{\Gamma})$ *converging to* φ *in* $L^2_{\mathcal{W}}(\widehat{\Gamma})$. *Then* $\varphi \in L^2_{\mathcal{W},m}(\widehat{\Gamma}) \subseteq \mathbb{D}^{1,2}$ *and*

$$\lim_{k\to\infty} D\varphi_k = D\varphi.$$

Proof (a) According to Theorem 17.2.1, φ_k has a chaos decomposition $\varphi_k = \sum_{n=0}^{\infty} I_n(f_{n,k})$. Then

$$\begin{aligned}
0 &= \lim_{k,l\to\infty} \|D\varphi_k - D\varphi_l\|^2_{\widehat{\Gamma\otimes\nu}} \\
&= \lim_{k,l\to\infty} \sum_{n=1}^{\infty} \left\| I_{n-1,1}(f_{n,k} - f_{n,l}) \right\|^2_{\widehat{\Gamma\otimes\nu}} \\
&= \lim_{k,l\to\infty} \sum_{n=1}^{\infty} \int_{T^{n-1}\times T} \left\| f_{n,k} - f_{n,l}) \right\|^2_{\mathbb{H}^n} d\widehat{\nu^n}.
\end{aligned}$$

Since all $f_{n,k}$ are symmetric, $\left(f_{n,k}\right)_{k\in\mathbb{N}}$ is a Cauchy sequence in $L^2_{\mathcal{L}^n}\left(\widehat{\nu^n}, \mathbb{H}^{\otimes n}\right)$ for all $n \in \mathbb{N}$. Let $\lim_{k\to\infty} f_{n,k} = f_n$ in $L^2_{\mathcal{L}^n}\left(\widehat{\nu^n}, \mathbb{H}^{\otimes n}\right)$. It follows that

$$\lim_{k\to\infty} I_{n-1,1}(f_{n,k}) = I_{n-1,1}(f_n) \text{ in } L^2_{\mathcal{W}\otimes\mathcal{L}^1}\left(\widehat{\Gamma\otimes\nu}, \mathbb{H}\right)$$

and
$$\lim_{k\to\infty} I_n(f_{n,k}) = I_n(f_n) \text{ in } L^2_{\mathcal{W}}\left(\widehat{\Gamma}\right).$$

By Proposition 5.6.5, $\sum_{n=1}^{\infty} I_n(f_n), \sum_{n=1}^{\infty} I_{n-1,1}(f_n)$ converge in $L^2_{\mathcal{W}}\left(\widehat{\Gamma}\right)$, $L^2_{\mathcal{W}\otimes\mathcal{L}^1}\left(\widehat{\Gamma\otimes\nu},\mathbb{H}\right)$, respectively, and

$$\lim_{k\to\infty}\left(\mathbb{E}\varphi_k + \sum_{n=1}^{\infty} I_n(f_{n,k})\right) = \left(\lim_{k\to\infty}\mathbb{E}\varphi_k\right) + \sum_{n=1}^{\infty} I_n(f_n) =: \varphi$$

in $L^2_{\mathcal{W}}\left(\widehat{\Gamma}\right)$, and

$$\lim_{k\to\infty}\sum_{n=1}^{\infty} I_{n-1,1}(f_{n,k}) = \sum_{n=1}^{\infty} I_{n-1,1}(f_n) = D\varphi$$

in $L^2_{\mathcal{W}\otimes\mathcal{L}^1}\left(\widehat{\Gamma\otimes\nu},\mathbb{H}\right)$

(b) Obviously, $\varphi \in L^2_{\mathcal{W},m}(\widehat{\Gamma})$. If $\varphi = \sum_{n=0}^{m}\varphi_n$ and $\varphi_k = \sum_{n=0}^{m}\varphi_{n,k}$, then, by Lemma 18.1.1,

$$\begin{aligned}\|D\varphi - D\varphi_k\|^2_{\widehat{\Gamma\otimes\nu}} &= \sum_{n=1}^{m} n\left\|\varphi_n - \varphi_{n,k}\right\|^2_{\widehat{\Gamma}} \\ &\le m\sum_{n=1}^{m}\left\|\varphi_n - \varphi_{n,k}\right\|^2_{\widehat{\Gamma}} \le m\left\|\varphi - \varphi_k\right\|^2_{\widehat{\Gamma}} \to_{k\to\infty} 0.\end{aligned}$$

□

18.6 The domain of the Malliavin derivative is a Hilbert space with respect to the norm $\|\cdot\|_{1,2}$

We shall see that the domain $\mathbb{D}^{1,2}$ of the Malliavin operator is a Hilbert space.

Proposition 18.6.1 *The domain $\mathbb{D}^{1,2}$ of the Malliavin derivative is a Hilbert space with respect to the norm $\|\cdot\|_{1,2}$, where the associated scalar product is given by*

$$\langle\varphi,\psi\rangle_{1,2} := \langle\varphi,\psi\rangle_{\widehat{\Gamma}} + \langle D\varphi, D\psi\rangle_{\widehat{\Gamma\otimes\nu}}.$$

Proof It is easy to see that $\langle\cdot,\cdot\rangle_{1,2}$ is a (positive definite) scalar product. It remains for us to show that $\mathbb{D}^{1,2}$ is complete: Fix a Cauchy sequence (φ_k) in $\mathbb{D}^{1,2}$ with respect to $\|\cdot\|_{1,2}$. Since

$$\|\varphi\|_{\widehat{\Gamma}}, \|D\varphi\|_{\widehat{\Gamma\otimes\nu}} \le \|\varphi\|_{1,2},$$

(φ_k) is a Cauchy sequence in $L^2_{\mathcal{W}}(\widehat{\Gamma})$ with respect to $\|\cdot\|_{\widehat{\Gamma}}$, converging to, say, α, and $(D\varphi_k)$ is a Cauchy sequence in $L^2_{\mathcal{W}\otimes\mathcal{L}^1}(\widehat{\Gamma\otimes\nu},\mathbb{H})$ with respect to $\|\cdot\|_{\widehat{\Gamma\otimes\nu}}$, converging to, say, β. By Theorem 18.5.1, $\alpha\in\mathbb{D}^{1,2}$ and $D\alpha=\beta$. Therefore, (φ_k) converges to α in $\mathbb{D}^{1,2}$ with respect to $\|\cdot\|_{1,2}$. □

Examples 18.6.2

(i) Let (φ_n) be a Cauchy sequence in the finite chaos level $L^2_{\mathcal{W},k}(\widehat{\Gamma})$ with respect to the norm $\|\cdot\|_{\widehat{\Gamma}}$ in $L^2_{\mathcal{W}}(\widehat{\Gamma})$. Since, by Theorem 18.5.1, (φ_n) converges to an element φ in $L^2_{\mathcal{W},k}(\widehat{\Gamma})\subseteq\mathbb{D}^{1,2}$ and $(D\varphi_n)$ converges to $D\varphi$ in $L^2_{\mathcal{W}\otimes\mathcal{L}^1,k}(\widehat{\Gamma\otimes\nu},\mathbb{H})$, we see that (φ_n) converges to φ with respect to $\|\cdot\|_{1,2}$.

(ii) Suppose that $\varphi=\sum_{n=0}^{\infty}I_n(f_n)\in\mathbb{D}^{1,2}$. Since $\lim_{k\to\infty}\sum_{n=0}^{k}I_n(f_n)=\varphi$ in $L^2_{\mathcal{W}}(\widehat{\Gamma})$ and $\lim_{k\to\infty}D\sum_{n=0}^{k}I_n(f_n)=D\varphi$ in $L^2_{\mathcal{W}\otimes\mathcal{L}^1}(\widehat{\Gamma\otimes\nu},\mathbb{H})$, we see that $\left(\sum_{n=0}^{k}I_n(f_n)\right)$ converges to φ with respect to $\|\cdot\|_{1,2}$.

We introduce the following notation: if $\psi\in L^2_{\mathcal{W}}(\widehat{\Gamma})$ has chaos decomposition $\psi=\sum_{n=0}^{\infty}I_n(g_n)$, we set $\psi^{\upharpoonright m}:=\sum_{n=0}^{m}I_n(g_n)$. Since $\mathbb{D}\subseteq\mathrm{Fin}L^2_{\mathcal{W}}(\widehat{\Gamma})$ (see Exercises 17.5 and 17.6), $\mathbb{D}\subseteq\mathbb{D}^{1,2}$. Note that $\psi^{\upharpoonright m}\in\mathbb{D}$ if $\psi\in\mathbb{D}$.

Proposition 18.6.3 *$\mathbb{D}$ is a dense subspace of the space* $\left(\mathbb{D}^{1,2},\|\cdot\|_{1,2}\right)$.

Proof Fix $\varphi\in\mathbb{D}^{1,2}$ with chaos decomposition $\varphi=\sum_{n=0}^{\infty}I_n(f_n)$. By Example 18.6.2 (i), $\lim_{m\to\infty}\left\|\varphi-\sum_{n=0}^{m}I_n(f_n)\right\|_{1,2}=0$. Fix $m\in\mathbb{N}$. Since $\mathbb{D}$ is dense in $L^2_{\mathcal{W}}(\widehat{\Gamma})$, there exists a sequence $(\varphi_k)_{k\in\mathbb{N}}$ in $\mathbb{D}$ with $\lim_{k\to\infty}\varphi_k=\sum_{n=0}^{m}I_n(f_n)$ in $L^2_{\mathcal{W}}(\widehat{\Gamma})$. Note that $\lim_{k\to\infty}\varphi_k^{\upharpoonright m}=\sum_{n=0}^{m}I_n(f_n)$ in $L^2_{\mathcal{W}}(\widehat{\Gamma})$. Since

$$\left\|D\varphi_k^{\upharpoonright m}-D\sum_{n=0}^{m}I_n(f_n)\right\|_{\widehat{\Gamma\otimes\nu}}\leq m\left\|\varphi_k^{\upharpoonright m}-\sum_{n=0}^{m}I_n(f_n)\right\|_{\widehat{\Gamma}}\to_{k\to\infty}0,$$

$\lim_{k\to\infty}\varphi_k^{\upharpoonright m}=\sum_{n=0}^{m}I_n(f_n)$ with respect to the norm $\|\cdot\|_{1,2}$. □

18.7 The range of the Malliavin derivative is closed

The domain of the Malliavin derivative D is not a closed subspace of $L^2_{\mathcal{W}}(\widehat{\Gamma})$, but we have the following.

Proposition 18.7.1 *The range $D[\mathbb{D}^{1,2}]$ of D is a closed subspace of* $L^2_{\mathcal{W}\otimes\mathcal{L}^1}(\widehat{\Gamma\otimes\nu},\mathbb{H})$.

Proof Let $(D\varphi_k)$ be a Cauchy sequence in $D[\mathbb{D}^{1,2}]$ with respect to the norm $\|\cdot\|_{\widehat{\Gamma\otimes\nu}}$ on $L^2_{\mathcal{W}\otimes\mathcal{L}^1}(\widehat{\Gamma\otimes\nu},\mathbb{H})$, converging to, say, $\beta\in L^2_{\mathcal{W}\otimes\mathcal{L}^1}(\widehat{\Gamma\otimes\nu},\mathbb{H})$. We may assume that $\mathbb{E}\varphi_k = 0$ for each $k \in \mathbb{N}$. By Theorem 18.5.1, (φ_k) converges to, say, $\alpha \in \mathbb{D}^{1,2}$ and $D\alpha = \beta$. Thus, β belongs to the range of D. □

18.8 A commutation rule for the directional derivative

Now we will prove a simple, but important, formula for the interplay of the Malliavin derivative and the directional derivative.

Proposition 18.8.1 *Fix a Malliavin differentiable $\varphi \in L^2_{\mathcal{W}}(\widehat{\Gamma})$ and $g \in L^2_{\mathcal{L}^1}(\widehat{\nu}, \mathbb{H})$ with $\|g\|_{\widehat{\nu}} = 1$. Let $\psi \in L^2_{\mathcal{W}}(\widehat{\Gamma})$. Then*

$$\langle D\varphi, \psi \otimes g \rangle_{\widehat{\Gamma \otimes \nu}} = \langle \psi, D\varphi_{\to g} \rangle_{\widehat{\Gamma}}.$$

Proof Let $\varphi = \mathbb{E}\varphi + \sum_{n=1}^{\infty} {}^{\circ}I_n(F_n)$ and $\psi = \mathbb{E}\varphi + \sum_{n=1}^{\infty} {}^{\circ}I_n(K_n)$ according to Theorem 17.2.1. Then

$$\begin{aligned}
\langle D\varphi, \psi \otimes g \rangle_{\widehat{\Gamma \otimes \nu}} &= \int_{\Omega} \int_{T} \langle D\varphi(\cdot, t), g(t) \cdot \psi \rangle_{\mathbb{H}} \, d\widehat{\nu}(t) d\widehat{\Gamma} \\
&= \int_{\Omega} \psi(X) \cdot \langle D\varphi(X, \cdot), g \rangle_{\widehat{\nu}} \, d\widehat{\Gamma}(X) \\
&= \langle \psi, D\varphi_{\to g} \rangle_{\widehat{\Gamma}}.
\end{aligned}$$

□

18.9 Product and chain rules for the Malliavin derivative

We choose $\varphi, \psi \in L^2_{\mathcal{W}}\left(\widehat{\Gamma}\right)$ with liftings $\Phi = \sum_{n=0}^{M} I_n(F_n), \Psi = \sum_{n=0}^{M} I_n(G_n)$, according to Theorem 17.3.1. Recall that $F_n, G_n \in SL^2\left(\nu^n, \mathbb{F}^{\otimes n}\right)$ are symmetric. Use the notation in Section 17.6; in particular, let $\mathfrak{E}$ be an ONB of $\mathbb{F}$.

First assume that φ and ψ belong to finite chaos levels. Then $M \in \mathbb{N}$ and we can assume that $D\Phi, D\Psi, D(\Phi \cdot \Psi) \in SL^2(\Gamma \otimes \nu, \mathbb{F})$ are liftings of $D\varphi, D\psi, D(\varphi \cdot \psi)$. In Section 17.6 we saw that the kernels of the chaos decomposition of $\varphi \cdot \psi$ are standard parts of $K_m : T^m_< \to \mathbb{F}^{\otimes m}$, where, for each $(a_1, \ldots, a_m) \in \mathfrak{E}^m$,

$$K_m(r_1, \ldots, r_m)(a_1, \ldots, a_m) = \sum_{k=0}^{m} \sum_{\sigma : k \uparrow m} \sum_{n=0}^{M^2} \sum_{t \in T^n_<} \sum_{\mathfrak{e} \in \mathfrak{E}^n} A \cdot B \cdot \frac{1}{H^n}$$

with

$$A = F_{n+k}(t, r_{\sigma_1}, \ldots, r_{\sigma_k}, \mathfrak{e}, a_{\sigma_1}, \ldots, a_{\sigma_k}),$$

$$B = G_{n+m-k}(t, r_{\overline{\sigma}_1}, \ldots, r_{\overline{\sigma}_{m-k}}, \mathfrak{e}, a_{\overline{\sigma}_1}, \ldots, a_{\overline{\sigma}_k}).$$

Since the F_n and G_n are symmetric, the K_m are symmetric on $T^m_{\neq}$. Note that for $s \in T$, $b \in \mathfrak{E}$ and $r_1 < \ldots < r_{m-1}$ the functions $K_m(\cdot, s, \cdot, b)$ with

$$K_m(r_1, \ldots, r_{m-1}, s)(a_1, \ldots, a_{m-1}, b)$$
$$= H^{m-1}\mathbb{E}_\Gamma\left(D(\Phi \cdot \Psi)_s(b) \cdot \langle X_{r_1}, a_1\rangle \cdot \ldots \cdot \langle X_{r_{m-1}}, a_{m-1}\rangle\right)$$

form the kernels of $D(\Phi \cdot \Psi)_s(b)$. Note that

$$K_m(r_1, \ldots, r_{m-1}, s, a_1, \ldots, a_{m-1}, b) = \left(K_m^{D\Phi \cdot \Psi} + K_m^{\Phi \cdot D\Psi}\right)(r, s, a, b)$$

with

$$K_m^{D\Phi \cdot \Psi}(r_1, \ldots, r_{m-1}, s, a_1, \ldots, a_{m-1}, b)$$
$$= \sum_{k=0}^{m-1} \sum_{\sigma : k \uparrow m-1} \sum_{n=0}^{M^2} \sum_{t \in T^n_<} \sum_{\mathfrak{e} \in \mathfrak{E}^n} \frac{1}{H^n} F_{n+k+1}(t, r_{\sigma_1}, \ldots, r_{\sigma_k}, s, \mathfrak{e}, a_{\sigma_1}, \ldots, a_{\sigma_k}, b)$$
$$\cdot G_{n+m-1-k}(t, r_{\overline{\sigma}_1}, \ldots, r_{\overline{\sigma}_{m-1-k}}, \mathfrak{e}, a_{\overline{\sigma}_1}, \ldots, a_{\overline{\sigma}_{m-1-k}})$$

and

$$K_m^{\Phi \cdot D\Psi}(r_1, \ldots, r_{m-1}, s, a_1, \ldots, a_{m-1}, b)$$
$$= \sum_{k=0}^{m-1} \sum_{\sigma : k \uparrow m-1} \sum_{n=0}^{M^2} \sum_{t \in T^n_<} \sum_{\mathfrak{e} \in \mathfrak{E}^n} \frac{1}{H^n} F_{n+k}(t, r_{\sigma_1}, \ldots, r_{\sigma_k}, \mathfrak{e}, a_{\sigma_1}, \ldots, a_{\sigma_k})$$
$$\cdot G_{n+m-k}(t, r_{\overline{\sigma}_1}, \ldots, r_{\overline{\sigma}_{m-1-k}}, s, \mathfrak{e}, a_{\overline{\sigma}_1}, \ldots, a_{\overline{\sigma}_{m-1-k}}, b).$$

Note that

$$K_m^{D\Phi \cdot \Psi}(r_1, \ldots, r_{m-1}, s, a_1, \ldots, a_{m-1}, b)$$
$$= H^{m-1}\mathbb{E}_\Gamma\left(D\Phi_s(b) \cdot \Psi \cdot \langle X_{r_1}, a_1\rangle \cdot \ldots \cdot \langle X_{r_{m-1}}, a_{m-1}\rangle\right)$$

and

$$K_m^{\Phi \cdot D\Psi}(r_1, \ldots, r_{m-1}, s, a_1, \ldots, a_{m-1}, b)$$
$$= H^{m-1}\mathbb{E}_\Gamma\left(\Phi \cdot D\Psi_s(b) \cdot \langle X_{r_1}, a_1\rangle \cdot \ldots \cdot \langle X_{r_{m-1}}, a_{m-1}\rangle\right).$$

It follows that the standard parts of the $K_m^{D\Phi \cdot \Psi}(\cdot, s, \cdot, b)$ form the kernels of $D\varphi_s(b) \cdot \psi$ and the standard parts of the $K_m^{\Phi \cdot D\Psi}(\cdot, s, \cdot, b)$ form the kernels of $\varphi \cdot D\psi_s(b)$. We obtain the following.

Proposition 18.9.1 *Suppose that $\varphi, \psi \in L^2_{\mathcal{W}}(\widehat{\Gamma})$ belong to finite chaos levels. Then*

$$D(\varphi \cdot \psi)_s = D\varphi_s \cdot \psi + \varphi \cdot D\psi_s \text{ in } L^2(\widehat{\Gamma} \otimes \lambda, \mathbb{H}).$$

Now let $\varphi = \sum_{n=0}^{\infty} I_n({}^\circ F_n), \psi = \sum_{n=0}^{\infty} I_n({}^\circ G_n) \in L^2_{\mathcal{W}}(\widehat{\Gamma})$. We set $\varphi_m = \sum_{n=0}^{m} I_n({}^\circ F_n), \psi_m = \sum_{n=0}^{m} I_n({}^\circ G_n)$. Using Proposition 18.9.1 and Theorem 18.5.1 and the relationship between Loeb and Lebesgue measures on $\mathcal{L}^n$ we obtain the following.

Theorem 18.9.2 (Product rule) *Assume that $\left(\mathbb{E}_{\widehat{\Gamma}}(\varphi_m \cdot \psi_m)\right)_{m\in\mathbb{N}}$ converges in $\mathbb{R}$.*

(A) *Suppose that the sequences $(D\varphi_m \cdot \psi_m)_{m\in\mathbb{N}}$, $(\varphi_m \cdot D\psi_m)_{m\in\mathbb{N}}$ converge in $L^2(\widehat{\Gamma} \otimes \lambda, \mathbb{H})$. Then $(D(\varphi_m \cdot \psi_m))_{m\in\mathbb{N}}$ converges in $L^2(\widehat{\Gamma} \otimes \lambda, \mathbb{H})$ and $\varphi \cdot \psi$ is Malliavin differentiable and*

$$(D(\varphi \cdot \psi))_s = (D\varphi)_s \cdot \psi + \varphi \cdot (D\psi)_s \text{ in } L^2(\widehat{\Gamma} \otimes \lambda, \mathbb{H}).$$

(B) *Suppose that $(D(\varphi_m \cdot \psi_m))_{m\in\mathbb{N}}$ converges in $L^2(\widehat{\Gamma} \otimes \lambda, \mathbb{H})$. Then $\varphi \cdot \psi$ is Malliavin differentiable and we have $\widehat{\Gamma} \otimes \lambda$-a.e.*

$$(D(\varphi \cdot \psi))_s = (D\varphi)_s \cdot \psi + \varphi \cdot (D\psi)_s.$$

The equation $D\left(\varphi_m^k\right)_s = k \cdot \varphi_m^{k-1} \cdot D(\varphi_m)_s$ implies the following.

Theorem 18.9.3 (Chain rule) *Fix $g : \mathbb{R}^n \to \mathbb{R}$ and Malliavin differentiable $\varphi_1, \ldots, \varphi_n$. Assume that the partial derivatives of g exist and that there are polynomials q_j in n variables with $\lim q_j = g$ and $\lim \partial_i q_j = \partial_i g$ for $i = 1, \ldots, n$.*

(A) *Fix $m \in \mathbb{N}$. Suppose that $\left(D\left(q_j\left(\varphi_{1,m}, \ldots, \varphi_{n,m}\right)\right)\right)_{j\in\mathbb{N}}$ converges in $L^2(\widehat{\Gamma} \otimes \lambda, \mathbb{H})$ and $\left(\mathbb{E}_{\widehat{\mu}}\left(q_j\left(\varphi_{1,m}, \ldots, \varphi_{n,m}\right)\right)\right)_{j\in\mathbb{N}}$ converges in $\mathbb{R}$. Then $g\left(\varphi_{1,m}, \ldots, \varphi_{n,m}\right)$ is Malliavin differentiable and we have $\widehat{\Gamma} \otimes \lambda$-a.e.*

$$\left(D(g\left(\varphi_{1,m}, \ldots, \varphi_{n,m}\right))\right)_s = \sum_{i=1}^{n} (\partial_i g)\left(\varphi_{1,m}, \ldots, \varphi_{n,m}\right) \cdot \left(D\varphi_{i,m}\right)_s.$$

(B) *Assume that (A) is true for all $m \in \mathbb{N}$, and g and $\partial_i g$ are continuous. Moreover, let $\left(D\left(g\left(\varphi_{1,m}, \ldots, \varphi_{n,m}\right)\right)\right)_{m\in\mathbb{N}}$, $\mathbb{E}_{\widehat{\mu}}\left(g\left(\varphi_{1,m}, \ldots, \varphi_{n,m}\right)\right)_{m\in\mathbb{N}}$ converge in $L^2(\widehat{\Gamma} \otimes \lambda, \mathbb{H})$, and in $\mathbb{R}$, respectively. Then $g(\varphi_1, \ldots, \varphi_n)$ is Malliavin differentiable and we have $\widehat{\Gamma} \otimes \lambda$-a.e.*

$$(D(g(\varphi_1, \ldots, \varphi_n)))_s = \sum_{i=1}^{n} (\partial_i g)(\varphi_1, \ldots, \varphi_n) \cdot (D\varphi_i)_s.$$

Exercises

18.1 Prove that $DI_n(F) = I_{n-1,1}(F)$ for symmetric $F : T^n_{\neq} \to \mathbb{F}^{\otimes n}$, where D is the derivative (see the beginning of Section 18.1).

18.2 Prove Theorem 18.3.1.

18.3 Prove that $\langle \cdot, \cdot \rangle_{1,2}$ is a positive definite scalar product.

18.4 Prove Theorem 18.9.2.

18.5 Prove Theorem 18.9.3.

19

The Skorokhod integral

In the preceding chapter we saw that the Malliavin derivative D is closely related to the usual derivative in finite-dimensional Euclidian spaces. In order to see that the Skorokhod integral can be defined as the standard part of a Riemann–Stieltjes integral, we will first prove that each $\psi \in L^2_{\mathcal{W}\otimes\mathcal{L}^1}(\widehat{\Gamma\otimes\nu},\mathbb{H})$ has the orthogonal expansion $\psi = \sum_{n=0}^{\infty} {}^\circ I_{n,1}(F_n)$, where $F_n \in SL^2_{\neq}(\nu^{n+1},\mathbb{F}^{\otimes(n+1)})$ is symmetric in the first n variables and nearstandard.

Then for elements ψ of a dense subspace $\Delta \subseteq L^2_{\mathcal{W}\otimes\mathcal{L}}(\widehat{\Gamma\otimes\nu},\mathbb{H})$ the Skorokhod integral δ is simply defined by Riemann–Stieltjes integration under the standard part map and under the sum, i.e. (see Section 19.1),

$$\delta\psi(X) := \sum_{n=0}^{\infty}\left({}^\circ\int^{V} I_{n,1}(F_n)\Delta B\right)(X) := \sum_{n=0}^{\infty} {}^\circ \sum_{t\in T} I_n(F_n(\cdot,t))(X_t).$$

19.1 Decomposition of processes

In order to define the Skorokhod integral, which is the reversal of the derivative from several points of view, and to define Skorokhod integral processes, we now provide a suitable decomposition of the functionals in $L^2_{\mathcal{W}\otimes\mathcal{L}^1}(\widehat{\Gamma\otimes\nu},\mathbb{H})$. Recall the definition of $\widetilde{F}$ for functions $F: T^{n+1}_{\neq} \to \mathbb{F}^{\otimes(n+1)}$, which are symmetric in the first n arguments (see Section 13.2). Note that $\widetilde{F} \in SL^2(\nu^{n+1},\mathbb{F}^{\otimes(n+1)})$ if $F \in SL^2(\nu^{n+1},\mathbb{F}^{\otimes(n+1)})$ and note that $\widetilde{F}$ is symmetric. The function $\widetilde{f}$ for $f \in L^2(\widehat{\nu^{n+1}},\mathbb{H}^{\otimes(n+1)})$ is defined in the same way. Moreover, by Corollary 13.4.3, ${}^\circ I_{n+1}(\widetilde{F}) = I_{n+1}({}^\circ\widetilde{F}) \in L_{\mathcal{W}}(\widehat{\Gamma})$ if ${}^\circ F$ exists and belongs to $L^2_{\mathcal{L}^{n+1}}(\widehat{\nu^{n+1}},\mathbb{H}^{\otimes(n+1)})$. We define for each $r \in T$

$$\left(\mathbf{1}_{T_r} \cdot_{n+1} F\right)(t_1,\ldots,t_{n+1}) := \mathbf{1}_{T_r}(t_{n+1}) \cdot F(t_1,\ldots,t_{n+1}).$$

Since $\nu^n\left(T^n \setminus T^n_{\neq}\right) \approx 0$, we may assume, in order to study S-integrability, that $T^n_{\neq}$ is the support of internal functions, defined on T^n. Then

$$\begin{aligned}\int^V I_{n,1}(\mathbf{1}_{T_r} \cdot_{n+1} F)\Delta B(X) &= \sum_{t \le r}\sum_{t_1<\ldots<t_n} F(t_1,\ldots,t_n,t)(X_{t_1},\ldots,X_{t_n},X_t)\\ &= I_{n+1}\left(\widetilde{\mathbf{1}_{T_r} \cdot_{n+1} F}\right).\end{aligned}$$

Theorem 19.1.1 *Let $\varphi \in L^2_{\mathcal{W}\otimes\mathcal{L}^1}(\widehat{\Gamma\otimes\nu},\mathbb{H})$. Then there exists a sequence $(F_n)_{n\in\mathbb{N}_0}$ of internal functions $F_n : T^{n+1} \to \mathbb{F}^{\otimes(n+1)}$ with the following five properties.*

(a) *F_n is symmetric in the first n variables, $F_n \in SL^2_{\neq}(\nu^{n+1},\mathbb{F}^{\otimes(n+1)})$ and ${}^\circ F_n \in L^2_{\mathcal{L}^{n+1}}(\widehat{\nu^{n+1}},\mathbb{H}^{\otimes(n+1)})$.*
(b) *$I_{n,1}(F_n) \in SL^2(\Gamma\otimes\nu,\mathbb{F})$ and ${}^\circ I_{n,1}(F_n) \in L^2_{\mathcal{W}\otimes\mathcal{L}^1}(\widehat{\Gamma\otimes\nu},\mathbb{H})$.*
(c) *$I_{n+1}(\widetilde{F}_n) \in SL^2(\Gamma)$ and ${}^\circ I_{n+1}(\widetilde{F}_n) \in L^2_{\mathcal{W}}(\widehat{\Gamma})$.*
(d) *$t \mapsto I_{n+1}\left(\widetilde{\mathbf{1}_{T_t} \cdot_{n+1} F_n}\right) \in SL^2(\Gamma\otimes\nu)$ and $t \mapsto {}^\circ I_{n+1}\left(\widetilde{\mathbf{1}_{T_t} \cdot_{n+1} F_n}\right)$ is in $L^2_{\mathcal{W}\otimes\mathcal{L}^1}(\widehat{\Gamma\otimes\nu})$.*
(e) *$\varphi = \sum_{n=0}^\infty {}^\circ I_{n,1}(F_n)$ converges in $L^2_{\mathcal{W}\otimes\mathcal{L}^1}(\widehat{\Gamma\otimes\nu},\mathbb{H})$.*

Uniqueness: If $\varphi = \sum_{n=0}^\infty {}^\circ I_{n,1}(K_n)$ converges in $L^2(\widehat{\Gamma\otimes\nu},\mathbb{H})$ and $K_n \in SL^2(\nu^{n+1},\mathbb{F}^{\otimes(n+1)})$ is symmetric in the first n variables, then $F_n \approx_{\mathbb{F}^{n+1}} K_n$ $\widehat{\nu^{n+1}}$-a.e.

This decomposition also yields a corresponding decomposition of functionals in $L^2_{\mathcal{W}}(\widehat{\Gamma}\otimes\lambda,\mathbb{H})$ or in $L^2(W\otimes\lambda,\mathbb{H})$.

Proof Let M be the set of all $\varphi \in L^2_{\mathcal{W}\otimes\mathcal{L}^1}(\widehat{\Gamma\otimes\nu},\mathbb{H})$ such that there exists a sequence $(F_n)_{n\in\mathbb{N}_0}$ of internal functions $F_n : T^{n+1}_{\neq} \to \mathbb{F}^{\otimes n+1}$ with the properties (a)–(e). We shall prove that $M = L^2_{\mathcal{W}\otimes\mathcal{L}^1}(\widehat{\Gamma\otimes\nu},\mathbb{H})$.

First assume that $\varphi := \mathbf{1}_{B\times C}\otimes a = \mathbf{1}_B \otimes \mathbf{1}_C \otimes a$ with $B \in \mathcal{W}$, $C \in \mathcal{L}^1$ and $a \in \mathbb{H}$. Note that for all $X \in \Omega$, $t \in T$ and $x \in \mathbb{H}$

$$\varphi(X,t)(x) = \mathbf{1}_B(X)\cdot\mathbf{1}_C(t)\cdot\langle a,x\rangle.$$

We will prove that $\varphi \in M$. According to Theorem 17.2.1, $\mathbf{1}_B$ has the decomposition

$$\mathbf{1}_B = \sum_{n=0}^\infty {}^\circ I_n(G_n) = \sum_{n=0}^\infty I_n({}^\circ G_n) \text{ in } L^2_{\mathcal{W}}(\widehat{\Gamma}) \text{ with } {}^\circ I_0(G_0) = \mathbb{E}(\mathbf{1}_B). \tag{59}$$

Since $C \in \mathcal{L}^1 \subseteq L_\nu(T)$, there exists an internal subset $A \subset T$ such that $\widehat{\nu}(A \triangle C) = 0$. For $n \in \mathbb{N}_0$ we define $F_n : T^n \times T \to \mathbb{F}$, setting

$$F_n := G_n \otimes \mathbf{1}_A \otimes a : (t,s,y) \mapsto G_n(t) \cdot \mathbf{1}_A(s) \cdot \langle {}^*a, y \rangle.$$

If $t \notin T_{\neq}^{n+1}$, set $F_n(t) := 0$. By Lemma 10.12.2 (b), $F_n \in SL^2_{\neq}(\nu^{n+1}, \mathbb{F}^{\otimes(n+1)})$. Thus, $I_{n+1}(\widetilde{F}_n) \in SL^2(\Gamma)$ (Corollary 13.1.4) and $I_{n,1}(F_n) \in SL^2(\Gamma \otimes \nu, \mathbb{F})$ (Theorem 13.1.6 (a)). Since ${}^\circ G_n$ is $\mathcal{L}^n$-measurable and $\mathbf{1}_A$ is $\mathcal{L}^1$-measurable, (a) is true. Since $\mathbf{1}_A \otimes a \in L^2_{\mathcal{L}^1}(\widehat{\nu}, \mathbb{H})$, we obtain ${}^\circ I_{n,1}(F_n) \in L^2_{\mathcal{W} \otimes \mathcal{L}^1}(\widehat{\Gamma \otimes \nu}, \mathbb{H})$, and, by Corollary 13.4.3, ${}^\circ I_{n+1}(\widetilde{F}_n) \in \mathcal{L}^2_{\mathcal{W}}(\widehat{\Gamma})$. Thus, (b) and (c) are true. In a similar way we see that (d) is true. Because of (59),

$$\begin{aligned}
\mathbf{1}_B \otimes \mathbf{1}_C \otimes a &= (\text{in } L^2_{\mathcal{W} \otimes \mathcal{L}^1}(\widehat{\Gamma \otimes \nu}, \mathbb{H}))\ \mathbf{1}_B \otimes \mathbf{1}_A \otimes a \\
&= \widehat{\Gamma}(B) \cdot \mathbf{1}_A \otimes a + \sum_{n=1}^{\infty} {}^\circ I_{n,1}(F_n) \text{ converges in } L^2_{\mathcal{W} \otimes \mathcal{L}^1}(\widehat{\Gamma \otimes \nu}, \mathbb{H}),
\end{aligned}$$

thus (e) is true. This shows that $\varphi = \mathbf{1}_B \otimes \mathbf{1}_C \otimes a \in M$. Since M is a linear space, it remains for us to show that M is complete: Let (φ_k) be a Cauchy sequence in M with $\varphi_k = \sum_{n=0}^{\infty} {}^\circ I_{n,1}(F_{n,k})$ in $L^2_{\mathcal{W} \otimes \mathcal{L}^1}(\widehat{\Gamma \otimes \nu}, \mathbb{H})$ such that (a)–(d) are true with $F_{n,k}$ instead of F_n. Then, by Proposition 13.1.7,

$$\begin{aligned}
\|\varphi_k - \varphi_l\|^2_{\widehat{\Gamma \otimes \nu}} &= \left\| \sum_{n=0}^{\infty} ({}^\circ I_{n,1}(F_{n,k} - F_{n,l})) \right\|^2_{\widehat{\Gamma \otimes \nu}} \\
&= \sum_{n=0}^{\infty} \|{}^\circ I_{n,1}(F_{n,k} - F_{n,l})\|^2_{\widehat{\Gamma \otimes \nu}} \\
&= \sum_{n=0}^{\infty} \int_{T^n_{\neq} \times T} \|{}^\circ F_{n,k} - {}^\circ F_{n,l}\|^2_{\mathbb{H}^{n+1}} \, d\widehat{\nu^{n+1}}.
\end{aligned}$$

Since $(F_{n,k})$ is symmetric in the first n variables, $({}^\circ F_{n,k})_{k \in \mathbb{N}}$ is a Cauchy sequence in $L^2(\widehat{\nu^{n+1}}, \mathbb{H}^{\otimes(n+1)})$ converging to an $f_n \in L^2_{\mathcal{L}^{n+1}}(\widehat{\nu^{n+1}}, \mathbb{H}^{\otimes(n+1)})$. In the same way, $\left({}^\circ I_{n,1}(F_{n,k})\right)_{k \in \mathbb{N}}$ converges to a $g_n \in L^2_{\mathcal{W} \otimes \mathcal{L}^1}(\widehat{\Gamma \otimes \nu}, \mathbb{H})$, $\left({}^\circ I_{n+1}(\widetilde{F}_{n,k})\right)_{k \in \mathbb{N}}$ converges to an $h_n \in L^2_{\mathcal{W}}(\widehat{\Gamma})$, $({}^\circ I_{n+1}(\mathbf{1}_{T_t} \cdot_{n+1} \widetilde{F_{n,k}}))_{k \in \mathbb{N}}$ converges to an $i_n \in L^2_{\mathcal{W} \otimes \mathcal{L}^1}(\widehat{\Gamma \otimes \nu})$. By Corollary 10.9.3, we may assume that $f_n = {}^\circ F_n$, $g_n = {}^\circ I_{n,1}(F_n)$, $h_n = {}^\circ I_{n+1}(\widetilde{F}_n)$, $i_n = t \mapsto {}^\circ I_{n+1}(\mathbf{1}_{T_t} \cdot_{n+1} \widetilde{F_n})$, and that the internal functions behind ${}^\circ$ fulfil the required S-integrability conditions. Moreover, we may assume that F_n is symmetric in the first n components and F_n is defined on $T^{n+1}_{\neq}$. This proves that M is complete.

We will now prove the uniqueness. Using similar computations, we obtain

$$0 = \left\| \sum_{n=0}^{\infty} {}^\circ I_{n,1}(F_n - K_n) \right\|^2_{\widehat{\Gamma \otimes \nu}} = \sum_{n=0}^{\infty} {}^\circ \int_{T^n_< \times T} \|F_n - K_n\|^2_{\mathbb{F}^{n+1}} \, d\nu^{n+1}.$$

It follows that $\int_{T^n_< \times T} \|F_n - K_n\|^2_{\mathbb{F}^{n+1}} \, d\nu^{n+1} \approx 0$. Since F_n and K_n are symmetric in the first n variables, $F_n \approx_{\mathbb{F}^{n+1}} K_n$ $\widehat{\nu^{n+1}}$-a.e. □

According to Theorem 17.3.1, Corollary 10.9.3 implies the following lifting result.

Corollary 19.1.2 *Let* $\varphi = \sum_{n=0}^{\infty} {}^\circ I_{n,1}(F_n) \in L^2_{\mathcal{W}\otimes\mathcal{L}^1}(\widehat{\Gamma \otimes \nu}, \mathbb{H})$ *according to Theorem 19.1.1. Choose an internal extension* $(F_n)_{n \in {}^*\mathbb{N}_0}$ *of* $(F_n)_{n \in \mathbb{N}_0}$. *Then there exists an unlimited* $K \in {}^*\mathbb{N}$ *such that for each unlimited* $M \leq K$, $M \in {}^*\mathbb{N}$

$$\Phi := \sum_{n=0}^{M} I_{n,1}(F_n) \in SL^2(\Gamma \otimes \nu, \mathbb{F}) \text{ and } \Phi \text{ is a lifting of } \varphi.$$

We may assume that all F_n *are symmetric in the first n components and* F_n *is defined on* $T^{n+1}_{\neq}$.

Moreover, according to Theorem 17.4.1, we obtain liftings of $\mathcal{W} \otimes \mathcal{L}^1$-measurable processes.

Theorem 19.1.3 *Let* $\varphi : \Omega \times [0, \infty[\to \mathbb{H}$ *be* $\mathcal{W} \otimes \mathcal{L}^1$*-measurable. Then* φ *has a lifting* Φ *of the form* $\Phi = \sum_{n=0}^{M} I_{n,1}(F_n)$.

19.2 Malliavin derivative of processes

Let $\varphi = \sum_{n=0}^{\infty} {}^\circ I_{n,1}(F_n) \in L^2_{\mathcal{W}\otimes\mathcal{L}^1}(\widehat{\Gamma \otimes \nu}, \mathbb{H})$, according to Theorem 19.1.1. Then the Malliavin derivative $D\varphi$ is an element of the Hilbert space $L^2_{\mathcal{W}\otimes\mathcal{L}^2}(\widehat{\Gamma \otimes \nu}, \mathbb{H}^{\otimes 2})$, defined by

$$D\varphi = \sum_{n=1}^{\infty} {}^\circ I_{n-1,2}(F_n)$$

if this series converges in $L^2_{\mathcal{W}\otimes\mathcal{L}^2}(\widehat{\Gamma \otimes \nu^2}, \mathbb{H}^{\otimes 2})$. Then φ is called **Malliavin differentiable**. It is left to the reader to study $D(D\varphi)$ and higher derivatives $D^{(k)}\varphi$.

19.3 The domain of the Skorokhod integral

Now we are able to define the Skorokhod integral. Let $\varphi \in L^2_{\mathcal{W}\otimes\mathcal{L}^1}(\widehat{\Gamma\otimes\nu}, \mathbb{H})$ with $\varphi = \sum_{n=0}^{\infty} {}^\circ I_{n,1}(F_n)$, where $F_n : T^{n+1}_{\neq} \to \mathbb{F}^{\otimes(n+1)}$ fulfils the conditions (a), (b), (c), (d) in Theorem 19.1.1; in particular, ${}^\circ I_{n+1}(\widetilde{F}_n) \in L^2_{\mathcal{W}}(\widehat{\Gamma})$. We define

$$\delta\varphi = \sum_{n=0}^{\infty} \left({}^\circ \int^V I_{n,1}(F_n)\Delta B \right) : X \mapsto \sum_{n=0}^{\infty} {}^\circ \sum_{t\in T} (I_n(F_n(\cdot,t))(X))\,(X_t)$$

if this series converges in $L^2_{\mathcal{W}}(\widehat{\Gamma})$. Note that

$$\sum_{t\in T} (I_n(F_n(\cdot,t))(X))\,(X_t) = I_{n+1}(\widetilde{F}_n)(X).$$

Therefore,

$$\delta\varphi = \sum_{n=0}^{\infty} {}^\circ I_{n+1}(\widetilde{F}_n),$$

and, since ${}^\circ I_{n+1}(\widetilde{F}_n)$ is $\mathcal{W}$-measurable, $\delta\varphi$ exists if and only if the series $\sum_{n=0}^{\infty} \mathbb{E}({}^\circ I_{n+1}(\widetilde{F}_n))^2$ converges. The linear operator δ is called the **Skorokhod integral**. Since for each finite sum $\psi = \sum_{n=0}^{k} {}^\circ I_{n,1}(F_n) \in L^2_{\mathcal{W}\otimes\mathcal{L}_1}(\widehat{\Gamma\otimes\nu}, \mathbb{H})$ the Skorokhod integral $\delta\psi$ of ψ always exists, the **domain** Δ of δ is a dense subspace of $L^2_{\mathcal{W}\otimes\mathcal{L}^1}(\widehat{\Gamma\otimes\nu}, \mathbb{H})$. A function $\varphi \in L^2_{\mathcal{W}\otimes\mathcal{L}^1}(\widehat{\Gamma\otimes\nu}, \mathbb{H})$ is called **Skorokhod integrable** (see [105]) if $\varphi \in \Delta$.

Since the spaces $L^2_{\mathcal{W}\otimes\mathcal{L}^1}(\widehat{\Gamma\otimes\nu}, \mathbb{H})$ and $L^2(W\otimes\lambda, \mathbb{H})$ can be identified, and also $L^2(W)$ and $L^2_{\mathcal{W}}(\widehat{\Gamma})$, the Skorokhod integral is densely defined for functionals in $L^2(W\otimes\lambda, \mathbb{H})$ and takes its values in $L^2(W)$.

Our aim now is to present a condition under which the Skorokhod integral of $\varphi \in L^2_{\mathcal{W}\otimes\mathcal{L}^1}(\widehat{\Gamma\otimes\nu}, \mathbb{H})$ exists. We assume that φ has the form

$$\varphi = \sum_{n=0}^{\infty} {}^\circ I_{n,1}(F^n) \text{ converging in } L^2_{\mathcal{W}\otimes\mathcal{L}^1}(\widehat{\Gamma\otimes\nu}, \mathbb{H}),$$

where the $F^n := F_n$ fulfil the conditions in Theorem 19.1.1.

Theorem 19.3.1 *Fix an orthonormal basis $\mathfrak{E}$ of $\mathbb{F}$. The functional $\varphi \in L^2_{\mathcal{W}\otimes\mathcal{L}^1}(\widehat{\Gamma\otimes\nu}, \mathbb{H})$ is Skorokhod integrable if and only if the series*

$$\sum_{n=1}^{\infty} {}^\circ \int_{T^2_{\neq}} \sum_{\mathfrak{e}\in\mathfrak{E}^2} \int_{T^{n-1}_{<}} \langle F^n_{(\cdot,s,t)}(\cdot,\mathfrak{e}_1,\mathfrak{e}_2) \cdot F^n_{(\cdot,t,s)}(\cdot,\mathfrak{e}_2,\mathfrak{e}_1)\rangle_{\mathbb{F}^{n-1}} d\nu^{n-1} d\nu(t,s)$$

converges in $\mathbb{R}$.

Proof Since $\sum_{n=0}^{\infty} \mathbb{E}_{\widehat{\Gamma}} \left({}^{\circ}I_{n+1}(\widetilde{F^n})\right)^2$ converges in $\mathbb{R}$ if and only if φ is Skorokhod integrable, we compute, using Lemma 11.1.3 again,

$$\mathbb{E}_{\Gamma}\left(I_{n+1}(\widetilde{F^n})^2\right) = \mathbb{E}_{\Gamma}\left(\sum_{t\in T}\sum_{t_1<\ldots<t_n} F^n_{(t_1,\ldots,t_n,t)}(X_{t_1},\ldots,X_{t_n},X_t)\right)^2 = \mathbb{E}_{n,1} + \mathbb{E}_{n,2},$$

where

$$\begin{aligned}
\mathbb{E}_{n,1} &= \mathbb{E}_{\Gamma}\sum_{t\in T}\sum_{\mathfrak{e}\in\mathfrak{E}} \langle X_t,\mathfrak{e}\rangle^2 \left(\sum_{t_1<\ldots<t_n} F^n_{(t_1,\ldots,t_n,t)}(X_{t_1},\ldots,X_{t_n},\mathfrak{e})\right)^2 \\
&= \sum_{\mathfrak{e}\in\mathfrak{E}^{n+1}}\sum_{t\in T}\sum_{t_1<\ldots<t_n} \left(F^n_{(t_1,\ldots,t_n,t)}\right)^2(\mathfrak{e})\frac{1}{H^{n+1}} \\
&= \int_{T^n_<\times T} \left\|F^n\right\|^2_{\mathbb{F}^{n+1}} d\nu^{n+1} = \int_{\Omega\times T} \left\|I_{n,1}(F^n)\right\|^2_{\mathbb{F}} d\Gamma\otimes\nu,
\end{aligned}$$

and with the shorthand $X_t^{\mathfrak{e}} := \langle X_t, \mathfrak{e}\rangle$

$$\begin{aligned}
&\mathbb{E}_{n,2} \\
&= \mathbb{E}_{\Gamma}\sum_{(s,t)\in T^2_{\neq}}\sum_{\mathfrak{e}\in\mathfrak{E}^2} X_t^{e_1}X_s^{\mathfrak{e}_2} I_n(F^n_{(\cdot,t)})(X,\mathfrak{e}_1)\cdot I_n(F^n_{(\cdot,s)})(X,\mathfrak{e}_2) \\
&= \mathbb{E}_{\Gamma}\sum_{(s,t)\in T^2_{\neq}}\sum_{\mathfrak{e}\in\mathfrak{E}^2} \left(X_t^{\mathfrak{e}_1}\right)^2\left(X_s^{\mathfrak{e}_2}\right)^2 I_{n-1}(F^n_{(\cdot,s,t)})(X,\mathfrak{e}_2,\mathfrak{e}_1)\cdot\, I_{n-1}(F^n_{(\cdot,t,s)})(X,\mathfrak{e}_1,\mathfrak{e}_2) \\
&= \int_{T^2_{\neq}}\sum_{\mathfrak{e}\in\mathfrak{E}^2}\int_{T^{n-1}_<} \langle F^n_{(\cdot,s,t)}(\cdot,\mathfrak{e}_2,\mathfrak{e}_1)\cdot F^n_{(\cdot,t,s)}(\cdot,\mathfrak{e}_1,\mathfrak{e}_2)\rangle_{\mathbb{F}^{n-1}} d\nu^{n-1} d\nu(t,s).
\end{aligned}$$

From $\varphi \in L^2_{\mathcal{W}\otimes\mathcal{L}^1}(\widehat{\Gamma\otimes\nu},\mathbb{H})$ it follows that $\sum_{n=0}^{\infty} {}^{\circ}\mathbb{E}_{n,1}$ converges. Therefore, $\delta\varphi = \sum_{n=0}^{\infty} {}^{\circ}I_{n+1}(\widetilde{F}_n)$ converges if and only if $\sum_{n=0}^{\infty} {}^{\circ}\mathbb{E}_{n,2}$ converges. This proves the theorem. □

19.4 A lifting theorem for the integral

We have the following lifting theorem for Skorokhod integrable stochastic processes, according to Theorem 18.3.1.

Proposition 19.4.1 *Let $\varphi = \sum_{n=0}^{\infty} {}^{\circ}I_{n,1}(F_n)$ be Skorokhod integrable. Choose an internal extension $(F_n)_{n\in{}^*\mathbb{N}_0}$ of $(F_n)_{n\in\mathbb{N}_0}$. Then there exists an unlimited*

$K \in {}^*\mathbb{N}$ such that for each unlimited $M \leq K$, $M \in {}^*\mathbb{N}$,

$$\Phi := \sum_{n=0}^{M} I_{n,1}(F_n) \in SL^2(\Gamma \otimes \nu, \mathbb{F}) \text{ and } \Phi \approx_{\mathbb{F}} \varphi \ \widehat{\Gamma \otimes \nu}\text{-a.e.},$$

$$\delta\Phi : X \mapsto \sum_{s \in T} \Phi(X,s)(X_s) = \int^{V} \Phi \Delta B(X) = \sum_{n=0}^{M} I_{n+1}(\widetilde{F}_n)(X) \in SL^2(\Gamma)$$

and

$$\delta\Phi \approx \delta\varphi \ \widehat{\Gamma}\text{-a.s. in } \mathbb{R}.$$

We may assume that all the F_n are symmetric in the first n variables.

19.5 The Itô integral is a special case of the Skorokhod integral

In this section we shall see that the Itô integral, established in Section 12.2, is a special case of the Skorokhod integral.

In order to apply the chaos decomposition, one has to restrict the filtration $(\mathfrak{b}_r)_{r \in [0,\infty[}$ (the standard part of $(\mathcal{B}_t)_{t \in T}$) to the coarser filtration $(\mathfrak{b}_r^{\mathcal{W}})_{r \in [0,\infty[} := (\mathfrak{b}_r \cap \mathcal{W})_{r \in [0,\infty[}$. Denote by $\mathfrak{P}^{\mathcal{W}}$ the σ-algebra on $\Omega \times [0,\infty[$, generated by the $(\mathfrak{b}_r^{\mathcal{W}})_{r \in [0,\infty[}$-predictable rectangles, augmented by the $\widehat{\Gamma} \otimes \lambda$-nullsets.

An internal function $F : T^{n+1} \to \mathbb{F}^{\otimes(n+1)}$ is called **non-anticipating** if $F(t_1, \ldots, t_n, t) \neq 0$ implies $t_1, \ldots, t_n < t$. We have the following expansion result for functions in $L^2_{\mathfrak{P}^{\mathcal{W}}}(\widehat{\Gamma \otimes \nu}, \mathbb{H})$, corresponding to a result of Bouleau and Hirsch [16] for the classical Wiener space. We extensively use the identification of a $\mathcal{W} \otimes \mathcal{L}^1$-measurable function f with its equivalent $\mathcal{W} \otimes$Leb-measurable g, i.e., $f(X,t) = g(X, {}^\circ t)$ for $\widehat{\Gamma \otimes \nu}$-almost all (X,t).

Theorem 19.5.1 *Fix $\varphi \in L^2_{\mathfrak{P}^{\mathcal{W}}}(\widehat{\Gamma} \otimes \lambda, \mathbb{H})$. Then there exists a sequence $(F_n)_{n \in \mathbb{N}_0}$ of functions $F_n : T_{\neq}^{n+1} \to \mathbb{F}^{\otimes(n+1)} \in SL^2(\nu^{n+1}, \mathbb{F}^{\otimes(n+1)})$ with the following four properties:*

(a) *F_n is symmetric in the first n variables and ${}^\circ F_n \in L^2_{\mathcal{L}^{n+1}}(\widehat{\nu^{n+1}}, \mathbb{H}^{n+1})$.*
(b) *F_n is non-anticipating.*
It follows that $I_{n,1}(F_n)$ is $(\mathcal{B}_{k^-})_{k \in T}$-adapted and $I_{n+1}(\widetilde{F}_n) = I_{n+1}(F_n)$.
(c) *${}^\circ I_{n,1}(F_n) \in L^2_{\mathfrak{P}^{\mathcal{W}}}(\widehat{\Gamma \otimes \nu}, \mathbb{H})$.*
(d) *$\varphi = \sum_{n=0}^{\infty} {}^\circ I_{n,1}(F_n)$ converges in $L^2_{\mathfrak{P}^{\mathcal{W}}}(\widehat{\Gamma \otimes \nu}, \mathbb{H})$.*

Proof Let M be the space of all $\varphi \in L^2_{\mathfrak{P}^{\mathcal{W}}}(\widehat{\Gamma \otimes \nu}, \mathbb{H})$ such that there exists a sequence (F_n) with the properties (a)–(d). We have to show that $M = L^2_{\mathfrak{P}^{\mathcal{W}}}(\widehat{\Gamma \otimes \nu}, \mathbb{H})$. Of course, M is a linear space. Let $B \times C$ be a $(\mathfrak{b}_r^{\mathcal{W}})_{r\in[0,\infty[}$-predictable rectangle with $B \in \mathfrak{b}_r^{\mathcal{W}}$ and $C =]r,s]$. Then there exist a $t_r \approx r$ in T and a $\widehat{\Gamma}$-approximation $A \in \mathcal{B}_{t_r}$ of B. Fix $t_s \approx s$ in T and $h \in \mathbb{H}$. By Theorem 17.2.1, $\mathbf{1}_B$ has the expansion $\mathbf{1}_B = \sum_{n=0}^{\infty} {}^\circ I_n(K_n)$ in $L^2_{\mathcal{W}}(\widehat{\Gamma})$. Since $B \in \mathcal{B}_{t_r} \vee \mathcal{N}_{\widehat{\Gamma}}$, we obtain, using Theorem 10.10.1,

$$\mathbf{1}_B = \mathbb{E}^{\mathcal{B}_{t_r} \vee \mathcal{N}_{\widehat{\Gamma}}} \mathbf{1}_B = \sum_{n=0}^{\infty} \mathbb{E}^{\mathcal{B}_{t_r} \vee \mathcal{N}_{\widehat{\Gamma}}} {}^\circ I_n(K_n) = \sum_{n=0}^{\infty} {}^\circ \mathbb{E}^{\mathcal{B}_{t_r}} I_n(K_n) = \sum_{n=0}^{\infty} {}^\circ I_n(G_n)$$

where $G_n = \mathbf{1}_{\{(t_1,\ldots,t_n) | t_1,\ldots,t_n \leq t_r\}} \cdot K_n$. Thus, $\mathbb{E}^{\mathcal{B}_{t_r} \vee \mathcal{N}_{\widehat{\Gamma}}} {}^\circ I_n(K_n) = {}^\circ I_n(G_n) = {}^\circ I_n(K_n)$ in $L^2(\widehat{\Gamma})$ and ${}^\circ G_n = {}^\circ K_n$ in $L^2(\widehat{\nu^n}, \mathbb{H}^{\otimes n})$, by Proposition 17.2.2. It follows that ${}^\circ I_n(G_n)$ is $\mathfrak{b}_r^{\mathcal{W}}$-measurable and ${}^\circ G_n$ is $\mathcal{L}^n$-measurable. Set $D :=]t_r, t_s] \cap T$. Then $\widehat{\nu}(D \Delta C) = 0$. Since $C \in \mathcal{L}^1$ and $C = D$ $\widehat{\nu^n}$-a.e., $D \in \mathcal{L}^1$. Set $F_n := G_n \otimes \mathbf{1}_D \otimes h$. Then (a) and (b) are true. (c) is true, because

$${}^\circ I_{n,1}(F_n) = {}^\circ I_n(G_n) \otimes \mathbf{1}_D \otimes h \text{ is } \mathfrak{P}^{\mathcal{W}}\text{-measurable.}$$

The proof for the other predictable rectangles is similar.

It remains to show that M is complete. Fix a Cauchy sequence $(\varphi_k)_{k\in\mathbb{N}}$ in M with $\varphi_k = \sum_{n=0}^{\infty} {}^\circ I_{n,1}(F_{n,k})$, where $F_{n,k}$ has the properties (a)–(c) with $F_{n,k}$ instead of F_n. Note that (see the proof of Theorem 19.1.1)

$$\sum_{n=0}^{\infty} \int_{T^n_< \times T} \| {}^\circ F_{n,k} - {}^\circ F_{n,k} \|^2_{\mathbb{H}^{n+1}} \, d\widehat{\nu^{n+1}} = \int_{\Omega \times T} \| \varphi_k - \varphi_l \|^2_{\mathbb{H}} \, d\widehat{\Gamma \otimes \nu} \to_{k,l\to\infty} 0.$$

Since the $F_{n,k}$ are symmetric in the first n arguments, $({}^\circ F_{n,k})_{k\in\mathbb{N}}$ is a Cauchy sequence in $L^2_{\mathcal{L}^{n+1}}(\widehat{\nu^{\otimes(n+1)}}, \mathbb{H}^{\otimes(n+1)})$ converging to an ${}^\circ F_n$ in the space $L^2(\widehat{\nu^{n+1}}, \mathbb{H}^{\otimes(n+1)})$. Since we may apply saturation to construct F_n, we can assume that F_n is non-anticipating and symmetric in the first n arguments (see Corollary 10.9.3). Moreover, ${}^\circ I_{n,1}(F_{n,k})$ converges to ${}^\circ I_{n,1}(F_n)$ in $L^2_{\mathfrak{P}^{\mathcal{W}}}(\widehat{\Gamma} \otimes \lambda, \mathbb{H})$. By Proposition 5.6.5, (φ_k) converges to $\varphi =: \sum_{n=0}^{\infty} {}^\circ I_{n,1}(F_n)$ in $L^2_{\mathfrak{P}^{\mathcal{W}}}(\widehat{\Gamma} \otimes \lambda, \mathbb{H})$. This proves the completeness of M and the theorem, by Proposition 5.6.1. □

Now fix $\varphi = \sum_{n=0}^{\infty} {}^\circ I_{n,1}(F_n)$ in $L^2_{\mathfrak{P}^{\mathcal{W}}}(\widehat{\Gamma} \otimes \lambda, \mathbb{H})$, according to the theorem. By Theorem 19.3.1, φ is Skorokhod integrable, because F_n is non-anticipating, and therefore

$$\int_{T_<^{n-1}} \sum_{(i,j)\in\omega^2} \int_{T_{\neq}^2} \langle F_n(\cdot,s,t)(\cdot,\mathfrak{e}_i,\mathfrak{e}_j), F_n(\cdot,t,s)(\cdot,\mathfrak{e}_j,\mathfrak{e}_i)\rangle_{\mathbb{F}^{n-1}} d\nu^2(s,t) d\nu^{n-1} = 0.$$

for each $n \in \mathbb{N}$.

By saturation, we obtain the following lifting result, according to Proposition 19.4.1.

Proposition 19.5.2 *Fix $\varphi = \sum_{n=0}^{\infty} {}^\circ I_{n,1}(F_n) \in L^2_{\mathfrak{P}\mathcal{W}}(\widehat{\Gamma\otimes\nu}, \mathbb{H})$, according to the previous theorem. Then there exists an internal extension $(F_n)_{n\in{}^*\mathbb{N}_0}$ of $(F_n)_{n\in\mathbb{N}_0}$ and an unlimited $K \in {}^*\mathbb{N}$ such that for each unlimited $M \le K$, $M \in {}^*\mathbb{N}$,*

$$\Phi := \sum_{n=0}^{M} I_{n,1}(F_n) \in SL^2(\Gamma\otimes\nu, \mathbb{F}) \quad \text{and} \quad \Phi \approx_{\mathbb{F}} {}^*\varphi \; \widehat{\Gamma\otimes\nu}\text{-a.e.}$$

where each F_n, $n \le M$, is non-anticipating, thus Φ is $(\mathcal{B}_{n^-})_{n\in T}$-adapted, and

$$\delta\Phi : X \mapsto \sum_{s\in T} \Phi(X,s)(X_s) = \int^V \Phi\Delta B(X) = \sum_{n=0}^{M} I_{n+1}(F_n)(X) \in SL^2(\Gamma)$$

and

$$\delta\Phi \approx \delta\varphi \; \widehat{\Gamma}\text{-a.s.}$$

Corollary 19.5.3 *If $\varphi \in L^2_{\mathfrak{P}\mathcal{W}}(\widehat{\Gamma\otimes\nu}, \mathbb{H})$, then $\delta\varphi = \int^V \varphi db_{\mathbb{B}}$, the Itô integral of φ as a random variable (see Chapter 12).*

Exercises

19.1 In the proof of Theorem 19.1.1 prove that $\left({}^\circ I_{n,1}(F_{n,k})\right)$ converges to a g_n in $L^2_{\mathcal{W}\otimes\mathcal{L}^1}(\widehat{\Gamma\otimes\nu}, \mathbb{H})$.

19.2 Prove Corollary 19.1.2.

19.3 Let $\varphi \in L^2_{\mathcal{W}\otimes\mathcal{L}^1}(\widehat{\Gamma\otimes\nu}, \mathbb{H})$ be Malliavin differentiable (see Section 19.2). Prove that φ is Skorokhod integrable.

Since the domain of the Skorokhod integral is quite messy, one often takes the coarser domain $\mathbb{L}_{1,2}$ of Malliavin differentiable processes.

19.4 Prove Proposition 19.4.1.

19.5 Prove Proposition 19.5.2.

19.6 Prove Corollary 19.5.3.

20

The interplay between derivative and integral

In this chapter we study some basic facts on the connection between the Malliavin derivative and the Skorokhod integral.

20.1 The integral is the adjoint operator of the derivative

We will prove this result, using appropriate liftings of the entities involved. In this way we avoid convergence arguments.

Theorem 20.1.1 *Suppose that $\varphi \in L^2_{\mathcal{W}}(\widehat{\Gamma})$ is Malliavin differentiable and $\psi \in L^2_{\mathcal{W}\otimes\mathcal{L}^1}(\widehat{\Gamma\otimes\nu},\mathbb{H})$ is Skorokhod integrable. Then*

$$\langle\varphi,\delta\psi\rangle_{\widehat{\Gamma}} = < D\varphi,\psi >_{\widehat{\Gamma\otimes\nu}}.$$

Proof By Theorems 18.3.1 and 19.4.1, we may assume that

$$\varphi = {}^{\circ}\sum_{n=0}^{M} I_n(F_n),\ \delta\psi = {}^{\circ}\sum_{n=0}^{M} I_{n+1}(\widetilde{G_n}) \text{ in } L^2(\widehat{\Gamma}),$$

$$\psi = {}^{\circ}\sum_{n=0}^{M} I_{n,1}(G_n),\ D\varphi = {}^{\circ}\sum_{n=1}^{M} I_{n-1,1}(F_n) \text{ in } L^2(\widehat{\Gamma\otimes\nu},\mathbb{H}),$$

where $\sum_{n=0}^{M} I_n(F_n)$ and $\sum_{n=0}^{M} I_{n+1}(\widetilde{G_n})$ belong to $SL^2(\Gamma)$ and $\sum_{n=0}^{M} I_{n,1}(G_n)$ and $\sum_{n=1}^{M} I_{n-1,1}(F_n)$ belong to $SL^2(\Gamma\otimes\nu,\mathbb{F})$. Moreover, we may assume that the F_n are symmetric and the G_n are symmetric in the first n arguments. By the pairwise orthogonality of the $I_n(F_n)$ and $I_{n,1}(G_n)$, we obtain

$$
\begin{aligned}
\langle\varphi,\delta\psi\rangle_{\widehat{\Gamma}} &= \mathbb{E}_{\widehat{\Gamma}}\left({}^{\circ}\sum_{n=0}^{M} I_n(F_n)\cdot{}^{\circ}\sum_{n=0}^{M} I_{n+1}(\widetilde{G}_n)\right)\\
&= {}^{\circ}\mathbb{E}_{\Gamma}\left(\sum_{n=0}^{M} I_n(F_n)\cdot\sum_{n=0}^{M} I_{n+1}(\widetilde{G}_n)\right)\\
&= {}^{\circ}\mathbb{E}_{\Gamma}\left(\sum_{n=1}^{M} I_n(F_n)\cdot\sum_{n=0}^{M-1} I_{n+1}(\widetilde{G}_n)\right)\\
&= {}^{\circ}\mathbb{E}_{\Gamma}\left(\sum_{n=0}^{M-1} I_{n+1}(F_{n+1})\cdot\sum_{n=0}^{M-1} I_{n+1}(\widetilde{G}_n)\right)\\
&= {}^{\circ}\sum_{n=0}^{M-1} \mathbb{E}_{\Gamma}\left(I_{n+1}(F_{n+1})\cdot I_{n+1}(\widetilde{G}_n)\right)\\
&= {}^{\circ}\sum_{n=0}^{M-1}\int_{T^n_{<}\times T}\langle F_{n+1},G_n\rangle_{\mathbb{F}^{n+1}}\,d\nu^{n+1}\\
&= {}^{\circ}\sum_{n=0}^{M-1}\langle I_{n,1}(F_{n+1}),I_{n,1}(G_n)\rangle_{\Gamma\otimes\nu}\\
&= {}^{\circ}\left\langle\sum_{n=0}^{M-1} I_{n,1}(F_{n+1}),\sum_{n=0}^{M-1} I_{n,1}(G_n)\right\rangle_{\Gamma\otimes\nu}\\
&= {}^{\circ}\left\langle\sum_{n=1}^{M} I_{n-1,1}(F_n),\sum_{n=0}^{M-1} I_{n,1}(G_n)\right\rangle_{\Gamma\otimes\nu}\\
&= {}^{\circ}\left\langle\sum_{n=1}^{M} I_{n-1,1}(F_n),\sum_{n=0}^{M} I_{n,1}(G_n)\right\rangle_{\Gamma\otimes\nu}\\
&= \left\langle{}^{\circ}\sum_{n=1}^{M} I_{n-1,1}(F_n),{}^{\circ}\sum_{n=0}^{M} I_{n,1}(G_n)\right\rangle_{\widehat{\Gamma\otimes\nu}}\\
&= < D\varphi,\psi >_{\widehat{\Gamma\otimes\nu}}.
\end{aligned}
$$

□

Let O be a **densely defined** operator between two Hilbert spaces X and Y, i.e., the domain E of O is a dense subspace of X. Then O is called **closed** if, for each sequence (ψ_n) in E, whenever (ψ_n) converges to $\psi\in E$ and $(O\psi_n)$ converges to an element $\varphi\in Y$, then $O\psi=\varphi$.

Corollary 20.1.2

(a) The Skorokhod integral operator δ is closed.
(b) The Malliavin derivative operator D is closed.

Proof (a) Let $\psi_n, \psi \in \Delta$ with $\lim_{n\to\infty} \psi_n = \psi$ in $L^2_{\mathcal{W}\otimes\mathcal{L}^1}(\widehat{\Gamma\otimes\nu}, \mathbb{H})$ and suppose that $\lim_{n\to\infty} \delta(\psi_n) = \varphi$ in $L^2_{\mathcal{W}}(\widehat{\Gamma})$. We obtain for all $\gamma \in \mathbb{D}_{1,2}$

$$\langle \gamma, \delta(\psi)\rangle_{\widehat{\Gamma}} = \langle D\gamma, \psi\rangle_{\widehat{\Gamma\otimes\nu}} = \lim_{n\to\infty} \langle D\gamma, \psi_n\rangle_{\widehat{\Gamma\otimes\nu}} = \lim_{n\to\infty} \langle \gamma, \delta(\psi_n)\rangle_{\widehat{\Gamma}} = \langle \gamma, \varphi\rangle_{\widehat{\Gamma}}.$$

This proves that $\delta(\psi) = \varphi$ in $L^2_{\mathcal{W}}(\widehat{\Gamma})$. The proof of (b) is very similar. □

20.2 A Malliavin differentiable function multiplied by square-integrable deterministic functions is Skorokhod integrable

In general a function $\varphi \otimes a : \Omega \times T \ni (X,t) \mapsto \varphi(X)\cdot a \in \mathbb{H}$, where $\varphi \in L^2_{\mathcal{W}}(\widehat{\Gamma})$ and $a \in \mathbb{H}$, is not necessarily Skorokhod integrable (note that φ does not depend on t). But we have the following first version of the 'integration by parts' formula.

Proposition 20.2.1 *Fix a Malliavin differentiable φ. Fix $g \in L^2_{\mathcal{L}^1}(\widehat{\nu}, \mathbb{H})$ with $\|g\|_{\widehat{\nu}} = 1$. Then $\varphi \otimes g$ is Skorokhod integrable and*

$$\delta(\varphi \otimes g) = \varphi \cdot I(g) - D\varphi_{\to g} \text{ in } L^2_{\mathcal{W}}(\widehat{\Gamma}).$$

Recall that $D\varphi_{\to g} = \langle D\varphi, g\rangle_{\widehat{\nu}}$ in $L^2_{\mathcal{W}}(\widehat{\Gamma})$ (see Section 18.4).

Proof Obviously, $\varphi \otimes g \in L^2_{\mathcal{W}\otimes\mathcal{L}_1}(\widehat{\Gamma\otimes\nu})$. Let $G \in SL^2(\nu, \mathbb{F})$ be a lifting of g and let $\varphi = \sum_{n=0}^{\infty} {}^\circ I_n(F_n)$ according to Theorem 17.2.1. Let us write F^n instead of F_n. In order to prove that $\varphi \otimes {}^\circ G$ is Skorokhod integrable, by Theorem 19.3.1, we have to show that for any internal orthonormal basis $\mathfrak{E}$ of $\mathbb{F}$

$$\sum_{n=1}^{\infty}{}^\circ \int_{T^{n-1}_< \times T^2_{\neq}} \sum_{\mathfrak{e}\in\mathfrak{E}^2} \left\langle F^n_{(\mathfrak{o},s)}(\cdot,\mathfrak{e}_1)G_t(\mathfrak{e}_2), F^n_{(\mathfrak{o},t)}(\cdot,\mathfrak{e}_2)G_s(\mathfrak{e}_1)\right\rangle_{\mathbb{F}^{n-1}} d\nu^{n+1}(\mathfrak{o},s,t)$$

converges in $\mathbb{R}$. By the Cauchy–Schwarz inequality, we obtain for all $r \in T_{<}^{n-1}$

$$\left| \int_{T_{\neq}^2} \sum_{\mathfrak{e} \in \mathfrak{E}^2} \langle F_{(r,s)}^n(\cdot, \mathfrak{e}_1) G_t(\mathfrak{e}_2) F_{(r,t)}^n(\cdot, \mathfrak{e}_2) G_s(\mathfrak{e}_1) \rangle_{\mathbb{F}^{n-1}} \, d\nu^2(s,t) \right|$$

$$= \left| \int_{T_{\neq}^2} \sum_{\mathfrak{b} \in \mathfrak{E}^{n-1}} \sum_{\mathfrak{e} \in \mathfrak{E}^2} F_{(r,s)}^n(\mathfrak{b}, \mathfrak{e}_1) G_t(\mathfrak{e}_2) \cdot F_{(r,t)}^n(\mathfrak{b}, \mathfrak{e}_2) G_s(\mathfrak{e}_1) d\nu^2(s,t) \right|$$

$$\leq \int_{T_{\neq}^2} \sum_{\mathfrak{b} \in \mathfrak{E}^{n-1}} \sum_{\mathfrak{e} \in \mathfrak{E}^2} \left(F_{(r,s)}^n(\mathfrak{b}, \mathfrak{e}_1) G_t(\mathfrak{e}_2) \right)^2 \cdot d\nu^2(s,t)$$

$$\leq \int_T \| F_{(r,s)}^n \|_{\mathbb{F}^n}^2 \, d\nu(s) \int_T \| G_t \|_{\mathbb{F}}^2 \, d\nu(t).$$

It follows that

$$\left| \int_{T_{<}^{n-1} \times T_{\neq}^2} \sum_{\mathfrak{e} \in \mathfrak{E}^2} \langle F_{(\mathfrak{o},s)}^n(\cdot, \mathfrak{e}_1) G_t(\mathfrak{e}_2), F_{(\mathfrak{o},t)}^n(\cdot, \mathfrak{e}_2) G_s(\mathfrak{e}_1) \rangle_{\mathbb{F}^{n-1}} \, d\nu^{n+1}(\mathfrak{o}, s, t) \right|$$

$$\leq \int_{T^{n-1} \times T} \| F^n \|_{\mathbb{F}^n}^2 \, d\nu^n \cdot \int_T \| G_t \|_{\mathbb{F}}^2 \, d\nu(t).$$

Since φ is Malliavin differentiable,

$$\sum_{n=1}^{\infty} {}^{\circ} \| I_{n-1,1}(F^n) \|_{\mathbb{F}}^2 = \sum_{n=1}^{\infty} \int_{T_{<}^{n-1} \times T} \| {}^{\circ} F^n \|_{\mathbb{F}^n}^2 \, d\widehat{\nu^n}$$

$$= \sum_{n=1}^{\infty} {}^{\circ} \int_{T_{<}^{n-1} \times T} \| F^n \|_{\mathbb{F}^n}^2 \, d\nu^n$$

converges. This proves that $\varphi \circ g$ is Skorokhod integrable. Now, by Theorem 13.2.1,

$$\left(\sum_{n=0}^{\infty} {}^{\circ} I_n(F^n) \right) I(g) = \sum_{n=0}^{\infty} \left({}^{\circ} \left(I_n(F^n) \cdot I(G) \right) \right)$$

$$= \sum_{n=0}^{\infty} \left({}^{\circ} I_{n+1}(\widetilde{F^n \circ G}) + {}^{\circ} \int_T \langle I_{n-1,1}(F^n), G \rangle_{\mathbb{F}} \, d\nu \right)$$

$$= \sum_{n=0}^{\infty} \left({}^{\circ} I_{n+1}(\widetilde{F^n \circ G}) + \int_T \langle {}^{\circ} I_{n-1,1}(F^n), {}^{\circ} G \rangle_{\mathbb{H}} \, d\widehat{\nu} \right)$$

$$= \delta(\varphi \otimes {}^{\circ} G) + D\varphi_{\to {}^{\circ} G} = \delta(\varphi \otimes g) + D\varphi_{\to g}.$$

□

20.3 The duality between the domains of D and δ

The following result is often a useful tool in order to decide whether a functional belongs to the domain of the Skorokhod integral or to the domain of the Malliavin derivative. Recall that $\mathrm{Fin}L^2_{\mathcal{W}}(\widehat{\Gamma})$ denotes the linear space of functions in $L^2_{\mathcal{W}}(\widehat{\Gamma})$ of finite chaos level. We define

$$L^2_{\mathcal{W}\otimes\mathcal{L}^1,k}(\widehat{\Gamma\otimes\nu},\mathbb{H}) := \left\{\sum_{n=0}^{\infty} I_{n,1}(F_n) \in L^2_{\mathcal{W}\otimes\mathcal{L}^1}(\widehat{\Gamma\otimes\nu},\mathbb{H}) \mid F_l = 0 \text{ for all } l > k\right\}$$

and

$$\mathrm{Fin}L^2_{\mathcal{W}\otimes\mathcal{L}^1}(\widehat{\Gamma\otimes\nu},\mathbb{H}) := \bigcup_{k\in\mathbb{N}_0} L^2_{\mathcal{W}\otimes\mathcal{L}^1,k}(\widehat{\Gamma\otimes\nu},\mathbb{H}).$$

Then $\mathrm{Fin}L^2_{\mathcal{W}\otimes\mathcal{L}^1}(\widehat{\Gamma\otimes\nu},\mathbb{H})$ is a subspace of the domain Δ of the Skorokhod integral, and is dense in $L^2_{\mathcal{W}\otimes\mathcal{L}^1}(\widehat{\Gamma\otimes\nu},\mathbb{H})$. Recall that $\mathrm{Fin}L^2_{\mathcal{W}}(\widehat{\Gamma}) \subseteq \mathbb{D}^{1,2}$ is a dense subspace of $L^2_{\mathcal{W}}(\widehat{\Gamma})$.

Proposition 20.3.1 (a) *Let $\mathbb{D}^{1,2}_0$ be the set of all $\varphi \in L^2_{\mathcal{W}}(\widehat{\Gamma})$ such that the linear form*

$$C_\varphi : \mathrm{Fin}L^2_{\mathcal{W}\otimes\mathcal{L}^1}(\widehat{\Gamma\otimes\nu},\mathbb{H}) \ni \alpha \mapsto \langle \delta\alpha, \varphi\rangle_{\widehat{\Gamma}} \in \mathbb{R} \textit{ is continuous.}$$

Then $\mathbb{D}^{1,2}_0 = \mathbb{D}^{1,2}$.

(b) *Let Δ_0 be the set of all $\varphi \in L^2_{\mathcal{W}\otimes\mathcal{L}^1}(\widehat{\Gamma\otimes\nu},\mathbb{H})$ such that the linear form*

$$C_\varphi : \mathrm{Fin}L^2_{\mathcal{W}}(\widehat{\Gamma}) \ni \alpha \mapsto \langle D\alpha, \varphi\rangle_{\widehat{\Gamma\otimes\nu}} \in \mathbb{R} \textit{ is continuous.}$$

Then $\Delta_0 = \Delta$.

Proof We will only prove (a). The proof of (b) is similar.

Suppose that $\varphi = \sum_{n=0}^{\infty} I_n(f_n) \in \mathbb{D}^{1,2}_0$. Since C_φ is continuous, there is a constant c such that $|C_\varphi(\alpha)| \le c\,\|\alpha\|_{\widehat{\Gamma\otimes\nu}}$ for all $\alpha \in \mathrm{Fin}L^2_{\mathcal{W}\otimes\mathcal{L}^1}(\widehat{\Gamma\otimes\nu},\mathbb{H})$. In order to show that $D\varphi$ converges in $L^2_{\mathcal{W}\otimes\mathcal{L}^1}(\widehat{\Gamma\otimes\nu},\mathbb{H})$, it suffices to show that

$\sum_{n=1}^{k} \|I_{n-1,1}(f_n)\|^2_{\widehat{\Gamma\otimes\nu}} \leq c^2$ for each $k \in \mathbb{N}$. This is true, because

$$\begin{aligned}
\sum_{n=1}^{k} \|I_{n-1,1}(f_n)\|^2_{\widehat{\Gamma\otimes\nu}} &= \left\|\sum_{n=1}^{k} I_{n-1,1}(f_n)\right\|^2_{\widehat{\Gamma\otimes\nu}} \\
&= \left\langle \sum_{n=1}^{k} I_{n-1,1}(f_n), \sum_{n=1}^{k} I_{n-1,1}(f_n)\right\rangle_{\widehat{\Gamma\otimes\nu}} \\
&= \left\langle \sum_{n=0}^{k} I_n(f_n), \delta\left(\sum_{n=1}^{k} I_{n-1,1}(f_n)\right)\right\rangle_{\widehat{\Gamma}} \quad \text{(by 20.1.1)} \\
&= \left\langle \varphi, \delta\left(\sum_{n=1}^{k} I_{n-1,1}(f_n)\right)\right\rangle_{\widehat{\Gamma}} \leq c \cdot \left\|\sum_{n=1}^{k} I_{n-1,1}(f_n)\right\|_{\widehat{\Gamma\otimes\nu}}.
\end{aligned}$$

It follows that $\varphi \in \mathbb{D}^{1,2}$.

Now assume that $\varphi \in \mathbb{D}^{1,2}$. Since for all $\alpha \in \mathrm{Fin}L^2_{\mathcal{W}\otimes\mathcal{L}^1}(\widehat{\Gamma\otimes\nu}, \mathbb{H})$

$$|C_\varphi(\alpha)| = |\langle \delta\alpha, \varphi\rangle_{\widehat{\Gamma}}| = |\langle \alpha, D\varphi\rangle_{\widehat{\Gamma\otimes\nu}}| \leq \|D\varphi\|_{\widehat{\Gamma\otimes\nu}} \|\alpha\|_{\widehat{\Gamma\otimes\nu}},$$

C_φ is continuous. This proves that $\varphi \in \mathbb{D}_0^{1,2}$. □

20.4 $L^2_{\mathcal{W}\otimes\mathcal{L}^1}(\widehat{\Gamma\otimes\nu}, \mathbb{H})$ is the orthogonal sum of the range of D and the kernel of δ

In an application of the preceding duality theorems, we will show that $L^2_{\mathcal{W}\otimes\mathcal{L}^1}(\widehat{\Gamma\otimes\nu}, \mathbb{H})$ is the orthogonal sum of the range of the Malliavin derivative and the kernel of the Skorokhod integral.

Proposition 20.4.1 $L^2_{\mathcal{W}\otimes\mathcal{L}^1}(\widehat{\Gamma\otimes\nu}, \mathbb{H}) = D[\mathbb{D}^{1,2}] \oplus \{\psi \in \Delta \mid \delta\psi = 0\}$.

Proof Since $D[\mathbb{D}^{1,2}]$ is a closed subspace of $L^2_{\mathcal{W}\otimes\mathcal{L}^1}(\widehat{\Gamma\otimes\nu}, \mathbb{H})$ (see Proposition 18.7.1), by the projection theorem, each $\varphi \in L^2_{\mathcal{W}\otimes\mathcal{L}^1}(\widehat{\Gamma\otimes\nu}, \mathbb{H})$ is the sum $\varphi = \psi + u$ of an element $\psi \in D[\mathbb{D}^{1,2}]$ and $u \in L^2_{\mathcal{W}\otimes\mathcal{L}^1}(\widehat{\Gamma\otimes\nu}, \mathbb{H})$ such that $u \perp D[\mathbb{D}^{1,2}]$. Since

$$\mathbb{D}^{1,2} \ni \alpha \mapsto \langle D\alpha, u\rangle_{\widehat{\Gamma\otimes\nu}} = 0,$$

this function is continuous. By Proposition 20.3.1 (b), u is Skorokhod integrable.

Since for all $\alpha \in \mathbb{D}^{1,2}$

$$\langle \delta u, \alpha \rangle_{\widehat{\Gamma}} = \langle u, D\alpha \rangle_{\widehat{\Gamma \otimes \nu}} = 0,$$

$\delta u = 0$. □

20.5 Integration by parts

The next aim is to show the integration by parts formula. Recall from Section 18.4 that $D\varphi_{\to g} \in L^2_{\mathcal{W}}(\widehat{\Gamma})$ is the directional derivative of $\varphi \in \mathbb{D}^{1,2}$ into the direction $g \in L^2_{\mathcal{L}^1}(\widehat{\nu}, \mathbb{H})$.

Theorem 20.5.1 *Suppose that φ and ψ are Malliavin differentiable. Let $g \in L^2_{\mathcal{L}^1}(\widehat{\nu}, \mathbb{H})$ with $\|g\|_{\widehat{\nu}} = 1$. Then*

$$\mathbb{E}_{\widehat{\Gamma}}(\varphi\psi I(g)) = \mathbb{E}_{\widehat{\Gamma}}\left(\psi \cdot D\varphi_{\to g} + \varphi \cdot D\psi_{\to g}\right).$$

Proof Note that

$$\begin{aligned}
&\mathbb{E}_{\widehat{\Gamma}}\left(\psi \cdot D\varphi_{\to g} + \varphi \cdot D\psi_{\to g}\right) \\
&\quad = \left\langle \psi, D\varphi_{\to g}\right\rangle_{\widehat{\Gamma}} + \langle \psi, \delta(\varphi \otimes g)\rangle_{\widehat{\Gamma}} \quad \text{(by 20.1.1 and 20.2.1)} \\
&\quad = \langle \psi, \varphi \cdot I(g)\rangle_{\widehat{\Gamma}} \quad \text{(by 20.2.1)} \\
&\quad = \mathbb{E}_{\widehat{\Gamma}}(\varphi\psi I(g)).
\end{aligned}$$

□

Exercises

20.1 Prove Corollary 20.1.2 (b).
20.2 Prove Proposition 20.3.1 (b).

21

Skorokhod integral processes

In this chapter we study the Skorokhod integral process operator δ^{pr}, densely defined on $L^2_{\mathcal{W}\otimes\mathcal{L}^1}(\widehat{\Gamma\otimes\nu},\mathbb{H})$ with values in $L^2_{\mathcal{W}\otimes\mathcal{L}^1}(\widehat{\Gamma\otimes\nu})$. Fix $\varphi\in L^2_{\mathcal{W}\otimes\mathcal{L}^1}(\widehat{\Gamma\otimes\nu},\mathbb{H})=L^2_{\mathcal{W}\otimes\mathrm{Leb}[0,\infty[}(\widehat{\Gamma}\otimes\lambda,\mathbb{H})$ with decomposition

$$\varphi=\sum_{n=0}^{\infty}{}^{\circ}I_{n,1}(F_n)=\sum_{n=0}^{\infty}I_{n,1}({}^{\circ}F_n),$$

according to Theorem 19.1.1. Recall that the range of the Skorokhod integral is a set of random variables. We have defined

$$\delta\left(\sum_{n=0}^{\infty}{}^{\circ}I_{n,1}(F_n)\right):X\mapsto\sum_{n=0}^{\infty}{}^{\circ}\left(\int^{V}I_{n,1}(F_n)\Delta B(X)\right)$$

if this series converges in $L^2_{\mathcal{W}}(\widehat{\Gamma})$. Recall from Section 19.1 that

$$\int^{V}I_{n,1}(F_n)\Delta B(X)=I_{n+1}(\widetilde{F}_n).$$

In analogy to the Itô integral we want the Skorokhod integral of φ to be dependent on time. The resulting stochastic process is called the Skorokhod integral process of φ and is denoted by $\delta^{pr}(\varphi)$. It is natural to address the question about continuous versions of $\delta^{pr}(\varphi)$ for non-adapted processes φ. This problem is the topic of this chapter.

21.1 The Skorokhod integral process operator

In analogy to the definition of the Malliavin derivative and the Skorokhod integral the **Skorokhod integral process** $\delta^{pr}(\varphi):\Omega\times T\to\mathbb{R}$ of φ is defined by internal Riemann–Stieltjes integration under the standard part map and the sum, in the following way (see Section 12.1). Using the notation $\cdot_{n+1}$ from Section 19.1, we define

$$\delta^{pr}(\varphi)(X,r) := \sum_{n=0}^{\infty} {}^{\circ}\left(\int I_{n,1}(F_n)\Delta B(X,r)\right) = \sum_{n=0}^{\infty} {}^{\circ}I_{n+1}(\widetilde{\mathbf{1}_{T_r} \cdot_{n+1} F_n})(X)$$

if this series converges in $L^2_{\mathcal{W}\otimes\mathcal{L}^1}(\widehat{\Gamma\otimes\nu})$. By Theorem 19.1.1 (d), it is sufficient to require that this series converges in $L^2(\widehat{\Gamma\otimes\nu})$.

Since $\varphi \in L^2_{\mathcal{W}\otimes\mathcal{L}^1}(\widehat{\Gamma\otimes\nu},\mathbb{H})$ and $\delta^{pr}(\varphi) \in L^2_{\mathcal{W}\otimes\mathcal{L}^1}(\widehat{\Gamma\otimes\nu})$ (if $\delta^{pr}(\varphi)$ converges), these functions are equivalent to functionals ψ in $L^2_{\mathcal{W}}(\widehat{\Gamma}\otimes\lambda,\mathbb{H}) = L^2(W\otimes\lambda,\mathbb{H})$, $\delta^{pr}(\psi)$ in $L^2_{\mathcal{W}}(\widehat{\Gamma}\otimes\lambda) = L^2(W\otimes\lambda)$, respectively. We may define for $r\in[0,\infty[$:

$$\left(\int \psi\, db_{\mathbb{B}}\right)(\cdot,r) := \int_0^r \psi\, db_{\mathbb{B}} := \delta^{pr}(\psi)(\cdot,r).$$

This integral is well defined $\widehat{\Gamma}\otimes\lambda$-a.e.; in fact, we have for $\widehat{\Gamma\otimes\nu}$ almost all $(X,t)\in\Omega\times T$:

$$\psi(X,{}^{\circ}t) = \varphi(X,t) \quad \text{and} \quad \int_0^{{}^{\circ}t} \psi\, db_{\mathbb{B}} = \delta^{pr}(\varphi)(\cdot,t).$$

The proof of the following theorem is similar to the proof of Theorem 19.3.1, where $\mathfrak{E}$ is an ONB of $\mathbb{F}$.

Theorem 21.1.1 *Let us write F^n instead of F_n. Then $\delta^{pr}(\varphi)$ converges if and only if $\sum_{n=1}^{\infty} {}^{\circ}A^n$ converges in $\mathbb{R}$, where $A^n =$*

$$\int_T \int_{T^2_{r,\neq}} \sum_{\mathfrak{e}\in\mathfrak{E}^2} \int_{T^{n-1}_<} \langle F^n_{(\cdot,s,t)}(\cdot,\mathfrak{e}_1,\mathfrak{e}_2)\cdot F^n_{(\cdot,t,s)}(\cdot,\mathfrak{e}_2,\mathfrak{e}_1)\rangle_{\mathbb{F}^{n-1}} d\nu^{n-1} d\nu^2(t,s) d\nu(r).$$

21.2 On continuous versions of Skorokhod integral processes

Following [84] and also [90], we will use Theorem 13.7.2 to construct a continuous version of $\delta^{pr}(\varphi)$ if φ belongs to finite chaos levels. This is an extension of a result due to Imkeller [48], who proved it for the classical Wiener space. Our proof is more elementary.

We have to prove that $(X,r) \mapsto I_{n+1}(\widetilde{\mathbf{1}_{T_r} \cdot_{n+1} F})(X)$ has a uniquely determined continuous version, where $F : T_{\neq}^{n+1} \to \mathbb{F}^{\otimes(n+1)} \in SL^2\left(\nu^{n+1}, \mathbb{F}^{\otimes(n+1)}\right)$. We do not need that F is a lifting of a standard function $f \in L^2\left(\lambda^{n+1}, \mathbb{H}^{\otimes(n+1)}\right)$ (see [90]). Fix limited $s < r$ in $T \cup \{0\}$ and define $G_i : T^{n+1} \to \mathbb{F}^{\otimes(n+1)}$ for $i = 1, \ldots, n+1$ by setting for $a := (a_1, \ldots, a_{n+1}) \in \mathbb{F}^{n+1}$ and $t := (t_1, \ldots, t_{n+1}) \in T_{\neq}^{n+1}$

$$G_{i,t}(a) := \mathbf{1}_{]s,r]} \cdot_{n+1} F_{(t_1,\ldots,t_{i-1},t_{i+1},\ldots,t_{n+1},t_i)}(a_1, \ldots, a_{i-1}, a_{i+1}, \ldots, a_{n+1}, a_i),$$

where $]s,r] := \{t \in T \mid s < t \leq r\}$. Then

$$\begin{aligned}&\left(\mathbb{E}_\Gamma\left(I_{n+1}(\widetilde{\mathbf{1}_{T_r} \cdot_{n+1} F}) - I_{n+1}(\widetilde{\mathbf{1}_{T_s} \cdot_{n+1} F})\right)^4\right)^{\frac{1}{4}} \\ &\leq \sum_{i=1}^{n+1} \left(\mathbb{E}_\Gamma\left((I_{n+1}(G_i))^4\right)\right)^{\frac{1}{4}}.\end{aligned}$$

By Theorem 13.1.3,

$$\begin{aligned}&\mathbb{E}_\Gamma\left((I_{n+1}(G_i))^4\right) \\ &\leq 3^{n+2}\left(\sum_{t \in T_<^{n+1}} \|G_{i,t}\|^2_{\mathbb{F}^{n+1}} \frac{1}{H^{n+1}}\right)^2 \\ &\leq 3^{n+2}\left(\sum_{t \in T_<^n} \sum_{x \in]s,r]} \|F(t,x)\|^2_{\mathbb{F}^{n+1}} \frac{1}{H^{n+1}}\right)^2 \\ &= 3^{n+2}\left(\sum_{t \in T_<^n} \sum_{x \in T_r} \|F(t,x)\|^2_{\mathbb{F}^{n+1}} \frac{1}{H^{n+1}} - \sum_{t \in T_<^n} \sum_{x \in T_s} \|F(t,x)\|^2_{\mathbb{F}^{n+1}} \frac{1}{H^{n+1}}\right)^2.\end{aligned}$$

Set $\alpha(k) := \sum_{t \in T_<^n} \sum_{x \in T_k} \|F(t,x)\|^2_{\mathbb{F}^{n+1}} \frac{1}{H^{n+1}}$ for $k \in T$. Then α is S-continuous on T, because $\|F\|^2_{\mathbb{F}^{n+1}} \in SL^2\left(\nu^{n+1}\right)$. Therefore we may set $\alpha(^\circ k) := {}^\circ\alpha(k))$ for all limited $k \in T$ and obtain:

$$\begin{aligned}&{}^\circ\mathbb{E}_\Gamma\left(I_{n+1}(\widetilde{\mathbf{1}_{T_r} \cdot_{n+1} F}) - I_{n+1}(\widetilde{\mathbf{1}_{T_s} \cdot_{n+1} F})\right)^4 \\ &\leq (n+1)^4 \cdot 3^{n+2} \cdot (\alpha(^\circ r) - \alpha(^\circ s))^2.\end{aligned}$$

By Theorem 13.7.2, we obtain the following result, due to [84].

Theorem 21.2.1 *The Skorokhod integral process $\delta^{pr}(\varphi)$ has a continuous version if φ belongs to finite chaos levels.*

Exercises

21.1 Prove Theorem 21.1.1.

22

Girsanov transformations

In this part we apply the smooth representation of Wiener functionals to study time-anticipating and non-time-anticipating Girsanov transformations on abstract Wiener spaces in the style of finite-dimensional analysis.

Fix again an abstract Wiener–Fréchet space $\mathbb{B}$ over $\mathbb{H}$. Let γ_1 be the Gaussian measure of variance 1 on the Borel σ-algebra of $\mathbb{B}$, constructed from the Gaussian measure γ_1 on the algebra of cylinder sets of $\mathbb{H}$ (see Section 4.3). In view of Proposition 4.8.1 this measure is called the **Wiener measure on** $\mathbb{B}$.

Each measurable mapping $\psi : \mathbb{B} \to \mathbb{H}$ defines a shift $\sigma : \mathbb{B} \to \mathbb{B}$ by setting

$$\sigma(f) := f + \psi(f).$$

The following questions have been studied extensively in the standard literature (see for example [18], [38], [57], [58], [79], [115]) and are the topics of this chapter.

(A) When is the image measure of the Wiener measure on $\mathbb{B}$ under σ absolutely continuous (or equivalent) to the Wiener measure?

(B) Is it possible to find a new measure P on the Borel σ-algebra of $\mathbb{B}$ that is absolutely continuous (or equivalent) to the Wiener measure and such that the Wiener measure is the image measure of P under σ?

In the results on these questions, it is often presumed that ψ is Skorokhod integrable or locally $\mathbb{L}_{1,2}$ or that ψ is $\mathbb{H}$-continuously differentiable. Recall that the space $\mathbb{L}_{1,2}$ is a subspace of the domain of the Skorokhod integral, consisting of the Malliavin differentiable processes.

Here we construct smooth liftings of ψ, defined on our *finite-dimensional space, and then use the substitution rule of finite-dimensional analysis.

In order to take into account the timeline $[0,\infty[$, we study the abstract Wiener space $C_{\mathbb{B}}$ (see Section 4.8). Fix a $\mathcal{B}(C_{\mathbb{B}}) \otimes \mathrm{Leb}[0,\infty[$-measurable process φ_{st} :

$C_{\mathbb{B}} \times [0,\infty[\to \mathbb{H}$ such that $\varphi_{\mathrm{st}}(f,\cdot)$ is locally in $L^1(\lambda,\mathbb{H})$ for W-almost all f. Recall that W is the Wiener measure on $C_{\mathbb{B}}$. It is the image measure of $\widehat{\Gamma}$ by $X \mapsto b_{\mathbb{B}}(X,\cdot)$. Let ψ be a mapping from $C_{\mathbb{B}}$ into the space of absolutely continuous $\mathbb{H}$-valued functions defined on $[0,\infty[$ by setting

$$\psi(f)(r) := \int_0^r \varphi_{\mathrm{st}}(f,s)d\lambda(s).$$

The integral here is the Bochner integral of $\varphi_{\mathrm{st}}(f,\cdot)$. The shift $\sigma : C_{\mathbb{B}} \to C_{\mathbb{B}}$, mentioned above, has the form

$$(\sigma(f))(r) := f(r) + \int_0^r \varphi_{\mathrm{st}}(f,s)d\lambda(s).$$

Since $C_{\mathbb{B}}$ is the set of trajectories of the Brownian motion $b := b_{\mathbb{B}}$ and φ_{st} can be identified with the function $\varphi : \Omega \times [0,\infty[\to \mathbb{H}$, $(X,r) \mapsto \varphi_{\mathrm{st}}(b(X,\cdot),r)$, we see that σ is a shift of b, given by

(GT) $(\sigma \circ b)(X,r) := b(X,r) + \int_0^r \varphi(X,s)d\lambda(s)$ with $(X,r) \in \Omega \times [0,\infty[$.

Let us address the following questions which are essentially the questions asked under (A) and (B) above: When does there exist a probability measure P on $L_\Gamma(\mathcal{B})$, absolutely continuous or even equivalent to $\widehat{\Gamma}$, such that

(a) $\sigma \circ b$ is again a Brownian motion with respect to P,

or, if this is not possible,

(b) $\sigma \circ b$ follows at least the law of Brownian motion with respect to P ?

In a remark we will point out that the answers can be transferred to the Wiener measure on $C_{\mathbb{B}}$. The results in this sections are slight extensions of results in [84].

I would like to mention that we obtain similar answers to the preceding question about continuous 'shifts' $\sigma : \Omega \times [0,\infty[\to \mathbb{B}$ with

$$\sigma(X,r) := b(X,r) + \int_0^r \varphi(X,s)d\lambda(s), \quad (X,r) \in \Omega \times [0,\infty[,$$

where φ is not necessarily $\mathcal{W}$-measurable in the first component (see Section 22.5 below).

To give answers to questions (a) and (b), we shall proceed as follows. Using the chaos decomposition of φ, we construct an internally smooth lifting, defined on $\Omega \times T$ with values in $\mathbb{F}$, of the $\mathbb{H}$-valued process $(X,r) \mapsto \int_0^r \varphi(X,s)d\lambda(s)$. Since the internal Brownian motion B is an internally smooth lifting of b, σ is infinitely close to an internally smooth mapping Φ from Ω into Ω. In order to obtain positive answers to the questions (a) and (b), the conditions on φ have to guarantee that the substitution rule in elementary analysis can be applied to the internal representation Φ of σ.

It should be mentioned that, instead of $\Omega = \mathbb{F}^T$, we can take $\Omega = \mathbb{F}^{T_\sigma}$ for some $\sigma \in T$, in particular, for a limited σ.

22.1 From standard to internal shifts

Recall that we have fixed a $\mathcal{W}\otimes\text{Leb}[0,\infty[$-measurable process $\varphi : \Omega \times [0,\infty[\to \mathbb{H}$ such that $\varphi(X,\cdot)$ is locally in $L^1(\lambda,\mathbb{H})$ for $\widehat{\Gamma}$-almost all X. We may identify φ with the $\mathcal{W}\otimes\mathcal{L}^1$-measurable function $(X,t)\mapsto\varphi(X,{}^\circ t)$. Then $\varphi(X,\cdot)$ is locally in $L^1_{\mathcal{L}^1}(\widehat{\nu},\mathbb{H})$ for $\widehat{\Gamma}$-almost all X.

There exists a sequence $(\varphi_n)_{n\in\mathbb{N}}$ of functions $\varphi_n \in L^2_{\mathcal{W}\otimes\mathcal{L}_1}(\widehat{\Gamma\otimes\nu},\mathbb{H})$ converging to φ in measure. We may also assume that, for all $r\in\mathbb{N}$, $\lim_{n\to\infty}\int_{T_r}\|\varphi_n-\varphi\|_{\mathbb{H}}\,d\widehat{\nu}=0$ in measure. Let $\Psi:\Omega\times T\to\mathbb{F}$ be a lifting of φ and let $\Phi_n\in SL^2(\Gamma\otimes\nu,\mathbb{F})$ be a lifting of φ_n, according to Corollary 19.1.2. Note that $({}^\circ\|\Phi_n-\Psi\|_{\mathbb{F}}) = (\|\varphi_n-\varphi\|_{\mathbb{H}})$ converges to 0 and $\lim_{n,m\to\infty}\left({}^\circ\int_{T_r}\|\Phi_n-\Phi_m\|_{\mathbb{F}}\,d\nu\right)=0$, both in measure, for all $r\in\mathbb{N}$. Using techniques established in Corollary 10.9.3, we can find a lifting Φ of φ such that $\Phi(X,\cdot)$ is locally in $SL^1(\nu,\mathbb{F})$ for $\widehat{\Gamma}$-almost all $X\in\Omega$. Moreover, we may assume that Φ inherits from the Φ_n their internal properties; in particular, Φ has the form

$$\Phi(X,t)=\sum_{k=0}^{M}\sum_{t_1<\ldots<t_k}F_k(t_1,\ldots,t_k,t)\left(X_{t_1},\ldots,X_{t_k},\cdot\right)\in\mathbb{F}'=\mathbb{F},$$

where $M\in{}^*\mathbb{N}$ and $F_k:T^{k+1}_{\neq}\to\mathbb{F}^{\otimes(k+1)}$.

(+) If φ is $(\mathfrak{b}_r)_{r\in[0,\infty[}$-adapted, we may assume that all F_k are non-anticipating, thus, Φ is $(\mathcal{B}_{t^-})_{t\in T}$-adapted (see Proposition 19.5.2).

(++) If $\varphi\in L^2_{\mathcal{W}\otimes\mathcal{L}^1}\left(\widehat{\Gamma\otimes\nu},\mathbb{H}\right)$, we may assume that $\Phi\in SL^2(\Gamma\otimes\nu,\mathbb{F})$.

Since $\mathbb{H}\subseteq\mathbb{F}$ and $\frac{a}{\|a\|_{\mathbb{F}}}$ can be extended to an internal ONB of $\mathbb{F}$ for each $a\in\mathbb{H}$, $a\neq0$, we see that for $\widehat{\Gamma}$-almost all X

$$\Phi(X,\cdot)(a)\text{ is locally in }SL^1(\nu)\quad\text{and}\quad\Phi(X,\cdot)(a)\approx\varphi(X,\cdot)(a)\text{ for each }a\in\mathbb{H}.$$

Here we have identified $a \in \mathbb{H}$ with ${}^*a \in \mathbb{F}$. Define the function $\widetilde{\Phi} : \Omega \to \Omega$ by

$$\widetilde{\Phi}(X) := \left(X_t + \Phi(X,t)\frac{1}{H} \right)_{t \in T}.$$

Since there exists a set U of $\widehat{\Gamma}$-measure 1 such that, for all $X \in U$, all $a \in \mathbb{H}$ and all limited $t \in T$,

$$b(X, {}^\circ t) \approx_{\mathbb{B}} B(X,t) \quad \text{and} \quad \int_0^{{}^\circ t} \varphi(X,s)(a) d\lambda(s) \approx \int_0^t \Phi(X,s)(a) d\nu(s),$$

we obtain an internal representation of σ, as follows.

Proposition 22.1.1 *For $\widehat{\Gamma}$-almost all X,*

$$a(\sigma \circ b(X, {}^\circ t)) \approx \langle B_t \circ \widetilde{\Phi}(X), a \rangle$$

for each $a \in \mathbb{B}'$ and each limited $t \in T$.

22.2 The Jacobian determinant of the internal shift

We need the Jacobian determinant of $\widetilde{\Phi}$. To this end let $F : \mathbb{G} \to \mathbb{K}$ be an internal mapping between *finite-dimensional internal Euclidean spaces $\mathbb{G}$ and $\mathbb{K}$. By the transfer principle, F is **differentiable in** $a \in \mathbb{G}$ if there exists an internal linear mapping $F'_a : \mathbb{G} \to \mathbb{K}$ such that

$$\lim_{h \to 0, h \in \mathbb{G}} \frac{1}{\|h\|_{\mathbb{G}}} \left(F(a+h) - F(a) - F'_a(h) \right) = 0.$$

Then F'_a is called the **derivative of** F in a. Since $\mathbb{K} = \mathbb{K}'$, we may equivalently assume that $F'_a : \mathbb{G} \times \mathbb{K} \to {}^*\mathbb{R}$ is a bilinear form, thus it is an element of the tensor product $\mathbb{G} \otimes \mathbb{K}$ of $\mathbb{G}$ and $\mathbb{K}$. The internal Hilbert–Schmidt norm on $\mathbb{G} \otimes \mathbb{K}$ is denoted by $\|\cdot\|_{\mathbb{G} \times \mathbb{K}}$.

F' is called **continuously differentiable** if $F' : \mathbb{G} \to \mathbb{G} \otimes \mathbb{K}$, $a \mapsto F'_a$ exists and is continuous (in the sense of the model $\mathfrak{W}$).

In analogy to the standard theory we call an internal function $G : \mathbb{G} \to \mathbb{G}$ **regular** if

(1) G is bijective and
(2) G is continuously differentiable internally with $\det G'_a \neq 0$ for all $a \in \mathbb{G}$.

Let $(\mathfrak{e}_i)_{i \le \rho}$ be an ONB of $\mathbb{G}$. The Lebesgue measure on $\mathbb{G}$ is the image measure of the internal Lebesgue measure on ${}^*\mathbb{R}^\rho$ by the mapping $(\alpha_i)_{i \le \rho} \mapsto$

$\sum_{i=1}^{\rho} \alpha_i \mathfrak{e}_i$. From the proof of Lemma 4.2.1 it follows that this image measure does not depend on the ONB of $\mathbb{G}$. By the change of variables rule for the finite-dimensional analysis and the transfer principle, we obtain the following rule:

Proposition 22.2.1 *Suppose that $F:\mathbb{G}\to\mathbb{K}$ is Borel measurable and $G:\mathbb{G}\to\mathbb{G}$ is regular. Then, for each $B\in\mathcal{B}(\mathbb{G})$,*

$$\int_B F\circ G\left|\det G'\right| d\lambda^{\rho} = \int_{G[B]} F d\lambda^{\rho}$$

if the integral of either side exists.

Since $\widetilde{\Phi}:\Omega\to\Omega$ is a *finite sum of multilinear forms, $\widetilde{\Phi}$ is continuously differentiable. Let $\widetilde{\Phi}_t:\Omega\to\mathbb{F}$ be the tth component of $\widetilde{\Phi}$. We identify $\widetilde{\Phi}_t'(X):\Omega\to\mathbb{F}$ with its representation matrix $\left(\widetilde{\Phi}_t'(X,s)\right)_{s\in T}$, where $\widetilde{\Phi}_t'(X,s)$ is the partial derivative of $\widetilde{\Phi}_t$ for the variable X_s in X. Since $F_k(t)=0$ for $t\notin T_{\neq}^{k+1}$, $\widetilde{\Phi}_t'(X,s)$ is the identity map for $s=t$ and for $s\neq t$

$$\widetilde{\Phi}_t'(X,s)=\sum_{k=1}^{M}\sum_{t_1<\ldots<t_{k-1}} F_k\left(t_1,\ldots,t_{k-1},s,t\right)\left(X_{t_1},\ldots,X_{t_{k-1}},\cdot,\cdot\right)\frac{1}{H},$$

which is a bilinear form on $\mathbb{F}\times\mathbb{F}$, thus a linear mapping from $\mathbb{F}$ into $\mathbb{F}$.

Let $M_{t,s}^{\mathfrak{E}}(X)$ be the representation matrix of $\widetilde{\Phi}_t'(X,s)$ with respect to an ONB $\mathfrak{E}:=(\mathfrak{e}_i)_{i\in\omega}$ of $\mathbb{F}$. Let $L^{\mathfrak{E}}(X)$ denote the $\omega\cdot H^2\times\omega\cdot H^2$-matrix, which results from the $H^2\times H^2$-matrix $\left(\widetilde{\Phi}_t'(X,s)\right)_{s,t\in T}$ by replacing $\widetilde{\Phi}_t'(X,s)$ with $M_{t,s}^{\mathfrak{E}}(X)$. Then $L^{\mathfrak{E}}(X)$ is the representation matrix of $\Phi_X':\Omega\to\Omega$ with respect to the canonical ONB basis $\left((\mathfrak{e}_{i,s})_{i\in\omega}\right)_{s\in T}$ of Ω over $\mathfrak{E}$, where

$$\mathfrak{e}_{i,s}:T\ni t\mapsto\begin{cases}\mathfrak{e}_i & \text{if } t=s\\ 0 & \text{otherwise.}\end{cases}$$

Note that $L^{\mathfrak{E}}(X)$ has the number 1 in the diagonal and infinitesimals for $\widehat{\Gamma}$-almost all $X\in\Omega$ at the other places. Therefore, the matrix $L^{\mathfrak{E}}(X)$, which is called the **Carleman–Fredholm matrix**, can be seen as an interference matrix of the identity.

22.3 Time-anticipating Girsanov transformations

First we will give necessary and sufficient conditions for the fact that a time-anticipating shift of Brownian motion results in a process that follows the law of Brownian motion with respect to a measure which is absolutely continuous or equivalent to $\widehat{\Gamma}$. Define an internal *real function $\Theta_\Phi : \Omega \to {}^*\mathbb{R}$ by

$$\Theta_\Phi(X) := e^{-\sum_{s\in T}\left(\Phi(X,s)(X_s)+\frac{1}{2}\|\Phi(X,s)\|^2_{\mathbb{F}}\frac{1}{H}\right)}\left|\det \widetilde{\Phi}'_X\right|.$$

The function Φ is called **adequate** if the following three conditions hold:

(A) $\widetilde{\Phi}$ is regular.
(B) $\det \widetilde{\Phi}' \not\approx 0\, \widehat{\Gamma}$-a.s.
(C) $\Theta_\Phi \in SL^1(\Gamma)$. Condition (C) is called the **Novikov condition**.

If in addition to (A), (B) and (C) also

(D) $e^{-\sum_{s\in T}\left(\Phi(X,s)(X_s)+\frac{1}{2}\|\Phi(X,s)\|^2_{\mathbb{F}}\frac{1}{H}\right)} \not\approx 0$ for $\widehat{\Gamma}$ almost all $X \in \Omega$

holds, Φ is called **strongly adequate**.

We shall see in the following section that for predictable φ we can choose the lifting Φ of φ such that conditions (A) and (B) are true. By Proposition 19.4.1 if φ is Skorokhod integrable, then we can choose Φ such that (D) holds and the internal function $X \mapsto e^{-\sum_{s\in T}\left(\Phi(X,s)(X_s)+\frac{1}{2}\|\Phi(X,s)\|^2_{\mathbb{F}}\frac{1}{H}\right)}$ is a lifting of $X \mapsto e^{-\delta\varphi-\frac{1}{2}\int_T\|\varphi(X,s)\|^2_{\mathbb{H}}d\widehat{\nu}(s)}$.

The following lemma, part (a) (iii), shows that, assuming (A), Condition (C) is equivalent to $\widehat{\Theta_\Phi\Gamma}$ being absolutely continuous to $\widehat{\Gamma}$. Note that for two internal measures ρ and ς on $\mathcal{B}$ with limited $\rho(\Omega)$, $\varsigma(\Omega)$, the Loeb measure $\widehat{\varsigma}$ is absolutely continuous to $\widehat{\rho}$ if and only if $\rho(B) \approx 0$ implies $\varsigma(B) \approx 0$ for each $B \in \mathcal{B}$.

Lemma 22.3.1

(a) *Suppose (A) is true. Then*
  (i) $\Theta_\Phi\Gamma : \mathcal{B} \ni A \mapsto \int_A \Theta_\Phi d\Gamma$ *is the internal image measure of* Γ *by* $\widetilde{\Phi}^{-1}$*; in particular, it is a probability measure. It follows that for all* $\mathcal{B}$-*measurable* $F : \Omega \to {}^*\mathbb{R}$

$$\int_\Omega F \circ \widetilde{\Phi} \cdot \Theta_\Phi d\Gamma = \int_\Omega F d\Gamma,$$

  if the integral of one of both sides exists.
  (ii) *For each* $a \in \mathbb{H}$, $e^{\lambda\langle B_n \circ \widetilde{\Phi}, a\rangle} \in SL^1(\Theta_\Phi\Gamma)$ *for each limited* $\lambda \in {}^*\mathbb{R}$ *and each limited* $n \in T$.
  (iii) $\Theta_\Phi \in SL^1(\Gamma)$ *(i.e., (C) holds) if and only if* $\widehat{\Theta_\Phi\Gamma}$ *is absolutely continuous to* $\widehat{\Gamma}$.

By (i), $\mathbb{E}\Theta_\Phi = 1$, thus, by Corollary 10.8.2, $\mathbb{E}^\circ\Theta_\Phi \leq 1$ and $\mathbb{E}^\circ\Theta_\Phi = {}^\circ\mathbb{E}\Theta_\Phi = 1$ if and only if $\Theta_\Phi \in SL^1(\Gamma)$.

(b) *Assume that (B) and (D) are true. Then $\widehat{\Gamma}$ is absolutely continuous with respect to $\widehat{\Theta_\Phi\Gamma}$.*

Proof Recall that we identify X_s with the projection $\pi_s : X \mapsto X_s$.

(a) We set $g(X) := e^{-\frac{H}{2}\sum_{i\in\omega, s\in T}\langle X_s, \mathfrak{e}_i\rangle^2} w$ with $w := \sqrt{\frac{H}{2\pi}}^{\omega\cdot H^2}$.

(i) Fix $B \in \mathcal{B}$. Then, by Proposition 22.2.1,

$$\begin{aligned}
&\Gamma\left(\widetilde{\Phi}[B]\right)\\
&= \int_{\widetilde{\Phi}[B]} g(X)dX = \int_B g\circ\widetilde{\Phi}(X)\cdot\left|\det\widetilde{\Phi}'_X\right|dX\\
&= w\cdot\int_B e^{-\frac{H}{2}\sum_{i\in\omega, s\in T}\left\langle X_s + \frac{1}{H}\Phi(X,s), \mathfrak{e}_i\right\rangle^2}\left|\det\widetilde{\Phi}'_X\right|dX\\
&= w\int_B e^{\sum_{i\in\omega, s\in T}\left(-\frac{H}{2}\langle X_s, \mathfrak{e}_i\rangle^2 - \langle X_s, \mathfrak{e}_i\rangle\Phi(X,s)(\mathfrak{e}_i) - \frac{1}{2}\Phi(X,s)^2(\mathfrak{e}_i)\frac{1}{H}\right)}\left|\det\widetilde{\Phi}'_X\right|dX\\
&= w\int_B e^{\sum_{i\in\omega, s\in T} -\frac{H}{2}\langle X_s, \mathfrak{e}_i\rangle^2 - \sum_{s\in T}\left(\Phi(X,s)(X_s) + \frac{1}{2}\|\Phi(X,s)\|^2_{\mathbb{F}}\frac{1}{H}\right)}\left|\det\widetilde{\Phi}'_X\right|dX\\
&= \int_B \Theta_\Phi d\Gamma.
\end{aligned}$$

(ii) Suppose that λ is limited. We may assume that $\|a\| = 1$ and we may choose $\mathfrak{e}_1 := a$ in an orthonormal basis $\mathfrak{E} = (\mathfrak{e}_i)_{i\in\omega}$ of $\mathbb{F}$. Then, by (i),

$$\begin{aligned}
&\mathbb{E}_\Gamma e^{\lambda\langle B_n\circ\widetilde{\Phi}, a\rangle}\Theta_\Phi = \mathbb{E}_\Gamma e^{\lambda\langle B_n, \mathfrak{e}_1\rangle}\\
&= \int_{{}^*\mathbb{R}^{H\cdot n}} e^{\lambda\sum_{s\leq Hn} x_s} e^{-\frac{H}{2}\sum_{s\leq Hn} x_s^2}\sqrt{\frac{H}{2\pi}}^{H\cdot n} d(x_s)_{s\leq Hn}\\
&= \int_{{}^*\mathbb{R}^{H\cdot n}} e^{\sum_{s\leq Hn}\left(-\frac{H}{2}\left(x_s - \frac{\lambda}{H}\right)^2 + \frac{\lambda^2}{2}\frac{1}{H}\right)}\sqrt{\frac{H}{2\pi}}^{H\cdot n} d(x_s)_{s\leq Hn}\\
&= e^{\frac{1}{2}\lambda^2 n} \text{ is limited.}
\end{aligned}$$

This proves (ii).

(iii) Suppose that $\Theta_\Phi \in SL^1(\Gamma)$ and $\Gamma(N) \approx 0$ for $N \in \mathcal{B}$. Then

$$\Theta_\Phi\Gamma(N) = \int_N \Theta_\Phi d\Gamma \approx 0.$$

This proves that $\widehat{\Theta_\Phi\Gamma}$ is absolutely continuous to $\widehat{\Gamma}$.

Conversely, suppose that $\widehat{\Theta_\Phi \Gamma}$ is absolutely continuous to $\widehat{\Gamma}$. It follows that $\int_A \Theta_\Phi d\Gamma \approx 0$ for each $\widehat{\Gamma}$-nullset $A \in \mathcal{B}$. Moreover, $\int_\Omega \Theta_\Phi d\Gamma = 1$ is limited. It follows that $\Theta_\Phi \in SL^1(\Gamma)$.

(b) We have to show that $\Theta_\Phi \Gamma(A) \approx 0$ implies $\Gamma(A) \approx 0$ for each $A \in \mathcal{B}$. Let $\Theta_\Phi \Gamma(A) \approx 0$, thus $\mathbb{E}\mathbf{1}_A \cdot e^{-\sum_{s\in T}\left(\Phi(X,s)(X_s)+\frac{1}{2}\|\Phi(X,s)\|^2_{\mathbb{F}}\frac{1}{H}\right)}\left|\det \widetilde{\Phi}'(X)\right| \approx 0$. Therefore, $\mathbf{1}_A \cdot e^{-\sum_{s\in T}\left(\Phi(X,s)(X_s)+\frac{1}{2}\|\Phi(X,s)\|^2_{\mathbb{F}}\frac{1}{H}\right)}\left|\det \widetilde{\Phi}'(X)\right| \approx 0$ $\widehat{\Gamma}$-a.s. By (B), we obtain $\mathbf{1}_A \cdot e^{-\sum_{s\in T}\left(\Phi(X,s)(X_s)+\frac{1}{2}\|\Phi(X,s)\|^2_{\mathbb{F}}\frac{1}{H}\right)} \approx 0\,\widehat{\Gamma}$-a.s. By (D), $\mathbf{1}_A \approx 0\,\widehat{\Gamma}$-a.s., thus $\Gamma(A) \approx 0$. □

Corollary 22.3.2 *Assume that Φ is strongly adequate. Then the probability measures $\widehat{\Theta_\Phi \Gamma}$ and $\widehat{\Gamma}$ are equivalent. If Φ is adequate, then $\widehat{\Theta_\Phi \Gamma}$ is absolutely continuous to $\widehat{\Gamma}$.*

Remark 22.3.3 We assume that Φ is (strongly) adequate. By Corollary 22.3.2, and since the **Wiener measure** W is the image measure of $\widehat{\Gamma}$ by $\kappa : X \mapsto b_{\mathbb{B}}(X,\cdot)$, the image measure $\widehat{\Theta_\Phi \Gamma}_\kappa$ of $\widehat{\Theta_\Phi \Gamma}$ on $\mathcal{B}(C_{\mathbb{B}})$ by κ is absolutely continuous (equivalent) to W. Therefore, $\widehat{\Theta_\Phi \Gamma}$ has a density with respect to the Wiener measure. This density is equivalent to the conditional expectation $\mathbb{E}^{\mathcal{W}}({}^\circ\Theta_\Phi)$ of ${}^\circ\Theta_\Phi$.

In our first application we will show that the shift $\sigma \circ b$ follows the law of Brownian motion w.r.t. $\widehat{\Theta_\Phi \Gamma}$ if Φ is adequate.

Theorem 22.3.4 *Assume that Φ is (strongly) adequate. Fix $r, t \in [0, \infty[$ with $r < t$. Then for each $a \in \mathbb{B}'$ with $\|a\| = 1$, $a(\sigma \circ b(\cdot,t) - \sigma \circ b(\cdot,r))$ is $\mathcal{N}(0, t-r)$-distributed with respect to the measure $\widehat{\Theta_\Phi \Gamma}$ which is absolutely continuous (equivalent) to $\widehat{\Gamma}$.*

Proof We may assume that $a = \mathfrak{e}_1$ in an orthonormal basis $\mathfrak{E} = (\mathfrak{e}_i)_{i\in\omega}$ of $\mathbb{F}$. Choose $n, m \in T$ such that $n \approx t$ and $m \approx r$. Then, using (i) and (ii), we have, for each $\lambda \in \mathbb{R}$,

$$\int_\Omega e^{\lambda a(\sigma \circ b(\cdot,t) - \sigma \circ b(\cdot,r))} d\widehat{\Theta_\Phi \Gamma} \approx \int_\Omega e^{\lambda\left\langle B_n \circ \widetilde{\Phi} - B_m \circ \widetilde{\Phi}, a\right\rangle} d\Theta_\Phi \Gamma$$
$$= \int_\Omega e^{\lambda\langle B_n - B_m, a\rangle} d\Gamma \approx \int_\Omega e^{\lambda a(b(\cdot,t) - b(\cdot,r))} d\widehat{\Gamma}.$$

Since $a(b(\cdot,t) - b(\cdot,r))$ is $\mathcal{N}(0, t-s)$-distributed under the measure $\widehat{\Gamma}$, the proof is finished. □

22.4 Adapted Girsanov transformation

We use the notation of the preceding section. However, we now assume that φ is $(\mathfrak{b}_r \cap \mathcal{W})_{r\in[0,\infty[}$-predictable. Let Φ be a lifting of φ, obtained in Section 22.1 under (+).

The next result shows that under condition (C) Φ is adequate. Recall that (C) is equivalent to $\widehat{\Theta_\Phi \Gamma}$ being absolutely continuous to $\widehat{\Gamma}$.

Lemma 22.4.1 *Assume that condition (C) is fulfilled for Φ. Then Φ is adequate. If, in addition, $\varphi \in L^2_{\mathfrak{P}\mathcal{W}}(\widehat{\Gamma \otimes \nu}, \mathbb{H})$ (recall the notation in Section 19.5), then we can choose Φ strongly adequate.*

Proof Recall that Φ has the form $\Phi = \sum_{n=0}^{M} I_{n,1}(F_n)$ where each F_n is non-anticipating, thus Φ is $(\mathcal{B}_{t^-})_{t\in T}$-adapted. Therefore, we can assume that for each $X \in \Omega$ and each $t \in T$

$$\Phi(X,t) = \Phi((X_{\frac{1}{H}},\ldots,X_{t-\frac{1}{H}},0,\ldots,0),t).$$

We have to show that the conditions (A) and (B) are true.

(A) Fix $X, Y \in \Omega$ with $X \neq Y$. Set $i := \min\{s \in T \mid X_s \neq Y_s\}$. By (+), we obtain $\Phi(X,i) = \Phi(Y,i)$. Therefore, $X_i + \Phi(X,i)\frac{1}{H} \neq Y_i + \Phi(Y,i)\frac{1}{H}$, thus $\widetilde{\Phi}(X) \neq \widetilde{\Phi}(Y)$. In order to show that $\widetilde{\Phi}$ is surjective, fix $Z \in \Omega$. Define by internal induction:

$$X_{\frac{1}{H}} := Z_{\frac{1}{H}} - \Phi\left((0,\ldots,0), \frac{1}{H}\right)\frac{1}{H},$$

$$X_{s+\frac{1}{H}} := Z_{s+\frac{1}{H}} - \Phi\left(\left(X_{\frac{1}{H}},\ldots,X_s,0,\ldots,0\right), s+\frac{1}{H}\right)\frac{1}{H}.$$

Since $\Phi((X_{\frac{1}{H}},\ldots,X_s,0,\ldots,0),s+\frac{1}{H}) = \Phi((X_{\frac{1}{H}},\ldots,X_H),s+\frac{1}{H})$ for each $s \in T \cup \{0\}$ and each $X \in \Omega$, we obtain $\widetilde{\Phi}(X) = Z$.

(B) Since $\widetilde{\Phi}'_X$ is represented by a triangle matrix with 1 in the diagonal, we obtain $\det \widetilde{\Phi}'_X = 1$, thus (B) is true.

If $\varphi \in L^2_{\mathfrak{P}\mathcal{W}}(\widehat{\Gamma \otimes \nu}, \mathbb{H})$, we can choose Φ such that the internal function $\Omega \ni X \mapsto \sum_{s\in T} \Phi(X,s)(X_s)$ lifts $\delta\varphi$, thus $\sum_{s\in T} \Phi(X,s)(X_s)$ is limited for $\widehat{\Gamma}$-almost all $X \in \Omega$. Since, by Keisler's Fubini theorem,

$$\sum_{s\in T} \|\Phi(X,s)\|^2_{\mathbb{F}} \frac{1}{H} \approx \int_T \|\varphi(X,\cdot)\|^2_{\mathbb{H}} \, d\widehat{\nu} \text{ is limited}$$

for $\widehat{\Gamma}$-almost all X, condition (D) is true. □

Now we obtain the classical Girsanov theorem for abstract Wiener spaces:

Theorem 22.4.2 *Suppose that Φ fulfils condition (C). Then $\sigma \circ b$ is a $\mathbb{B}$-valued Brownian motion on the adapted space $(\Omega, \mathcal{W}, \widehat{\Theta_\Phi \Gamma}, (\mathfrak{b}_r^{\mathcal{W}})_{r\in[0,\infty[})$. Recall that $\widehat{\Theta_\Phi \Gamma}$ is absolutely continuous to $\widehat{\Gamma}$. Moreover, both measures are equivalent if condition (D) also holds.*

Proof First note that $\sigma \circ b(X,\cdot)$ is continuous and $\sigma \circ b(X,0) = 0$ for $\widehat{\Theta_\Phi \Gamma}$-almost all X. In order to show that $\sigma \circ b(X,\cdot)$ is a Brownian motion, it suffices to show that for each $a,b \in \mathbb{B}'$ with $\|a\| = \|b\| = 1$ and $a \perp b$:

(a) $\left(e^{\lambda a(\sigma\circ b(\cdot,t)) - \frac{1}{2}\lambda^2 t}\right)_{t\in[0,\infty[}$ is a $(\mathfrak{b}_t^{\mathcal{W}})_{t\in[0,\infty[}$-martingale with respect to $\widehat{\Theta_\Phi \Gamma}$ for each $\lambda \in \mathbb{R}$ (see Theorem 3.5.2);

(b) $(a(\sigma \circ b(\cdot,t)), b(\sigma \circ b(\cdot,t)))$ is independent for all $t \in [0,\infty[$.

To prove (a) fix $r,t \in [0,\infty[$ with $r<t$ and $B \in \mathfrak{b}_r$. There exist $m \in T$ and $A \in \mathcal{B}_m$ such that $m \approx r$ and $\widehat{\Gamma}(A \Delta B) = 0$. Fix $n \in T$ with $n \approx t$. Note that $\widetilde{\Phi}[A] \in \mathcal{B}_m \subset \mathfrak{b}_r$. We obtain, using (i), and the martingale property of $\left(e^{\lambda a(b_r) - \frac{1}{2}\lambda^2 r}\right)_{r\in[0,\infty[}$ with respect to the measure $\widehat{\Gamma}$,

$$
\begin{aligned}
\int_B e^{\lambda a(\sigma\circ b(\cdot,t)) - \frac{1}{2}\lambda^2 t} d\widehat{\Theta_\Phi \Gamma} &\approx \int_A e^{\lambda \langle B_n \circ \widetilde{\Phi}, a\rangle - \frac{1}{2}\lambda^2 n} d\Theta_\Phi \Gamma \\
&= \int_{\widetilde{\Phi}[A]} e^{\lambda \langle B_n, a\rangle - \frac{1}{2}\lambda^2 n} d\Gamma \\
&\approx \int_{\widetilde{\Phi}[A]} e^{\lambda a(b_t) - \frac{1}{2}\lambda^2 t} d\widehat{\Gamma} = \int_{\widetilde{\Phi}[A]} e^{\lambda a(b_r) - \frac{1}{2}\lambda^2 r} d\widehat{\Gamma} \\
&= \int_B e^{\lambda a(\sigma\circ b(\cdot,r)) - \frac{1}{2}\lambda^2 r} d\widehat{\Theta_\Phi \Gamma}.
\end{aligned}
$$

The proof of (b) is similar to the proof of Theorem 11.7.7. □

22.5 †Extension of abstract Wiener spaces

We now emphasize the following advantage of poly-saturated models. It is not only possible to simulate infinite (dimensional) entities by finite (dimensional) ones within the new model, but also new structures come up. Loeb spaces are an example. In recent papers, stimulated by Sun [114], new product spaces in probability theory have been studied systematically (see Keisler and Sun [54]; see also [72] and [9]).

Here we indicate how the notion 'admissible sequence' can be used to develop Malliavin calculus on many spaces different from $L^2(W) = L^2_{\mathcal{W}}(\widehat{\Gamma})$ and $L^2(W\otimes\lambda) = L^2_{\mathcal{W}\otimes\mathcal{L}^1}\left(\widehat{\Gamma\otimes\nu}\right)$. In particular, we obtain Malliavin calculus on spaces strictly larger than $L^2(W)$ and $L^2(W\otimes\lambda)$. The results are based upon chaos decompositions following [90].

Fix a vector space $\mathfrak{H}^V \subseteq SL^2(\nu,\mathbb{F})$ over $\mathbb{R}$ such that the standard part $^\circ F$ of F exists for each $F\in\mathfrak{H}^V$, thus $\|^*(^\circ F(t)) - F(t)\|_{\mathbb{F}} \approx 0$ for $\widehat{\nu}$-almost all $t\in T$. It is not required that $^\circ F$ is $\mathcal{L}^1$-measurable. In Remark 12.1.1 we constructed a simple nearstandard function $F\in SL^2(\nu,\mathbb{F})$ that is not $\mathcal{L}^1$-measurable.

Define σ-subalgebras $\mathcal{H}_1$, $\mathcal{H}$ of $L_\nu(T)$, $L_\Gamma(\mathcal{B})$, respectively, depending on $\mathfrak{H}^V$, by setting

$$\mathcal{H}_1 := \sigma\left\{^\circ F \mid F\in\mathfrak{H}^V\right\}\vee\mathcal{N}_{\widehat{\nu}},$$
$$\mathcal{H} := \sigma\left\{^\circ I(F) \mid F \text{ is } \mathcal{H}_1\text{-measurable}\right\}\vee\mathcal{N}_{\widehat{\Gamma}}.$$

A sequence $(\mathfrak{H}_n)_{n\in\mathbb{N}_0}$ is called an **admissible sequence over** $\mathfrak{H}^V$ if each $\mathfrak{H}_n$ fulfils the following closure properties (EW 1)–(EW 6).

(EW 1) $\mathfrak{H}_0 = \mathbb{R}$.
(EW 2) $\mathfrak{H}_n$ is a linear subspace of nearstandard elements of $SL^2(\nu^n\upharpoonright T^n_<,\mathbb{F}^{\otimes n})$.
(EW 3) $^\circ I_n(F)$ is $\mathcal{H}$-measurable for each $F\in\mathfrak{H}_n$.
(EW 4) $^\circ F$ is $(\mathcal{H}_1)^{\otimes n}\vee\mathcal{N}_{\widehat{\nu^n}}$-measurable for each $F\in\mathfrak{H}_n$.
(EW 5) $^\circ\mathfrak{H}_n$ is closed.
(EW 6) If $F\in SL^2(\nu,\mathbb{F})$ such that $^\circ F$ is $\mathcal{H}_1$-measurable, then $\mathbf{1}_{T^n_<}\cdot F^{\otimes n}\in\mathfrak{H}_n$. It follows that $\mathfrak{H}_1 = \left\{F\in SL^2(\nu) \mid {}^\circ F \text{ is } \mathcal{H}_1\text{-measurable}\right\}$.

Set

$$L^2_{\mathcal{H}}(\widehat{\Gamma},\mathbb{H}^{\otimes k}) := L^2(\Omega,\mathcal{H},\widehat{\Gamma},\mathbb{H}^{\otimes k})$$

and

$$L^2_{\mathcal{H}_1^n}(\widehat{\nu^n},\mathbb{H}^{\otimes k}) := L^2(T^n,(\mathcal{H}_1)^{\otimes n}\vee\mathcal{N}_{\widehat{\nu^n}},\widehat{\nu^n},\mathbb{H}^{\otimes k}).$$

Recall that $\widehat{\nu^n} = \widehat{\nu}^n$ on $(\mathcal{H}_1)^{\otimes n}\vee\mathcal{N}_{\widehat{\nu^n}}$.

Examples 22.5.1

(i) Set $\mathfrak{H}_n := \left\{F\in SL^2(\nu^n\upharpoonright T^n_<,\mathbb{F}^{\otimes n}) \mid {}^\circ F\in L^2_{\mathcal{L}^n}(\widehat{\nu^n})\right\}$ and $\mathfrak{H}^V := \mathfrak{H}_1$. Then $(\mathfrak{H}_n)$ is an admissible sequence over $\mathfrak{H}_1$, and $L^2_{\mathcal{H}}(\widehat{\Gamma}) = L^2_{\mathcal{W}}(\widehat{\Gamma})$ and $L^2_{\mathcal{H}_1}(\widehat{\nu^n}) = L^2_{\mathcal{L}^n}(\widehat{\nu^n})$.
(ii) Let $\mathfrak{H}_n := \left\{F\in SL^2(\nu^n\upharpoonright T^n_<,\mathbb{F}^{\otimes n}) \mid {}^\circ F\in L^2(\widehat{\nu}^n)\right\}$ and $\mathfrak{H}^V := \mathfrak{H}_1$. Then $(\mathfrak{H}_n)$ is an admissible sequence over $\mathfrak{H}_1$, and $L^2_{\mathcal{H}}(\widehat{\Gamma},\mathbb{H}^{\otimes k})$ is a strict extension of $L^2_{\mathcal{W}}(\widehat{\Gamma},\mathbb{H}^{\otimes k})$ and $L^2_{\mathcal{H}_1^n}(\widehat{\nu^n},\mathbb{H}^{\otimes k})$ is a strict extension of $L^2_{\mathcal{L}^n}(\widehat{\nu^n},\mathbb{H}^{\otimes k})$. By a

result of Y. Sun, $L^2(\widehat{\nu}^n, \mathbb{H}^{\otimes k})$ is a strict subspace of $L^2(\widehat{\nu^n}, \mathbb{H}^{\otimes k})$. Note that, by Proposition 10.4.1, $L^2_{\mathcal{H}_1^n}(\widehat{\nu^n}, \mathbb{H}^{\otimes k})$ and $L^2(\widehat{\nu}^n, \mathbb{H}^{\otimes k})$ can be identified.

The proofs of the following two theorems are similar to the proofs of Theorems 17.2.1 and 19.1.1.

Theorem 22.5.2 *Let $(\mathfrak{H}_n)$ be an admissible sequence over $\mathfrak{H}^V$. Then for each $\varphi \in L^2_{\mathcal{H}}(\widehat{\Gamma})$ there exists a sequence $(F_n)_{n\in\mathbb{N}_0}$ with $F_n \in \mathfrak{H}_n$ such that*

$$\varphi = \sum_{n=0}^{\infty} {}^\circ I_n(F_n) \text{ in } L^2_{\mathcal{H}}(\widehat{\Gamma}).$$

Moreover, ${}^\circ I_0(F_0) = {}^\circ F_0 = \mathbb{E}(\varphi)$.

Fix an admissible sequence $(\mathfrak{H}_n)$ over $\mathfrak{H}^V$ and use the previous notation. Define

$$L^2_{\mathcal{H}\otimes\mathcal{H}_1}(\widehat{\Gamma\otimes\nu}, \mathbb{H}) := \left(\Omega \times T, (\mathcal{H}\otimes\mathcal{H}_1) \vee \mathcal{N}_{\widehat{\Gamma\otimes\nu}}, \widehat{\Gamma\otimes\nu}, \mathbb{H}\right).$$

Theorem 22.5.3 *Let $\varphi \in L^2_{\mathcal{H}\otimes\mathcal{H}_1}(\widehat{\Gamma\otimes\nu}, \mathbb{H})$. Then there exists a sequence $(F_n)_{n\in\mathbb{N}_0}$ of functions $F_n : T^{n+1} \to \mathbb{F}^{\otimes(n+1)} \in SL^2_{\neq}(\nu^{n+1}, \mathbb{F}^{\otimes(n+1)})$ with the following four properties:*

(a) *F_n is symmetric in the first n variables and ${}^\circ F_n \in L^2_{\mathcal{H}_1^{n+1}}(\widehat{\nu^{n+1}}, \mathbb{H}^{\otimes(n+1)})$.*
(b) *${}^\circ I_{n,1}(F_n) \in L^2_{\mathcal{H}\otimes\mathcal{H}_1}(\widehat{\Gamma\otimes\nu}, \mathbb{H})$.*
(c) *${}^\circ I_{n+1}(\widetilde{F}_n) \in L^2_{\mathcal{H}}(\widehat{\Gamma})$.*
(d) *$\varphi = \sum_{n=0}^{\infty} {}^\circ I_{n,1}(F_n)$ converges in $L^2_{\mathcal{H}\otimes\mathcal{H}_1}(\widehat{\Gamma\otimes\nu}, \mathbb{H})$.*

Using these decompositions, it is possible to define the Malliavin derivative on a dense subspace of $L^2_{\mathcal{H}}(\widehat{\Gamma})$ and the Skorokhod integral on a dense subspace of $L^2_{\mathcal{H}\otimes\mathcal{H}_1}(\widehat{\Gamma\otimes\nu}, \mathbb{H})$ in the same manner as has been done in the earlier chapters.

Exercises

Use the notation at the beginning of Section 22.1.

22.1 Prove that $({}^\circ \|\Phi_n - \Psi\|_{\mathbb{F}}) = (\|\varphi_n - \varphi\|_{\mathbb{H}})$ converges to 0 in measure.
22.2 Prove that $\lim_{n,m\to\infty} \left({}^\circ \int_{T_r} \|\Phi_n - \Phi_m\|_{\mathbb{F}} \, d\nu\right) = 0$ in measure for all $r \in \mathbb{N}$.
22.3 Prove that there exists a lifting Φ of φ such that $\Phi(X, \cdot)$ is locally in $SL^1(\nu, \mathbb{F})$ for $\widehat{\Gamma}$-almost all $X \in \Omega$.

22.4 Let $(\Omega, \mathcal{E}, \zeta)$ and $(\Omega, \mathcal{E}, \rho)$ be two internal probability spaces. Prove that $\widehat{\zeta}$ is absolutely continuous to $\widehat{\rho}$ if and only if $\rho(D) \approx 0$ implies $\zeta(D) \approx 0$ for all $D \in \mathcal{E}$.

22.5 Prove Theorem 22.5.2.

22.6 Prove Theorem 22.5.3.

23

Malliavin calculus for Lévy processes

Following [89], we finally deal with Malliavin calculus for a large class of Lévy processes, based on stochastic integration in Chapters 15 and 16. Recall that we assume that the internal Borel measure μ^1 on ${}^*\mathbb{R}$ has the following properties: $H\mathbb{E}_{\mu^1}x^n$ is limited for all $n \in \mathbb{N}$ and $B : \Omega \times T \to {}^*\mathbb{R}$; $(X, t) \mapsto \sum_{s \leq t} X_s$ is μ-limited.

The standard theory of Malliavin calculus for Lévy processes with applications can be found, for example, in the book by Di Nunno, Oksendal and Proske [32] and in the articles [30], [29], [31].

23.1 Chaos

Admissible sequences have been used several times to prove certain closure properties of the summands in a chaos decomposition. We call $(\mathfrak{H}_n)_{n \in \mathbb{N}_0}$ an **admissible sequence** if the following six conditions hold.

(W 1) $\mathfrak{H}_0 = \mathbb{R}$.
(W 2) $\mathfrak{H}_1 = \{F : T \to {}^*\mathbb{R} \in SL^2(\nu) \mid {}^\circ F \text{ exists in } L^2(\lambda)\}$.
(W 3) $\mathfrak{H}_n$ is a linear space over $\mathbb{R}$.
(W 4) $\mathfrak{H}_n \subseteq \{F : T^n_< \to {}^*\mathbb{R} \in SL^2(\nu^n) \mid {}^\circ F \text{ exists in } L^2(\lambda^n)\}$.
(W 5) ${}^\circ\mathfrak{H}_n := \{{}^\circ F \mid F \in \mathfrak{H}_n\}$ is a closed subspace of $L^2(\lambda^n)$.
(W 6) For all restricted functions $F_1, \ldots, F_n$ in $\mathfrak{H}_1$, the n-ary function $\mathbf{1}_{T^n_<} \cdot (F_1 \otimes \ldots \otimes F_n)$ is an element of $\mathfrak{H}_n$. Recall that the F_i are bounded with support in T_σ for some $\sigma \in \mathbb{N}$.

Let $\overleftrightarrow{F}$ denote the symmetric extension of $F \in \mathfrak{H}_n$, according to Remark 17.1.1 (c). We obtain examples which are similar to Examples 17.1.2. The first one guarantees that at least one admissible sequence exists. The second one is the key for the Clark–Ocone formula. The proofs use results from Chapter 13, in particular on iterated integrals, similar to the proofs of the assertion in Examples 17.1.2.

Examples 23.1.1

(i) Set $\mathfrak{H}_n := \{F : T^n_< \to {}^*\mathbb{R} \in SL^2(\nu^n) \mid {}^\circ F \text{ exists in } L^2(\lambda^n)\}$ for $n \geq 1$ and $\mathfrak{H}_0 = \mathbb{R}$. Then $(\mathfrak{H}_n)_{n\in\mathbb{N}_0}$ is admissible.

(ii) Let $\mathfrak{G}_n$ be the set of all $F \in \mathfrak{H}_n$ such that for all $k \in \mathbb{N}^{n-1}_\mu$ we have in $L^2_{\mathcal{D}\otimes\mathcal{L}}\left(\widehat{\mu\otimes\nu}\right)$

$$^\circ I_{k,1}(F) = \left(r \mapsto \mathbb{E}^{\mathfrak{b}_{^\circ r}}\ {}^\circ I_{k,1}(\overleftrightarrow{F})(\cdot, r)\right).$$

Then $(\mathfrak{G}_n)_{n\in\mathbb{N}_0}$ is admissible. It follows that $\mathbb{E}^{\mathfrak{b}_{^\circ r}}\ {}^\circ I_{k,1}(\overleftrightarrow{F})(\cdot, r)$ is $\mathcal{D}$-measurable for $\widehat{\nu}$-almost all r. Moreover, $I_{k,1}(F)$ is a $(\mathcal{B}_{t^-})_{t\in T}$-adapted lifting in $SL^2(\mu\otimes\nu)$ of $r \mapsto \mathbb{E}^{\mathfrak{b}_r}\ {}^\circ I_{k,1}(\overleftrightarrow{F})(\cdot, r)$.

Proof We only prove part (ii). Obviously, (W 1)–(W 4) are true. To prove (W 5), fix a Cauchy sequence $({}^\circ F^m)_{m\in\mathbb{N}}$ with $F^m \in \mathfrak{G}_n$. There exists an $F \in \mathfrak{H}_n$ with $\lim_{m\to\infty} {}^\circ F^m = {}^\circ F$ in $L^2\left(\widehat{\nu^n}\right)$. Then

$$\lim_{m\to\infty} \int_{\Omega\times T} \left({}^\circ I_{k,1}(F^m) - {}^\circ I_{k,1}(F)\right)^2 d\widehat{\mu\otimes\nu} = 0$$

and, by Jensen's inequality,

$$\lim_{m\to\infty} \int_{\Omega\times T} \left(\mathbb{E}^{\mathfrak{b}_{^\circ r}} \left({}^\circ I_{k,1}(\overleftrightarrow{F^m})(\cdot, r) - {}^\circ I_{k,1}(\overleftrightarrow{F})\right)(\cdot, r)\right)^2 d\widehat{\mu\otimes\nu}(\cdot, r) = 0.$$

Together with ${}^\circ I_{k,1}(F^m) = \mathbb{E}^{\mathfrak{b}_{^\circ r}}\ {}^\circ I_{k,1}(\overleftrightarrow{F^m})(\cdot, r)$, this equality is also true for F instead of F^m. To prove (W 6), fix restricted $F_1, \ldots, F_n \in \mathfrak{G}_1$ and set $F := \mathbf{1}_{T^n_<} \cdot (F_1 \otimes \ldots \otimes F_n)$. We use the following notation: for $k \in \mathbb{N}^{n-1}_\mu$ and $t \in T^{n-1}_<$ set

$$p_{k,t}(X) := p_{k_1}(X_{t_1}) \cdot \ldots \cdot p_{k_{n-1}}(X_{t_{n-1}})$$

and

$$G_i(t) := F_1(t_1) \cdot \ldots \cdot F_{i-1}(t_{i-1}) \cdot F_{i+1}(t_i) \cdot \ldots \cdot F_n(t_{n-1}).$$

Then we have for all limited $r \in T$

$$\mathbb{E}^{\mathfrak{b}_{^\circ r}}\ {}^\circ I_{k,1}(\overleftrightarrow{F})(\cdot, r) = \sum_{i=1}^{n} {}^\circ F_i(r) \cdot \mathbb{E}^{\mathfrak{b}_{^\circ r}}\ {}^\circ \beta_i(\cdot, r)$$

with $\beta_i(X,r) := \sum_{t_1<\ldots<t_{i-1}<r<t_i<\ldots<t_{n-1}} G_i(t)p_{k,t}(X)$. Mimic the proof of condition (W 6) in Examples 17.1.2 (ii) to see that $\mathbb{E}^{\mathfrak{b}\circ_r}\ {}^\circ\beta_i(\cdot,r)=0$ for $i<n$. It follows that $\mathbb{E}^{\mathfrak{b}\circ_r}\ {}^\circ I_{k,1}(\overleftrightarrow{F})(\cdot,r) = {}^\circ I_{k,1}(F)(\cdot,r)$. □

Here is the first version of the chaos representation theorem for Lévy processes.

Proposition 23.1.2 *Fix $\varphi \in L^2_{\mathcal{D}}(\widehat{\mu})$, i.e., $\varphi \in L^2(\widehat{\mu})$ and φ is $\mathcal{D}$-measurable. Fix an admissible sequence $(\mathfrak{H}_n)_{n\in\mathbb{N}_0}$. Then there exists a sequence $\left(F_{(k_1,\ldots,k_n)}\right)_{n\in\mathbb{N}_0,(k_1,\ldots,k_n)\in\mathbb{N}^n_\mu}$ of functions $F_{(k_1,\ldots,k_n)} \in \mathfrak{H}_n$ with the following properties:*

(a)

$$\varphi = \sum_{n\in\mathbb{N}_0} \sum_{(k_1,\ldots,k_n)\in\mathbb{N}^n_\mu} I_{(k_1,\ldots,k_n)}({}^\circ F_{(k_1,\ldots,k_n)}) \text{ converges in } L^2(\widehat{\mu}).$$

(b) *If $g_{(k_1,\ldots,k_n)} : [0,\infty[^n_\le \to \mathbb{R}$ is Lebesgue measurable and*

$$\varphi = \sum_{n\in\mathbb{N}_0} \sum_{(k_1,\ldots,k_n)\in\mathbb{N}^n_\mu} I_{(k_1,\ldots,k_n)}(g_{(k_1,\ldots,k_n)}) \text{ converges in } L^2(\widehat{\mu}),$$

then ${}^\circ F_{(k_1,\ldots,k_n)} = g_{(k_1,\ldots,k_n)}$ λ^n-a.e. for all $n\in\mathbb{N}_0$ and all $(k_1,\ldots,k_n)\in\mathbb{N}^n_\mu$.

Proof (a) Let M be the set of all $\varphi \in L^2_{\mathcal{D}}(\widehat{\mu})$ having such a decomposition. Then M is a linear subspace of $L^2_{\mathcal{D}}(\widehat{\mu})$. Condition (W 5), Theorem 16.2.3, the pairwise orthogonality of the $I_k(f)$ and Proposition 5.6.5 tell us that M is closed. In order to prove that $M = L^2_{\mathcal{D}}(\widehat{\mu})$, fix $\varphi \in L^2_{\mathcal{D}}(\widehat{\mu})$ with $\varphi \perp M$. It suffices to prove $\varphi = 0$ in $L^2_{\mathcal{D}}(\widehat{\mu})$. Fix a finite sum $C = \sum_{j=1}^m I_{(l_j)}(f_j)$, with restricted functions $f_j \in L^2(\lambda)$ and $l_j \in \mathbb{N}_\mu$. By Theorem 16.5.1 and condition (W 6) we have for all polynomials p: $p(C)$ is a linear combination of iterated integrals with kernels in $\bigcup_{n\in\mathbb{N}_0} {}^\circ\mathfrak{H}_n$, thus $\varphi \perp p(C)$ for all polynomials p. Let $\widehat{\mu}^+, \widehat{\mu}^-$ be the measures with density φ^+ (the positive part of φ), φ^- (the negative part of φ), respectively. We have to prove that $\widehat{\mu}^+ = \widehat{\mu}^-$. Since $\mathcal{D}$ is generated by the $\widehat{\mu}$-integrals $I_{(k)}(f)$, it suffices to prove that,

$$\int_\Omega e^{i\cdot\sum_{j=1}^m I_{(l_j)}f_j}\,d\widehat{\mu}^+ = \int_\Omega e^{i\cdot\sum_{j=1}^m I_{(l_j)}f_j}\,d\widehat{\mu}^-.$$

Set $J := \sum_{j=1}^m I_{(l_j)}f_j$. Then

$$\int_\Omega e^{i\cdot J}\,d\widehat{\mu}^+ = \sum_{n=0}^\infty \mathbb{E}_{\widehat{\mu}} \frac{i^n\cdot J^n}{n!}\varphi^+ = \sum_{n=0}^\infty \mathbb{E}_{\widehat{\mu}} \frac{i^n\cdot J^n}{n!}\varphi^- = \int_\Omega e^{i\cdot J}\,d\widehat{\mu}^-.$$

The proof of part (a) is finished.

(b) The uniqueness follows from the orthogonality of the summands (see Proposition 16.3.3). □

In order to obtain the Malliavin derivative and the Skorokhod integral, we use a slight modification of Proposition 23.1.2.

Theorem 23.1.3 *Fix $\varphi \in L^2_{\mathfrak{D}}(\widehat{\mu})$ and an admissible sequence $(\mathfrak{H}_n)_{n\in\mathbb{N}_0}$.*

(A) *There exists a sequence $(F_n)_{n\in\mathbb{N}_0}$ of internal symmetric functions $F_n : M^n_\mu \times T^n_{\neq} \to {}^*\mathbb{R}$ with the following properties:*
 (a) *$F_n \in SL^2(c^n \otimes \nu^n)$, ${}^\circ F_n$ exists and belongs to $L^2(c^n \otimes \lambda^n)$.*
 (b) *$\mathbf{1}_{T^n_<}(\cdot)F_n(k_1,\ldots,k_n,\cdot) \in \mathfrak{H}_n$ for all $(k_1,\ldots,k_n) \in \mathbb{N}^n_\mu$.*
 (c) *$\varphi = \sum_{n=0}^\infty I_n({}^\circ F_n) = \mathbb{E}_{\widehat{\mu}}\varphi + \sum_{n=1}^\infty I_n({}^\circ F_n)$ and*

$$I_n({}^\circ F_n) = {}^\circ I_n(F_n) \text{ with } I_n(F_n) = \sum_{k\in M^n_\mu}\sum_{t\in T^n_<} F_n(k,t)\prod_{i=1}^n p_{k_i}(X_{t_i})$$

(see Theorem 16.4.1).

(B) *If $g_n : \mathbb{N}^n_\mu \times [0,\infty[^n \to \mathbb{R}$ is symmetric and $\varphi = \sum_{n=0}^\infty I_n(g_n)$, then $g_n = f_n$ in $L^2(c^n \otimes \lambda^n)$.*

Proof We have $\varphi = \sum_{n\in\mathbb{N}_0}\sum_{k\in\mathbb{N}^n_\mu} I_k(f_n(k,\cdot))$, according to Proposition 23.1.2, where $f_n : \mathbb{N}^n_\mu \times [0,\infty[^n_\le \to \mathbb{R}$ with $f_n(k,\cdot) := f_k \approx \widetilde{F}(k,\cdot) =: \widetilde{F}_n(k,\cdot)$ for a certain $\widetilde{F}(k,\cdot) \in \mathfrak{H}_n$. Let $\widetilde{F}_n : M^n_\mu \times T^n_< \to {}^*\mathbb{R}$ be an internal extension of $\widetilde{F}_n : \mathbb{N}^n_\mu \times T^n_< \to {}^*\mathbb{R}$ (see Proposition 16.1.1). Since $\sum_{k\in\mathbb{N}^n_\mu} I_k(f_n(k,\cdot))$ converges in $L^2(\widehat{\mu})$, we may assume that $\widetilde{F}_n \in SL^2(c^n \otimes \nu^n)$. We convert $\widetilde{F}_n$ to a symmetric function $F_n : M^n_\mu \times T^n_{\neq} \to {}^*\mathbb{R}$ by setting

$$F_n(k_1,\ldots,k_n,t_1,\ldots,t_n) := \widetilde{F}_n(k_{\sigma_1},\ldots,k_{\sigma_n},t_{\sigma_1},\ldots,t_{\sigma_n}) \text{ if } t_{\sigma_1} < \ldots < t_{\sigma_n}.$$

This proves (A). Part (B) follows from the orthogonality of the multiple integrals. □

The functions ${}^\circ F_n$ are called the **kernels** of φ. By the following method we can compute the kernels of the chaos decomposition, corresponding to Theorem 17.5.1; the proof here is similar to the proof there.

Theorem 23.1.4 *Fix $\varphi \in L^2_{\mathfrak{D}}(\widehat{\mu})$ and a lifting $\Phi \in SL^2(\mu)$ of φ. Define for all $n \in {}^*\mathbb{N}_0$, $l \in M^n_\mu$ and $s \in T^n_<$*

$$\Phi_n(l_1,\ldots,l_n,s_1,\ldots,s_n) := H^n \cdot \mathbb{E}_\mu\left(\Phi \cdot p_{l_1}(X_{s_1}) \cdot \ldots \cdot p_{l_n}(X_{s_n})\right).$$

Then $\Phi_n \in SL^2(c^n \otimes \nu^n)$ for all $n \in \mathbb{N}_0$, and there exists an unlimited $M \in {}^*\mathbb{N}$ such that $\sum_{n \in {}^*\mathbb{N}_0, n \leq M} I_n(\Phi_n) \in SL(\mu)$. Moreover,

$$\varphi = \sum_{n \in \mathbb{N}_0} {}^\circ I_n(\Phi_n) = \sum_{n \in \mathbb{N}_0} I_n({}^\circ \Phi_n).$$

Proof Let $\varphi = \sum_{n \in \mathbb{N}_0} {}^\circ I_n(F_n)$, according to Theorem 23.1.3. According to Theorem 17.3.1, φ, has a lifting of the form $\Xi := \sum_{n \leq M} I_n(F_n) \in SL^2(\mu)$, where $M \in {}^*\mathbb{N}$. Set

$$U := \left\{ \sum_{n \leq M} I_n(G_n) \mid G_n : M_\mu^n \times T_<^n \to {}^*\mathbb{R} \in L^2\left(c^n \otimes \nu^n\right) \right\}.$$

Since U is an internally closed subspace of $L^2(\mu)$, $\Phi = \Psi + \sum_{n \leq M} I_n(G_n)$ with $\Psi \in U^\perp$ and $\sum_{n \leq M} I_n(G_n) \in U$. Since $X \mapsto p_{l_1}(X_{s_1}) \cdot \ldots \cdot p_{l_n}(X_{s_n}) \in U$ for $n \leq M$, $l_1, \ldots, l_n \in M_\mu^n$, $s_1 < \ldots < s_n$, we obtain

$$\mathbb{E}_\mu\left(\Phi \cdot p_{l_1}(X_{s_1}) \cdot \ldots \cdot p_{l_n}(X_{s_n})\right) = G_n(l_1, \ldots, l_n, s_1, \ldots, s_n) \cdot \frac{1}{H^n}.$$

This proves that $G_n = \Phi_n$ for $n \leq M$. Now continue as in the proof of Theorem 17.5.1. □

23.2 Malliavin derivative

Let $\varphi \in L^2_{\mathcal{D}}(\widehat{\mu})$ with decomposition $\varphi = \sum_{n=0}^\infty I_n(f_n)$, according to Theorem 23.1.3. We may assume that $I_n(f_n) = \sum_{k \in \mathbb{N}_\mu^n} I_k {}^\circ F_n(k, \cdot)$, with $\mathbf{1}_{T_<^n}(\cdot) F_n(k, \cdot) \in \mathfrak{G}_n$ (see Example 23.1.1 (ii)). Then the **Malliavin derivative** $D\varphi$ of φ belongs to $L^2(c \otimes \widehat{\mu \otimes \nu})$, where for all $l \in \mathbb{N}_\mu$ and $\widehat{\mu \otimes \nu}$-almost all (X, t)

$$D\varphi(l, X, t) = \sum_{n=1}^\infty I_{n-1}(f_n(\cdot, l, \cdot, t))(X) = \sum_{n=1}^\infty \sum_{k \in \mathbb{N}_\mu^{n-1}} I_{k,1}(f_n(k, l, \cdot))(X, t)$$

if this series converges in $L^2(c \otimes \widehat{\mu \otimes \nu})$. The Malliavin derivative D is defined on a dense subspace of $L^2_{\mathcal{D}}(\widehat{\mu})$, because $D\varphi$ exists if and only if $\sum_{n=1}^\infty \sqrt{n} I_n(f_n)$ converges in $L^2(\widehat{\mu})$. Then φ is called **Malliavin differentiable**.

By Theorem 16.5.2, we can convert this derivative $D\varphi$ to an equivalent process in $L^2(c \otimes \widehat{\mu} \otimes \lambda)$ (see Corollary 10.9.1). There exists a $\mathcal{D} \otimes \mathrm{Leb}[0, \infty[$-measurable function $D^{\mathrm{st}}\varphi : \mathbb{N}_\mu \times \Omega \times [0, \infty[\to \mathbb{R}$ equivalent to $D\varphi$, i.e.,

$D^{\mathrm{st}}\varphi(k,X,{}^{\circ}t) = D\varphi(k,X,t)$ for all $k \in \mathbb{N}_{\mu}^{n}$ and $\widehat{\mu\otimes\nu}$-almost all $(X,t) \in \Omega \times T$. The densely defined operator D^{st} from $L^2_{\mathcal{D}}(\widehat{\mu})$ to $L^2_{\mathcal{D}\otimes\mathrm{Leb}}(c\otimes\widehat{\mu}\otimes\lambda)$ is called the **standard Malliavin derivative**. Let us identify $D\varphi$ and $D^{\mathrm{st}}\varphi$.

We have the following lifting result for Malliavin differentiable functions, according to Theorem 18.3.1.

Theorem 23.2.1 *Suppose that $\varphi = \sum_{n=0}^{\infty} {}^{\circ}I_n(F_n)$ in $L^2_{\mathcal{D}}(\widehat{\mu})$, according to Theorem 23.1.3. Then φ has a lifting $\Phi \in SL^2(\mu)$ of the form*

$$\Phi := \sum_{n=0}^{M} I_n(F_n),$$

where M is any unlimited number below some unlimited $M_{\infty} \in {}^{*}\mathbb{N}$. If φ is Malliavin differentiable, we may assume that

$$D\Phi : (k,X,t) \mapsto \sum_{n=1}^{M} I_{n-1,1}(F_n)(\cdot,k,\cdot,t)(X)) \in SL^2(c\otimes\mu\otimes\nu),$$

and

$$D\Phi \approx D\varphi \; c\otimes\widehat{\mu\otimes}\nu\text{-a.e.}$$

23.3 The Clark–Ocone formula

The Clark–Ocone formula has great importance in finance development, driven by Lévy processes (see Aase *et al.* [1]). It is an Itô integral representation of Lévy functionals.

Theorem 23.3.1 *Fix a Lévy functional $\varphi \in L^2_{\mathcal{D}}(\widehat{\mu})$ with decomposition $\varphi = \mathbb{E}_{\widehat{\mu}}\varphi + \sum_{n=1}^{\infty} {}^{\circ}I_n(F_n)$, according to Theorem 23.1.3. By Example 23.1.1 (ii), we may assume that $\mathbf{1}_{T^n_<} \cdot F_n(m,\cdot) \in \mathfrak{G}_n$ for all $m \in \mathbb{N}^n_{\mu}$. Then φ can be written as the stochastic integral*

$$\varphi = \mathbb{E}_{\widehat{\mu}}\varphi + \int^{V} (l,\cdot,r) \mapsto \mathbb{E}^{\mathfrak{b}_r} \sum_{n=1}^{\infty} I_{n-1}({}^{\circ}F_n(\cdot,l,\cdot,r))p.$$

If φ is Malliavin differentiable, we obtain

$$\varphi(X) = \mathbb{E}_{\widehat{\mu}}\varphi + \int^{V} (l,\cdot,r) \mapsto \mathbb{E}^{\mathfrak{b}_r} D\varphi(l,\cdot,r)p.$$

Since $\mathbb{E}^{\mathfrak{b}_r} I_{n-1}({}^{\circ}F_n(\cdot,l,\cdot,r))$ is $\mathcal{D}$-measurable, we may replace $\mathfrak{b}_r$ by $\mathfrak{b}_r \cap \mathcal{D}$.

Proof Since $\int \cdot p_l$ is a bounded linear operator, we see that

$$\begin{aligned}
&\varphi - \mathbb{E}_{\widehat{\mu}}(\varphi) \\
&= \sum_{n=1}^{\infty} \sum_{k \in \mathbb{N}_\mu^n} {}^\circ \sum_{t \in T_<^n} F_n(k,t) p_{k_1}(X_{t_1}) \ldots p_{k_n}(X_{t_n}) \\
&= \sum_{l \in \mathbb{N}_\mu} \sum_{n=1}^{\infty} \sum_{k \in \mathbb{N}_\mu^{n-1}} {}^\circ \sum_{s \in T} \sum_{t \in T_<^{n-1}, t<s} F_n(k,l,t,s) p_{k_1}(X_{t_1}) \ldots p_{k_{n-1}}(X_{t_{n-1}}) p_l(X_s) \\
&= \sum_{l \in \mathbb{N}_\mu} \sum_{n=1}^{\infty} \sum_{k \in \mathbb{N}_\mu^{n-1}} {}^\circ \int^V s \mapsto I_{k,1}(\mathbf{1}_{T_<^n}(\circ) \cdot F_n(k,l,\circ)(\cdot,s) p_l \\
&= \sum_{l \in \mathbb{N}_\mu} \sum_{n=1}^{\infty} \sum_{k \in \mathbb{N}_\mu^{n-1}} \int^V r \mapsto \mathbb{E}^{\mathfrak{b}_r} I_{k,1}({}^\circ F_n(k,l,\circ))(\cdot,r) p_l \\
&= \sum_{l \in \mathbb{N}_\mu} \int^V r \mapsto \mathbb{E}^{\mathfrak{b}_r} \sum_{n=1}^{\infty} \sum_{k \in \mathbb{N}_\mu^{n-1}} I_{k,1}({}^\circ F_n(k,l,\circ))(\cdot,r) p_l \\
&= \sum_{l \in \mathbb{N}_\mu} \int^V r \mapsto \mathbb{E}^{\mathfrak{b}_r} \sum_{n=1}^{\infty} {}^\circ I_{n-1}(F_n(\cdot,l,\cdot,r)) p_l \\
&= \int^V (l,\cdot,r) \mapsto \mathbb{E}^{\mathfrak{b}_r} \sum_{n=1}^{\infty} I_{n-1}({}^\circ F_n(\cdot,l,\cdot,r)) p \\
&= \int^V (l,\cdot,r) \mapsto \mathbb{E}^{\mathfrak{b}_r} D\varphi(l,\cdot,r) p \text{ (if } \varphi \text{ is differentiable).}
\end{aligned}$$

□

23.4 Skorokhod integral processes

In order to define the Skorokhod integral and Skorokhod integral processes, we need a suitable decomposition of functionals in $L^2_{\mathcal{D}\otimes\mathrm{Leb}}(c \otimes \widehat{\mu} \otimes \lambda)$. We also use the following common notation. Suppose that $F : M_\mu^{n+1} \times T_{\neq}^{n+1} \to {}^*\mathbb{R}$ is internal and symmetric in the first n components. Then $\widetilde{F} : M_\mu^{n+1} \times T_{\neq}^{n+1} \to {}^*\mathbb{R}$ is the symmetric function derived from F by setting for $t = (t_1, \ldots, t_{n+1})$ and $k = (k_1, \ldots, k_{n+1})$,

$$\widetilde{F}(k,t) := \sum_{i=1}^{n+1} F_{(k_1,\ldots,k_{i-1},k_{i+1},\ldots,k_{n+1},k_i)}(t_1,\ldots,t_{i-1},t_{i+1},\ldots,t_{n+1},t_i).$$

Recall the definition: $\left(\mathbf{1}_{T_\beta} \cdot_{n+1} F\right)(k,t) := \mathbf{1}_{T_\beta}(t_{n+1}) \cdot F(k,t)$ with $\beta \in T$. We may assume that the domain of $F(k,\cdot)$ is $T_{\neq}^{n+1}$. Then

$$\int I_{n,1}(\mathbf{1}_{T_\beta} \cdot_{n+1} F)p := \int_{\frac{1}{H}}^{\beta} I_{n,1}(F)p$$

$$:= \sum_{l \in M_\mu} \sum_{r \leq \beta} \left(\sum_{k \in M_\mu^n} \sum_{t \in T_<^n} F(k,l,t,r) p_{k_1}(X_{t_1}) \cdot \ldots \cdot p_{k_n}(X_{t_n}) \right) p_l(X_r)$$

$$= I_{n+1}\left(\widetilde{\mathbf{1}_{T_\beta} \cdot_{n+1} F}\right).$$

Theorem 23.4.1 *Fix $\varphi \in L^2_{\mathcal{D}\otimes \mathrm{Leb}}(c \otimes \widehat{\mu} \otimes \lambda)$. Then there exists a sequence $(F_n)_{n\in\mathbb{N}_0}$ of internal functions $F_n : M_\mu^{n+1} \times T_{\neq}^{n+1} \to {}^*\mathbb{R}$ such that ${}^\circ F_n$ exists in $L^2\left(\widehat{c^{n+1} \otimes \nu^{n+1}}\right)$ with the following properties:*

(a) *$F_n \in SL^2\left(c^{n+1} \otimes \nu^{n+1}\right)$ and F_n is symmetric in the first n arguments.*
(b) *$I_{n+1}({}^\circ\widetilde{F}_n) \in L^2_{\mathcal{D}}(\widehat{\mu})$.*
(c) *$s \mapsto I_{n+1}\left(\widetilde{\mathbf{1}_{T_s} \cdot_{n+1} F_n}\right) \in SL^2(\mu \otimes \nu)$ and $s \mapsto {}^\circ I_{n+1}\left(\widetilde{\mathbf{1}_{T_s} \cdot_{n+1} F_n}\right)$ is in $L^2_{\mathcal{D}\otimes\mathcal{L}}(\widehat{\mu \otimes \nu})$.*
(d) *$\varphi(m,X,{}^\circ s) = \sum_{n=0}^{\infty} {}^\circ I_n(F_n(\cdot,m,\cdot,s))(X)$ in $L^2_{\mathcal{D}\otimes\mathrm{Leb}}(\widehat{c \otimes \mu \otimes \nu})$.*

Proof Let M be the set of all functions φ in $L^2_{\mathcal{D}\otimes\mathrm{Leb}}(c \otimes \widehat{\mu} \otimes \lambda)$ having such a decomposition. Using Corollary 10.9.3, one can see that M is a complete linear space. Thus, it suffices to prove the result for $\varphi = \mathbf{1}_{\{l\}} \otimes \mathbf{1}_B \otimes \mathbf{1}_C$, where $l \in \mathbb{N}_\mu$, $B \in \mathcal{D}$ and $C \subset [0,\sigma]$ for some $\sigma \in \mathbb{N}$ and Lebesgue measurable. Now $\mathbf{1}_B$ has a decomposition $\mathbf{1}_B = \sum_{n=0}^{\infty} I_n({}^\circ G_n)$ according to Theorem 23.1.3. For C there exists an internal $A \subseteq T$ such that $\widehat{\nu}(A \Delta st^{-1}[C]) = 0$. Define for $k \in M_\mu^{n+1}$, $t \in T_{\neq}^{n+1}$

$$F_n(k,t) := \mathbf{1}_{\{l\}}(k_{n+1}) \cdot G_n(k_1,\ldots,k_n,t_1,\ldots,t_n) \cdot \mathbf{1}_A(t_{n+1}).$$

Note that (a)–(d) are true. □

The proof of the following result is similar to the proof of the preceding theorem and is left to the reader.

Theorem 23.4.2 *Fix $\varphi \in L^2_{\mathcal{D}\otimes\mathrm{Leb}}(\widehat{\mu} \otimes \lambda)$. Then there exists a sequence $(F_n)_{n\in\mathbb{N}_0}$ of internal functions $F_n : M_\mu^n \times T_{\neq}^{n+1} \to {}^*\mathbb{R}$ such that ${}^\circ F_n$ exists in $L^2\left(c^n \otimes \lambda^{n+1}\right)$ with the following properties:*

(a) *$F_n \in SL^2\left(c^n \otimes \nu^{n+1}\right)$ and F_n is symmetric in the first n arguments.*

(b) $(X,s) \mapsto I_n(F_n(\cdot,s))(X) \in SL^2(\mu \otimes \nu)$, *where*

$$I_n(F_n(\cdot,s))(X) = \sum_{k \in M^n_\mu} \sum_{t \in T^n_<} F_n(k,t,s) \prod_{i=1}^{k} p_{k_i}(X_{t_i}).$$

(c) $\varphi(X, {}^\circ s) = \sum_{n=0}^\infty {}^\circ I_n(F_n(\cdot,s))(X)$ *in* $L^2_{\mathcal{D}\otimes\mathrm{Leb}}(\widehat{\mu \otimes \nu})$.

Using Theorem 23.4.1, we are able to define the **Skorokhod integral** δ as a densely defined operator from $L^2_{\mathcal{D}\otimes\mathrm{Leb}}(c \otimes \widehat{\mu} \otimes \lambda)$ into $L^2_{\mathcal{D}}(\widehat{\mu})$. Suppose that φ has the decomposition according to Theorem 23.4.1. Set

$$\delta\varphi := \sum_{n=0}^\infty {}^\circ I_{n+1}\left(\widetilde{F}_n\right)$$

for those $\varphi \in L^2_{\mathcal{D}\otimes\mathrm{Leb}}(c \otimes \lambda \otimes \widehat{\mu})$ such that $\delta\varphi$ converges in $L^2_{\mathcal{D}}(\widehat{\mu})$, in which case φ is called **Skorokhod integrable**. The proof of the following result is left to the reader. It is similar to the proof of Theorem 19.5.1.

Theorem 23.4.3 *If* $\varphi \in L^2_{\mathcal{D}\otimes\mathrm{Leb}}(c \otimes \widehat{\mu} \otimes \lambda)$ *and each* φ_k *is* $(\mathfrak{b}_r \cap \mathcal{D})_{r \in [0,\infty[}$-*predictable. Then* φ *is Skorokhod integrable and* $\delta\varphi = \int \varphi\, p$.

Now we introduce the **Skorokhod integral process** as operator δ^{pr}, densely defined on $L^2_{\mathcal{D}\otimes\mathrm{Leb}}(c \otimes \widehat{\mu} \otimes \lambda)$ with values in $L^2_{\mathrm{Leb}\otimes\mathcal{D}}(\widehat{\mu} \otimes \lambda)$. Fix $\varphi \in L^2_{\mathcal{D}\otimes\mathrm{Leb}}(c \otimes \widehat{\mu} \otimes \lambda)$ with decomposition according to Theorem 23.4.1. Suppose that $r \mapsto \sum_{n=0}^\infty {}^\circ I_{n+1}\left(\widetilde{\mathbf{1}_{T_r} \cdot_{n+1} F_n}\right)$ converges in $L^2_{\mathcal{D}\otimes\mathcal{L}}(\widehat{\mu \otimes \nu})$. Then define $\delta^{\mathrm{pr}}(\varphi) : \Omega \times [0,\infty[\to \mathbb{R}$ by

$$\delta^{\mathrm{pr}}(\varphi)(X, {}^\circ r) := \sum_{n=0}^\infty {}^\circ I_{n+1}\left(\widetilde{\mathbf{1}_{T_r} \cdot_{n+1} F_n}\right)(X).$$

Recall that the process ${}^\circ I_{n+1}\left(\widetilde{\mathbf{1}_{T_\cdot} \cdot_{n+1} F_n}\right)$ is equivalent to a process in $L^2_{\mathcal{D}\otimes\mathrm{Leb}}(\widehat{\mu} \otimes \lambda)$.

23.5 Smooth representations

Using chaos decompositions, we can find S-square-integrable internally smooth liftings of square-integrable random variables or processes. In analogy to Theorem 17.3.1, Corollary 19.1.2 and Theorems 18.3.1 and 19.4.1, we have the following.

Theorem 23.5.1 *Suppose that, according to Theorems 23.1.3, 23.4.1,* $\varphi \in L^2_{\mathcal{D}}(\widehat{\mu})$, $\psi \in L^2_{\mathcal{D}\otimes\mathcal{L}}(\widehat{c\otimes\mu\otimes\nu})$ *have expansions*

$$\varphi = \sum_{n=0}^{\infty} {}^{\circ}I_n(F_n), \quad \psi(k,X,t) = \sum_{n=0}^{\infty} {}^{\circ}I_n(G_n(\cdot,k,\cdot,t))(X).$$

Let $(F_n)_{n\in{}^*\mathbb{N}_0}$, $(G_n)_{n\in{}^*\mathbb{N}_0}$ *be internal extensions of* $(F_n)_{n\in\mathbb{N}_0}$, $(G_n)_{n\in\mathbb{N}_0}$, *respectively, such that* F_n *is symmetric and* G_n *is symmetric in the first n arguments. Then there exists an unlimited* $K_\infty \in {}^*\mathbb{N}$ *such that for all unlimited* $K \in {}^*\mathbb{N}$ *with* $K \le K_\infty$ *the following hold.*

(a) $\Phi_K := \sum_{n=0}^{K} I_n(F_n)$ *belongs to* $SL^2(\mu)$ *and is a lifting of* φ. *If* φ *is Malliavin differentiable, then we may choose* K_∞ *such that*

$$D\Phi_K : (k,X,t) \mapsto \sum_{n=1}^{K} {}^{\circ}I_{n-1}(F_n(\cdot,k,\cdot,t)(X) \in SL^2(c\otimes\mu\otimes\nu)$$

and is a lifting of $D\varphi$.

(b) $\Psi_K : (l,X,r) \mapsto \sum_{n=0}^{K} I_n(G_n(\cdot,l.\cdot,r))(X)$ *belongs to* $SL^2(c\otimes\mu\otimes\nu)$ *and is a lifting of* ψ. *If* ψ *is Skorokhod integrable, then we may choose* K_∞ *such that*

$$\delta(\Psi_K) : X \mapsto \sum_{n=0}^{K} I_{n+1}(\widetilde{G_n})(X) \in SL^2(\mu)$$

and is a lifting of $\delta\psi$.

(c) *Moreover, for* $\psi \in L^2_{\mathcal{D}\otimes\mathcal{L}}(\widehat{\mu\otimes\nu})$, *according to Theorem 23.4.2,* ψ *has a lifting* $\Psi \in SL^2(\mu\otimes\nu)$ *of the form*

$$\Psi(X,r) = \sum_{n=0}^{K} I_n(G_n(\cdot,r))(X).$$

The next result shows that δ is the adjoint operator of the Malliavin derivative, thus δ and D are closed operators. The straightforward proof is left to the reader (see Theorem 20.1.1).

Theorem 23.5.2 *Let* φ *be Skorokhod integrable and* ψ *be Malliavin differentiable. Then*

$$\langle \varphi, D\psi \rangle_{c\otimes\widehat{\mu}\otimes\lambda} = \langle \delta\varphi, \psi \rangle_{\widehat{\mu}}.$$

Moreover, we obtain lifting results for non-integrable functions in analogy to Theorems 17.4.1 and 19.1.3.

Theorem 23.5.3 *Fix $\mathcal{D}$-measurable $\varphi : \Omega \to \mathbb{R}$ and $\mathcal{D} \otimes \mathcal{L}$-measurable $\psi : \mathbb{N}_\mu \times \Omega \times T \to \mathbb{R}$ ($\psi : \Omega \times T \to \mathbb{R}$). Then φ and ψ have liftings of the form*

$$\Phi : X \mapsto \sum_{n=0}^{K} I_n(F_n) \text{ with internal symmetric } F_n : M_\mu^n \times T_{\neq}^n \to^* \mathbb{R},$$

$$\Psi : (l, X, r) \mapsto \sum_{n=0}^{K} I_n(F_n(\cdot, l, \cdot, r)(X)) \left(\Psi : (X, r) \mapsto \sum_{n=0}^{K} I_n(F_n(\cdot, r)(X) \right),$$

where $F_n : M_\mu^{n+1} \times T_{\neq}^{n+1} \to {}^*\mathbb{R}$ $\left(F_n : M_\mu^n \times T_{\neq}^{n+1} \to {}^*\mathbb{R}\right)$ are internal and symmetric in the first n variables.

Proof The proof uses the fact that φ and ψ can be approximated in measure by square-integrable functions. Then use liftings of these approximations, according to Theorem 23.5.1. To obtain the desired liftings of φ and ψ, use techniques, established in Corollary 10.9.3. □

23.6 A commutation rule for derivative and limit

The following result and its proof are similar to Theorem 18.5.1 and the proof there. The proof is left to the reader.

Theorem 23.6.1 *Suppose that $(\varphi_i)_{i\in\mathbb{N}}$ is a sequence of Malliavin differentiable functions such that $(D\varphi_i)_{i\in\mathbb{N}}$ converges in $L^2_{\mathcal{D}\otimes\mathcal{L}}(c \otimes \widehat{\mu \otimes \nu})$ and suppose that $(\mathbb{E}_{\widehat{\mu}}\varphi_i)_{i\in\mathbb{N}}$ converges in the real numbers. Then $(\varphi_i)_{i\in\mathbb{N}}$ converges to a Malliavin differentiable function and D ($\lim \varphi_i$) = $\lim D\varphi_i$.*

23.7 The product rule

Following [92], product and chain rules for general Lévy processes will be studied now in a quite constructive way, using somehow technical but elementary finite combinatorics. I would recommend the reader to study first the main lemma (Lemma 23.7.2) in the case $m = 3$.

Both rules are more complicated than the corresponding rules for Brownian motion. The reasons are: the set $\mathbb{N}_\mu$ is often different from $\{1\}$ and the multiple integrals are only square-integrable, in general. It follows that, in contrast to Brownian motion, the product of two Lévy functionals in finite chaos levels is not necessarily square integrable. The most challenging difference, however, is the fact that $\mathbb{E}_\mu p_1^3 \neq 0$, for example, in the case of Poisson processes.

In order to overcome the difficulties caused by the loss of moments larger than 2 for the multiple integrals, we use restricted functions. For each $m \in \mathbb{N}$ define for $\varphi = \sum_{n \in N_o} I_n(f_n)$

$$\varphi_m := \varphi \upharpoonright m := \sum_{n \in \mathbb{N}_0} I_n(f_n \upharpoonright m),$$

where $(f_n \upharpoonright m)(k,r) := f_n(k,r)$ if $n \leq m$, $|f_n(k,r)| \leq m$, $k \in \{1,\ldots,m\}^n$ and $r \leq m$. Otherwise, $(f_n \upharpoonright m)(k,r) := 0$. If $\varphi = \varphi_m$ or $f = f \upharpoonright m$, then we say that φ, f, respectively, are **restricted by** m.

Restricted $\varphi \in L^2_{\mathcal{D}}(\widehat{\mu})$ are Malliavin differentiable. Fix $\varphi \in L^2_{\mathcal{D}}(\widehat{\mu})$. Since I_n is a continuous operator and the $(I_n(f_n))_{n \in \mathbb{N}}$ are pairwise orthogonal, we have $\lim \varphi_m = \varphi$ in $L^2(\widehat{\mu})$, and if φ is Malliavin differentiable, then $\lim D\varphi_m = D\varphi$ in $L^2(c \otimes \widehat{\mu} \otimes \lambda)$. It is always assumed that $\mathbb{N}$ is an initial segment of $\mathbb{N}$ depending on the Lévy process L. The following terms are crucial: for $l, \kappa, \widetilde{\kappa} \in \mathbb{N}_\mu$ Set

$$\sigma(\kappa, \widetilde{\kappa}, l) := H\mathbb{E}_{\mu^1} p_\kappa \cdot p_{\widetilde{\kappa}} \cdot p_l \quad \text{and} \quad \alpha(\kappa, \widetilde{\kappa}, l) := {}^\circ\sigma(\kappa, \widetilde{\kappa}, l).$$

Theorem 23.7.1 (Product rule) *Let $\mathbb{N}_\mu$ be finite. Fix Malliavin differentiable $\varphi, \psi \in L^2_{\mathcal{D}}(\widehat{\mu})$ such that $\left(\mathbb{E}_{\widehat{\mu}}(\varphi_m \cdot \psi_m)\right)_{m \in \mathbb{N}}$ converges in $\mathbb{R}$.*

(A) *Suppose that the sequences $(D\varphi_m \cdot \psi_m)_{m \in \mathbb{N}}$, $(\varphi_m \cdot D\psi_m)_{m \in \mathbb{N}}$ converge in $L^2(c \otimes \widehat{\mu} \otimes \lambda)$. Then $(D(\varphi_m \cdot \psi_m))_{m \in \mathbb{N}}$ converges in $L^2(c \otimes \widehat{\mu} \otimes \lambda)$ if and only if $\left((l,r,X) \mapsto \sum_{\kappa, \widetilde{\kappa} \in \mathbb{N}_\mu} \alpha(\kappa, \widetilde{\kappa}, l) \cdot D(\varphi_m)_{\kappa,r}(X) \cdot D(\psi_m)_{\widetilde{\kappa},r}(X)\right)_{m \in \mathbb{N}}$ converges in $L^2(c \otimes \widehat{\mu} \otimes \lambda)$, in which case $\varphi \cdot \psi$ is Malliavin differentiable and*

$$(D(\varphi \cdot \psi))_{(l,r)} = (D\varphi)_{(l,r)} \cdot \psi + \varphi \cdot (D\psi)_{(l,r)} + \sum_{\kappa, \widetilde{\kappa} \in \mathbb{N}_\mu} \alpha(\kappa, \widetilde{\kappa}, l) \cdot D(\varphi)_{\kappa,r} \cdot D(\psi)_{\widetilde{\kappa},r}$$

in $L^2(c \otimes \widehat{\mu} \otimes \lambda)$. In the case $\mathbb{N}_\mu = \{1\}$, we have for $\alpha = \alpha(1,1,1)$

$$(D(\varphi \cdot \psi))_r = \frac{(\varphi + \alpha \cdot (D\varphi)_r) \cdot (\psi + \alpha \cdot (D\psi)_r) - \varphi \cdot \psi}{\alpha} \quad \text{if } \alpha \neq 0,$$

in $L^2(\widehat{\mu} \otimes \lambda)$. If $\alpha = 0$, then $(D(\varphi \cdot \psi))_r = (D\varphi)_r \cdot \psi + \varphi \cdot (D\psi)_r$.

(B) *Suppose that $(D(\varphi_m \cdot \psi_m))_{m \in \mathbb{N}}$ converges in $L^2(c \otimes \widehat{\mu} \otimes \lambda)$. Then $\varphi \cdot \psi$ is Malliavin differentiable and for all $l \in \mathbb{N}_\mu$ we have $\widehat{\mu} \otimes \lambda$-a.e.*

$$(D(\varphi \cdot \psi))_{l,r} = (D\varphi)_{l,r} \cdot \psi + \varphi \cdot (D\psi)_{l,r} + \sum_{\kappa, \widetilde{\kappa} \in \mathbb{N}_\mu} \alpha(\kappa, \widetilde{\kappa}, l) \cdot (D\varphi)_{\kappa,r} \cdot (D\psi)_{\widetilde{\kappa},r}.$$

If L is the Brownian motion, then $\mathbb{N}_\mu = \{1\}$ and $\alpha = 0$. For the Poisson process with rate β we have $\mathbb{N}_\mu = \{1\}$ and $\alpha = \beta^{-\frac{1}{2}}$. In the case of symmetrized Poisson processes, we have $\mathbb{N}_\mu = \{1,2\}$.

In the proof of this result we use the computation of the kernels (see Theorem 23.1.4) of the terms in the theorem, like $\varphi_m \cdot \psi_m, D(\varphi_m \cdot \psi_m), \ldots$

We use the following notation. An internal function $\Phi : \Omega \to {}^*\mathbb{R}$ is called a **polynomial restricted by** $S \in {}^*\mathbb{N}$ if

$$\Phi(X) = \sum_{n \in {}^*\mathbb{N}_0} \sum_{k \in M_\mu^n} \sum_{t \in T_<^n} F_n(k,t) \prod_{i=1}^{n} p_{k_i}(X_{t_i}),$$

where $F_n : M_\mu^n \times T_{\neq}^n \to {}^*\mathbb{R}$ is internal and symmetric, and Φ and therefore also the F_n are restricted by S. By Theorem 23.1.3, we can assume that each $\varphi : \Omega \to \mathbb{R}$ in $L^2_{\mathcal{D}}(\widehat{\mu})$ has a polynomial lifting $\Phi : \Omega \to {}^*\mathbb{R} \in SL^2(\mu)$. If φ is Malliavin differentiable, then we can, in addition, assume that

$$D\Phi : (l,r,X) \mapsto \sum_{n \in {}^*\mathbb{N}} \sum_{k \in M_\mu^n} \sum_{t \in T_<^{n-1}} F_n(k,l,t,r) \prod_{i=1}^{n-1} p_{k_i}(X_{t_i})$$

belongs to $SL^2(c \otimes \nu \otimes \mu)$ and is a lifting of the Malliavin derivative of φ.

In analogy to the Brownian motion case we define the following. Fix $m \in \mathbb{N}$, $\rho \in m \cup \{0\}$, a strictly increasing ρ-tuple $\beta_1 < \ldots < \beta_\rho$ in m and $i \in \{\rho, \ldots, m\}$. Let $\tau : i - \rho \uparrow m \setminus \{\beta_1, \ldots, \beta_\rho\}$ be a strictly monotone increasing function from $i - \rho$ into $m \setminus \{\beta_1, \ldots, \beta_\rho\}$. Then $\overline{\tau}$ denotes the **complement** of τ, i.e., $\overline{\tau}$ is the uniquely determined strictly monotone increasing function from $m - i$ onto $m \setminus (\mathrm{range}(\tau) \cup \{\beta_1, \ldots, \beta_\rho\})$. Note that τ depends on m. Here is the key to the product rule and also to the chain rule:

Lemma 23.7.2 *Suppose that $\varphi, \psi \in L^2_{\mathcal{D}}(\widehat{\mu})$ are restricted by some standard $S \in \mathbb{N}$. Then we have in $L^2(c \otimes \widehat{\mu} \otimes \lambda)$:*

$$(D(\varphi \cdot \psi))_{(l,r)} = (D\varphi)_{(l,r)} \cdot \psi + \varphi \cdot (D\psi)_{(l,r)} + \sum_{\kappa, \widetilde{\kappa} \in \mathbb{N}_\mu} \alpha(\kappa, \widetilde{\kappa}, l)$$
$$\cdot D(\varphi)_{\kappa,r} \cdot D(\psi)_{\widetilde{\kappa},r} \cdot$$

Proof By Theorem 23.1.3, φ and ψ have polynomial liftings Φ and Ψ. Since φ, ψ are restricted by S, we can assume that Φ and Ψ are also restricted by S. Therefore $\Phi \cdot \Psi$ belongs to $SL^2(\mu)$ and is a lifting of $\varphi \cdot \psi \in L^2_{\mathcal{D}}(\widehat{\mu})$. Moreover, $D(\Phi \cdot \Psi), D\Phi \cdot \Psi, \Phi \cdot D\Psi \in SL^2(c \otimes \mu \otimes \nu)$ are liftings of $D(\varphi \cdot \psi), D\varphi \cdot$

$\psi, \varphi \cdot D\psi$, respectively, and $\sum_{\kappa,\widetilde{\kappa}\in\mathbb{N}_\mu} \sigma(\kappa,\widetilde{\kappa},\cdot)\cdot D\Phi_\kappa \cdot D\Psi_{\widetilde{\kappa}} \in SL^2(c\otimes\mu\otimes\nu)$ is a lifting of $\sum_{\kappa,\widetilde{\kappa}\in\mathbb{N}_\mu} \alpha(\kappa,\widetilde{\kappa},l)\cdot D\varphi_\kappa \cdot D\psi_{\widetilde{\kappa}}$. By Theorem 23.1.4, the kernels of $\varphi\cdot\psi$ are the standard parts of the kernels $K_m^{\Phi\cdot\Psi}$ under $\Phi\cdot\Psi$, given by $K_0^{\Phi\cdot\Psi} = \mathbb{E}_\mu(\Phi\cdot\Psi)$ and, for $m \geq 1$,

$$K_m^{\Phi\cdot\Psi}(l,r) = H^m \mathbb{E}_\mu\left(\Phi\cdot\Psi\cdot p_{l_1}(X_{r_1})\cdot\ldots\cdot p_{l_m}(X_{r_m})\right)$$

with $l \in \mathbb{N}_\mu^m$, $r \in T_<^m$. Let F_n, G_n, be the kernels of Φ, Ψ, respectively. Finite combinatorics tells us that $K_m^{\Phi\cdot\Psi}(l,r)$ is the finite sum:

$$K_m^{\Phi\cdot\Psi}(l,r) = \sum_{\rho=0}^{m}\sum_{\kappa,\widetilde{\kappa}\in\mathbb{N}_\mu^\rho}\sum_{\beta\in m_<^\rho}\sum_{i=\rho}^{m}\sum_{\tau:i-\rho\uparrow m\smallsetminus\{\beta_1,\ldots,\beta_\rho\}}\sum_{n\in\mathbb{N}_0}\sum_{k\in\mathbb{N}_\mu^n}\sum_{t\in T_<^n}\frac{1}{H^n}\cdot\Pi$$

with

$$\begin{aligned}\Pi = {} & F_{n+i}(k,\varkappa,l_{\tau_1},\ldots l_{\tau_{i-\rho}},t,r_{\beta_1},\ldots r_{\beta_\rho},r_{\tau_1},\ldots r_{\tau_{i-\rho}}) \\ & \cdot G_{n+m-i+\rho}(k,\widetilde{\varkappa},l_{\overline{\tau}_1},\ldots l_{\overline{\tau}_{m-i}},t,r_{\beta_1},\ldots r_{\beta_\rho},r_{\overline{\tau}_1},\ldots r_{\overline{\tau}_{m-i}}) \\ & \cdot \sigma(\kappa_1,\widetilde{\kappa}_1,l_{\beta_1})\cdot\ldots\cdot\sigma(\kappa_\rho,\widetilde{\kappa}_\rho,l_{\beta_\rho}).\end{aligned}$$

Note that

$$K_m^{\Phi\cdot\Psi}(l,r) = A + B + C(l_m, r_m)$$

with

$$A = \sum_{\rho=0}^{m}\sum_{\kappa,\widetilde{\kappa}\in\mathbb{N}_\mu^\rho}\sum_{\beta\in(m-1)_<^\rho}\sum_{i=\rho}^{m}\sum_{\tau:i-\rho\uparrow m\smallsetminus\{\beta_1,\ldots,\beta_\rho\},\tau(i-\rho)=m}\sum_{n\in\mathbb{N}_0}\sum_{k\in\mathbb{N}_\mu^n}\sum_{t\in T_<^n}\frac{1}{H^n}\cdot\Pi,$$

$$B = \sum_{\rho=0}^{m}\sum_{\kappa,\widetilde{\kappa}\in\mathbb{N}_\mu^\rho}\sum_{\beta\in(m-1)_<^\rho}\sum_{i=\rho}^{m}\sum_{\tau:i-\rho\uparrow m\smallsetminus\{\beta_1,\ldots,\beta_\rho\},\overline{\tau}(m-i)=m}\sum_{n\in\mathbb{N}_0}\sum_{k\in\mathbb{N}_\mu^n}\sum_{t\in T_<^n}\frac{1}{H^n}\cdot\Pi,$$

$$C(l_m,r_m) = \sum_{\rho=0}^{m}\sum_{\kappa,\widetilde{\kappa}\in\mathbb{N}_\mu^\rho}\sum_{\beta\in m_<^\rho,\beta_\rho=m}\sum_{i=\rho}^{m}\sum_{\tau:i-\rho\uparrow m\smallsetminus\{\beta_1,\ldots,\beta_\rho\}}\sum_{n\in\mathbb{N}_0}\sum_{k\in\mathbb{N}_\mu^n}\sum_{t\in T_<^n}\frac{1}{H^n}\cdot\Pi.$$

In the same way, computing the kernel $K_{m-1}^{D\Phi_{l_m,r_m}\cdot\Psi}$ under $D\Phi_{l_m,r_m}\cdot\Psi$, we obtain for $l = (l_1,\ldots,l_{m-1}), r = (r_1,\ldots r_{m-1})$,

$$\begin{aligned}K_{m-1}^{D\Phi_{l_m,r_m}\cdot\Psi}(l,r) &= H^{m-1}\mathbb{E}_\mu\left(D\Phi_{l_m,r_m}\cdot\Psi\cdot p_{l_1}(X_{r_1})\cdot\ldots\cdot p_{l_{m-1}}(X_{r_{m-1}})\right) \\ &= \sum_{\rho=0}^{m-1}\sum_{\kappa,\widetilde{\kappa}\in\mathbb{N}_\mu^\rho}\sum_{\beta\in m-1_<^\rho}\sum_{i=\rho}^{m-1}\sum_{\tau:i-\rho\uparrow m-1\smallsetminus\{\beta_1,\ldots,\beta_\rho\}}\sum_{n\in\mathbb{N}_0}\sum_{k\in\mathbb{N}_\mu^n}\sum_{t\in T_<^n}\frac{1}{H^n}\end{aligned}$$

$$\cdot F_{n+i+1}(k,\varkappa,l_{\tau_1},\ldots l_{\tau_{i-\rho}},l_m,t,r_{\beta_1},\ldots r_{\beta_\rho},r_{\tau_1},\ldots r_{\tau_{i-\rho}},r_m)$$
$$\cdot G_{n+m-1-i+\rho}(k,\widetilde{\varkappa},l_{\overline{\tau}_1},\ldots l_{\overline{\tau}_{m-1-i}},t,r_{\beta_1},\ldots r_{\beta_\rho},r_{\overline{\tau}_1},\ldots r_{\overline{\tau}_{m-1-i}})$$
$$\cdot \sigma(\kappa_1,\widetilde{\kappa}_1,l_{\beta_1})\cdot\ldots\cdot\sigma(\kappa_\rho,\widetilde{\kappa}_\rho,l_{\beta_\rho}).$$

Note that $K_{m-1}^{D\Phi_{l_m,r_m}\cdot\Psi}(l,r)=A$ and

$$B=K_{m-1}^{\Phi\cdot D\Psi_{l_m,r_m}},\quad C(l_m,r_m)=\sum_{\eta,\widetilde{\eta}\in\mathbb{N}_\mu}K_{m-1}^{D\Phi_{\eta,r_m}\cdot D\Psi_{\widetilde{\eta},r_m}}\sigma(\eta,\widetilde{\eta},l_m),$$

where $K_{m-1}^{\Phi\cdot D\Psi_{l_m,r_m}}$ and $K_{m-1}^{D\Phi_{\eta,r_m}\cdot D\Psi_{\widetilde{\eta},r_m}}$ are the kernels under $\Phi\cdot D\Psi_{l_m,r_m}$ and under $D\Phi_{\eta,r_m}\cdot D\Psi_{\widetilde{\eta},r_m}$, respectively. This proves that for all $l\in\mathbb{N}_\mu$ and $\widehat{\nu}$-almost all $r\in T$

$$(D(\Phi\cdot\Psi))_{(l,r)}\approx(D\Phi)_{(l,r)}\cdot\Psi+\Phi\cdot(D\Psi)_{(l,r)}+\sum_{\kappa,\widetilde{\kappa}\in\mathbb{N}_\mu}\sigma(\kappa,\widetilde{\kappa},l)\cdot D\Phi_{\kappa,r}\cdot D\Psi_{\widetilde{\kappa},r}.$$

Taking standard parts, we obtain the desired result. □

The product rule now follows from Lemma 23.7.2 and Theorem 23.6.1.

23.8 The chain rule

In order to prove the chain rule, which is an extension of the product rule for the case $\mathbb{N}_\mu=1$, we need the following useful lemma.

Lemma 23.8.1 *Suppose that $\mathbb{N}_\mu=\{1\}$. Fix $g:\mathbb{R}^n\to\mathbb{R}$ with $g(x_1,\ldots,x_n)=x_1^{k_1}\cdot\ldots\cdot x_n^{k_n}$ and $\varphi_1,\ldots,\varphi_n\in L^2_{\mathcal{D}}(\widehat{\mu})$ restricted by some $S\in\mathbb{N}$. Then as an identity in $L^2(\lambda\otimes\widehat{\mu})$*

$$D(g(\varphi_1,\ldots,\varphi_n))$$
$$=\begin{cases}\frac{1}{\alpha}(g(\varphi_1+\alpha\cdot D\varphi_1,\ldots,\varphi_n+\alpha\cdot D\varphi_n)-g(\varphi_1,\ldots,\varphi_n)) & \text{if } \alpha\neq0\\ \sum_{i=1}^n(\partial_i g)(\varphi_1,\ldots,\varphi_n)\cdot D\varphi_i & \text{if } \alpha=0\end{cases}.$$

Proof By induction on n, using the product rule. In the case $n=1$ apply induction on k_1. □

Theorem 23.8.2 (Chain rule) *Suppose that $\mathbb{N}_\mu=\{1\}$. Fix $g:\mathbb{R}^n\to\mathbb{R}$ and Malliavin differentiable functions $\varphi_1,\ldots,\varphi_n$. Assume that the partial derivatives* *of g exist and that there are polynomials q_j in n variables with* $\lim q_j = g$ *and* $\lim \partial_i q_j = \partial_i g$ *for* $i = 1, \ldots, n$.

(A) *Fix* $m \in \mathbb{N}$. *Suppose that* $\left(D\left(q_j\left(\varphi_{1,m}, \ldots, \varphi_{n,m}\right)\right)\right)_{j \in \mathbb{N}}$ *converges in* $L^2(\widehat{\mu} \otimes \lambda)$ *and* $\left(\mathbb{E}_{\widehat{\mu}}\left(q_j\left(\varphi_{1,m}, \ldots, \varphi_{n,m}\right)\right)\right)_{j \in \mathbb{N}}$ *converges in* $\mathbb{R}$.
Then $g\left(\varphi_{1,m}, \ldots, \varphi_{n,m}\right)$ *is Malliavin differentiable and we have* $\widehat{\mu} \otimes \lambda$*-a.e.,*

$$\begin{aligned}&\left(D\left(g\left(\varphi_{1,m}, \ldots, \varphi_{n,m}\right)\right)\right)_r \\ &= \frac{1}{\alpha}\left(g\left(\varphi_{1,m} + \alpha \cdot \left(D\varphi_{1,m}\right)_r, \ldots, \varphi_{n,m} + \alpha \cdot \left(D\varphi_{n,m}\right)_r\right) - g\left(\varphi_{1,m}, \ldots, \varphi_{n,m}\right)\right),\end{aligned}$$

where this fraction is equal to $\sum_{i=1}^{n}\left(\partial_i g\right)\left(\varphi_{1,m}, \ldots, \varphi_{n,m}\right) \cdot \left(D\varphi_{i,m}\right)_r$ *if* $\alpha = 0$.

(B) *Assume that (A) is true for all* $m \in \mathbb{N}$, *and* g *and* $\partial_i g$ *are continuous. Moreover, let* $\left(D\left(g\left(\varphi_{1,m}, \ldots, \varphi_{n,m}\right)\right)\right)_{m \in \mathbb{N}}$, $\mathbb{E}_{\widehat{\mu}}\left(g\left(\varphi_{1,m}, \ldots, \varphi_{n,m}\right)\right)_{m \in \mathbb{N}}$ *converge in* $L^2(\widehat{\mu} \otimes \lambda)$, *in* $\mathbb{R}$, *respectively. Then* $g\left(\varphi_1, \ldots, \varphi_n\right)$ *is Malliavin differentiable and we have* $\widehat{\mu} \otimes \lambda$*-a.e.*

$$\begin{aligned}&\left(D\left(g\left(\varphi_1, \ldots, \varphi_n\right)\right)\right)_r \\ &= \frac{g\left(\varphi_1 + \alpha \cdot \left(D\varphi_1\right)_r, \ldots, \varphi_n + \alpha \cdot \left(D\varphi_n\right)_r\right) - g\left(\varphi_1, \ldots, \varphi_n\right)}{\alpha},\end{aligned}$$

where this fraction is equal to $\sum_{i=1}^{n}\left(\partial_i g\right)\left(\varphi_1, \ldots, \varphi_n\right) \cdot \left(D\varphi_i\right)_r$ *if* $\alpha = 0$.

In the work of Di Nunno et al. [31] on pure jump processes the product rule has the form

$$(D(\varphi \cdot \psi))_l = (D\varphi)_l \cdot \psi + \varphi \cdot (D\psi)_l + D\left(\varphi_m\right)_l \cdot D\left(\psi_m\right)_l.$$

In case of the chain rule, they prove a corresponding formula via the Wick product similar to the formula above for $\alpha = 0$. Moreover, in that work and also in the work of Solé et al. [112], the set $\mathbb{R} \times \Omega \times [0, \infty[$ is the domain of the Malliavin derivative $D\varphi$ of a Lévy functional φ, where the measure on $\mathbb{R}$ depends on the Lévy process.

In our approach $D\varphi$ is defined on $\mathbb{N}_\mu \times \Omega \times [0, \infty[$, the measure on $\mathbb{N}_\mu \times [0, \infty[$ is the product of the counting measure on $\mathbb{N}_\mu$ and the Lebesgue measure on $[0, \infty[$. In their work, Nualart and Schoutens [80] take the power jump processes of a Lévy process to prove a chaos representation result for Lévy functionals. Léon et al. [62] use their approach to define the directional Malliavin derivative and the directional Skorokhod integral.

23.9 Girsanov transformations

Girsanov transformations, in particular for Lévy processes, are important tools in finance mathematics and are used to find solutions to stochastic differential equations. They have been studied for Brownian motion in a vast numbers of papers. We only refer to Nualart's book [79] and the references in it. The literature on Girsanov transformations of jump processes is less extensive, in particular for time-anticipating transformations. Privault has studied anticipative Girsanov transformations for Poisson space in [95]. Léandre [60] investigates Girsanov transformations for Poisson processes in semi-group theory. Here we study more general Lévy processes, following [89].

It is generally understood that the transformations have to fulfil certain differentiability conditions. In our setting differentiability often comes for free, thanks to the smooth representations of Lévy functionals in [85]. We study, according to Chapter 22, shifts σ of a large class of Lévy processes L of the form

$$(\sigma \circ L)(\cdot, r) = L(\cdot, r) + \int_0^r \varphi(\cdot, s) d\lambda(s), \tag{60}$$

where λ is the Lebesgue measure on $[0, \infty[$ and $\varphi : \Omega \times [0, \infty[\to \mathbb{R}$ is a $\widehat{\mu}$ almost surely locally Lebesgue integrable stochastic process. These transformations often define shifts on the whole space of càdlàg functions. According to Girsanov transformations in the Gaussian case, answers are given to the following question.

> When does there exist a probability measure P, absolutely continuous or even equivalent to the original probability measure, such that $\sigma \circ L$ follows the law of L with respect to P?

Answers to this question are quite simple if φ is $(\mathfrak{b}_r)_{r \in [0,\infty[}$- adapted, because the Carleman–Fredholm determinant is equal to 1. This is, in general, not true if φ is time-anticipating. Since the underlying probability space Ω is *finite-dimensional and Lévy functionals can be represented by smooth functions, we can again use the substitution rule in elementary analysis. Similar ideas can be found in the work of Cutland [23] and Cutland and Ng [24] for finite-dimensional Brownian motion.

The internal concept

It suffices to study transformations of the standard part $L = {}^\circ B$ of a μ-limited Lévy process B. Since L lives on the *finite-dimensional space Ω, the substitution rule in finite-dimensional analysis can be applied. In the Gaussian case similar transformations have been used in Chapter 14 to construct continuous solutions to Langevin equations in infinite dimensions. In the remaining subsections of Section 23.9 the results are used to give answers to the above question.

Let us assume in this section that ρ^1 is an internal Borel probability measure on ${}^*\mathbb{R}$, which has a density δ with respect to *Lebesgue measure.

Fix an internal process $\Psi : \Omega \times T \to {}^*\mathbb{R}$ such that $\Psi(\cdot, t)$ is continuously differentiable for all $t \in T$. Define $\widetilde{\Psi} : \Omega \to \Omega$, $X \mapsto (X_t + \Psi(X,t))_{t\in T}$ and $\Theta_\Psi : \Omega \to {}^*\mathbb{R}$ by setting

$$\Theta_\Psi(X) := \prod_{t\in T} \frac{\delta\,(X_t + \Psi(X,t))}{\delta(X_t)} \left|\det \widetilde{\Psi}'(X)\right|.$$

This function Θ_Ψ is ρ-a.s. well defined and is called the **Cameron–Martin formula** of Ψ. The $H^2 \times H^2$-matrix $\widetilde{\Psi}'$ ($\widetilde{\Psi}'$ denotes the Jacobian of $\widetilde{\Psi}$) is called the **Carleman–Fredholm matrix** of Ψ. Let us call Ψ **adequate** if the following three conditions hold:

(A) $\widetilde{\Psi}$ is **regular**, i.e., $\widetilde{\Psi}$ is bijective and $\det \widetilde{\Psi}'(X) \neq 0$ for all $X \in \Omega$.
(B) ${}^\circ\det \widetilde{\Psi}' \neq 0$ $\widehat{\rho}$-a.s.
(C) $\mathbb{E}_{\widehat{\rho}}\, {}^\circ\Theta_\Psi = 1$. This condition is called the **Novikov condition**.

If in addition to (A), (B) and (C) also

(D) ${}^\circ \prod_{t\in T} \frac{\delta(X_t+\Psi(t,X))}{\delta(X_t)} \neq 0$ for $\widehat{\rho}$ almost all $X \in \Omega$
holds, Φ is called **strongly adequate**.

We have the following general example. The proof is similar to the proof of Lemma 22.4.1.

Lemma 23.9.1 *If Ψ is $(\mathcal{B}_{t^-})_{t\in T}$-adapted, then $\widetilde{\Psi}$ is regular and* $\det \widetilde{\Psi}' = 1$, *thus (A) and (B) are true.*

Let $\Theta_\Psi \rho$ be the measure with density Θ_Ψ with respect to ρ. The following result shows that the sequence $\left(\widetilde{\Psi}(X)_t\right)_{t\in T}$ has the same behaviour with respect to $\Theta_\Psi \rho$ as X has with respect to ρ if condition (A) is true.

Lemma 23.9.2 *Assume that condition (A) holds.*

(a) *$\Theta_\Psi \rho$ is the image measure of ρ by $\widetilde{\Psi}^{-1}$. It follows that $\Theta_\Psi \rho$ is a probability measure and that for all $\mathcal{B}$-measurable $F : \Omega \to {}^*\mathbb{R}$*

$$\int_\Omega F \circ \widetilde{\Psi} \cdot \Theta_\Psi d\rho = \int_\Omega F d\rho$$

if the integral of one of both sides exists. It follows that $\mathbb{E}_\rho\Theta_\Psi = 1$ *and, therefore,* $\mathbb{E}_{\widehat{\rho}}\,{}^\circ\Theta_\Psi \leq 1$. *By Corollary 10.8.2 (b), the Novikov condition is equivalent to the fact that* Θ_Ψ *is* S_ρ*-integrable.*

(b) $\widetilde{\Psi}(X)_t$ *has the same distribution with respect to* $\Theta_\Psi\rho$ *as* X_t *has with respect to* ρ.

(c) $\left(\widetilde{\Psi}(X)_t\right)_{t\in T}$ *is independent.*

Proof (a) follows from the substitution rule as follows. For each internal Borel set D in Ω we have

$$\begin{aligned}
\Theta_\Psi\rho(D) &= \int_D \prod_{t\in T} \frac{\delta(X_t + \Psi(X,t))}{\delta(X_t)} \left|\det \widetilde{\Psi}'(X)\right| d\rho \\
&= \int_D \prod_{t\in T} \frac{\delta(X_t + \Psi(X,t))}{\delta(X_t)} \left|\det \widetilde{\Psi}'(X)\right| \prod_{t\in T} \delta(X_t) dX \\
&= \int_D \prod_{t\in T} \delta(X_t + \Psi(X,t)) \left|\det \widetilde{\Psi}'(X)\right| dX \\
&= \int_{\widetilde{\Psi}[D]} \prod_{t\in T} \delta(X_t)\, dX = \rho\left(\widetilde{\Psi}[D]\right).
\end{aligned}$$

(b) Let α be the image measures of $\Theta_\Psi\rho$ by $X \mapsto \widetilde{\Psi}(X)_t$ and let β be the image measure of ρ by $X \mapsto X_t$. By (a) we obtain for all $\lambda \in {}^*\mathbb{R}$

$$\int_{^*\mathbb{R}} e^{i\lambda x} d\alpha = \int_\Omega e^{i\lambda\widetilde{\Psi}(\cdot)_t} d\Theta_\Psi\rho = \int_\Omega e^{i\lambda X_t} d\rho = \int_{^*\mathbb{R}} e^{i\lambda x} d\beta.$$

(c) The proof of (c) is left to the reader. □

Finally we present conditions under which $\widehat{\Theta_\Psi\rho}$ is absolutely continuous to $\widehat{\rho}$ or even equivalent to $\widehat{\rho}$, using Lemma 23.9.2.

Lemma 23.9.3 *Assume that (A) is true.*

(a) *Condition (C) is true if and only if* $\widehat{\Theta_\Psi\rho}$ *is absolutely continuous to* $\widehat{\rho}$.

(b) *If (B) and (D) are true, then* $\widehat{\rho}$ *is absolutely continuous to* $\widehat{\Theta_\Psi\rho}$.

Proof We identify X_s with the projection $\pi_s : X \mapsto X_s$.

(a) Suppose that condition (C) is true and ${}^\circ\rho(N) = 0$ for $N \in \mathcal{B}$. Then

$${}^\circ\Theta_\Psi\rho(N) = {}^\circ\int_N \Theta_\Psi d\rho = 0,$$

because Θ_Ψ is S_ρ-integrable. This proves that $\widehat{\Theta_\Psi \rho}$ is absolutely continuous to $\widehat{\rho}$. Conversely, suppose that $\widehat{\Theta_\Psi \rho}$ is absolutely continuous to $\widehat{\rho}$. It follows that $^\circ \int_A \Theta_\Psi d\rho = 0$ for each $\widehat{\rho}$-nullset $A \in \mathcal{B}$. Moreover, $\int_\Omega \Theta_\Psi d\rho = 1$ is limited. It follows that condition (C) is true.

(b) Let $^\circ\Theta_\Psi \rho(A) = 0$, thus, by (B), $\mathbf{1}_A \cdot {}^\circ \prod_{t \in T} \frac{\delta(X_t + \Psi(t,X))}{\delta(X_t)} = 0$ $\widehat{\rho}$-a.s. By (D), $\mathbf{1}_A = 0$ $\widehat{\rho}$-a.s., thus $^\circ\rho(A) = 0$. Therefore, $\widehat{\rho}$ is absolutely continuous to $\widehat{\Theta_\Psi \rho}$. □

Shifts by continuous functions

Now we go back to our measure μ^1 and assume that μ^1 has a density δ with respect to *Lebesgue measure. We assume that in our shift σ the process φ is $\mathcal{D} \otimes$Leb-measurable and $\varphi(X, \cdot) \upharpoonright [0, S] \in L^1(\lambda)$ for $\widehat{\mu}$-almost all $X \in \Omega$ and all $S \in \mathbb{N}$. Since $r \mapsto \int_0^r \varphi(\cdot, s) d\lambda(s)$ is continuous $\widehat{\mu}$-a.s., $\sigma(L)$ has the same jumps as L $\widehat{\mu}$-a.s. The following result shows that in most cases the shift σ defines a shift on the space D of càdlàg functions.

Proposition 23.9.4 *Suppose that φ is $\mathcal{D}_L \otimes$Leb-measurable, where $\mathcal{D}_L$ is the σ-algebra generated by L. For each càdlàg function $f : [0, \infty[\to \mathbb{R}$ define*

$$\sigma(f) := \sigma(L)(X, \cdot),$$

if $f = L(X, \cdot)$. The shift σ is well defined and $\sigma(f)$ is càdlàg.

Proof By Proposition 15.5.4, for càdlàg f there is an $X \in \Omega$ with $f = L(X, \cdot)$. If $f = L(X, \cdot) = L(Y, \cdot)$, then $\varphi(X, \cdot) = \varphi(Y, \cdot)$, because φ is $\mathcal{D}_L \otimes$Leb-measurable. □

Note that $\mathcal{D}$ is a slight extension of $\mathcal{D}_L$. In the remainder of this chapter we assume that the process $\varphi : \Omega \times [0, \infty[\to \mathbb{R}$ is $\mathcal{D} \otimes$Leb-measurable, $\widehat{\mu}$-almost surely locally integrable and represents the shift σ in equation (60). Recall that we identify φ with its equivalent function defined on $\Omega \times T$.

Characterization of the shifts by internal shifts

Using the chaos decomposition, we can find a smooth lifting of φ which leads to a smooth representation of our shift σ. The key is the following lemma.

Lemma 23.9.5 *There exists an internal function $\Phi : \Omega \times T \to {}^*\mathbb{R}$ such that*

(a) $^\circ\Phi(X, t) = \varphi(X, {}^\circ t)$ *for $\widehat{\mu \otimes \nu}$-almost all (X, t),*

(b) $X \mapsto \left(X_t + \Phi(X,t)\frac{1}{H}\right)_{t\in T}$ *is continuously differentiable,*
(c) $\int_0^{{}^\circ t} \varphi(\cdot,s)d\lambda(s) = {}^\circ\sum_{s\le t}\Phi(\cdot,s)\frac{1}{H}$ *for all limited* $t \in T$ $\widehat{\mu}$*-a.s.*

Proof We can proceed as in Section 22.1. There exists a sequence $(\varphi_n)_{n\in\mathbb{N}}$ of functions $\varphi_n \in L^2_{\mathcal{D}\otimes\mathcal{L}_1}(\widehat{\mu\otimes\nu})$ converging to φ in measure. We may also assume that, for all $r\in\mathbb{N}$, $\lim_{n\to\infty}\int_{T_r}|\varphi_n-\varphi|\,d\widehat{\nu}=0$ in measure. Let $\Psi:\Omega\times T\to {}^*\mathbb{R}$ be a lifting of φ and let $\Phi_n \in SL^2(\mu\otimes\nu)$ be a lifting of φ_n, according to Theorem 23.5.1. Note that $({}^\circ|\Phi_n-\Psi|) = (|\varphi_n-\varphi|)$ converges to 0 in measure and $\lim_{n,m\to\infty}\left({}^\circ\int_{T_r}|\Phi_n-\Phi_m|\,d\nu\right)=0$ in measure for all $r\in\mathbb{N}$. Using techniques established in Corollary 10.9.3, we can find a lifting Φ of φ such that $\Phi(X,\cdot)$ is locally in $SL^1(\nu)$ for $\widehat{\mu}$-almost all $X\in\Omega$. Moreover, we may assume that Φ inherits from the Φ_n their internal properties; in particular, Φ has the form

$$\Phi(X,s)=\sum_{n=0}^{M}\sum_{k\in M_\mu^n}\sum_{t\in T_<^n}F_n(k,t,s)p_{k_1}(X_{t_1})\cdot\ldots\cdot p_{k_n}(X_{t_n}),$$

where $M\in{}^*\mathbb{N}$ and $F_n: M_\mu^n\times T_{\neq}^{k+1}\to{}^*\mathbb{R}$. Then (a), (b) and (c) are true. □

Using Lemma 16.4.1 and saturation, we obtain the following internal representation of σ.

Proposition 23.9.6 *For each* $r\in[0,\infty[$ *there exists a set* U_r *of* $\widehat{\mu}$*-measure* 1 *and a* $t\in T$ *with* ${}^\circ t=r$ *such that, for all* $X\in U_r$,

$$\sigma\circ L(X,r)\approx\left(B(X,t)+\sum_{s\le t}\Phi(X,s)\frac{1}{H}\right)=:\sigma\circ B(X,t).$$

We define $\Psi:=\Phi\cdot\frac{1}{H}$ and and use the notation $\widetilde{\Psi}$, Θ_Ψ and $\rho=\mu$ in Section 23.9.1. The Carleman–Fredholm matrix now can be seen as the interference matrix of the identity in the following sense: if $\frac{\partial\Phi(t,X)}{\partial X_j}$ is limited, then in the diagonal there are numbers infinitely close to 1 and outside there are infinitesimals.

Time-anticipating transformations

The preceding results can be applied to give necessary and sufficient conditions for the fact that a time-anticipating shift of a Lévy process L results in a process that follows the law of L with respect to a measure which is absolutely continuous or equivalent to $\widehat{\mu}$.

Theorem 23.9.7 *Assume that φ has a (strongly) adequate lifting Φ. Fix $r, \widetilde{r} \in [0, \infty[$ with $r < \widetilde{r}$. Then $A := (\sigma \circ L(\cdot, r) - \sigma \circ L(\cdot, \widetilde{r}))$ has the same distribution with respect to $\widehat{\Theta_\Psi \mu}$ as $C := L(\cdot, r) - L(\cdot, \widetilde{r})$ has with respect to $\widehat{\mu}$. Moreover, the measure $\widehat{\Theta_\Psi \mu}$ is (equivalent) absolutely continuous to $\widehat{\mu}$.*

Proof It suffices to prove that the Fourier transformations of A and C coincide. To this end choose $t, \widetilde{t} \in T$ with $t \approx r$ and $\widetilde{t} \approx \widetilde{r}$, and $L(\cdot, r) \approx B(\cdot, t)$ and $L(\cdot, \widetilde{r}) \approx B(\cdot, \widetilde{t})$ $\widehat{\mu}$-a.s. By Lemma 23.9.2, we have for each $\lambda \in \mathbb{R}$

$$
\begin{aligned}
&\int_\Omega e^{i\lambda(L(\cdot,r) - L(\cdot,\widetilde{r}))} d\widehat{\mu} \\
&= {}^\circ \int_\Omega e^{i\lambda(B(\cdot,t) - B(\cdot,\widetilde{t}))} d\mu \\
&= {}^\circ \int_\Omega e^{i\lambda(\sigma \circ B(\cdot,t) - \sigma \circ L(\cdot,\widetilde{t}))} d\Theta_\Psi \mu \\
&= \int_\Omega e^{i\lambda(\sigma \circ L(\cdot,r) - \sigma \circ L(\cdot,\widetilde{r}))} d\widehat{\Theta_\Phi \mu}.
\end{aligned}
$$

□

Adapted Girsanov transformation

In this section we sketch the much simpler case of $(\mathfrak{b}_r \cap \mathcal{D})_{r \in [0,\infty[}$-adapted processes φ. By Theorem 12.3.1 with $\widehat{\mu}$ instead of $\widehat{\Gamma}$, there exists an internal $(\mathcal{B}_{t^-})_{t \in T}$-adapted $\Phi : \Omega \times T \to {}^*\mathbb{R}$ such that $\Phi(X, t) \approx \varphi(X, {}^\circ t)$ for $\widehat{\mu \otimes \nu}$-almost all (X, t). We may assume that Φ is locally in $SL^1(\nu)$ $\widehat{\mu}$-a.s. and smooth. Then, by Lemma 23.9.1, $\widetilde{\Psi}$ is regular and $\det \widetilde{\Psi}' = 1$. Since conditions (A) and (B) are true in this example, Ψ is adequate if and only if condition (C) is fulfilled. Now we obtain the classical Girsanov theorem:

Theorem 23.9.8 *Suppose that condition (C), the Novikov condition, holds for Φ. Then $\sigma \circ L$ has the same law with respect to $\widehat{\Theta_\Psi \mu}$ as L has with respect to $\widehat{\mu}$. Recall that $\widehat{\Theta_\Psi \mu}$ is absolutely continuous to $\widehat{\mu}$. Moreover, both measures are equivalent if condition (D) also holds.*

Girsanov transformations for Poisson processes

Here we apply the preceding results to Poisson processes. Let $d\rho^1 := \beta e^{-\beta x} dx$ on ${}^*[0, \infty[$ with $\beta \in \mathbb{R}^+$ and let ρ be the internal product of ρ^1 on Ω. Let Ψ and $\widetilde{\Psi}$ be defined as in Section 23.9.1 and assume that $\widetilde{\Psi}$ is regular. The independence of $(X_t)_{t \in T}$, the exponential distribution of the X_t with rate β under ρ and Lemma 23.9.2 imply that the $X_t + \Psi(t, X)$ are exponentially distributed with rate β and that $(X_t + \Psi(t,X))_{t\in T}$ is independent now under the measure $\Theta_\Psi \rho$. By Exercises 23.9 and 23.10, the standard part of

$$N : \Omega \times T \to^* \mathbb{R}, (X,t) \mapsto \max \left\{ n \in {}^*\mathbb{N}_0 \mid \sum_{s \leq \frac{n}{H}} X_t \leq t \right\}$$

is a Poisson process under $\widehat{\rho}$ with rate β. Therefore, the standard part of

$$N_\Psi : \Omega \times T \to \ {}^*\mathbb{R}, (X,t) \mapsto \max \left\{ n \in {}^*\mathbb{N}_0 \mid \sum_{s \leq \frac{n}{H}} X_t + \Psi(t,X) \leq t \right\}$$

is also a Poisson processes with rate β under $\widehat{\Theta_\Psi \rho}$. Lemma 23.9.3 gives information about the absolute continuity between $\widehat{\rho}$ and $\widehat{\Theta_\Psi \rho}$.

Exercises

23.1 Prove Lemma 23.9.2 part (c).

23.2 Proof Theorem 23.4.2

23.3 Proof Theorem 23.5.3

23.4 Prove Theorem 23.6.1.

23.5 Find conditions under which Skorokhod integral processes have continuous versions.

23.6 $L^2_{\mathcal{L}\otimes\mathcal{D}}(\widehat{c \otimes \mu \otimes \nu})$ is the orthogonal sum of the range of D and the kernel of δ.

Let $\mathrm{Fin}L^2_{\mathcal{D}}(\widehat{\mu}), \mathrm{Fin}L^2_{\mathcal{L}\otimes\mathcal{D}}(\widehat{c \otimes \mu \otimes \nu})$ denote the space of restricted chaos levels of $L^2_{\mathcal{D}}(\widehat{\mu})$, $L^2_{\mathcal{L}\otimes\mathcal{D}}(\widehat{c \otimes \mu \otimes \nu})$, respectively.

Let $\mathbb{D}^{1,2}_0$ be the set of all $\varphi \in L^2_{\mathcal{D}}(\widehat{\mu})$ such that the linear form

$$C_\varphi : \mathrm{Fin}L^2_{\mathcal{L}\otimes\mathcal{D}}(\widehat{c \otimes \mu \otimes \nu}) \ni \alpha \mapsto \langle \delta\alpha, \varphi \rangle_{\widehat{\mu}} \in \mathbb{R} \text{ is continuous.}$$

23.7 Prove that $\mathbb{D}^{1,2}_0$ is the domain of the Malliavin derivative.

23.8 Let Δ_0 be the set of all $\varphi \in L^2_{\mathcal{L}\otimes\mathcal{D}}(\widehat{c \otimes \mu \otimes \nu})$ such that the linear form

$$C_\varphi : \mathrm{Fin}L^2_{\mathcal{D}}(\widehat{\mu}) \ni \alpha \mapsto \langle D\alpha, \varphi \rangle_{\widehat{c\otimes\nu\otimes\mu}} \in \mathbb{R} \text{ is continuous.}$$

Prove that Δ_0 is the domain of the Skorokhod integral.

Use the notation of Section 23.9.6.

23.9 Prove that $\rho(N_t = n) = e^{-\beta t}\frac{(\beta t)^n}{n!}$ for all $n \in {}^*\mathbb{N}$ with $n < H^2$ and $\rho\left(N_t = H^2\right) = \sum_{n \geq H^2} e^{-\beta t}\frac{(\beta t)^n}{n!}$.

23.10 Prove that we have $\widehat{\rho}$-a.s.
$N_t \in \mathbb{N}_0$, $N_t - N_s \leq 1$ for all limited $s < t$ in T and N_t is SDJ.

23.11 Prove that the recursively defined **Laguerre polynomials** $p_n(x) := \frac{\beta^n}{n!}x^n - \sum_{i=0}^{n-1} p_i(x)$ form an ONB of $L^2(\rho)$.

Appendices: Existence of poly-saturated models

Here we will give the full details on the construction of poly-saturated models of mathematics and prove the existence of such models. Compare these appendices with Chapter 7.

Appendix A. Poly-saturated models

In this appendix the reader can find a precise axiomatic introduction to poly-saturated models. It is a common practice in mathematics to work in naive set theory together with the axiom of choice, where sets are used and the usual element-relation $\in$ between them.

A.1 Weak models and models of mathematics

Here is the precise definition of the phrase 'model of mathematics'. First, a sequence $\mathfrak{V} := (\mathfrak{V}_n)_{n\in\mathbb{N}_0}$ of sets is called a **weak model over** $\mathfrak{V}_0$ if

(W 1) $\mathfrak{V}_0 \neq \emptyset$,

(W 2) $\mathfrak{V}_n \subset \mathcal{P}(\mathfrak{V}_0 \cup \ldots \cup \mathfrak{V}_{n-1})$ for each $n \geq 1$. Therefore, the elements of $\mathfrak{V}_n$ are subsets of $\mathfrak{V}_0 \cup \ldots \cup \mathfrak{V}_{n-1}$. However, and this is important for the following, it is not required that each subset of $\mathfrak{V}_0 \cup \ldots \cup \mathfrak{V}_{n-1}$ is an element of $\mathfrak{V}_n$.

The elements of $\mathfrak{V}_0$ are called **urelements** or **individuals in** $\mathfrak{V}$ and the elements of $\bigcup_{1\leq n} \mathfrak{V}_n$ are called the **(internal) sets in** $\mathfrak{V}$. A subset $E \subseteq \mathfrak{V}_0 \cup \ldots \cup \mathfrak{V}_{n-1}$ is called an **external set in** $\mathfrak{V}$ if E is not a set in $\mathfrak{V}$, i.e., $E \notin \bigcup_{1\leq n} \mathfrak{V}_n$.

It is inevitable for us that urelements in $\mathfrak{V}$ are different from the sets in $\mathfrak{V}$; see Proposition B.1.1 below. Moreover, the urelements of $\mathfrak{V}$ should be empty relative to $\mathfrak{V}$. Therefore, we have the following definition. A weak model $\mathfrak{V} = (\mathfrak{V}_n)_{n\in\mathbb{N}_0}$ over $\mathfrak{V}_0$ is called a **model over** $\mathfrak{V}_0$ if

(M 3) $\mathfrak{V}_0 \cap \bigcup_{1\leq n} \mathfrak{V}_n = \emptyset$, i.e., urelements are different from sets,

(M 4) $u \cap \bigcup_{n\in\mathbb{N}_0} \mathfrak{V}_n = \emptyset$ for each $u \in \mathfrak{V}_0$, i.e., elements of urelements don't belong to the model.

Let $S := (S_n)_{n\in\mathbb{N}_0}$ be a sequence of sets. We use the following notation:

$$S_{<\infty} := \bigcup_{n\in\mathbb{N}_0} S_n \quad \text{and} \quad S_{\leq n} := S_0 \cup \ldots \cup S_n.$$

A.2 From weak models to models

Following [86], we will now prove that the urelements of a weak model $\mathfrak{V}$ can be renamed in such a way that $\mathfrak{V}$ becomes a model.

Proposition A.2.1 *For each set $Y \neq \emptyset$ there exists a set X and a bijection c from Y onto X such that each weak model over X is already a model over X.*

So we rename the elements y of Y by $c(y)$. The code $c(y)$ is quite technical, but we never use its special form. It is only important that some coding of the elements of Y exists. If the definition of $c(y)$ bothers you, try to come up with a better one.

Proof Set

$$\widetilde{Y} := \{\{y\} \mid y \in Y\}.$$

Then $\operatorname{card}(\widetilde{Y}) = \operatorname{card}(Y)$. Define by recursion:

$$Y_0 := \widetilde{Y}, \quad Y_{n+1} := \mathcal{P}(Y_n) \quad \text{and} \quad Y_\infty := \bigcup_{n \in \mathbb{N}_0} Y_n,$$

and

$$X := \left\{\{\{a, Y_\infty\}\} \mid a \in \widetilde{Y}\right\}.$$

Now define for all $y \in Y$

$$c(y) := \{\{\{y\}, Y_\infty\}\}.$$

Note that

(i) Y_∞ is an infinite set, $\operatorname{card}(a) = 1$ and $\operatorname{card}(\{a, Y_\infty\}) = 2$ for each $a \in \widetilde{Y} \cup X$.

It follows that

(ii) the function $i : a \mapsto \{\{a, Y_\infty\}\}$ defines a bijection from $\widetilde{Y}$ onto X. Moreover, $Y_0 = \widetilde{Y} \notin X$. Since $\emptyset \in Y_n$ for each $n \geq 1$, $Y_n \notin X$ for each $n \in \mathbb{N} \cup \{\infty\}$.

Let $\mathfrak{V} := (\mathfrak{V}_n)_{n \in \mathbb{N}_0}$ be a weak model over X. By induction over $n \in \mathbb{N}_0$, we see that $Y_n \notin \mathfrak{V}_k$ for each $n \in \mathbb{N}_0$ and each $k \in \{0, \ldots, n\}$. It follows that

(iii) $Y_\infty \notin \mathfrak{V}_k$ for each $k \in \mathbb{N}_0$.

Now we prove that $\mathfrak{V}$ is a model over X:

(iv) We first show that $X \cap \bigcup_{n \geq 1} \mathfrak{V}_n = \emptyset$. Assume that $k \geq 1$ and $\{\{a, Y_\infty\}\} \in \mathfrak{V}_k$ for some $a \in \widetilde{Y}$. Then $\{a, Y_\infty\} \in \mathfrak{V}_i$ for some $i < k$. By (i), $\{a, Y_\infty\} \notin X = \mathfrak{V}_0$. Therefore, there exists a $j < i$ with $Y_\infty \in \mathfrak{V}_j$, contradicting (iii).

(v) Finally, we prove that $b \cap \mathfrak{V}_n = \emptyset$ for each $n \in \mathbb{N}$ and for each $b \in \mathfrak{V}_0$. Fix $b = \{\{a, Y_\infty\}\} \in X$. Assume that there exist $n \in \mathbb{N}$ and $x \in \mathfrak{V}_n$ with $x \in b$.

Then $x = \{a, Y_\infty\}$. By (i), $x \notin X = V_0$, thus $n \geq 1$. It follows that there exists a $k < n$ with $Y_\infty \in \mathfrak{V}_k$, contradicting (iii). □

Now we would like to make a remark about the usual notion of global urelements. These are objects which are not sets (in particular, they are different from the empty set), and do not contain any elements. Because of the extensionality axiom, global urelements don't exist in Zermelo–Fraenkel (ZF) set theory. In many books on poly-saturated models the existence of such global urelements is presupposed. In [86] we suggested using 'local' urelements instead. Local urelements are urelements relative to the model according to equations (M 3) and (M 4). These local urelements are sufficient to lay the foundation for the theory of poly-saturated models and they exist in ZF set theory.

A model $\mathfrak{V} = (\mathfrak{V}_n)_{n \in \mathbb{N}_0}$ is called a **standard model** if it has no external sets, i.e., $\mathfrak{V}_n = \mathcal{P}(\mathfrak{V}_0 \cup \ldots \cup \mathfrak{V}_{n-1})$ for all $n \geq 1$.

A.3 Languages for models

In this section we want to make the phrase 'mathematical property' precise. Therefore, we define languages strong enough to formalize every mathematical statement.

Given a model $\mathfrak{V} := (\mathfrak{V}_n)_{n \in \mathbb{N}_0}$, the alphabet of the language $\mathcal{L}_\mathfrak{V}$ has the following symbols:

Logical symbols: $\vee$, $\neg$, $\exists$, $\dot{=}$, $\dot{\in}$
Variables: $x, y, z, u, v, x_1, \ldots, x_n, \ldots$.
Parameters: The elements of $\mathfrak{V}_{<\infty}$ are the parameters in $\mathcal{L}_\mathfrak{V}$.
Auxiliary symbols: Parentheses '(', ')', point '.' and comma ','.

We assume that all these symbols are different.

A **sentence in** $\mathcal{L}_\mathfrak{V}$ is built up inductively from these rules:

(i) Fix $a, b \in \mathfrak{V}_{<\infty}$. Then $(a \dot{=} b)$ is a sentence in $\mathcal{L}_\mathfrak{V}$ and, if b is a set in $\mathfrak{V}$, $(a \dot{\in} b)$ is a sentence in $\mathcal{L}_\mathfrak{V}$.

(ii) If A and B are sentences in $\mathcal{L}_\mathfrak{V}$, then $(A \vee B)$ and $(\neg A)$ are sentences in $\mathcal{L}_\mathfrak{V}$.

(iii) Let A be a sentence in $\mathcal{L}_\mathfrak{V}$ and let a, b be parameters in $\mathcal{L}_\mathfrak{V}$. If a is a set in $\mathfrak{V}$ and x is a variable not occurring in A, then $(\exists x \dot{\in} a A_b(x))$ is a sentence in $\mathcal{L}_\mathfrak{V}$. Here $A_b(x)$ is the string of signs of the alphabet of $\mathcal{L}_\mathfrak{V}$ that results from A by replacing b, where b occurs in A, with x.

(iv) The set of sentences of $\mathcal{L}_\mathfrak{V}$ is the smallest set having the properties (i), (ii) and (iii).

A **formula in** $\mathcal{L}_{\mathfrak{V}}$ results from a sentence in $\mathcal{L}_{\mathfrak{V}}$ by replacing some parameters $b_1,\dots,b_k$ with variables $x_1,\dots,x_k$ not occurring in the sentence. We then say that $x_1,\dots,x_k$ are **free** in the formula. We shall write $A(x_1,\dots,x_k)$, to indicate that at most $x_1,\dots,x_k$ are free in the formula A.
As usual, we use the following abbreviations:

$$(A \Rightarrow B) \text{ for } ((\neg A) \vee B),$$
$$(A \wedge B) \text{ for } (\neg((\neg A) \vee (\neg B))),$$
$$(A \Leftrightarrow B) \text{ for } ((A \Rightarrow B) \wedge (B \Rightarrow A)),$$
$$(\forall x \,\dot{\in}\, a A(x)) \text{ for } (\neg(\exists x \,\dot{\in}\, a(\neg A(x)))).$$

In order to save parentheses, we agree that $\neg$, $\exists$, $\forall$ bind stronger than $\wedge$, $\wedge$ binds stronger than $\vee$, $\vee$ binds stronger than $\Rightarrow$, and $\Rightarrow$ binds stronger than $\Leftrightarrow$. The relation 'binds stronger' is transitive. Moreover, we will use the following shorthand

$$\exists x_1,\dots,x_k \,\dot{\in}\, a A \text{ for } \exists x_1 \,\dot{\in}\, a \dots \exists x_k \,\dot{\in}\, a A,$$
$$\forall x_1,\dots,x_k \,\dot{\in}\, a A \text{ for } \forall x_1 \,\dot{\in}\, a \dots \forall x_k \,\dot{\in}\, a A.$$

A.4 Interpretation of the language

Fix a model $\mathfrak{V} := (\mathfrak{V}_n)_{n \in \mathbb{N}_0}$. Here is the **truth predicate** for sentences in $\mathcal{L}_{\mathfrak{V}}$:

(a) Fix $a, b \in \mathfrak{V}_{<\infty}$.
  (i) The sentence $a \dot{=} b$ is true in $\mathfrak{V}$ if $a = b$ in the sense of common set theory.
  (ii) The sentence $a \,\dot{\in}\, b$ is true in $\mathfrak{V}$ if $a \in b$ in the sense of common set theory.

(b) Let A, B be sentences of $\mathcal{L}_{\mathfrak{V}}$.
  (i) $\neg A$ is true in $\mathfrak{V}$ if A is not true in $\mathfrak{V}$.
  (ii) $A \vee B$ is true in $\mathfrak{V}$ if A is true in $\mathfrak{V}$ or B is true in $\mathfrak{V}$.
  (iii) Let a be a set in $\mathfrak{V}$. $\exists x \,\dot{\in}\, a A(x)$ is true in $\mathfrak{V}$ if there exists a $c \in a$ such that $A(x)(c)$ is true in $\mathfrak{V}$. Here the sentence $A(x)(c)$ results from the formula $A(x)$ by replacing x with c. Instead of $A(x)(c)$ we will write $A(c)$. Note that, if a is a set in $\mathfrak{V}$, then $a \subseteq \mathfrak{V}_{<\infty}$.

Following a common practice in model theory, we will write $\mathfrak{V} \models A$ to denote that A is true in $\mathfrak{V}$. Sometimes, we also say that A is a **theorem of** $\mathfrak{V}$ if A is true in $\mathfrak{V}$.

A.5 Models closed under definition

We will now study models that have nice closure properties. A model $\mathfrak{V} := (\mathfrak{V}_n)_{n\in\mathbb{N}_0}$ is called **closed under definition** if for each formula $A(x)$ in $\mathcal{L}_{\mathfrak{V}}$ in which at most the variable x occurs free, and each $n \in \mathbb{N}_0$,

$$\left\{a \in \mathfrak{V}_{\leq n} \mid \mathfrak{V} \models A(x)(a)\right\} \in \mathfrak{V}_{n+1}.$$

Here are several examples.

Proposition A.5.1 *Suppose that $\mathfrak{V}$ is closed under definition and the positive integers are urelements in $\mathfrak{V}$. (According to Proposition A.2.1, we assume that the elements in $\mathfrak{V}_0$ are coded in such a way that they become urelements.) Fix $n \geq 1$. Then the following hold.*

(a) $\mathfrak{V}_1 \subseteq \ldots \subseteq \mathfrak{V}_n \subseteq \ldots$, *thus* $\mathfrak{V}_{\leq n} = \mathfrak{V}_0 \,\dot\cup\, \mathfrak{V}_n$.
(b) *If F is a finite subset of $\mathfrak{V}_{\leq n-1}$, then $F \in \mathfrak{V}_n$.*
(c) *If $a, b \in \mathfrak{V}_{\leq n-1}$, then $\langle a,b\rangle := \{\{a,b\},\{b\}\} \in \mathfrak{V}_{n+1}$.*
(d) *Fix $a_1,\ldots,a_k \in \mathfrak{V}_{\leq n-1}$. Then $(a_1,\ldots,a_n) := \{\langle 1,a_1\rangle,\ldots,\langle k,a_k\rangle\} \in \mathfrak{V}_{n+2}$.*
(e) *Fix $a_1,\ldots,a_k \in \mathfrak{V}_n$. Then $a_1 \cup \ldots \cup a_k \in \mathfrak{V}_n$ and $a_1 \cap \ldots \cap a_k \in \mathfrak{V}_n$.*
(f) *If $a, b \in \mathfrak{V}_n$, then $a \setminus b \in \mathfrak{V}_n$.*
(g) $\emptyset \in \mathfrak{V}_n$.
(h) $\mathfrak{V}_{\leq n-1} \in \mathfrak{V}_n$.
(i) $\mathfrak{V}_{n-1} \in \mathfrak{V}_n$.
(j) *Fix $a_1,\ldots,a_k \in \mathfrak{V}_n$. Then*

$$a_1 \times \ldots \times a_k = \{(\alpha_1,\ldots,\alpha_k) \mid \alpha_1 \in a_1 \text{ and } \ldots \text{ and } \alpha_k \in a_k\} \in \mathfrak{V}_{n+3}.$$

Proof (a) Fix $b \in \mathfrak{V}_n$ with $1 \leq n$. Then

$$b = \left\{a \in \mathfrak{V}_{\leq n} \mid \mathfrak{V} \models a \,\dot\in\, b\right\} \in \mathfrak{V}_{n+1}.$$

(b)–(g) follow from the Examples (iv), (vi), (v) in Section 7.1.

(h) $\mathfrak{V}_{\leq n-1} = \left\{a \in \mathfrak{V}_{\leq n-1} \mid \mathfrak{V} \models a \doteq a\right\} \in \mathfrak{V}_n$.

(i) For $n = 1$ the result follows from (h). Let $n \geq 2$. Then, by (a), (h) and (f),

$$\mathfrak{V}_{n-1} = \mathfrak{V}_{\leq n-1} \setminus \mathfrak{V}_0 \in \mathfrak{V}_n.$$

(j) Using (d), we obtain:

$$a_1 \times \ldots \times a_k =$$

$$\left\{a \in \mathfrak{V}_{\leq n+2} \mid \mathfrak{V} \models \exists x_1 \,\dot\in\, a_1 \ldots \exists x_k \,\dot\in\, a_k \,(a \doteq (x_1,\ldots,x_k))\right\} \in \mathfrak{V}_{n+3},$$

where we have used the following abbreviations:

$a \doteq (x_1,\ldots,x_k)$ for $\forall x \,\dot{\in}\, \mathfrak{V}_{\leq n+1} \, (x \,\dot{\in}\, a \Leftrightarrow x \doteq \langle 1,x_1\rangle \vee \ldots \vee x \doteq \langle k,x_k\rangle)$,
$x \doteq \langle i,x_i\rangle$ for $\forall z \,\dot{\in}\, \mathfrak{V}_{\leq n}(z \,\dot{\in}\, x \Leftrightarrow z \doteq \{i,x_i\} \vee z \doteq \{x_i\})$,
$z \doteq \{i,x_i\}$ for $\forall y \,\dot{\in}\, \mathfrak{V}_{\leq n-1}(y \,\dot{\in}\, z \Leftrightarrow y \doteq i \vee y \doteq x_i)$,
$z = \{x_i\}$ for $\forall y \,\dot{\in}\, \mathfrak{V}_{\leq n-1}(y \,\dot{\in}\, z \Leftrightarrow y \doteq x_i)$. □

Remark A.5.2 We shall identify $\langle a,b\rangle$ with (a,b).

A.6 Elementary embeddings

Fix a model $\mathfrak{V} = (\mathfrak{V}_n)_{n\in\mathbb{N}_0}$ closed under definition and a second model $\mathfrak{W} = (\mathfrak{W}_n)_{n\in\mathbb{N}_0}$. Suppose * is a mapping from $\mathfrak{V}_{<\infty}$ into $\mathfrak{W}_{<\infty}$. If A is a sentence (formula) in $\mathcal{L}_{\mathfrak{V}}$, then the *-transform $^*A$ of A is the sentence (formula) in $\mathcal{L}_{\mathfrak{W}}$ obtained by replacing each parameter a in A by $^*(a)$. We will write $^*a$ instead of $^*(a)$. The mapping $^* : \mathfrak{V}_{<\infty} \to \mathfrak{W}_{<\infty}$ is called an **elementary embedding** from $\mathfrak{V}$ into $\mathfrak{W}$ if the following conditions (E 1) and (E) are true.

(E 1) For each $n \in \mathbb{N}_0$, $^*\mathfrak{V}_n = \mathfrak{W}_n$. It follows that $\mathfrak{W}_n \in \mathfrak{W}_{<\infty}$.

(E) (Transfer principle) For each sentence A in $\mathcal{L}_{\mathfrak{V}}$

$$\mathfrak{V} \models A \Leftrightarrow \mathfrak{W} \models {}^*A.$$

The transfer principle means that * preserves each property which holds in $\mathfrak{V}$.

Proposition A.6.1 *Let * be an elementary embedding from $\mathfrak{V}$ into $\mathfrak{W}$ and assume that the positive integers are urelements in $\mathfrak{V}$ (see Proposition A.2.1). We obtain the following for each $n \in \mathbb{N}_0$.*

(a) *The restriction $^* \restriction \mathfrak{V}_n$ maps $\mathfrak{V}_n$ into $\mathfrak{W}_n$, and $\mathfrak{W}_n$ is a set in $\mathfrak{W}$.*
(b) *The map * is injective and a* ***homomorphism****, i.e., for $a,b \in \mathfrak{V}_{<\infty}$,*

$$a \in b \Leftrightarrow {}^*a \in {}^*b.$$

Fix sets a,b in $\mathfrak{V}$.

(c) *$a \subseteq b$ if and only if $^*a \subseteq {}^*b$.*
(d) *$^*(a \cup b) = {}^*a \cup {}^*b$. Similar results hold for intersections and differences of sets.*
(e) *$^*(\mathfrak{V}_{\leq n}) = \mathfrak{W}_{\leq n} \in \mathfrak{W}_{n+1}$.*
(f) *$\mathfrak{W}$ is closed under definition.*

(g) *Let $A(x)$ be a formula in $\mathcal{L}_{\mathfrak{V}}$. Then*

$$^*\left\{a \in \mathfrak{V}_{\le n} \mid \mathfrak{V} \models A(x)(a)\right\} = \left\{b \in \mathfrak{W}_{\le n} \mid \mathfrak{W} \models \left({}^*(A(x))\right)(b)\right\}.$$

(h) *If $\{a_1,\ldots,a_k\}$ is a finite subset of $\mathfrak{V}_{<\infty}$, then $^*\{a_1,\ldots,a_k\} = \{^*a_1,\ldots,{}^*a_k\}$, in particular, $^*\emptyset = \emptyset$. It follows that $^*(a_1,\ldots,a_k) = (^*a_1,\ldots,{}^*a_k)$ for all $a_1,\ldots,a_k \in \mathfrak{V}_{<\infty}$.*

(i) *Let $a_1,\ldots,a_k,a$ be sets in $\mathfrak{V}$. Let f be a set in $\mathfrak{V}$ and a mapping from $a_1 \times \ldots \times a_k$ into a. Then $^*f$ is a mapping from $^*a_1 \times \ldots \times {}^*a_k$ into $^*a$ such that, for all $(b_1,\ldots,b_k) \in a_1 \times \ldots \times a_k$,*

$$^*(f(b_1,\ldots,b_k)) = {}^*f(^*b_1,\ldots,{}^*b_k).$$

If, in addition, f is surjective onto a or injective or bijective, then $^*f$ is surjective onto $^*a$, injective or bijective, respectively.

Proof (a) By Proposition A.5.1 (i), we obtain

$$a \in \mathfrak{V}_n \Rightarrow \mathfrak{V} \models a \mathrel{\dot\in} \mathfrak{V}_n \Rightarrow \mathfrak{W} \models {}^*a \mathrel{\dot\in} {}^*\mathfrak{V}_n \Rightarrow {}^*a \in {}^*\mathfrak{V}_n = \mathfrak{W}_n.$$

Moreover, $\mathfrak{V}_n \in \mathfrak{V}_{n+1}$, thus $\mathfrak{W}_n = {}^*\mathfrak{V}_n \in {}^*\mathfrak{V}_{n+1} = \mathfrak{W}_{n+1}$. This proves that $\mathfrak{W}_n$ is a set in $\mathfrak{W}$.

(b) follows from (xii) and (xiii) in Section 7.1.

(c) $a \subseteq b \Leftrightarrow \mathfrak{V} \models \forall x \mathrel{\dot\in} a\,(x \mathrel{\dot\in} b) \Leftrightarrow \mathfrak{W} \models \forall x \mathrel{\dot\in} {}^*a\,(x \mathrel{\dot\in} {}^*b) \Leftrightarrow {}^*a \subseteq {}^*b$.

(d) The sentence $\forall x \mathrel{\dot\in} {}^*(a \cup b)\,(x \mathrel{\dot\in} {}^*a \vee x \mathrel{\dot\in} {}^*b)$ is true in $\mathfrak{W}$, because the same sentence without $*$ is true in $\mathfrak{V}$. This proves that $^*(a \cup b) \subseteq {}^*a \cup {}^*b$. By (c) $^*a \subseteq {}^*(a \cup b)$ and $^*b \subseteq {}^*(a \cup b)$, thus $^*a \cup {}^*b \subseteq {}^*(a \cup b)$.

(e) By Proposition A.5.1 (i) and (b), $\mathfrak{W}_0 = {}^*\mathfrak{V}_0 \in {}^*\mathfrak{V}_1 = \mathfrak{W}_1$. Let $n \ge 1$. Then $^*(\mathfrak{V}_{\le n}) = {}^*(\mathfrak{V}_0 \cup \mathfrak{V}_n) = {}^*\mathfrak{V}_0 \cup {}^*\mathfrak{V}_n = \mathfrak{W}_0 \cup \mathfrak{W}_n \subseteq \mathfrak{W}_{\le n}$. Since, for all $i = 0,\ldots,n$, $\mathfrak{W}_i = {}^*\mathfrak{V}_i \subseteq {}^*(\mathfrak{V}_{\le n})$, we have $\mathfrak{W}_{\le n} \subseteq {}^*(\mathfrak{V}_{\le n})$. Since $\mathfrak{V}_{\le n} \in \mathfrak{V}_{n+1}$, $^*\left(\mathfrak{V}_{\le n}\right) \in {}^*\mathfrak{V}_{n+1} = \mathfrak{W}_{n+1}$.

(f) Let $A(x)$ be a formula in $\mathcal{L}_{\mathfrak{W}}$ and let $n \in \mathbb{N}_0$. There is an $m \in \mathbb{N}_0$ such that the parameters $b_1,\ldots,b_k$ occurring in $A(x)$ belong to $\mathfrak{W}_{\le m}$. Let $A(x,x_1,\ldots,x_k)$ result from $A(x)$ by replacing b_i with x_i, $i = 1,\ldots,k$. We may assume that $x_1,\ldots,x_k$ do not occur in $A(x)$ and are different. Then $A(x,x_1,\ldots,x_k)$ is a formula in $\mathcal{L}_{\mathfrak{V}}$. Since $\mathfrak{V}$ is closed under definition,

$$\mathfrak{V} \models \forall x_1,\ldots,x_k \mathrel{\dot\in} \mathfrak{V}_{\le m} \exists z \mathrel{\dot\in} \mathfrak{V}_{n+1} \forall x \mathrel{\dot\in} \mathfrak{V}_{\le n}(x \mathrel{\dot\in} z \Leftrightarrow A(x,x_1,\ldots,x_k))).$$

Since $^*\mathfrak{V}_{\le n} = \mathfrak{W}_{\le n}$ and $^*\mathfrak{V}_n = \mathfrak{W}_n$ for all $n \in \mathbb{N}_0$, we see that

$$\mathfrak{W} \models \forall x_1,\ldots,x_k \mathrel{\dot\in} \mathfrak{W}_{\le m} \exists z \mathrel{\dot\in} \mathfrak{W}_{n+1} \forall x \mathrel{\dot\in} \mathfrak{W}_{\le n}(x \mathrel{\dot\in} z \Leftrightarrow A(x,x_1,\ldots,x_k))).$$

It follows that for $b_1,\dots,b_k$ there is a $B\in\mathfrak{W}_{n+1}$ such that, for $a\in\mathfrak{W}_{\leq n}$, $a\in B$ if and only if $\mathfrak{W}\models A(x)(a)$. Therefore,

$$B=\{a\in\mathfrak{W}_{\leq n}\mid\mathfrak{W}\models A(x)(a)\}\in\mathfrak{W}_{n+1}.$$

(g) Define $B:=\{a\in\mathfrak{V}_{\leq n}\mid\mathfrak{V}\models A(x)(a)\}$. Then $B\in\mathfrak{V}_{n+1}$ and

$$\mathfrak{V}\models\forall x\,\dot{\in}\mathfrak{V}_{\leq n}(x\,\dot{\in}\,B\Leftrightarrow A(x)).$$

Since $\mathfrak{W}\models\forall x\,\dot{\in}\mathfrak{W}_{\leq n}(x\,\dot{\in}\,{}^*B\leftrightarrow{}^*A(x))$ and ${}^*B\in\mathfrak{W}_{n+1}$, we obtain

$${}^*B=\{b\in\mathfrak{W}_{\leq n}\mid\ \mathfrak{W}\models{}^*A(x)(b)\}.$$

The proof of (h) is an exercise (see Exercise 7.7).

(i) We identify the k-ary functions f with binary relations, where the second argument is uniquely determined by the first argument, which is a k-tuple. Therefore, if f is a k-ary function, we can use $x_{k+1}=f(x_1,\dots,x_k)$ as a shorthand for $((x_1,\dots,x_k),x_{k+1})\in f$.

We may assume that there exists an $n\geq 1$ such that $a_1,\dots,a_k,a,f\in\mathfrak{V}_n$. Then $\mathfrak{V}\models A_1\wedge A_2\wedge A_3$, where

$$A_1:=\forall x_1\,\dot{\in}\,a_1\dots\forall x_k\,\dot{\in}\,a_k\exists y\,\dot{\in}\,a\,(((x_1,\dots,x_k),y)\,\dot{\in}\,f).$$

$$A_2:=\forall x_1,\dots,x_k,y\,\dot{\in}\mathfrak{V}_{\leq n}(((x_1,\dots,x_k),y)\,\dot{\in}\,f\Rightarrow x_1\,\dot{\in}\,a_1\wedge\dots\wedge x_k\,\dot{\in}\,a_k\wedge y\,\dot{\in}\,a).$$

$$A_3:=\forall x_1,\dots,x_k,y,y'\,\dot{\in}\mathfrak{V}_{\leq n}(((x_1,\dots,x_k),y)\,\dot{\in}\,f\wedge\big((x_1,\dots,x_k),y'\big)\,\dot{\in}\,f\Rightarrow y\doteq y').$$

A_1 means that domain$(f)\supseteq a_1\times\dots\times a_k$; A_2 means that range$(f)\subseteq a$ and domain$(f)\subseteq a_1\times\dots\times a_k$; A_3 means that f is a function. Since $\mathfrak{V}\models A_1\wedge A_2\wedge A_3$ and therefore $\mathfrak{W}\models{}^*A_1\wedge{}^*A_2\wedge{}^*A_3$, we see that ${}^*f$ is a mapping from ${}^*a_1\times\dots\times{}^*a_k$ into ${}^*a$. Assume that $f(b_1,\dots,b_k)=b$. Then $((b_1,\dots,b_k),b)\in f$ and $(({}^*b_1,\dots,{}^*b_k),{}^*b)\in{}^*f$. Therefore, ${}^*f({}^*b_1,\dots,{}^*b_k)={}^*b={}^*(f(b_1,\dots,b_k))$. The other parts of assertion (g) can be proved in a similar way. □

A.7 Poly-saturated models

In the whole book we have fixed a standard model $\mathfrak{V}:=(\mathfrak{V}_n)_{n\in\mathbb{N}_0}$ such that $\mathfrak{V}_0$ is an infinite set. Of course, $\mathfrak{V}$ is closed under definition. Recall the notions 'small' and 'deep' in Sections 7.1 and 7.2. A model $\mathfrak{W}:=(\mathfrak{W}_n)_{n\in\mathbb{N}_0}$, which is closed under definition, is called **poly-saturated for** $\mathfrak{V}$ if for each $n\in\mathbb{N}$ and each small and deep set $\mathcal{B} \subseteq \mathfrak{W}_n$ the intersection $\bigcap \mathcal{B}$ of all elements of $\mathcal{B}$ is non-empty. We have the following applications of saturation.

Proposition A.7.1 *Assume that $\mathfrak{W}$ is poly-saturated for $\mathfrak{V}$. Fix $n \in \mathbb{N}$.*

(a) *Let $C \subseteq \mathfrak{W}_{\leq n-1}$ be infinite and small. Then C is external. It follows that an internal set C in $\mathfrak{W}$ is either finite or very large.*
(b) *Let R be a set in $\mathfrak{V}$. (R could be a k-ary relation.) Then*

$$^*[R] \subseteq {}^*R \text{ and } (^*[R] \neq {}^*R \text{ if and only if } R \text{ is infinite}).$$

(c) *Let $\mathcal{B} \subseteq \mathfrak{W}_n$ be small and $C \in \mathfrak{W}_n$ with $C \subseteq \bigcup \mathcal{B}$. Then there exist finitely many $B_1, \ldots, B_n \in \mathcal{B}$ with $C \subseteq B_1 \cup \ldots \cup B_n$.*
(d) *If $(A_k)_{k \in \mathbb{N}}$ is a strictly increasing (decreasing) sequence in $\mathfrak{W}_n$, then $\bigcup_{k \in \mathbb{N}} A_k$ ($\bigcap_{k \in \mathbb{N}} A_k$) is external.*

Proof (a), (b), (c) and (d) follow from the proofs of Examples (xi), (xiii), (ix) and (x) in Section 7.1. □

If R is a k-ary relation and a set in $\mathfrak{V}$, then, by Proposition A.7.1 (b), $^*R$ can be understood as an extension of R in the following sense: if $a_1, \ldots, a_k$ are elements of the model $\mathfrak{V}$, then $(a_1, \ldots, a_k) \in R$ if and only if $^*(a_1, \ldots, a_k) = (^*a_1, \ldots, {}^*a_k) \in {}^*R$; in particular, if R is a relation on the set of urelements of $\mathfrak{V}$, then for urelements $a_1, \ldots, a_k$ in $\mathfrak{V}$, $(a_1, \ldots, a_k) \in R$ if and only if $(a_1, \ldots, a_k) \in {}^*R$ by identification of urelements a in $\mathfrak{B}$ with $^*a$ in $\mathfrak{W}$.

For n-ary functions f in $\mathfrak{V}$ (or in $\mathfrak{W}$; functions are special relations and $f(a_1, \ldots, a_n) = a$ is an abbreviation for $((a_1, \ldots, a_n), a) \in f$), we obtain in a similar way: If $a_1, \ldots, a_n$ are elements of $\mathfrak{V}$, then $f(a_1, \ldots, a_n) = a$ if and only if $^*f(^*a_1, \ldots, {}^*a_n) = {}^*a$, and for urelements $a_1, \ldots, a_n, a$ in $\mathfrak{V}$, $f(a_1, \ldots, a_n) = a$ if and only if $^*f(a_1, \ldots, a_n) = a$.

If R is a binary relation (for example, $R = <$), we will write aRb for $(a, b) \in R$, and if f is a binary function (for example, $f = +$), we will write $a f b = c$ for $((a, b), c) \in f$.

Appendix B. The existence of poly-saturated models

Following [86], we now present an elementary proof of Theorem 7.2.1, without using the existence of countably incomplete κ-good ultrafilters.

B.1 From pre-models to models

In order to deal with ultrapowers and limits of elementary chains under the same umbrella, we introduce the notion 'pre-model': A triple $((P_n)_{n\in\mathbb{N}_0}, \sim, E)$ is called a **pre-model** if the following conditions are true, where we use the shorthand $P_{1\leq}$ for $\bigcup_{n\geq 1} P_n$.

(PM 1) **Existence of urelements.** $P_0 \neq \emptyset$ and $P_0 \cap \bigcup_{n\geq 1} P_n = \emptyset$.

(PM 2) **Compatibility of $\sim$ with P_n.** The relation $\sim$ is an equivalence relation on $P_{<\infty}$ such that for all $a, b \in P_{<\infty}$ with $a \sim b$ and all $n \in \mathbb{N}_0$:

$$a \in P_n \text{ if and only if } b \in P_n.$$

(PM 3) **Compatibility of $\sim$ with E.** The relation E is a subset of $P_{<\infty} \times P_{1\leq}$ such that for $a, a', b, b' \in P_{<\infty}$ with $a' \sim a$ and $b' \sim b$:,

$$aEb \text{ if and only if } a'Eb'.$$

(PM 4) **Transitivity.** If $n \geq 1$ and $a \in P_n$, then

$$bEa \Rightarrow b \in P_{\leq n-1}.$$

(PM 5) **Extensionality.** Fix $a, b \in P_{1\leq}$. Then $a \sim b$ if $(cEa \Leftrightarrow cEb)$ for all $c \in P_{<\infty}$.

Later on we shall be concerned with two important examples of pre-models.

Fix a pre-model $P = ((P_n)_{n\in\mathbb{N}_0}, \sim, E)$. Using the so-called **Mostowski collapse**, we define inductively on $n \in \mathbb{N}_0$ sets $\mathfrak{V}_n$ and their elements $\overline{a}^n$ such that $(\mathfrak{V}_n)$ becomes a model generated by P.

For each $a \in P_0$ set $\widetilde{a}^0 := \{b \in P_0 | \; b \sim a\}$. According to Proposition A.2.1, we can rename each $\widetilde{a}^0$ to $\overline{a}^0$ such that the function $\widetilde{a}^0 \mapsto \overline{a}^0$ defines a bijection between $\{\widetilde{a}^0 \,|\, a \in P_0\}$ and $\{\overline{a}^0 \,|\, a \in P_0\}$, and such that each weak model over

$$\mathfrak{V}_0 := \{\overline{a}^0 \,|\, a \in P_0\}$$

is a model over $\mathfrak{V}_0$.

Now let $n \geq 1$ and assume that $\overline{a}^k$ is already defined for each $k < n$ and each $a \in P_k$. Moreover, assume that $\mathfrak{V}_k := \{\overline{a}^k \,|\, a \in P_k\}$ for each $k < n$. If $a \in P_n$, set

$$\overline{a}^n := \{\overline{c}^k | k < n, c \in P_k \text{ and } c\, E\, a\}$$

and

$$\mathfrak{V}_n := \{\overline{a}^n \,|\, a \in P_n\}.$$

Since $P_0 \neq \emptyset$, we have $\mathfrak{V}_0 \neq \emptyset$. Since for $n \geq 1$, $\overline{a}^n \subset \mathfrak{V}_{\leq n-1}$, we see that $\mathfrak{V}_n \subset \mathcal{P}(\mathfrak{V}_{\leq n-1})$. It follows that $\mathfrak{V} := (\mathfrak{V}_n)_{n\in\mathbb{N}_0}$ is a model over $\mathfrak{V}_0$. We say that the model $\mathfrak{V}$ is **generated by** $((P_n)_{n\in\mathbb{N}_0}, \sim, E)$.

In the following crucial result we need the fact that urelements are different from the sets in a model.

Proposition B.1.1 *Fix $n, m \in \mathbb{N}_0$ and $a \in P_n$ and $b \in P_m$. Then*

$$\overline{a}^n = \overline{b}^m \Leftrightarrow a \sim b.$$

Proof We prove this result by induction on $\mu_{m,n} := \max\{n, m\}$.

Let $\mu_{m,n} = 0$. Then, by the definition of $\overline{a}^0$, $\overline{a}^0 = \overline{b}^0 \Leftrightarrow \widetilde{a}^0 = \widetilde{b}^0 \Leftrightarrow a \sim b$. For the induction step let $\mu_{m,n} > 0$.

Assume that $\overline{a}^n = \overline{b}^m$. Since $\mathfrak{V}_0 \cap \mathfrak{V}_{1\leq} = \emptyset$, we have $1 \leq n, m$. Suppose that $c\, E\, a$ holds. By (PM 4), there exists an $i < n$ with $c \in P_i$. It follows that $\overline{c}^i \in \overline{a}^n = \overline{b}^m$. Therefore, there exists a $j < m$ and a $d \in P_j$ with $d\, E\, b$ and $\overline{d}^j = \overline{c}^i$. Since $\mu_{i,j} < \mu_{m,n}$, by the induction hypothesis, $d \sim c$. From (PM 3) it follows that $c\, E\, b$ holds. Therefore, $c\, E\, a$ implies $c\, E\, b$ and vice versa. By (PM 5), $a \sim b$.

Now assume that $a \sim b$. Since $m > 0$ or $n > 0$, by (PM 1) and (PM 2), $1 \leq m, n$. In order to show that $\overline{a}^n = \overline{b}^m$, let $\overline{c}^i \in \overline{a}^n$ with $i < n$, $c \in P_i$ and $c\, E\, a$. By (PM 3), $c\, E\, b$. By (PM 4), there exists a $j < m$ with $c \in P_j$. It follows that $\overline{c}^j \in \overline{b}^m$. Since $\mu_{i,j} < \mu_{m,n}$, by the induction hypothesis, $\overline{c}^i = \overline{c}^j \in \overline{b}^m$. This proves that $\overline{a}^n \subseteq \overline{b}^m$. The proof of $\overline{a}^n \supseteq \overline{b}^m$ is similar. □

In view of the previous proposition, we may define for each $a \in P_{<\infty}$

$$\overline{a} := \overline{a}^n \text{ if } a \in P_n.$$

We obtain the following

Corollary B.1.2 *Fix $a, b \in P_{<\infty}$. Then*

(a) $\overline{a} = \overline{b} \Leftrightarrow a \sim b$,
(b) $\overline{a} \in \overline{b} \Leftrightarrow a E b$.

Proof We only need to prove (b). Assume that $\overline{a} \in \overline{b}$. Since $\mathfrak{V}$ is a model, $\overline{b}$ is a set in $\mathfrak{V}$. Therefore, there exist $i \in \mathbb{N}$ and $a' \in P_i$ with $a' E b$ and $\overline{a'} = \overline{a}$. By (a), $a' \sim a$. By (PM 3), $a E b$. Now assume that $a E b$. Since $b \in P_n$ for some $n \geq 1$, by (PM 4), there exists an $i < n$ with $a \in P_i$. By the definition of $\overline{b}$, $\overline{a} \in \overline{b}$. □

B.2 Ultrapowers

The aim of this section is the proof of the compactness theorem. We assume that the reader is familiar with filters and ultrafilters. First we will construct a pre-model generating the ultrapower of a model $\mathfrak{V}$:

Fix a non-empty set I, an ultrafilter D on I and a model $\mathfrak{V} = (\mathfrak{V}_n)_{n \in \mathbb{N}_0}$ that is closed under definition. We will use the following abbreviations.

If $(a_i)_{i \in I}$ is an I-sequence, then we will write (a_i) instead of $(a_i)_{i \in I}$. If $\mathcal{F}(i)$ is an assertion about elements i of I, then we will write $\mathcal{F}(i)$ a.e. instead of $\{i \in I \mid \mathcal{F}(i)\} \in D$. For each $n \in \mathbb{N}_0$ set

$$F_n := \{(a_i) \mid a_i \in \mathfrak{V}_n \text{ a.e.}\}.$$

On $F_{<\infty} = \bigcup_{n \in \mathbb{N}_0} F_n$ we define a relation $\sim$ by setting

$$(a_i) \sim (b_i) :\Leftrightarrow a_i = b_i \text{ a.e.}$$

and we define a relation $E \subseteq F_{<\infty} \times F_{1\leq}$ by setting,

$$(a_i)\, E\, (b_i) :\Leftrightarrow a_i \in b_i \text{ a.e.}$$

Using the ultrafilter properties one can easily prove the following

Lemma B.2.1 *The triple $\big((F_n)_{n \in \mathbb{N}_0}, \sim, E\big)$ is a pre-model.*

The model generated by $((F_n)_{n \in \mathbb{N}_0}, \sim, E)$ is called the ***D*-ultrapower of** $\mathfrak{V}$ and is denoted by $\Pi_D(\mathfrak{V})$.

Now we will show that the D-ultrapower of $\mathfrak{V}$ has the same formal properties as $\mathfrak{V}$. Let A be a (sentence) formula in $\mathcal{L}_{\Pi_D(\mathfrak{V})}$. By $(A^i)_{i\in I}$ we denote one of the I-sequences of (sentences) formulas in $\mathcal{L}_{\mathfrak{V}}$ that result from A by replacing each parameter $\overline{(a_i)}$ in A by a_i. Note that if $(\widetilde{A}^i)_{i\in I}$ is another result, then $\widetilde{A}^i = A^i$ a.e.

Theorem B.2.2 (Theorem of Loś) *Fix a sentence A in $\mathcal{L}_{\Pi_D(\mathfrak{V})}$. Then*

$$\Pi_D(\mathfrak{V}) \models A \Leftrightarrow \mathfrak{V} \models A^i \text{ a.e.}$$

Proof By induction over the definition of the sentences in $\mathcal{L}_{\Pi_D(\mathfrak{V})}$.

(1) (a) Let A be the sentence $(\overline{(a_i)} \mathrel{\dot\in} \overline{(b_i)})$. Then

$$\Pi_D(\mathfrak{V}) \models \overline{(a_i)} \mathrel{\dot\in} \overline{(b_i)} \Leftrightarrow \overline{(a_i)} \in \overline{(b_i)} \Leftrightarrow a_i \in b_i \quad \text{a.s.} \Leftrightarrow \mathfrak{V} \models A^i \text{ a.e.}$$

(b) The proof for $A = (\overline{(a_i)} \doteq \overline{(b_i)})$ is similar to the proof of (a).

(2) (a) Let $A = B \vee C$ be a formula of $\mathcal{L}_{\Pi_D(\mathfrak{V})}$. Assume that the assertion is true for B and C.

Suppose that $\Pi_D(\mathfrak{V}) \models A$. Then $\Pi_D(\mathfrak{V}) \models B$ or $\Pi_D(\mathfrak{V}) \models C$. By the induction hypothesis, $\mathfrak{V} \models B^i$,a.e. or $\mathfrak{V} \models C^i$ a.e. Therefore, $\mathfrak{V} \models B^i \vee C^i$ a.e. The result follows, since A^i equals $B^i \vee C^i$ a.e. Suppose that $\mathfrak{V} \models B^i \vee C^i$ a.e. Since D is an ultrafilter, $\mathfrak{V} \models B^i$ a.e. or $\mathfrak{V} \models C^i$ a.e. By the induction hypothesis, $\Pi_D(\mathfrak{V}) \models B$ or $\Pi_D(\mathfrak{V}) \models C$, thus $\Pi_D(\mathfrak{V}) \models B \vee C$.

(b) Let $A = \neg B$. Assume that $\Pi_D(\mathfrak{V}) \models A$, thus $\Pi_D(\mathfrak{V}) \not\models B$. By the induction hypothesis, $\{\mathfrak{V} \models B^i\} \notin D$. Since D is an ultrafilter $\{\mathfrak{V} \not\models B^i\} \in D$, thus $\mathfrak{V} \models \neg B^i$ a.e.

Assume that $\mathfrak{V} \models \neg B^i$ a.e. Since $\emptyset \notin D$, $\{\mathfrak{V} \models B^i\} \notin D$. The induction hypothesis implies $\Pi_D(\mathfrak{V}) \models \neg B$.

(c) Let $A = \exists x \mathrel{\dot\in} \overline{(a_i)} B(x)$. If $\Pi_D(\mathfrak{V}) \models A$, then there is a $\overline{(b_i)} \in \overline{(a_i)}$ with $\Pi_D(\mathfrak{V}) \models B(x)(\overline{(b_i)})$. By the induction hypothesis, $\mathfrak{V} \models B(x)^i(b_i)$ a.e. Since $b_i \in a_i$ a.e., $\mathfrak{V} \models A^i$ a.e.

Suppose that $\mathfrak{V} \models A^i$ a.e., that is, $Y := \{\mathfrak{V} \models \exists x \mathrel{\dot\in} a_i (B(x)^i)\} \in D$. We choose an I-sequence (b_i) with $\mathfrak{V} \models b_i \mathrel{\dot\in} a_i \wedge B(x)^i(b_i)$ for each $i \in Y$ and set $b_i = \emptyset$ if $i \notin Y$. Since $\mathfrak{V} \models B(x)^i(b_i)$ a.e., by the induction hypothesis, $\Pi_D(\mathfrak{V}) \models B(x)(\overline{(b_i)})$. Since $b_i \in a_i$ a.e., $\overline{(b_i)} \in \overline{(a_i)}$. It follows that $\Pi_D(\mathfrak{V}) \models A$. □

An important consequence of Loś' theorem is the existence of an elementary embedding from $\mathfrak{V}$ into the ultrapower $\Pi_D(\mathfrak{V})$ of $\mathfrak{V}$.

For each $b \in \mathfrak{V}_{<\infty}$ let (b) be the constant I-sequence (b_i) with $b_i = b$ for each $i \in I$. The mapping $^* : \mathfrak{V}_{<\infty} \to \Pi_D(\mathfrak{V})_{<\infty}$ is defined by setting

$$^*b := \overline{(b)}.$$

Corollary B.2.3 *The function* * *is an elementary embedding from* $\mathfrak{V}$ *into* $\Pi_D(\mathfrak{V})$.

Proof We have to check the properties (E 1) and (E).

(E 1) Since $\mathfrak{V}_n \in \mathfrak{V}_{n+1}$, $(\mathfrak{V}_n) \in F_{n+1}$, thus $^*\mathfrak{V}_n = \overline{(\mathfrak{V}_n)} \in \Pi_D(\mathfrak{V})_{<\infty}$. By Corollary B.1.2 (b), and since $\mathfrak{V}_n$ is a set in $\mathfrak{V}$, we obtain

$$\overline{(a_i)} \in \overline{(\mathfrak{V}_n)} \Leftrightarrow (a_i)\ E\ (\mathfrak{V}_n) \Leftrightarrow \overline{(a_i)} \in \Pi_D(\mathfrak{V})_n.$$

This proves that $^*\mathfrak{V}_n = \Pi_D(\mathfrak{V})_n$.

(E) Let A be a sentence in $\mathcal{L}_{\mathfrak{V}}$. Then $^*A$ is a sentence in $\mathcal{L}_{\Pi_D(\mathfrak{V})}$ and $(^*A)^i = A$ a.e. By Theorem B.2.2, we obtain

$$\mathfrak{V} \models A \Leftrightarrow \mathfrak{V} \models (^*A)^i \text{ a.e.} \Leftrightarrow \Pi_D(\mathfrak{V}) \models {}^*A. \qquad \square$$

Recall from Proposition A.6.1 (f) that the model $\Pi_D(\mathfrak{V})$ is closed under definition. Using the ultrapower construction given above, we can prove the compactness theorem. The proof is the same as in first-order model theory (see Chang and Keisler [21]).

Theorem B.2.4 (The compactness theorem) *Let* $\mathfrak{V} = (\mathfrak{V}_n)_{n\in\mathbb{N}_0}$ *be a model that is closed under definition, let* $n \geq 1$ *and let* $\mathcal{B}$ *be a deep subset of* $\mathfrak{V}_n$. *Then there exists a model* $\mathfrak{W} = (\mathfrak{W}_n)_{n\in\mathbb{N}_0}$ *and an elementary embedding* * *from* $\mathfrak{V}$ *into* $\mathfrak{W}$, *such that* $\bigcap{}^*[\mathcal{B}] \neq \emptyset$. *(Note that* $\bigcap{}^*[\mathcal{B}] = \bigcap\{^*A \mid A \in \mathcal{B}\}$.)

Proof We may assume that $\mathcal{B}$ is closed under finite intersections. Let I be the set of finite non-empty subsets of $\mathcal{B}$. Since $\mathcal{B}$ is deep, for each $i \in I$ there exists an $a_i \in \mathfrak{V}_{\leq n-1}$ such that $a_i \in \bigcap i \in \mathcal{B}$. For each set $A \in \mathcal{B}$ define

$$\widetilde{A} := \{i \in I \mid a_i \in A\} \subset I$$

and

$$G := \{\widetilde{A} \mid A \in \mathcal{B}\} \subset \mathcal{P}(I).$$

Then $\emptyset \notin G$, because $\{A\} \in \widetilde{A}$ for each $A \in \mathcal{B}$. Since $\mathcal{B} \neq \emptyset$, $G \neq \emptyset$. Now let $A, B \in \mathcal{B}$. Then $\widetilde{A \cap B} \in G$ and it is easy to see that $\widetilde{A \cap B} \subset \widetilde{A} \cap \widetilde{B}$. It follows that $F(G) := \{B \subseteq I \mid \exists C \in G (C \subseteq B)\}$ is a filter. Fix an ultrafilter $D \supseteq F(G)$ on I. Let $\Pi_D(\mathfrak{V})$ be the D-ultrapower of $\mathfrak{V}$ and let $^* : \mathfrak{V}_{<\infty} \to \Pi_D(\mathfrak{V})_{<\infty}$ be the elementary embedding introduced previously. Since $a_i \in \mathfrak{V}_{\leq n-1}$ and D is an ultrafilter, $(a_i) \in F_{\leq n-1}$, thus $\overline{(a_i)} \in \Pi_D(\mathfrak{V})_{\leq n-1}$. In order to show that $\overline{(a_i)} \in \bigcap{}^*[\mathcal{B}]$, fix $A \in \mathcal{B}$. Then

$$\{i \in I \mid \mathfrak{V} \models a_i \dot{\in} A\} = \{i \in I \mid a_i \in A\} = \widetilde{A} \in G \subset F(G) \subset D.$$

By the Theorem B.2.2, $\Pi_D(\mathfrak{V}) \models \overline{(a_i)} \in \overline{(A)} = {}^*A$. □

B.3 Elementary chains and their elementary limits

Here we extend the notions 'elementary chain' and 'elementary limit' to our notion of a model. Let λ be an ordinal number different from 0 and let $(\mathfrak{V}^\alpha)_{\alpha<\lambda}$ be a λ-sequence of models $\mathfrak{V}^\alpha = (\mathfrak{V}^\alpha_n)_{n\in\mathbb{N}_0}$, which are closed under definition. Then the pair $\big((\mathfrak{V}^\alpha)_{\alpha<\lambda}, (*_{\alpha\beta})_{\alpha\le\beta<\lambda}\big)$ is called an **elementary chain** if for each $\alpha \le \beta \le \gamma < \lambda$

(EC 1) $*_{\alpha\beta}$ is an elementary embedding from $\mathfrak{V}^\alpha$ into $\mathfrak{V}^\beta$,
(EC 2) $*_{\alpha\alpha}(a) = a$ for each $a \in \mathfrak{V}^\alpha_{<\infty}$,
(EC 3) $*_{\beta\gamma} \circ *_{\alpha\beta} = *_{\alpha\gamma}$.

Starting from an elementary chain C we define a pre-model such that the generated model becomes the elementary limit of C. To this end fix an elementary chain $\big((\mathfrak{V}^\alpha)_{\alpha<\lambda}, (*_{\alpha\beta})_{\alpha\le\beta<\lambda}\big)$. We define for each $n \in \mathbb{N}_0$

$$P_n := \{(a,\alpha) | \alpha < \lambda \text{ and } a \in \mathfrak{V}^\alpha_n\}.$$

Two elements $(a,\alpha),(b,\beta) \in P_{<\infty} = \bigcup_{n\in\mathbb{N}_0} P_n$ are called **equivalent** if there exists a $\gamma < \lambda$ with $\alpha,\beta \le \gamma$ such that $*_{\alpha\gamma}(a) = *_{\beta\gamma}(b)$, in which case we shall write $(a,\alpha) \sim (b,\beta)$.

For $(a,\alpha) \in P_{<\infty}$ and $(b,\beta) \in P_{1\le}$ we set $(a,\alpha)\, E\, (b,\beta)$ if there exists a $\gamma < \lambda$ with $\alpha,\beta \le \gamma$ such that $*_{\alpha\gamma}(a) \in *_{\beta\gamma}(b)$. The simple proof of the next result is left to the reader.

Lemma B.3.1 *$((P_n)_{n\in\mathbb{N}_0}, \sim, E)$ is a pre-model.*

The model, generated by the pre-model $((P_n)_{n\in\mathbb{N}_0}, \sim, E)$, is called the **elementary limit** of the elementary chain $((\mathfrak{V}^\alpha)_{\alpha<\lambda}, (*_{\alpha\beta})_{\alpha\le\beta<\lambda})$ and is denoted by $\mathfrak{V}^\lambda = (\mathfrak{V}^\lambda_n)_{n\in\mathbb{N}_0}$.

We will now collect some immediate consequences of the construction of the elementary limit (Corollary B.1.2).

Proposition B.3.2 *Fix an elementary chain $((\mathfrak{V}^\alpha)_{\alpha<\lambda}, (*_{\alpha\beta})_{\alpha\le\beta<\lambda})$ and let $\mathfrak{V}^\lambda$ be its elementary limit. Fix $n \in \mathbb{N}_0$. Then the following hold.*

(a) $\mathfrak{V}^\lambda_n = \{\overline{(a,\alpha)} \mid (a,\alpha) \in P_n\} = \{\overline{(a,\alpha)} \mid \alpha < \lambda \text{ and } a \in \mathfrak{V}^\alpha_n\}$.
Fix $\alpha,\beta < \lambda, a \in \mathfrak{V}^\alpha_{<\infty}, b \in \mathfrak{V}^\beta_{<\infty}$.

(b) $\overline{(a,\alpha)} = \overline{(b,\beta)}$ *if and only if there is a* γ *with* $\alpha,\beta \leq \gamma < \lambda$ *and* $*_{\alpha\gamma}(a) = *_{\beta\gamma}(b)$. *It follows that* $\overline{(a,\alpha)} = \overline{(*_{\alpha\beta}(a),\beta)}$ *if* $\alpha \leq \beta < \lambda$.

(c) $\overline{(a,\alpha)} \in \overline{(b,\beta)}$ *if and only if there is a* $\gamma < \lambda$ *with* $\alpha,\beta \leq \gamma$ *and* $*_{\alpha\gamma}(a) \in *_{\beta\gamma}(b)$.

(d) $\overline{(\mathfrak{V}_n^\alpha,\alpha)} = \overline{(\mathfrak{V}_n^\beta,\beta)}$, *because for each* γ *with* $\alpha,\beta \leq \gamma < \lambda$

$$*_{\alpha\gamma}(\mathfrak{V}_n^\alpha) = \mathfrak{V}_n^\gamma = *_{\beta\gamma}(\mathfrak{V}_n^\beta).$$

Definition B.3.3 Fix an elementary chain $((\mathfrak{V}^\alpha)_{\alpha<\lambda},(*_{\alpha\beta})_{\alpha\leq\beta<\lambda})$ and its elementary limit $\mathfrak{V}^\lambda$. For each $\alpha \leq \lambda$ define a mapping $*_{\alpha\lambda} : \mathfrak{V}_{<\infty}^\alpha \to \mathfrak{V}_{<\infty}^\lambda$ by setting

$$*_{\alpha\lambda}(a) := \overline{(a,\alpha)} \text{ for } \alpha < \lambda \quad \text{and} \quad *_{\lambda\lambda}\left(\overline{(a,\alpha)}\right) := \overline{(a,\alpha)}.$$

Proposition B.3.4 *Fix* $\alpha \leq \lambda$. *Then* $*_{\alpha\lambda}$ *is an elementary embedding from* $\mathfrak{V}^\alpha$ *into* $\mathfrak{V}^\lambda$. *It follows that* $((\mathfrak{V}^\alpha)_{\alpha<\lambda+1},(*_{\alpha\beta})_{\alpha\leq\beta<\lambda+1})$ *is an elementary chain.*

Proof To prove property (E 1), fix $n \in \mathbb{N}_0$. We have to show that $\mathfrak{V}_n^\lambda = *_{\alpha\lambda}(\mathfrak{V}_n^\alpha)$.

'$\subseteq$' Fix $x \in \mathfrak{V}_n^\lambda$. Then there exist $\beta < \lambda$ and $b \in \mathfrak{V}_n^\beta$ with $x = \overline{(b,\beta)}$. By Proposition B.3.2 (c) and (d),

$$x = \overline{(b,\beta)} \in \overline{(\mathfrak{V}_n^\beta,\beta)} = \overline{(\mathfrak{V}_n^\alpha,\alpha)} = *_{\alpha\lambda}\mathfrak{V}_n^\alpha.$$

'$\supseteq$' Fix $x \in *_{\alpha\lambda}(\mathfrak{V}_n^\alpha) = \overline{(\mathfrak{V}_n^\alpha,\alpha)}$. Since $\mathfrak{V}_n^\alpha$ is a set in $\mathfrak{V}^\alpha$, $\overline{(\mathfrak{V}_n^\alpha,\alpha)}$ is a set in $\mathfrak{V}^\lambda$. Therefore, there exist $\beta < \lambda$ and $b \in \mathfrak{V}_{<\infty}^\beta$ with $x = \overline{(b,\beta)}$. By Corollary B.1.2 (b), $(b,\beta)\, E\, (\mathfrak{V}_n^\alpha,\alpha)$, thus there exists a $\gamma \geq \alpha,\beta$ with $*_{\beta\gamma}(b) \in *_{\alpha\gamma}(\mathfrak{V}_n^\alpha) = \mathfrak{V}_n^\gamma$. We obtain

$$x = \overline{(b,\beta)} = \overline{(*_{\beta\gamma}(b),\gamma)} \in \mathfrak{V}_n^\lambda.$$

This proves (E 1). To prove (E), we will prove by induction on the definition of the sentences A in $\bigcup_{\alpha<\lambda} \mathcal{L}_{\mathfrak{V}^\alpha}$ that, if A is in $\mathcal{L}_{\mathfrak{V}^\alpha}$, then

$$\mathfrak{V}^\alpha \models A \Leftrightarrow \mathfrak{V}^\lambda \models {}^{*_{\alpha\lambda}}A.$$

(1) (a) Let $A = (a \dot{\in} b)$. Then

$$\mathfrak{V}^\alpha \models A \Leftrightarrow a \in b \Leftrightarrow (a,\alpha)\, E\, (b,\alpha) \Leftrightarrow \overline{(a,\alpha)} \in \overline{(b,\alpha)}$$

$$\Leftrightarrow *_{\alpha\lambda}(a) \in *_{\alpha\lambda}(b) \Leftrightarrow \mathfrak{V}^\lambda \models {}^{*_{\alpha\lambda}}A.$$

The second '$\Leftarrow$' of the previous computation can be seen as follows. If $(a,\alpha)\, E\, (b,\alpha)$, then $*_{\alpha\gamma}(a) \in *_{\alpha\gamma}(b)$ for some $\gamma \geq \alpha$, thus $a \in b$.

(1) (b) If $A = (a \dot{=} b)$, then the proof is similar to that of (1) (a).

(2) If $A = (B \vee C)$ or $A = \neg B$, then the assertion follows immediately from the induction hypothesis.

Let $A = \exists x \dot{\in} a B(x)$.

Assume that $\mathfrak{V}^\alpha \models A$. Then there is a $b \in a$ with $\mathfrak{V}^\alpha \models B(x)(b)$. By the induction hypothesis, $\mathfrak{V}^\lambda \models {}^{*\alpha\lambda} B(x) \left(\overline{(b,\alpha)}\right)$. Since $\overline{(b,\alpha)} \in \overline{(a,\alpha)}$, $\mathfrak{V}^\lambda \models {}^{*\alpha\lambda} A$.

Assume that $\mathfrak{V}^\lambda \models {}^{*\alpha\lambda} A$. Then $\mathfrak{V}^\lambda \models {}^{*\alpha\lambda} B(x) \left(\overline{(b,\delta)}\right)$ for some $\overline{(b,\delta)} \in \overline{(a,\alpha)}$. Set $\gamma := \max\{\alpha, \delta\}$. Since $\overline{(*_{\delta\gamma}(b), \gamma)} = \overline{(b,\delta)}$, ${}^{*\gamma\lambda}({}^{*\alpha\gamma} B(x) (*_{\delta\gamma}(b)))$ is true in $\mathfrak{V}^\lambda$. By the induction hypothesis, ${}^{*\alpha\gamma} B(x) (*_{\delta\gamma}(b))$ is true in $\mathfrak{V}^\lambda$. Since $*_{\delta\gamma}(b) \in *_{\alpha\gamma}(a)$, we see that $\mathfrak{V}^\gamma \models \exists x \dot{\in} *_{\alpha\gamma}(a) {}^{*\alpha\gamma} B(x)$, that is, $\mathfrak{V}^\gamma \models {}^{*\alpha\gamma} \exists x \dot{\in} a B(x)$. Since the parameters of $\exists x \dot{\in} a B(x)$ belong to $\mathfrak{V}^\alpha_{<\infty}$, we obtain $\mathfrak{V}^\alpha \models \exists x \dot{\in} a B(x)$. □

B.4 Existence of poly-saturated models with the same properties as standard models

In order to finish the proof of Theorem 7.2.1, we need one more lemma, which is an iteration of the compactness theorem. First some notation: let id denote the identity map, i.e., $\mathrm{id}(x) := x$.

Lemma B.4.1 *Fix a model $\mathfrak{W} := (\mathfrak{W}_n)_{n \in \mathbb{N}_0}$ which is closed under definition. Then there exist a model U and an elementary embedding $*$ from $\mathfrak{W}$ into U such that, for each $n \geq 1$ and each deep $\mathcal{B} \subseteq \mathfrak{W}_n$, $\bigcap {}^*[\mathcal{B}] \neq \emptyset$.*

Proof There exists a cardinal number θ and a listing $(\mathcal{B}_\alpha)_{\alpha<\theta}$ of all deep subsets $\mathcal{B}_\alpha \subseteq \mathfrak{W}_n$ for some $n \geq 1$. By transfinite recursion, we define an elementary chain $((\mathfrak{W}^\alpha)_{\alpha<\theta}, (*_{\alpha\beta})_{\alpha\leq\beta<\theta})$ in the following way. Set

$$\mathfrak{W}^0 := \mathfrak{W}, \quad *_{00} := \mathrm{id} \upharpoonright \mathfrak{W}_{<\infty}.$$

Assume that $\lambda < \theta$ and that $\mathfrak{W}^\alpha, *_{\alpha\beta}$ are already defined for each $\alpha \leq \beta < \lambda$ such that the following conditions $(1,\lambda)$ and $(2,\lambda)$ hold:

$(1,\lambda)$ $((\mathfrak{W}^\alpha)_{\alpha<\lambda}, (*_{\alpha\beta})_{\alpha\leq\beta<\lambda})$ is an elementary chain,

$(2,\lambda)$ $\bigcap {}^{*0\alpha} [\mathcal{B}_\beta] \neq \emptyset$ for each $\beta < \alpha < \lambda$.

In order to construct $\mathfrak{W}^\lambda$ and $*_{\alpha\lambda}$ for each $\alpha < \lambda$, we have to consider two cases.

Case 1: $\lambda = \gamma + 1$ is a successor ordinal.

Since $\mathcal{B}_\gamma$ is deep and $*_{0\gamma}$ is an elementary embedding from $\mathfrak{W}^0$ into $\mathfrak{W}^\gamma$, ${}^{*0\gamma}[\mathcal{B}_\gamma]$ is deep. By the compactness theorem, there exist a model $\mathfrak{W}^\lambda$ and an elementary embedding $*_{\gamma\lambda}$ from $\mathfrak{W}^\gamma$ into $\mathfrak{W}^\lambda$, such that $\bigcap^{*_{\gamma\lambda}\circ *_{0\gamma}}[\mathcal{B}_\gamma] \neq \emptyset$. We now define for each $\alpha \leq \gamma$

$$*_{\alpha\lambda} := *_{\gamma\lambda} \circ *_{\alpha\gamma} \quad \text{and} \quad *_{\lambda\lambda} := \text{id} \restriction \mathfrak{W}^\lambda_{<\infty}.$$

Note that $(1, \lambda+1)$ and $(2, \lambda+1)$ are true.

Case 2: λ is a limit ordinal.

Then let $\mathfrak{W}^\lambda$ be the elementary limit of $((\mathfrak{W}^\alpha)_{\alpha<\lambda}, (*_{\alpha\beta})_{\alpha\leq\beta<\lambda})$. For each $\alpha < \lambda$, the elementary embedding $*_{\alpha\lambda}$ from $\mathfrak{W}^\alpha$ into $\mathfrak{W}^\lambda$ is, defined in Definition B.3.3. Set $*_{\lambda\lambda} := \text{id} \restriction \mathfrak{W}^\lambda_{<\infty}$. Note that $(1, \lambda+1)$ and $(2, \lambda+1)$ are true.

We thus obtain an elementary chain $((\mathfrak{W}^\lambda)_{\lambda<\theta}, (*_{\alpha\lambda})_{\alpha\leq\lambda<\theta})$. Let $U := \mathfrak{W}^\theta$ be its elementary limit. Set $* := *_{0\theta}$. By Proposition B.3.4, $*$ is an elementary embedding from $\mathfrak{W}$ into U. It is easy to check that $\bigcap^* [\mathcal{B}] \neq \emptyset$ if $\mathcal{B}$ is a deep subset of $\mathfrak{W}_n$ with $n \geq 1$. □

Now we are able to finish the proof of Theorem 7.2.1.

Proof Fix a standard model $\mathfrak{V} = (\mathfrak{V}_n)_{n\in\mathbb{N}_0}$. Let κ^+ be the smallest cardinal number greater than κ, where κ is the cardinality of $\mathfrak{V}_{<\infty}$. Then κ^+ is a **regular cardinal**, that is, for each $\rho < \kappa^+$ and each ρ-sequence $(\alpha_\beta)_{\beta<\rho}$ in κ^+, $\sup_{\beta<\rho} \alpha_\beta < \kappa^+$. By transfinite recursion, we construct again an elementary chain $((\mathfrak{V}^\alpha)_{\alpha<\kappa^+}, (*_{\alpha\beta})_{\alpha\leq\beta<\kappa^+})$. Set

$$\mathfrak{V}^0 := \mathfrak{V} \text{ and } *_{00} := \text{id} \restriction \mathfrak{V}_{<\infty}.$$

Fix an ordinal λ strictly between 0 and κ^+ and assume that there already exists an elementary chain $((\mathfrak{V}^\alpha)_{\alpha<\lambda}, (*_{\alpha\beta})_{\alpha\leq\beta<\lambda})$ such that

$$(\diamond\lambda) \quad \text{for } \alpha, \beta < \lambda, \alpha < \beta,\ n \geq 1 \text{ and each deep } \mathcal{B} \subseteq \mathfrak{V}^\alpha_n,\ \bigcap^{*_{\alpha\beta}} [\mathcal{B}] \neq \emptyset.$$

We now define $\mathfrak{V}^\lambda$ and $*_{\alpha\lambda}$ for each $\alpha \leq \lambda$.

First assume that $\lambda = \gamma + 1$ is a successor ordinal.

Then, by Lemma B.4.1, there exists a model $\mathfrak{V}^\lambda$ and an elementary embedding $*_{\gamma\lambda}$ from $\mathfrak{V}^\gamma$ into $\mathfrak{V}^\lambda$, such that for each $n \geq 1$ and each deep $\mathcal{B} \subseteq \mathfrak{V}^\gamma_n$, $\bigcap^{*_{\gamma\lambda}} [\mathcal{B}] \neq \emptyset$. For each $\alpha < \lambda$ set

$$*_{\alpha\lambda} := *_{\gamma\lambda} \circ *_{\alpha\gamma} \quad \text{and} \quad *_{\lambda\lambda} := \text{id} \restriction \mathfrak{V}^\lambda_{<\infty}.$$

Note that $((\mathfrak{V}^\alpha)_{\alpha<\lambda+1}, (*_{\alpha\beta})_{\alpha\leq\beta<\lambda+1})$ is an elementary chain and $(\diamondsuit(\lambda+1))$ is true.

Now assume that λ is a limit number.

Let $\mathfrak{V}^\lambda$ be the elementary limit of $((\mathfrak{V}^\alpha)_{\alpha<\lambda}, (*_{\alpha\beta})_{\alpha\leq\beta<\lambda})$. For $\alpha \leq \lambda$ let $*_{\alpha\lambda}$ be the elementary embedding from $\mathfrak{V}^\alpha$ into $\mathfrak{V}^\lambda$, defined under Definition B.3.3. Note that $((\mathfrak{V}^\alpha)_{\alpha<\lambda+1}, (*_{\alpha\beta})_{\alpha\leq\beta<\lambda+1})$ is an elementary chain and $(\diamondsuit(\lambda+1))$ is true.

Let $\mathfrak{W} := \mathfrak{V}^{\kappa^+}$ be the elementary limit of $((\mathfrak{V}^\alpha)_{\alpha<\kappa^+}, (*_{\alpha\beta})_{\alpha\leq\beta<\kappa^+})$. Set $^* := *_{0\kappa^+}$. Then * is an elementary embedding from $\mathfrak{V}$ into $\mathfrak{W}$. To prove that $\mathfrak{W}$ is κ^+-saturated, fix $n \geq 1$ and a deep set $\mathcal{B} \subseteq \mathfrak{W}_n$ with $\mathrm{card}(\mathcal{B}) < \kappa^+$. Since κ^+ is a regular cardinal, there exists a $\delta < \kappa^+$ such that

$$\mathcal{B} \subseteq \left\{\overline{(b,\beta)} \mid \beta < \delta \text{ and } b \in \mathfrak{V}_n^\beta\right\}.$$

Set $\mathcal{B}' := \left\{*^{\beta\delta}(b) \mid \beta < \delta, b \in \mathfrak{V}_n^\beta \text{ and } \overline{(b,\beta)} \in \mathcal{B}\right\}$. Note that $^{*_{\delta\kappa^+}}[\mathcal{B}'] = \mathcal{B}$ and that $\mathcal{B}'$ is deep. From the construction of $((\mathfrak{V}^\alpha)_{\alpha<\kappa^+}, (*_{\alpha\beta})_{\alpha\leq\beta<\kappa^+})$ it follows that $\bigcap^{*_{\delta(\delta+1)}}[\mathcal{B}'] \neq \emptyset$. Since $*_{(\delta+1)\kappa^+}$ is an elementary embedding from $\mathfrak{V}^{\delta+1}$ into $\mathfrak{V}^{\kappa^+} = \mathfrak{W}$, by (EC 3),

$$\bigcap \mathcal{B} = \bigcap{}^{*_{\delta\kappa^+}}[\mathcal{B}'] = \bigcap{}^{*_{(\delta+1)\kappa^+} \circ *_{\delta(\delta+1)}}[\mathcal{B}'] \neq \emptyset. \qquad \square$$

References

[1] K. Aase, B. Øksendal, N. Privault and J. Ubøe, White noise generalizations of the Clark–Hausmann–Ocone theorem with applications to mathematical finance, Prépublications du Département de Mathématiques, Université de la Rochelle (1999).

[2] S. Albeverio, J. E. Fenstad, R. Høegh Krohn and T. Lindstrøm, *Nonstandard Methods in Stochastic Analysis and Mathematical Physics* (New York: Academic Press, 1986).

[3] J. M. Aldaz and H. Render, Borel measure extensions of measures defined on sub-σ algebras, *Adv. Math.* **150** (2000), 233–63.

[4] R. M. Anderson, A nonstandard representation of Brownian motion and Itô integration, *Israel J. Math.* **25** (1976), 15–46.

[5] R. B. Ash, *Real Analysis and Probability* (New York: Academic Press, 1972).

[6] H. Bauer, *Wahrscheinlichkeitstheorie* (Berlin: Walter de Gruyter, 1974).

[7] J. Berger, private communication.

[8] J. Berger, An infinitesimal approach to stochastic analysis on abstract Wiener spaces, Dissertation, Ludwig Maximilians Universität München (2002).

[9] J. Berger, H. Osswald, Y. Sun and J. L Wu, On nonstandard product measure spaces, *Illinois J. Math.* **46** (2002), 319–30.

[10] J. Bertoin, *Lévy Processes* (Cambridge: Cambridge University Press, 1996).

[11] K. Bichteler, J. B. Gravereaux and J. Jacod, *Malliavin Calculus for Processes with Jumps*, Stochastics Monographs (New York, NJ: Gordon and Breach, 1987).

[12] P. Billingsley, *Convergence of Probability Measures* (New York: Wiley, 1968).

[13] P. Billingsley, *Probability and Measure* (New York: Wiley, 1979).

[14] V. I. Bogachev, *Gaussian Measures*, Mathematical Surveys and Monographs 62 (Providence, RI: American Mathematical Society, 1998).

[15] S. Boucheron, G . Lugosi and M. Massart, Concentration inequalities using the entropy method, *Ann. Prob.* **31**:3 (2003), 1583–614.

[16] N. Bouleau and F. Hirsch, *Dirichlet Forms and Analysis on Wiener Space*, de Gruyter Studies in Mathematics (New York: Walter de Gruyter, 1991).

[17] Z. Brzezniak and J. M. A. M. van Neerven, Stochastic convolution in separable Banach spaces and the stochastic linear Cauchy Problem, *Studia Math.* **143** (2000), 43–74.

[18] R. Buckdahn, Anticipative Girsanov transformations, *Probab. Theory Rel. Fields* **90** (1991), 223–40.

[19] D. L. Burkholder, B. J. Davis and R. F. Gandy, Integral inequalities of convex functions of operators on martingales. In *Proceedings of the 6th Berkeley Symposium* vol. 2 (Berkeley, CA: University of California Press, 1970), pp. 223–40.

[20] M. Capinski and N. Cutland, *Nonstandard Methods for Stochastic Fluid Mechanics*, (Singapore: World Scientific, 1995).

[21] C. C. Chang and H. J. Keisler, *Model Theory* (Amsterdam: North Holland, 1973).

[22] J. M. C. Clark, The representation of functionals of Brownian motion by stochastic integrals, *Ann. Math. Statist.* **41** (1970), 1282–95; **42** (1971), 1778.

[23] N. Cutland, Infinitesimals in action, *J. Lond. Math. Soc.* **35**:2 (1987), 202–17.

[24] N. Cutland and S.-A. Ng, A nonstandard approach to the Malliavin calculus. In *Advances in Analysis, Probability and Mathematical Physics (Blaubeuren 1992)*, ed. S. Albeverio, W. A. J. Luxemburg and M. P. H. Wolff. Mathematics and Its Applications 314 (Dordrecht: Kluwer, 1995), pp. 149–70.

[25] N. Cutland, Brownian motion on the Wiener sphere and the infinite-dimensional Ornstein–Uhlenbeck processes, *Stoch. Process. Appl.* **79** (1999), 95–107.

[26] N. Cutland, *Loeb Measure in Practice: Recent Advances*, Lecture Notes in Mathematics 1751 (Springer-Verlag: Berlin, 2000).

[27] G. Da Prato and J. Zabczyk, *Stochastic Equations in Infinite Dimensions* (Cambridge: Cambridge University Press, 1992).

[28] G. Da Prato, G. S. Kwapien and J. Zabczyk, Regularity of solutions of linear stochastic equations in Hilbert spaces, *Stochastics* **23** (1987), 1–23.

[29] G. Di Nunno, On orthogonal polynomials and the Malliavin derivative for Lévy random stochastic measures, *Stoch. and Stoch. Rep.* **76** (2004), 517–48.

[30] G. Di Nunno, B. Øksendal and F. Proske, White noise analysis for Lévy processes, *J. Funct. Anal.* **206**:1 (2004), 109–48.

[31] G. Di Nunno, Th. Meyer-Brandis, B. Øksendal and F. Proske, Malliavin calculus and anticipative Itô formulae for Lévy processes, *Infin. Dimens. Anal. Quantum Probab. Relat. Top.* **8** (2005), 235–58.

[32] G. Di Nunno, B. Øksendal and F. Proske, *Malliavin Calculus for Lévy Processes with Applications to Finance* (Berlin: Springer Verlag, 2009).

[33] J. L. Doob, *Stochastic Processes* (New York: Wiley, 1965).

[34] C. Dellacherie and P. A. Meyer, *Probabilities and Potential B*, Mathematics Studies (Amsterdam: North Holland, 1982).

[35] T. Duncan and P. Varaiya, On the solutions of a stochastic control system, *SIAM J. Control* **13**:5 (1975), 1077–92.

[36] T. Duncan, Fréchet valued martingales and stochastic integrals, *Stochastics* **1** (1976), 269–84.

[37] O. Enchev, Nonlinear transformations on the Wiener space, *Ann. Prob.* **21** (1993), 2169–88.

[38] O. Enchev and D. W. Strook, Anticipative diffusions and related change of measure, *J. Funct. Anal.* **116** (1993), 449–77.

[39] M. X. Fernique, Intégrabilité des vecteurs Gaussiens, *C. R. Acad. Sci. Paris Ser. A* **270** (1970), 1698–9.

[40] B. Gaveau, Noyau des probabilités de certains opérateurs d'Ornstein–Uhlenbeck dans l'espace de Hilbert, *C. R. Acad Sci. Paris Ser. I Math.* **293** (1981), 469–72.

[41] L. Gross, Measurable functions on Hilbert space, *Trans. Amer. Math. Soc.* **105** (1962), 372–90.

[42] L. Gross, Abstract Wiener spaces. In *Proceedings of the 5th Berkeley Symposium on Mathematical Statistics and Probability*, Vol. II (Berkeley, CA: University of California Press, 1965), pp. 31–41.

[43] P. R. Halmos, *Measure Theory* (Princeton, NJ: van Nostrand, 1959).

[44] C. W. Henson and L. Moore, Nonstandard analysis and the theory of Banach spaces. In *Nonstandard Analysis–Recent Developments*, ed. A. H. Hurd. Lecture Notes in Mathematics, 983 (Berlin: Springer, 1983).

[45] H. Heuser, *Lehrbuch der Analysis, Teil 2* (Stuttgart: Teubner Verlag, 1990).

[46] D. L. Hoover and E. A. Perkins, Nonstandard construction of the stochastic integral and applications to stochastic differential equations I and II, *Trans. Amer. Math. Soc.* **275** (1983), 1–58.

[47] P. Imkeller, Stochastische Analysis, Vorlesungsmanuskript, Universität München (1988).

[48] P. Imkeller, Regularity of Skorokhod integral processes based on integrands in a finite Wiener Chaos, *Probab. Theory Rel. Fields* **98** (1994), 137–42.

[49] I. Iscoe, M. B. Marcus, D. McDonald, M. Talagrand and J. Zinn, Continuity of l^2-valued Ornstein–Uhlenbeck processes, *Ann. Probab.* **18** (1990), 68–84.

[50] K. Itô, *Foundations of Stochastic Differential Equations in Infinite-dimensional Spaces* (Philadelphia, PA: SIAM, 1984).

[51] Y. Itô, Generalized Poisson functionals, *Probab. Theory Rel. Fields* **77** (1988), 1–28.

[52] I. Karatzas and S. E. Shreve, *Brownian Motion and Stochastic Calculus* (Berlin: Springer Verlag, 1988).

[53] H. J. Keisler, *An Infinitesimal Approach to Stochastic Analysis*, Memoirs of the American Mathematical Society 48 (Providence, RI: American Mathematical Society, 1984).

[54] H. J. Keisler and Y. N. Sun, A metric on probabilities, and products of Loeb spaces, *J. Lond. Math. Soc.* **69** (2004), 257–72.

[55] J. Kuelbs and R. Lepage, The law of the iterated logarithm for Brownian motion in a Banach space, *Trans. Amer. Math. Soc.* **185** (1973), 253–64.

[56] H. H. Kuo, *Gaussian Measures on Banach Spaces*, Lecture notes in Mathematics 463 (Berlin: Springer Verlag, 1975).

[57] S. Kusuoka, The non-linear transformation of Gaussian measure on Banach space and its absolute continuity (I), *J. Fak. Sci. Univ. Tokyo Sec. IA Math* **29** (1982), 567–97.

[58] S. Kusuoka, On the absolute continuity of the law of a system of multiple Wiener integrals, *J. Fak. Sci. Univ. Tokyo Sec. IA Math* **30** (1983), 191–7.

[59] D. Landers and L. Rogge, Universal Loeb-measurability of sets and of the standard part map with applications, *Trans. Amer. Math. Soc.* **304** (1987), 229–43.

[60] R. Léandre, Girsanov transformations for poisson processes in semi-group theory. Preprint, Université de Bourgogne, Dijon 2007.

[61] M. Ledoux and M. Talagrand, *Probability in Banach Spaces*, Ergebnisse der Mathematik und ihrer Grenzgebiete 23 (Berlin: Springer Verlag, 1991).

[62] J. A. Léon, J. L. Solé, F. Utzet and J. Vives, On Lévy processes, Malliavin calculus and market models with jumps, *Finance and Stochastics* **6**:2 (2002), 197–225.

[63] P. Lévy, *Théorie de l'Addition des Variables Aléatoires* (Paris: Gauthier-Villars, 1937).

[64] T. Lindstrøm, Hyperfinite stochastic integration I, II, III and Addendum, *Math. Scand.* **46** (1980), 265–333.

[65] T. Lindstrøm, A Loeb measure approach to theorems of Prohorov, Sazonov and Gross, *Trans. Amer. Math. Soc.* **269** (1982), 521–34.

[66] T. Lindstrøm, Anderson's random walk and the infinite-dimensional Ornstein–Uhlenbeck process. In *Advances in Analysis, Probability and Mathematical Physics (Blaubeuren 1992)*, ed. S. Albeverio, W. A. J. Luxemburg and M. P. H. Wolff. Mathematics and Its Applications 314 (Dordrecht: Kluwer, 1995) pp. 186–99.

[67] T. Lindstrøm, Hyperfinite Lévy processes, *Stochastics* **76**:6 (2004), 517–48.

[68] T. Lindstrøm, Stochastic integration in hyperfinite-dimensional linear spaces. In *Nonstandard Analysis–Recent Developments*, ed. A. H. Hurd (Berlin: Springer Verlag, 1983), pp. 134–61.

[69] P. A. Loeb, Conversion from nonstandard to standard measure spaces and applications in probability theory, *Trans. Amer. Math. Soc.* **211** (1975), 113–22.

[70] P. A. Loeb, A functional approach to nonstandard measure theory, *Contemp. Math.* **26** (1984), 251–61.

[71] P. A. Loeb and H. Osswald, Nonstandard integration theory in topological vector lattices, *Monatshefte für Mathematik* **124** (1997), 53–82.

[72] P. A. Loeb, H. Osswald, Y. Sun and Z. Zhang, Uncorrelatedness and orthogonality for vector-valued processes, *Trans. Amer. Math. Soc.* **356**:8 (2004), 3209–25.

[73] P. A. Loeb and M. Wolff, eds, *Nonstandard-Analysis for the Working Mathematician* (Dordrecht: Kluwer Scientific 2000).

[74] W. A. J. Luxemburg, A general theory of monads. In *Applications of Model Theory to Algebra, Analysis and Probability* (New York: Hold, Rinehart and Winston, 1969), pp. 18–86.

[75] P. Malliavin, Stochastic calculus of variations and hypoelliptic operators. In *Proceedings of the International Symposium on Stochastic Differential Equations, Kyoto, 1976* (New York: Wiley, 1978), pp. 529–47.

[76] P. R. Masani, *Norbert Wiener*, Vita Mathematica 5 (Berlin: Birkhäuser Verlag, 1990).

[77] A. Millet and W. Smoleński, On the continuity of Ornstein–Uhlenbeck processes in infinite-dimensions, *Probab. Theory Rel. Fields* **92** (1992), 529–47.

[78] S. Ng, *Hypermodels in Mathematical Finance Modelling via Infinitesimal Analysis* (River Edge, NJ: World Scientific, 2003).

[79] D. Nualart, *The Malliavin Calculus and Related Topics* (Berlin: Springer Verlag, 1995).

[80] D. Nualart and W. Schoutens, Chaotic and predictable representations for Lévy processes, *Stoch. Proc. Appl.* **90** (2000), 109–22.

[81] D. Ocone, Malliavin calculus and stochastic integral representation of diffusion processes, *Stochastics* **12** (1984), 161–85.

[82] H. Osswald and Y. Sun, Measure and probability theory with applications. In *Nonstandard Analysis for the Working Mathematician*, ed. P. Loeb and M. Wolff (Dordrecht: Kluwer Scientific, 2000), pp. 137–256.

[83] H. Osswald, Infinitesimals in abstract Wiener spaces. In *Stochastic Processes, Physics and Geometry: New Interplays, II (Leipzig, 1999)* ed. F. Gesztesy *et al.*, CMS Conf. Proc 29 (Providence, RI: American Mathematical Society, 2000), pp. 539–46.

[84] H. Osswald, Malliavin calculus in abstract Wiener spaces using infinitesimals, *Adv. Math.* **176** (2003), 1–37.

[85] H. Osswald, On the Clark–Ocone formula for the abstract Wiener space, *Adv. Math.* **176** (2003), 38–52.

[86] H. Osswald, The existence of poly-saturated models. In *Nonstandard Analysis for the Working Mathematician*, ed. P. Loeb and M. Wolff (Berlin: Kluwer Scientific, 2000), pp. 57–72.

[87] H. Osswald and J.-L. Wu, On infinite-dimensional continuous Ornstein–Uhlenbeck processes, *Acta Appl. Math.* **83** (2004), 289–312.

[88] H. Osswald, Malliavin calculus on product measures of $\mathbb{R}^{\mathbb{N}}$ based on chaos, *Stochastics* **77**:6 (2005), 501–14.

[89] H. Osswald, A smooth approach to Malliavin calculus for Lévy processes, *J. Theor. Prob.* **22** (2009), 441–73.

[90] H. Osswald, Malliavin calculus on extensions of abstract Wiener spaces, *J. Math. Kyoto Univ.* **8**:2 (2008), 239–62.

[91] H. Osswald, On anticipative Girsanov transformations, *J. Theor. Prob.* **22** (2009), 474–81.

[92] H. Osswald, Computation of the kernels of Lévy functionals and applications, *Illinois J. Math.* to appear.

[93] N. Privault, Absolute continuity in infinite-dimensions and anticipating stochastic calculus, *Potential Analysis* **8** (1998), 325–43.

[94] N. Privault, Calcul des variations stochastiques pour la mésure de densité uniforme, *Potential Analysis* **29** (2008), 327–49.

[95] N. Privault, Girsanov theorem for anticipative shifts, *Prob. Theory Rel. Fields* **104** (1996), 501–14.

[96] N. Privault, A transfer principle from Wiener to Poisson space and applications, *J. Funct. Anal.* **132** (1995), 335–60.

[97] P. Protter, *Stochastic Integration and Differential Equations*, 2nd edn (Berlin: Springer, 2004).

[98] M. Reed and B. Simon, *Methods of Modern Mathematical Physics, I: Functional Analysis* (San Diego, CA: Academic Press, 1980).

[99] B. Rehle, A Nonstandard approach to Lévy Processes, Bachelor Thesis, University of Munich, May 2011.

[100] A. Robinson, *Nonstandard Analysis* (Amsterdam: North-Holland, 1966).

[101] W. Rudin, *Functional Analysis* (New York: McGraw-Hill, 1973).

[102] J. Sacks, *Saturated Model Theory* (Reading, MA: W. A. Benjamin, 1972).

[103] K. Sato, *Lévy Processes and Infinitely Divisible Distributions* (Cambridge: Cambridge University Press, 1999).

[104] W. Schoutens, *Stochastic Processes and Orthogonal Polynomials*, Lecture Notes in Statistics (Berlin: Springer, 2000).

[105] A. V. Skorokhod, On a generalization of a stochastic integral, *Th. Prob. Appl.* **20** (1975), 219–33.

[106] I. Simao, Pinned Ornstein–Uhlenbeck process on an infinite-dimensional space. In *Stochastic Analysis and Applications*, ed. I. M. Davies, A. Truman and K. D. Elworthy (Singapore: World Scientific, 1996), pp. 401–7.
[107] W. Smoleński, Continuity of Ornstein–Uhlenbeck processes, *Bull. Polish Acad. Sci. Math.* **37** (1990), 203–6.
[108] O. G. Smolyanov and H. von Weizsäcker, Smooth probability measures and associated differential operators, *Inf.-dim. Anal., Quantum Prob. Rel. Topics* **2**:1 (1999), 51–78.
[109] E. M. Stein, *Singular Integrals and Differentiability of Functions* (Princeton, NJ: Princeton University Press, 1970).
[110] K. Stroyan and W. A. J. Luxemburg, *Introduction to the Theory of Infinitesimals* (New York: Academic Press, 1976).
[111] K. S. Stroyan and J. Bayod, *Foundation of Infinitesimal Stochastic Analysis* (Amsterdam: North Holland, 1985).
[112] J. L. Solé, F. Utzet and J. Vives, Canonical Lévy process and Malliavin calculus, *Stoch. Proc. Appl.* **117** (2007), 165–87.
[113] Y. N. Sun, A theory of hyperfinite processes: the complete removal of individual uncertainty via exact LLN, *J. Math. Econ.* **29** (1998), 419–503.
[114] Y. N. Sun, The almost equivalence of pairwise and mutual independence and the duality with exchangeability, *Probab. Theory Rel. Fields* **112** (1998), 425–56.
[115] A. S. Üstünel and M. Zakai, Transformations of Wiener measure under anticipative flows, *Probab. Theory Rel. Fields* **93** (1992), 91–136.
[116] A. S. Üstünel and M. Zakai, Embedding the abstract Wiener space in a probability space, *J. Func. Anal.* **171** (2000), 124–38.
[117] A. S. Üstünel and M. Zakai, *Transforms of Measure on a Wiener Space*, Springer Monographs in Mathematics (Berlin: Springer Verlag, 2000).
[118] J.-L. Wu, On the regularity of stochastic difference equations in hyperfinite-dimensional vector spaces and applications to $\mathcal{D}'$-valued stochastic differential equations, *Proc. R. Soc. Edin. A* **124** (1994), 1089–117.
[119] N. Wiener, The homogeneous chaos, *Amer. J. Math.* **60** (1938), 879–936.
[120] M. Zakai, The Malliavin calculus, *Acta Appl. Math.* **3** (1985), 175–207

Index

absolutely continuous, 67
absolutely continuous to, 7
abstract Wiener space, 56
adapted Loeb space, 143
adapted probability space, 9
adapted process, 12
admissible sequence
  for Brownian functionals, 294
  for Lévy functionals, 352
algebra, 5
almost everywhere, 6
almost surely, 6, 10
approximation, 131
  of the Dirac function, 73
approximation property, 72
atom, 260

Bochner integrable, 64
Borel sets, 5
Borel σ-algebra, 5
Borel measure, 6
Brownian motion
  infinite-dimensional, 66
  one-dimensional, 45

c-filtration, 12
càdlàg, 124
canonical, 194
canonical isometry, 79
canonical martingale, 10
Cauchy–Schwarz inequality, 4
centred Gaussian measure, 38
chaos level, 298
closed under definition, 98, 383
compact, 100
complete, 6
conditional expectation, 6
continuous in probability, 223
continuous iterated integral, 219
continuous standard part, 121
continuous version of, 224
convergence in measure, 65
critical
  for Lévy functionals, 272
  for sequences, 83
cylinder set, 92
  for continuous functions, 61
  in Fréchet spaces, 56
  in Hilbert spaces, 51
  in tensor products, 63
  simple, 58

Dedekind cut axiom, 102
deep, 99
dense, 5
density, 7
derivative, 67
derivative into the direction, 311
direction, 310
discrete Lévy process, 42
distance, 42
divides, 108
Doob–Meyer conditions, 12

elementary chain, 393
elementary embedding, 101, 384
elementary limit, 393
end extension, 108
equivalence
  in pre-models, 393

of L^p-spaces, 79, 141
of measurable sets, 101
of measures, 7
event, 6
expected value, 6
external, 97, 379

filtration, 9
filtration, generated by, 10
finite, 6
finite chaos level, 298
finite intersection property, 99
finitely additive, 5
formula, 382
Fourier coefficient, 5
Fréchet space, 4
free, 382

generated by, 389

Hahn–Banach theorem
for locally convex spaces, 5
for normed spaces, 5
Hilbert–Schmidt norm, 62
Hilbert–Schmidt operator, 55

image measure, 7
independent, 39
independent increments, 259
indicator function, 7
individual, 379
infinitely close
between tensor products, 122
for reals, 111
in topological spaces, 115
infinitesimal, 110, 111
infinitesimal increments, 266
integrability, 143
integrable, 10
integral
for Lévy processes, 277
for sequences, 84
for tensor products, 205
integral of order k
for Lévy processes, 276
for sequences, 84
internal
cardinality, 109
filtration, 142
iterated integral, 212
iterated integral process, 218
iterated integral with parameters, 212
object, 97, 379
stochastic integral, 197
intersection stable, 47
iterated integral, 215
of order k for sequences, 85
with parameters for Brownian functionals, 215
iterated integral of order k
for Lévy processes, 280
with parameters for Lévy processes, 280
iterated integral process of order *k*, 281

kernels
for sequences, 86
of Brownian functionals, 298
of Lévy functionals, 355

Lévy measure, 260
Lévy process, 258
Lévy triplet, 265
lifting
for processes, 150
for random variables, 135
for tensor products, 139
limit of a projective system, 52
limited, 111
by a measure, 251
limited increments, 248
locally S-integrable, 149
Loeb-measurable, 140
Loeb space, 133

Malliavin derivative
for Brownian functionals, 306
for Lévy functionals, 356
Malliavin differentiable
for Brownian functionals, 307
for Lévy functionals, 356
for sequences, 88
martingale, 12
associated to a Lévy process, 43
maximal, 4
mean, 39
measurable, 7
measurable semi-metric, 54
metric
generated by, 4
translation invariant, 4
model, 379
monad, 115

multiple integral
  for sequences, 87
  process for Lévy processes, 282
  with parameters for Lévy processes, 282

natom, 260
natural filtration, 84
nearstandard
  for reals, 113
  in tensor products, 123
  in topological spaces, 115
neighbourhood, 115
non-anticipating
  for Brownian functionals, 325
  for sequences, 89
normalized counting measure, 17
normally distributed, 39
Novikov condition, 344
nullset, 6, 131

open, 115
orthogonal projection, 5
orthogonal set, 4
orthonormal basis, 4
outer measure, 131

p-integrable, 10
parameter, 97
parameters, 381
Poisson process, 260
poly-saturated, 99, 386
power set, 5
pre-Fréchet space, 4
pre-model, 388
pre-S-continuous, 121, 161
predictable, 71
predictable rectangle, 71
probability measure, 6
probability space, 6
process, 12
product, 63
projection theorem, 4
projective system, 50
pure jump process, 266

quadratic variation, 11

rate, 260
regular, 119
regular cardinal, 396
regular conditional probability, 41
representation, 44
restricted, 283, 363
restricted increments, 43
Riesz lemma
  for Hilbert spaces, 4
  for Lebesgue spaces, 8
right continuous, 12, 142
right interval, 5
right rectangle, 5

S-continuous, 120
S-continuous on, 224
S-dense, 224
S-integrability, 149
S-integrable, 143
S-square-integrable, 143
same properties as, 101
sentence, 381
separable, 5
separating semi-norms, 3
σ-additive, 6
σ-algebra
  generated by, 5
σ-finite, 6
simple functions, 64
Skorokhod integral
  for Brownian functionals, 323
  for Lévy processes, 360
  for sequences, 89
Skorokhod integral process
  for Brownian functionals, 335
  for Lévy processes, 360
Skorokhod metric, 125
small by, 99
splitting infinitesimal, 257
standard model, 96, 381
standard part
  of an internal filtration, 143
  map, 113
  of a martingale, 169
  for reals, 113
  right continuous, 167
  in tensor products, 123
  in topological spaces, 117
*-continuous, 128
*-finite, 109
state of information, 10
stochastic process, 12
stopping time, 13
strongly adapted, 12
strongly predictable, 71
submartingale, 10

supremum semi-norm, 61
symmetric, 165
  for sequences, 87
symmetric difference, 5

tensor product, 123
theorem, 382
timeline, 9
topological dual, 4
transformation rule, 7
true, 98
truth predicate, 382

ultra power, 390
uncritical
  for Lévy processes, 272
  for sequences, 83
uniquely determined version, 224
unlimited, 108, 111
urelement, 379

variance, 39
version, 169, 221
very simple, 66

weak approximation property, 77
weak model, 379
weakly S-continuous, 225
Wiener integral, 206
Wiener measure, 67, 192, 291
Wiener–Lévy integral, 280
witness
  for integrability, 64
  for measurability, 54
  for S-integrability, 155
  for the SD-property, 128
  for the SDJ-property, 126

Printed in the United States
by Baker & Taylor Publisher Services

Printed in the United States
by Baker & Taylor Publisher Services